Understanding Neuromuscular Health and Disease: Advances in Genetics, Omics, and Molecular Function

Understanding Neuromuscular Health and Disease: Advances in Genetics, Omics, and Molecular Function

Editors

William Duddy
Stephanie Duguez

MDPI • Basel • Beijing • Wuhan • Barcelona • Belgrade • Manchester • Tokyo • Cluj • Tianjin

Editors

William Duddy	Stephanie Duguez
Biomedical Sciences Research Institute	Biomedical Sciences Research Institute
University of Ulster	University of Ulster
Derry/Londonderry	Derry/Londonderry
United Kingdom	United Kingdom

Editorial Office
MDPI
St. Alban-Anlage 66
4052 Basel, Switzerland

This is a reprint of articles from the Special Issue published online in the open access journal *Journal of Personalized Medicine* (ISSN 2075-4426) (available at: https://www.mdpi.com/journal/jpm/special_issues/Understanding_Neuromuscular).

For citation purposes, cite each article independently as indicated on the article page online and as indicated below:

LastName, A.A.; LastName, B.B.; LastName, C.C. Article Title. *Journal Name* **Year**, *Volume Number*, Page Range.

ISBN 978-3-0365-1622-6 (Hbk)
ISBN 978-3-0365-1621-9 (PDF)

Contents

About the Editors

William Duddy Dr Duddy is a lecturer in Bioinformatics at the University of Ulster. He works closely with colleagues in cell biology and translational medicine, blending computational analysis with wet-lab experimental approaches to understand and treat pathology at the molecular level. His interest in neuromuscular health stems from the experience of having a brother with the muscle wasting disease, Duchenne Muscular Dystrophy. His current research focus is on the neurodegenerative condition, Amyotrophic Lateral Sclerosis, where he has helped to identify abnormalities in the behaviour of patient's muscle cells, including the secretion of neurotoxic vesicles.

Stephanie Duguez Dr Duguez's research group at Ulster University is composed of biologists and bioinformaticians and explores the intercellular communication between muscle tissue and surrounding cells, including motor neurons, inflammatory cells, and fibroblasts. Her group is using human muscle stem cells and studying their secretome profiles by different OMICS approaches such as proteomic, metabolomic, and transcriptomic analyses. The aim is to decipher the role of intercellular communication in different physiological and pathological conditions such as Amyotrophic Lateral Sclerosis (ALS), Spinal and Bulbar Muscular Atrophy (SBMA), and Spinal Muscular Atrophy (SMA-III & IV). By characterizing the muscle secretome, she aims to identify biomarkers and to unravel new therapeutic strategies for neurodegenerative disorders.

Preface to "Understanding Neuromuscular Health and Disease: Advances in Genetics, Omics, and Molecular Function"

Recently, the field of neuromuscular research has seen considerable advances in the molecular and cellular understanding of muscle biology, and the treatment of neuro-muscular disease. These advances are at the forefront of modern molecular methodologies, often integrating across wet-lab cell and tissue models, dry-lab computational approaches, and clinical studies. The continuing development and application of multi-omics methods offer particular challenges and opportunities, not least in the potential for personalized medicine.

This compilation, entitled "Understanding Neuromuscular Health and Disease: Advances in Genetics, Omics, and Molecular Function", encompasses some 15 publications from colleagues working on diverse aspects of neuromuscular health and disease. It is structured according to the following broad themes:

- Therapeutics (Buscara et al., Joubert et al.)
- Genotype-phenotype correlations (Anwar et al., Lim et al., Sheikh et al., Connolly et al.)
- Cellular and molecular mechanisms (Sidlauskaite et al., Le Gall et al.)
- Biomarkers (Smeriglio et al., Tawalbeh et al., Heier et al., D'Cruz et al.)
- Omics and mathematical modelling (Speciale et al., Vignier et al., Vasilopoulou et al.)

More than 500 different genes are known to be associated with neuromuscular disorders (Buscara et al.), and the identification of causative mutations has allowed the development of personalized therapies (Buscara et al., Smeriglio et al.). However, even if great progress has been made during the past two decades in different subgroups of neuromuscular disorders, there are still numerous challenges to resolve, such as the optimization of therapeutic knock-down strategies (Joubert et al.), targeting specific muscles or tissues of the nervous system (Buscara et al.), identifying genetic modifiers that can impair a therapeutic strategy (Smeriglio et al.), targeting common pathways being affected in different patient subgroups for a given disease (Le Gall et al., Connolly et al.), or understanding the impact of neuromuscular disorders on other tissues that could be affected but may be understudied.

The neuromuscular pathologies considered in this book include Duchenne muscular dystrophy (Anwar et al., Lim et al., Tawalbeh et al., Sheikh et al.), facioscapulohumeral dystrophy (Joubert et al., Heier et al., Sidlauskaite et al.), amyotrophic lateral sclerosis (Le Gall et al., Connolly et al., Vasilopoulou et al.), spinal muscular atrophy (Smeriglio et al.), Emery–Dreifuss muscular dystrophy (Vignier et al.), and rheumatoid arthritis (D'Cruz et al.). Looking across diseases, several themes are recurrent, such as the efforts to identify genotype–phenotype correlations in DMD (Anwar et al., Lim et al., Sheikh et al.) and ALS (Le Gall et al., Connolly et al.), the quest for effective biomarkers in many neuromuscular conditions (Smeriglio et al., Tawalbeh et al., Heier et al., D'Cruz et al.), and the use of genomic and multi-omic approaches towards better ways to identify biomarkers and to understand disease (Heier et al., Vasilopoulou et al., Vignier et al.).

The search for genotype–phenotype correlations can be aimed at the improved understanding of disease (Le Gall et al., Connolly et al., Lim et al., Sheikh et al.), but may also be relevant to potential therapeutic outcomes (Anwar et al.). Of relevance to this are genotype–phenotype correlations in DMD (Anwar et al., Lim et al., Sheikh et al.) and in ALS (Le Gall et al., Connolly et al.). It is interesting to contrast the state of investigations in these two conditions, differences which are related to the underlying genetics: DMD being due to mutations at a single gene, and, therefore, correlations being sought between clinical outcomes and specific mutation patterns within that gene (Anwar et al., Lim et al., Sheikh et al.); ALS being a disease of unclear aetiology for the majority of patients and the focus of these genotype–phenotype investigations thus being on the relationship of different genes and their functional roles to the implicated mechanisms (Le Gall et al.) or clinical outcomes of the condition (Connolly et al.).

The use of genomics and multi-omics approaches is a theme which itself cuts across the aims of current research, from the overlay of multiple omics data to achieve a global perspective and new understanding of Emery–Dreifuss muscular dystrophy (Vignier et al.), to the identification of novel circulating miRNA and protein biomarkers for FSHD using multi-omics (Heier et al.), through to the application of machine learning to the genomics of ALS, which is aimed at understanding the molecular basis of this disease (and is also relevant to genotype–phenotype correlations) (Vasilopoulou et al.). Deciphering the pathways and gene mutations involved in neuromuscular diseases may allow for the development of computational models helping our understanding of muscle pathologies, which could enable preclinical studies of neuromuscular diseases in the context of personalized medicine (Speciale et al.).

Aside from therapeutic strategy development, the use of biomarkers may be critical as disease trackers for the development of effective therapeutics (for example, in FSHD, Heier et al.), but also to the personalized tailoring of existing treatments (such as in DMD, Tawalbeh et al., and in rheumatoid arthritis, D'Cruz et al.), and may prove useful in a broad sense for improved stratification, diagnosis, and treatment (e.g., in adult SMA, Smeriglio et al.).

We hope that studies such as these, that integrate modern molecular methodologies across cell and tissue models, computational approaches, and clinical studies, will continue to drive progress towards improved neuromuscular health and treatments for these often severe diseases.

William Duddy, Stephanie Duguez
Editors

Journal of
Personalized
Medicine

MDPI

Review

Of rAAV and Men: From Genetic Neuromuscular Disorder Efficacy and Toxicity Preclinical Studies to Clinical Trials and Back

Laurine Buscara [1], David-Alexandre Gross [1,2] and Nathalie Daniele [1,*]

1 Genethon, 91000 Evry, France; lbuscara@genethon.fr (L.B.); dagross@genethon.fr (D.-A.G.)
2 Université Paris-Saclay, Univ Evry, Inserm, Genethon, Integrare Research Unit UMR_S951, 91000 Evry, France
* Correspondence: daniele@genethon.fr; Tel.: +33-1-6947-2987

Received: 31 October 2020; Accepted: 23 November 2020; Published: 28 November 2020

Abstract: Neuromuscular disorders are a large group of rare pathologies characterised by skeletal muscle atrophy and weakness, with the common involvement of respiratory and/or cardiac muscles. These diseases lead to life-long motor deficiencies and specific organ failures, and are, in their worst-case scenarios, life threatening. Amongst other causes, they can be genetically inherited through mutations in more than 500 different genes. In the last 20 years, specific pharmacological treatments have been approved for human usage. However, these "à-la-carte" therapies cover only a very small portion of the clinical needs and are often partially efficient in alleviating the symptoms of the disease, even less so in curing it. Recombinant adeno-associated virus vector-mediated gene transfer is a more general strategy that could be adapted for a large majority of these diseases and has proved very efficient in rescuing the symptoms in many neuropathological animal models. On this solid ground, several clinical trials are currently being conducted with the whole-body delivery of the therapeutic vectors. This review recapitulates the state-of-the-art tools for neuron and muscle-targeted gene therapy, and summarises the main findings of the spinal muscular atrophy (SMA), Duchenne muscular dystrophy (DMD) and X-linked myotubular myopathy (XLMTM) trials. Despite promising efficacy results, serious adverse events of various severities were observed in these trials. Possible leads for second-generation products are also discussed.

Keywords: AAV; genetic neuromuscular disorders; gene therapy; clinical trials; toxicity; SMA; DMD; XLMTM

1. Introduction

Neuromuscular disorders are a group of heterogeneous rare diseases characterised by skeletal muscle dysfunction and caused primarily by motoneuron, peripheral nerve, motor end plate or muscle deficiencies. This family of pathologies encompasses a wide clinical spectrum, ranging from very weak and barely detectable clinical signs to extremely severe and life-shortening forms. Common symptoms include muscle-specific patterns of atrophy and weakness, occasionally associated with the involvement of additional organs, the most common complication being cardiac and/or respiratory failure. These diseases can be caused by many factors, notably autoimmunity; inflammation; poisoning; toxin accumulation; tumours; environmental agents; neurologic, metabolic or traumatic syndromes [1–3]; aging [1,4]; and genetic inheritance or spontaneous mutations in muscle or nerve-essential genes. The large majority of mutations are monogenic, with every nature of mutation and transmission mode possible. The classification of neuromuscular diseases based on their origins and phenotypical features is published every year at http://www.musclegenetable. The 2020 update reports 1042 neuromuscular disorders caused by mutations in 587 different genes, classified in 16 groups, and many remain to be discovered [5]. The pathogenic mechanisms are very diverse, as they depend on the gene involved, and proteins with very different functions and subcellular localisations are affected (enzymes, structural proteins, metabolic key-players, etc). In this review, we will focus our interest on genetic neuromuscular diseases currently under interventional clinical trials with whole-body delivery.

2. Marketed Pharmacological Treatments

Before the 1990s, treatment options were limited to supportive therapies aiming at improving life comfort and lengthening lifespan. Anti-inflammatory drugs proved very efficient in preventing muscle degeneration and mortality in inflammatory myopathies [6]. Corticosteroids are also commonly used and show limited success in Duchenne muscular dystrophy (DMD, OMIM 310200), a very severe and the most common form of degenerative muscular pathology. Long-term clinical trials showed that prednisolone/prednisone or deflazacort corticosteroids reduce chronic muscle inflammation, stabilise muscle function, prolong ambulation and improve respiratory function and patients' survival [7–9]. However, several side effects are associated with the prolonged usage of these immuno-modulators, the most severe being a drastic inhibition of the immune system's functionality, occasionally leading to life-threatening opportunistic infections.

More recently, specifically targeted treatments were developed and approved for human applications by the regulatory agencies. The USA Food and Drug Administration (FDA) and the European Medicines Agency (EMA) approved Myozyme® (α-glucosidase, Sanofi-Genzyme, Cambridge, MA, USA) for long-term enzyme replacement therapy in Pompe patients, who suffer from a severe metabolic myopathy caused by mutations in the α-glucosidase-encoding gene (glycogen storage disease type 2 or Pompe Disease, OMIM 232300) [10–12]. The treatment proved particularly efficient in improving lifespan and muscle, respiratory and cardiac functions in classical infantile-onset Pompe disease (<1 year of age, with cardiomyopathy) [13,14], with more contrasted results in late-onset forms of the disease [15,16]. Occasional infusion-associated reactions and adverse events related to the treatment were reported. Nearly all were resolved with an interruption or reduction of the infusion rate or symptomatic treatment. In almost every case, repeated bi-monthly intravenous injections of the product led to the generation of α-glucosidase-specific antibodies [13–16], although seldom showed evidence of in vitro inhibitory activity. Some patients developed anaphylactic shock [15].

Exondys 51® (Eteplirsen, Sarepta Therapeutics, Cambridge, USA), a drug targeting the DMD pathology, was granted accelerated approval by the FDA in 2016 on the grounds of phenotype stabilisation, making it the first FDA-approved drug for DMD [17]. Severe Duchenne myopathy is caused by a variety of mutations in the dystrophin-encoding *DMD* gene [18]. The large majority are out-of-frame mutations resulting in the total loss of dystrophin, while the expression of shorter forms of dystrophin caused by in-frame *DMD* mutations leads to the milder Becker phenotype [19]. This observation constitutes the proof of concept that expressing shorter forms of dystrophin could be a therapeutic option for ameliorating, if not curing, DMD symptoms. Eteplirsen is a 30-nucleotide-long phosphorodiamidate morpholino oligomer (PMO) designed to skip *DMD* exon 51 and restore a shorter but functional reading frame. The weekly intravenous injection of this drug restores partial dystrophin expression in skeletal fibres [20], prevents muscle loss of function [21–23] and protects pulmonary and cardiac functions [24,25]. This drug offers a very good safety profile, probably due to its uncharged chemical nature. Of note, Translarna® (Ataluren, PTC Therapeutics, South Plainfield, NJ, USA), a read-through RNA interference molecule targeting non-sense mutations in Duchenne, showed a weak benefit in DMD ambulatory patients in clinical trials [26,27] and was granted conditional approval for ambulatory patients by the EMA in 2014, but was refused by the FDA. The treatment proved safe and delayed ambulation loss in longer-term studies compared with a historical cohort [28].

Spinraza® (Nusinersen, Biogen, Cambridge, MA, USA) was the first curative drug for spinal muscular atrophy (SMA) in paediatric and adult patients to be approved by the FDA in 2016 and the following year by the EMA [29,30]. SMA, the most common motoneuron degenerative disease and the leading genetic cause of infant mortality, is due to hereditary bi-allelic mutations in the *SMN* gene [31]. Spinraza® is an antisense oligonucleotide interfering with the splicing of an alternative form of the gene (*SMN2*) and leading to the production of a functional SMN protein. Repeated intrathecal injections result in an increase in SMN proteins and meaningful improvement in motor development and function with the associated survival of the patients [32,33]. Just recently, in August 2020, the FDA approved Evrysdi® (Risdiplam, Genentech/Roche, San Francisco, CA, USA) as the first oral and at-home treatment for all SMA patients from 2 months of age [34]. Similarly to Spinraza, this *SMN2* splicing modifier increases the levels of SMN proteins and shows clinically meaningful improvements in survival and motor and respiratory functions in SMA patients [35–37]. However, while Spinraza requires four administrations in the spinal cord a year, Evrysdi is taken orally for systemic distribution once a day, widening the field of application to patients excluded from intrathecal injections because of scoliosis.

Even though these drugs ameliorate the patient's life and prognosis, they do not cure the diseases and necessitate constant re-dosing, a burdensome shortcoming for patients with an already altered quality of life. Long-term adverse events due to constant drug re-administration are also an important issue, especially as an immune response towards the treatment often develops with time, impeding its efficacy. Moreover, these personalised medicine treatments are generally highly specific for the targeted disease and mutation. Because of their wider range of application, a very intense research field is focused on developing gene replacement approaches. These strategies, which take advantage of the natural capacity of viruses to infect specific human cells, consist of inserting therapeutic genes in place of viral sequences in vectors devoid of replicative capacity. They offer the advantage of being usable regardless of the mutation type and position, at least for pathologies caused by losses of function. After a long period of difficulties linked mostly to the route of administration and to the production of the therapeutic vectors, the last ten years finally saw the translation of several proofs of concept into promising clinical trials. The vector favoured for the delivery of genes in neuromuscular tissues is derived from the adeno-associated virus (AAV). In the last ten years, several AAV-based treatments have been approved for human usage. In 2012, Glybera® (alipogene tiparvovec, UniQure, Lexington, KY, USA) was the first to be accepted by the EMA for the correction of a rare inherited metabolic disorder, substantiating AAV innocuousness and long-term efficacy [38,39]. Luxturna® (Voretigene Neparvovec, Spark Therapeutics, Philadelphia, PA, USA) was later approved for the local treatment of a rare retinal disease [40–42]. Very recently, regulatory agencies granted full (FDA) and conditional (EMA) approval to Zolgensma® (onasemnogene abeparvovec-xioi, AveXis/Novartis, Bannockburn, IL, USA), the first AAV-based treatment for the whole-body correction of SMA [43–45], paving the way for other myopathies.

3. The Therapeutic Toolbox for Muscle Gene Therapy

3.1. About Wild-Type AAV

The AAV virus is a 25 nm-diameter non-enveloped human parvovirus, with a simple architecture composed of a single-stranded 4.7 kb linear DNA genome encapsidated within an icosahedral protein capsid. The DNA bears four open-reading frames (ORFs) coding, respectively, for the four non-structural Rep proteins involved in the viral cell cycle (Rep 78, 68, 52 and 40); the three structural Cap proteins VP1, VP2 and VP3, assembling in a 1:1:10 ratio to constitute the 60 monomers of the capsid; the assembly-activating protein (AAP), promoting capsid assembly [46]; and the recently described membrane-associated accessory protein (MAAP) [47]. The ORFs are framed by two highly structured 145 bp palindromic inverted terminal repeats (ITRs) acting in cis as structural signals to drive AAV replication and genome packaging. AAV can infect both dividing and quiescent cells [48].

Various AAV serotypes of human and primate origin (AAV1 to AAV13) and more than a hundred natural variants have been identified (AAV1 [49], AAV2 and AAV3 [50], AAV4 [51], AAV5 [52], AAV6 [53], AAV7 and AAV8 [54], AAV9 [55], AAV10 and AAV11 [56], AAV12 [57] and AAV13 [58]). Based on VP1-capsid composition, the AAVs were phylogenetically classified into six clades, regrouped together according to genetic relatedness [59]. Although many display a broad tissue tropism, they generally show preferential infections of specific organs. The cell tropisms depend on many parameters, but subtle differences in the capsid's amino acid sequence and structure are one essential feature driving tissue targeting [55,60,61]. Once disseminated in the blood stream, AAVs have to overcome several barriers to deliver their DNA within host cells' nuclei. First, AAVs can be neutralised by pre-existing neutralising antibodies (NAbs), as seroprevalence resulting from natural infections with wild-type AAV is common in the general human population, with a high cross-reactivity between serotypes [62–64]. Second, AAVs have to attach to specific receptors before being internalised within host cells. AAV capsids were shown to interact with specific glycan moieties of host membrane proteoglycans: heparan sulfate for AAV2 [65]; heparin for AAV3 and AAV6 [66]; sialic acid for AAV1, AAV4, AAV5 and AAV6 [66–69]; and galactose for AAV9 [70–72]. Transmembrane receptors such as PDGFR for AAV5 [73] and the 37/67 kDa laminin receptor LamR for AAV8 [74] were also reported to be surface receptors. Their in vivo biodistribution correlates with and could account for virus tropism. Efficient virus endocytosis requires secondary binding events with membrane co-receptors. For AAV2, the most widely studied AAV, the hepatocyte growth factor receptor c-Met [75], $\alpha V\beta 5$ integrin [76] and fibroblast growth factor receptor 1 (FGFR1) [77] were demonstrated to increase AAV infectiosity and proposed as co-receptors. However, using a candidate approach based on genetic deletion and supplementation, Pilay et al. demonstrated the existence of a co-receptor common to all the tested serotypes (AAV1, 2, 3B, 5, 6, 8 and 9), the previously uncharacterised type I transmembrane protein KIAA0319L, renamed AAVR [78,79]. Within the cell cytoplasm, AAV undergoes intracellular trafficking via the microtubule network to reach the nucleus [80] and achieves endosomal escape, nuclear entry and capsid unfolding. The AAV lytic cycle needs co-infection with a helper virus such as adenovirus [49], herpesvirus [81] or cytomegalovirus [82] for replication to occur. In the absence of this helper virus, the AAV enters a latent state. Several reports have evidenced the preferential integration of the AAV genome into the transcriptionally active environment of the AAVS1 locus in the q13.4-Ter region of host chromosome 19 genomic DNA [83–87]. Other hotspots were evidenced in chromosome 5p13.3 (AAVS2) and chromosome 3p24.3 (AAVS3) [88]. However, these studies were performed in cell culture, and it was recently evidenced in vivo that AAV mainly persists as transcriptionally active episomal forms and sometimes integrates randomly in the host genome [89]. Clonal integration in six oncogenes in liver tissue associated with hepatic tumorigenesis was also identified [89,90]. No specific enrichment was found in major AAV targets previously identified in cell lines [89,90].

3.2. Of the Usage of Recombinant AAV for Central Nervous System (CNS) and Muscle-Specific Targeting

For neuromuscular diseases, AAV vectors stand out as the most promising tools for driving body-wide muscle gene expression, as their wild-type counterparts have not been associated with a pathologic condition, they target myocytes and they are relatively poorly immunogenic. Nonetheless, several factors limit their application, mainly their low packaging capacity (<5 kb), especially as many neuromuscular genes are larger. Another issue is the targeting specificity, as specific gene delivery is desirable to reduce the risk of toxic off-target effects.

AAV recombinant vectors (rAAVs) derived from wild-type viruses are devoid of viral genetic elements, apart from the two ITRs in between which the transgene of interest is inserted. Plasmids encoding Rep, Cap and a helper are brought in trans within an appropriate production cell line to achieve DNA packaging [91]. AAV2-based recombinant genomes have been packaged in many different capsid types, resulting in a wide collection of "pseudotyped vectors" (rAAV2/X, where X stands for the capsid serotype). In the absence of the Rep gene, in both murine models and cell lines, the rAAV genome mostly concatamerises and forms circular, transcriptionally active episomes unable to divide when host cells cycle [92–94], or integrates at a very low rate in the host genome, randomly [48,95–97] or in preferential regions: near chromosomal instability points or in CpG islands, active genes and regulatory sequences [98–101]. In cell lines, chromosomal rearrangements were observed near the AAV-host genome breaking points [102,103]. Importantly, in murine hepatocytes, rAAVs were also reported to integrate at a low frequency into chromosome 12, at the *Rian* locus (RNA imprinted and accumulated in the nucleus), upregulating neighbouring non-coding RNAs and genes [104,105]. This integration, suggested to participate in murine hepatocellular carcinogenesis, seems specific to neonate animals and to some genetic backgrounds, and was not seen in adult mice [106].

More than 20 years ago, rAAV2s were the first vectors to prove their efficacy for the efficient and persistent transduction of post-mitotic neuromuscular cells [95,107]. The local brain delivery of a reporter transgene placed under the control of a ubiquitous strong promoter resulted in neuron and, to a lower level, glial cell transduction in rodents [107]. The long-term transduction of muscle fibres was observed after the intramuscular injection of a reporter gene in wild-type mice and rhesus monkeys, pointing out the inter-species tropism of this vector [95]. However, rAAV2s have a preference for slow-twitch muscle fibres, which might restrict their therapeutic benefits [108]. Additionally, of all the serotypes identified to date, AAV2 is the most common target of pre-existing NAbs in human populations [62,63], which could potentially prevent effective transduction in most of the putative patients [109]. Finally, a side-by-side comparative study of rAAV1 to 9 carried out with a ubiquitously driven luciferase reporter transgene evidenced that rAAV2 is amongst the lowest for general and muscle-specific transduction after intravenous injection, the optimal administration route for myopathy [110]. Today, rAAV2 is mainly used for tissue-specific gene therapy, such as local brain injection in clinical trials aimed at CNS delivery for Parkinson's disease [111].

Recombinant AAV1, 7, 8 and 9 showed higher muscle transduction than rAAV2 after local injection in mice [54,55,112–114] and dogs [115]. Muscle targeting was also achieved, though with lower efficacy with AAV5 [113] and AAV6 [114].

Apart from very few diseases in which a specific group of muscles are affected and can be targeted by local delivery, whole-body muscle transfer has to be achieved for myopathy treatments, and the delivery is usually performed by systemic administration, or specific cerebro-spinal fluid delivery in the case of CNS-specific pathology. The body-wide intravascular delivery of rAAV packaged with reporter genes confirmed the widespread dissemination and highest muscle tropism of rAAV1, 7, 8 and 9 in mice [54,110,116–121], dogs [115,122,123] and monkeys [119,124]. Conflicting publications report on rAAV9's preferential tropism for fast fibres [120] or slow fibres [114], but as they were performed with different promotors and different murine genetic backgrounds, general conclusions cannot be drawn.

The vascular endothelium is a major barrier for rAAV tissue distribution. Its permeation through the use of vascular endothelium growth factor (VEGF) was once demonstrated to enhance tissue transduction with rAAV6, largely inefficient by the intravascular route [125], though this effect is lost at high doses of the vector, and ensuing attempts to use it failed [116]. Muscle ischemia, a feature associated with some myopathies [126], was also shown to improve muscle targeting, partly for the same reasons [127].

Skeletal muscles are composed of long-lived mature post-mitotic fibres and of satellite cells, a population of progenitors crucial for muscle regeneration. Ideally, for expression persistence, therapeutic vectors should target myofibres and satellite cells, but unfortunately, rAAVs are inefficient for satellite cell transduction [128]. However, even though the episomal DNA can be diluted by successive cell divisions during muscle growth or regeneration, the transgene genome was shown to be stable for years in terminally differentiated myocytes, leading to continuous transgene expression [95,124,129].

Heart targeting is essential for treating neuromuscular diseases with cardiomyopathic features. Interestingly, several reports showed that rAAV9s lead to the highest levels in heart muscle [110,118,119,130,131], although rAAV6's superior cardiac efficacy was once reported [132]. Differences in the vector doses and systemic routes of administration are likely to account for this discrepancy. The cardiotropic properties of rAAV9s might originate, at least partly, from their specific binding to galactose receptors [70,71]. Interestingly, the intravascular delivery of rAAV9 is also an appealing strategy for CNS targeting, as this serotype is the most efficient for crossing the blood–brain barrier. Indeed, motoneurons and glial cells were transduced in the spinal cords and brains of mice, cats and non-human primates [118,133–138]. This strategy is a safer alternative to local CNS delivery. This unique feature of rAAV9s could come from increased vascular permeability and/or from their attachment to specific receptors distinct from those of other serotypes, such as the galactose receptor [70–72]. Indeed, the crystallographic structure of AAV9 revealed the specificity of the capsid in regions associated with receptor attachment that could account for its unique cellular tropism [61].

Finally, serotypes AAV1–9 transduce the liver with very high efficacy in mice, dogs and primates [110,116–120,122,129,135,136], but apart from the transient elevation of the aminotransferase enzyme related to the expression of the GFP transgene [135], no serious adverse events (SAEs) related to the capsid were reported during the biodistribution studies. Capsid-specific NAbs commonly developed with both local and systemic injection, though the levels varied with the dose and route of administration, but no major immunotoxicity was evidenced [95,110,112,115,135,136].

3.3. Restricting Expression by Muscle and CNS-Specific Promoters

Apart from capsid choice, a careful selection of the transgene regulatory elements, especially the promoter, is essential for specific expression. Viral promoters, such as the cytomegalovirus (CMV) or the Rous sarcoma virus (RSV), have generally been used for proofs of concept in early muscular gene therapy development, as they allow broad and powerful transgene expression [139–142]. The CMV promoter is currently used in several clinical trials for Duchenne and Becker muscular dystrophies, sporadic inclusion body myositis and Pompe disease (NCT02354781, NCT01519349, NCT00428935 and NCT00976352) [143–147]. However, it is now known that eukaryotic cells progressively silence transgene expression driven by viral promoters as a result of an immune mechanism to shut off viral expression, limiting their use for gene therapy applications where the long-lasting expression of the transgene is crucial [148–151]. An alternative to limit transgene silencing is the use of eukaryotic constitutive promoters, such as the elongation factor 1α (EF-1α), phosphoglycerate kinase (PGK), ubiquitin C (UBC) or hybrid promoters such as the chicken β-actin promoter coupled with the CMV early enhancer (CAG promoter), which shows high levels of transgene expression [152]. Interestingly, the CAG promoter is currently being used in a gene therapy clinical trial aiming at treating SMA type 1 patients: it shows success in driving appropriate expression levels in target tissues, as clinically meaningful benefits are achieved [153–156]. Nevertheless, constitutive transgene expression, notably in antigen presenting cells (APCs), was reported to induce an immune response [157–159].

In order to minimise ectopic transgene expression, muscle-specific promoters such as muscle creatine kinase (MCK) [160], desmin (Des) [161] or α-myosin heavy chain (α-MHC) [162] have been developed, showing higher muscle specificity compared to constitutive promoters [161]. Transgene expression efficacy driven by the Des promoter was successfully demonstrated in 2014 in preclinical studies performed in murine and canine models of X-linked myotubular myopathy (XLMTM) [163] and has recently shown promising results in a clinical trial with meaningful improvements in neuromuscular and respiratory functions (NCT03199469) [164].

Despite specific gene expression, muscle-specific promoters usually do not allow a high level of transgene expression in muscle cells and have a large size, limiting the packaging capacity for the transgene. Therefore, different laboratories have developed truncated muscle-specific promoters by selecting specific regulatory sequences to optimise both promoter strength and muscle-specific expression [165–168]. In 2008, Wang et al. designed compact muscle-specific promoters by combining an 87 bp proximal basal MCK promoter with a double (dMCK) or a triple (tMCK) tandem of the modified MCK enhancer [169], leading to highly efficient shorter promoters of 509 bp and 720 bp, respectively [170]. These two hybrid promoters demonstrate high transgene expression in skeletal muscles (except for the diaphragm), with no expression in the brain or liver. Interestingly, the dMCK and tMCK promoters are not active in the heart, which could be an advantage for the gene therapy of muscular diseases without cardiomyopathies. Following a successful proof of principle of gene transfer efficacy using tMCK in mouse models of Charcot-Marie-Tooth neuropathy type 1A [171] and limb girdle muscular dystrophy (LGMD) type 2D [172], this promoter has moved forward to clinical trials for these diseases (NCT03520751, NCT01976091 and NCT00494195) [173,174]. However, both dMCK and tMCK were reported to show fast-twitch myofibre preferences, which could limit treatment efficacy depending on the pathology [170]. Inversely, the MHCK7 promoter (770 bp), based on the assembly of specific enhancer and promoter regions of MCK and α-MHC, was shown to direct high levels of transgene expression specifically in the skeletal and cardiac muscles, with the advantage of being expressed in both fibre types [175], and proved more efficient for muscle expression than MCK1 in a murine model of Pompe disease [176]. This promoter was shown to direct robust micro-dystrophin expression in a systemic gene replacement clinical trial for Duchenne muscular dystrophy [177]. Compact muscle-specific promoters were also designed by assembling multiple copies of myogenic regulatory elements of natural muscle promoters and enhancers. The synthetic muscle-specific C5.12 promoter was reported to present a 6- to 8-fold expression increase over the CMV promoter [178].

The use of tissue-restricted promoters has also revealed their ability to evade undesirable adaptive immune responses directed against the transgene product. A possible explanation for these results is the inhibition of transgene expression in transduced professional APCs. Cordier et al. previously showed that inserting the muscle-specific C5.12 promoter instead of the ubiquitous CMV promoter enables human γ-sarcoglycan expression in mice, probably impairing the anti-transgene immune response [179]. The same observation was made with α-sarcoglycan driven by the C5.12 promoter [180] or α-galactosidase A driven by the DC190 liver promoter for treating Fabry disease [181]. Another hybrid promoter, also based on the MCK enhancer and coupled with the SV40 promoter (MCK/SV40), resulted in the long-term sustainability of the transgene expression with a minimal cellular and humoral immune response compared to the ubiquitous CMV and CAG promoters, suggesting benefits for gene therapy applications with immunogenic transgenes [168].

Liver targeting is important to promote tolerance to the transgene product in order to lead to stable muscle expression [182,183]. In this context, Colella et al. designed a new tandem promoter enabling the expression of the transgene in the targeted muscle cells for treatment efficacy, as well as in hepatocytes to trigger immune tolerance to the transgenic protein [184]. To combine muscle-specific and hepatic transgene expression, both the apolipoprotein E enhancer (ApoE) and the human alpha-1 anti-trypsin promoter elements, known to allow tolerogenic transgene expression in the liver, were multiplexed with the muscle-specific C5.12 promoter. This approximately 1 kb hybrid promoter efficiently promotes transgene expression in muscles and prevents transgene immunity.

4. Translating Preclinical Studies into Clinical Trials

The achievable skeletal muscle, heart and CNS-specific targeting, together with the apparent safety of capsids, paved the way for the preclinical assessment of AAV-driven therapies for myopathies. Dozens of proofs of concept were made, but we will focus this discussion on the strategies that were translated into clinics for the whole-body treatment of SMA, DMD and XLMTM congenital myopathy. Nonetheless, Table 1 provides a general overview of the main clinical features and SAEs observed in all body-wide and CNS-targeted AAV-driven interventional clinical trials ongoing for neuromuscular disorders.

Table 1. Summary of current body-wide and central nervous system (CNS)-targeted adeno-associated virus (AAV)-mediated gene replacement therapy clinical trials for neuromuscular diseases, as reported in *ClinicalTrials.gov*.

	Product/Administration					Clinical Design				
Disease	**AAV Serotype**	**Promoter**	**Transgene**	**Name**	**Administration/Dose**	**Clinical Trial ID (Study Name) Sponsor/Collaborator**	**Study Phase/Status**	**Study Timelines (Clinical Follow-Up)**	**Age, Gender, Actual or Estimated/ Planned Number of Participants Enrolled**	**Serious Adverse Events**
DMD	AAV9	CK8	Micro-dystrophin	SGT-001	Intravenous 2 doses	NCT03368742 (IGNITE DMD) Solid Biosciences, LLC	Phase 1/2 active, not recruiting	2017–2024 (2 years)	4 to 17 years, males, n = 16/same as current	Complement activation kidney failure, platelet count drop ($n = 1$ at 5×10^{13} vg/kg) + cardiopulmonary insufficiency ($n = 1$ at 2×10^{14} vg/kg) [185,186]
DMD	AAVrh74	MHCK7	Micro-dystrophin	SRP-9001	Intravenous 2×10^{14} vg/kg	NCT03375164 Sarepta Therapeutics, Inc.	Phase 1/2 active, not recruiting	2018–2021 (3 years)	3 months to 7 years, males, n = 4/12	No serious adverse events [177]
DMD	AAVrh74	MHCK7	Micro-dystrophin	SRP-9001	Intravenous 1 dose	NCT03769116 Sarepta Therapeutics, Inc.	Phase 2 active, not recruiting	2018–2026 (5 years)	4 years to 7 years, males, n = 41/24	-
DMD	AAV9	Human muscle-specific	Mini-dystrophin	PF-06939926	Intravenous 1×10^{14} vg/kg 3×10^{14} vk/kg	NCT03362502 Pfizer	Phase 1B active, not recruiting	2018–2026 (5 years)	4 years and older, males, n = 30/12	Antibody response, complement activation, acute kidney injury, haemolysis, thrombocytopenia ($n = 1$ at 3×10^{14} vg/kg) [187]
SMA	AAV9	Hybrid CMV enhancer/chicken β-actin promoter	Human SMN	AVXS-101	Intravenous 6.7×10^{13} vg/kg 2×10^{14} vg/kg	NCT02122952 AveXis, Inc.	Phase 1 completed	2014–2017 (2 years)	Up to 6 months of age, males and female, n = 15/9	Elevated serum aminotransferase levels (>10× normal level) [44]
SMA	AAV9	Hybrid CMV enhancer/chicken β-actin promoter	Human SMN	AVXS-101	Intravenous therapeutic dose	NCT03306277 (STR1VE) AveXis, Inc.	Phase 3 completed	2017–2019 (18 months of age)	Up to 6 months of age, males and females, n = 22/15	-
SMA	AAV9	Hybrid CMV enhancer/chicken β-actin promoter	Human SMN	AVXS-101	Intrathecal 6×10^{13} vg 1.2×10^{14} vg 2.4×10^{14} vg	NCT03381729 (STRONG) AveXis, Inc.	Phase 1 suspended (on clinical hold pending further discussions regarding pre-clinical findings)	2017–2021 (15 months)	6 to 60 months of age, males and females, n = 51/27	SAE mainly related to the disease itself ($n = 7$). Transaminitis events probably related to treatment ($n = 2$). [188]

Table 1. *Cont.*

Product/Administration						Clinical Design				
Disease	**AAV Serotype**	**Promoter**	**Transgene**	**Name**	**Administration/Dose**	**Clinical Trial ID (Study Name) Sponsor/Collaborator**	**Study Phase/Status**	**Study Timelines (Clinical Follow-Up)**	**Age, Gender, Actual or Estimated/ Planned Number of Participants Enrolled**	**Serious Adverse Events**
SMA	AAV9	Hybrid CMV enhancer/chicken β-actin promoter	Human SMN	AVXS-101	Intravenous	NCT03461289 (STRIVE-EU) AveXis, Inc.	Phase 3 completed	2018–2020 (18 months of age)	Up to 6 months of age, males and females, $n = 33/30$	-
SMA	AAV9	Hybrid CMV enhancer/chicken β-actin promoter	Human SMN	AVXS-101	Intravenous 1.1×10^{14} vg/kg	NCT03505099 (SPR1NT) AveXis, Inc./PRA Health Sciences	Phase 3 active, not recruiting	2018–2021 (18 and 24 months of age)	Up to 42 days, males and females, $n = 30/44$	-
SMA	AAV9	Hybrid CMV enhancer/chicken β-actin promoter	Human SMN	AVXS-101	Intravenous single dose	NCT03837184 AveXis, Inc./PRA Health Sciences	Phase 3 active, not recruiting	2019–2021 (18 months of age)	Up to 6 months of age, males and females, $n = 2/6$	-
XLMTM	AAV8	Des	Human MTM1	AT132	Intravenous 1×10^{14} vg/kg 3×10^{14} vg/kg	NCT03199469 (ASPIRO) Audentes Therapeutics	Phase 1/2 active, not recruiting (FDA placed on clinical hold since June 2020)	2017–2024 (5 years)	Up to 5 years, males, $n = 24/12$	Progressive liver dysfunction, hyperbilirubinemia, death from sepsis or gastrointestinal bleeding ($n = 3/17$ at 3×10^{14} vg/kg) [189]
Pompe	AAV2/8	Liver-specific promoter	hGAA	ACTUS-101	Intravenous 2 doses	NCT03533673 Asklepios Biopharmaceuticals, INC./Duke University and National Institute of Arthritis and Musculoskelatal and Skin Diseases (NIAMS)	Phase 1/2 recruiting	2018–2022 (52 weeks)	18 years and older, males and females, $n = 8/6$	-
Pompe	AAV	Liver-specific promoter	hGAA	SPK-3006	Intravenous dose escalation	NCT04093349 (RESOLUTE) Spark Therapeutics	Phase 1/2 Recruiting	2020–2023 (52 weeks)	18 years and older, males and females, $n = 20$/same as current	-
Pompe	AAV8	Hybrid liver/desmin promoter	hGAA	AT845	Intravenous 2 doses	NCT04174105 (FORTIS) Audentes Therapeutics	Phase 1/2 Recruiting	2020–2027 (5 years)	18 to 80 years, males and females, $n = 8$/same as current	-
Danon	AAV9	CAG	hLAMP2B	RP-A501	Intravenous 2 doses	NCT03882437 Rocket Pharmaceuticals Inc.	Phase 1 recruiting	2019–2023 (3 years)	8 years to 14 years and 15 years and older, males, $n = 24$/same as current	-

Table 1. *Cont.*

Product/Administration						Clinical Design				
Disease	AAV Serotype	Promoter	Transgene	Name	Administration/Dose	Clinical Trial ID (Study Name) Sponsor/Collaborator	Study Phase/Status	Study Timelines (Clinical Follow-Up)	Age, Gender, Actual or Estimated/ Planned Number of Participants Enrolled	Serious Adverse Events
LGMD2E	scAAV rh74	MHCK7	SGCB	SRP-9003	Intravenous 5×10^{13} vg/kg	NCT03652259 Sarepta Therapeutics, Inc.	Phase 1/2 active, not recruiting	2018–2020 (3 years)	4 to 15 years, males and females, $n = 6/9$	Elevated liver enzymes associated with transient increase in bilirubin ($n = 1$) [190]
Batten disease	AAV2	CU	hCLN2	-	CNS administration 3×10^{12} vg	NCT00151216 Weill Medical College of Cornell University/Nathan's Battle Foundation	Phase 1 completed	2004–2019 (18 months)	3 to 18 years, males and females, $n = 10/11$	-
Batten disease	AAVrh.10	CU	hCLN2	-	Direct CNS administration 9×10^{11} vg/2.85×10^{11} vg	NCT01414985 Weill Medical College of Cornell University	Phase 1/2 completed	2010–2017 (18 months)	3 to 18 years, males and females, $n = 8/16$	-
Batten disease	AAVrh.10	CU	hCLN2	-	Direct CNS administration 9×10^{11} vg 2.85×10^{11} vg	NCT01161576 Weill Medical College of Cornell University/National Institute of Health	Phase 1 active, not recruiting	2010–2032 (18 months)	2 to 18 years, males and females, $n = 25/16$	-
Batten disease	scAAV9	CB	CLN6	AT-GTX-501	Intrathecal	NCT02725580 Amicus Therapeutics	Phase 1/2A active, not recruiting	2016–2021 (24 months)	1 year and older, males and females, $n = 13/6$	-
Batten disease	scAAV9	P546	CLN3	AT-GTX-502	Intrathecal 2 doses	NCT03770572 Amicus Therapeutics	Phase 1/2A active, not recruiting	2018–2023 (36 months)	3 to 10 years, males and females, $n = 7$/same as current	-
GSD1a	AAV8	Native promoter	G6Pase	DTX401	Intravenous 3 doses	NCT03517085 Ultragenyx Pharmaceutical INC	Phase 1/2 recruiting	2018–2020 (52 weeks)	18 years and older, males and females, $n = 18/9$	No treatment-related serious adverse events reported to date

DMD: Duchenne muscular dystrophy; SMA: Spinal muscular atrophy; XLMTM: X-linked myotubular myopathy; LGMD2E: Limb girdle muscular dystrophy type 2E; GSD1a: Glycogen storage disease type 1a.

4.1. SMA Trial

The ubiquitous SMN protein plays a key role in RNA regulation, and its deficiency in SMA is associated with cell-specific pre-mRNA splicing defects, possibly accounting for the tissue selectivity of the pathology [191]. Indeed, lower motoneurons are the cells primarily affected by degeneration in SMA, but other tissues, in particular, the heart, are also occasionally affected [192]. Local CNS delivery performed by the intrathecal administration of an rAAV9-h*SMN* vector proved efficient in correcting the motoneurons pathology in mice at doses in the 10^{13} vg/kg range [193]. Widespread distribution of the transgene was also demonstrated in non-human primate (NHP) spinal cord and brain motoneurons. However, local delivery might be clinically risky, and limits body-wide vector distribution and the correction of extra-CNS symptoms. Interestingly, taking advantage of the ability of AAV9 to cross the blood-brain barrier after systemic administration, three independent laboratories reported the preclinical safety and efficacy of an rAAV9-human *SMN* vector in different animal models [194–197]. All these studies used a self-complementary (sc) vector, which bears a DNA construct enabling the shunting of the transcription of the second DNA strand and hence leads to quicker gene expression than conventional single-stranded vectors [198]. A remarkable rescue of the phenotype was observed in mouse and cat models of SMA receiving intravenous doses ranging from 3×10^{13} to 3.3×10^{14} vg/kg of body weight (see Table 2 for details). The treatments rescued survival and all the major clinical manifestations, such as muscle atrophy and weakness, respiratory distress, weight loss and paralysis. The correction of murine cardiomyopathy was also reported [195]. An extensive motoneuron distribution was confirmed in cynomolgus macaques [194]. The product used by Barkats and collaborators has similar effects to the one used by Kaspar and collaborators at a 10-fold lower AAV dosage [194,196] (Table 2). This might be due to the codon-optimised enhancement of the transgene expression and/or by way of promoter regulation. Widespread motoneuron transduction was observed in the spinal cord, and the heart, skeletal muscles and liver were also highly transduced. Apart from the necrosis of the tails and ears seen in long-term survivors and attributed to the lack of SMN in these tissues [194,196,197], no safety issues were evidenced, and a clinical trial was initiated in 2014 [44]. Fifteen 0.9- to 7.9-month-old patients were treated intravenously with an rAAV9-h*SMN* product controlled by the hybrid CMV enhancer/chicken-β actin promoter (product referred to as AVXS-101), either with 6.7×10^{13} vg/kg (three patients) or 2×10^{14} vg/kg (twelve patients) (AveXis/Novartis, NCT02122952) [44]. To this day, all the patients are alive and show significant amelioration of motor, respiratory and nutritional functions [44,153]. The improvements are substantial as seen from a comparison with a natural history cohort [154]. The effect is dose-dependent and related to the time of initiation: the earlier the injection, the more efficient the treatment [156]. This treatment proved more efficient than Spinraza® [155]. It is also longer-lived and safer, as Spinraza® necessitates constant re-administration by the risky intrathecal route [199]. Additional benefits might come from the widespread correction of SMN-related defects in other organs, the most important being the cardiac tissue. To date, fifty-six SAEs have been reported, amongst which two are deemed treatment-related [44]. They are limited to elevated serum aminotransferase levels reaching more than 10 times the normal range, without any other liver enzyme abnormalities or clinical manifestations. This important elevation of hepatic enzymes was rescued by a short course of glucocorticoids in the first patient, treated with a low dose (prednisolone, 1 mg/kg/day for 30 days, starting one day before AAV injection), which was thereafter administered systematically one day before the treatment administration to prevent liver-related toxicity. Granted these excellent results, the FDA approved AVXS-101 for usage in SMA patients in the USA in May 2019 [43]. This new drug goes by the name of Zolgensma and is the third AAV-based gene therapy approved to date for genetic diseases.

Table 2. Preclinical animal studies in mice models of SMA.

Reference.	Promoter	Codon Optimisation	Dose vg/per Mouse	Dose vg/kg of Body Weight	Expression in CNS	Mean Survival (Days)	Adverse Events
[196]	PGK	Yes	4.5×10^{10}	3×10^{13}	SC: 80–140% of WT levels Brain: low	160 d (in 100% mice)	- Hyperactivity - Tail necrosis - Ear necrosis - Bilateral cataract
[194]	CBA	No	5×10^{11}	3.3×10^{14}	SC: 42% of WT levels	>250 d (*n* = 4, 1 death at d 97)	- Necrotic pinna
[197]	CMV	Yes	1×10^{11}	6.7×10^{13}	Lumbar SC: 66.5% MN Thoracic SC: 45% MN Cervical SC: 55% MN	69 d (in 80% of mice)	- Short tail - Ear necrosis - Moderate eyelid inflammation

All mice were from the SMN delta 7 strain ($SMN2^{+/+}$, $SMN\Delta7^{+/+}$, $smn^{-/-}$) and received an intravenous injection in the facial vein at p1 of an AAV9 solution. The doses expressed in vg/kg were calculated using a mouse body weight at p1 of 1.5 g (the average weight of a p1 SMN-Delta7 pup [193]).

Intriguingly, a preclinical report published after this trial's initiation demonstrates the acute toxicity of a closely related product composed of an identical CAG-h*SMN* cassette packaged in the rAAV9 variant AAVhu68 and injected into wild-type NHPs and piglets at 2×10^{14} vg/kg (the highest dose in the SMA trial) [200]. The biological abnormalities did not resolve in one out of three injected NHPs, leading to euthanasia at Day 5. No piglets died. The vector genome copy number was roughly 1000-fold higher in the liver than in other tissues. Intense dorsal root sensory neuron degeneration was evidenced in both the NHPs and piglets, with additional acute hepatocellular injury and liver failure, systemic inflammation and internal haemorrhage in monkeys. Because of the acute time course (abnormal parameters at Day 4–5), the toxic effects are not thought to be related to the activation of an adaptive immune response to the capsid or transgene and destruction of hepatocytes. They are more likely to result from the activation of an intracellular cellular stress pathway linked to genome or capsid overload in hepatocytes, together with the activation of systemic inflammation and the associated coagulopathy. In line with these findings, another AAV9-derived vector, AAV-PHP.P, coding for an unrelated GFP transgene and injected in NHPs at a slightly lower dose of 7.5×10^{13} vg/kg (nearly identical to the lower dose of 6.7×10^{13} vg/kg in the SMA trial) led to similar toxic effects on the liver and to thrombocytopenia and haemorrhage [201]. Here again, the time course of the acute symptoms is not consistent with an undesirable activation of the adaptive immune system. It is unclear whether liver damage or coagulopathy is the primary defect, but it is worth mentioning that the liver damage induced by some viral infections participates in lowering the platelet number, although this mechanism was not reported for AAV vectors [202]. Whether the toxic effects are related only to the dose or to the capsid used remains unclear. AAVhu68 and AAV PHP-B are closely related to AAV9 (two and seven amino acids of variation, respectively), but that could substantially change vector entry and processing. It would seem so, as high doses of rAAV9 ranging from 7.5×10^{13} vg/kg [201] to $1–3 \times 10^{14}$ vg/kg [138] did not lead to toxicity in NHPs. The toxicity might also be related to the species used (NHPs and piglets) and the health status (wild-type animals) and will not necessarily translate into human toxicity in patients, as vector processing might be substantially different. The toxicity could also relate to the un-unified mode of vector purification and contaminants, especially as the ratio of empty/full capsids varies according to protocols. Thus, the therapeutic window is probably quite narrow in the SMA trial, as high levels of vector are necessary to achieve therapeutic benefit.

4.2. DMD Trials

DMD is a devastating and the most common muscle degenerative pathology, and as such, it has been the subject of many therapeutic attempts. Respiratory and cardiac complications are common, and patients' lifespans are severely reduced. *DMD* is the largest human gene (≈14 kb cDNA, NM_004006.2), which impedes its encapsidation within an AAV. Dystrophin is composed of an N-terminal actin-binding domain, 24 spectrin-like repeats articulated by four hinge regions and, at the C-terminal extremity, a cysteine-rich domain and a specific C-terminal domain. This highly flexible molecule interacts with a membrane-bound molecular complex (DAPC for dystrophin-associated-protein complex) and with sarcomeric actin, ensuring the plasticity of the muscle structure and resistance to contraction-induced injury. The observation that the deletion of a large part of the central domain leads to a very mild phenotype in patients set the ground for a large number of therapeutic trials aiming at expressing dystrophin forms shortened in the spectrin-like region [19]. Two main paths have been followed: shortening the natural gene by inducing the skipping of specific exons and restoring the reading frame (mutation-specific therapies) or bringing in trans a reduced version of the gene (a therapy amenable to all forms of dystrophinopathies). A founder paper of Chamberlain's team established the importance of the different domains by investigating the phenotypes of transgenic mice in which various forms of shortened dystrophin, named micro (<30% of the full-length coding sequence)- or mini-dystrophin, were expressed [203]. Exon-skipping feasibility and efficacy was demonstrated by bringing the adequate oligonucleotide within myofibres using a U7-driven AAV [204–209]. The gene transfer of micro-dystrophin offers more versatile options and proved very efficient in improving muscle force, protecting muscle from contraction-induced lesions and improving heart function in murine and canine models of the disease using the rAAV serotypes 6 [125,210–212], 8 [213–216] and 9 [139,141,217–221] delivered by the intravascular route. The most commonly used micro-dystrophins are variants of the ΔR4-R23ΔCT form, but the inclusion of the R16-R17 spectrin-like domains, involved in linkage to the membrane-bound cell metabolism regulator nitric oxide synthase NOS [222], was proposed to have additional therapeutic benefits [140,218]. Based on these preclinical proofs of concept, three clinical trials using micro-dystrophin gene transfer have been in progress since 2017 (Pfizer [New York, NY, USA], NCT03362502; Sarepta Therapeutics: NCT03375164; Solid Biosciences [Cambridge, MA, USA], NCT03368742). The three trials use muscle-specific promoters, rAAV9 or rAAVrh74 serotypes, and high and comparable doses of vector (1×10^{14} to 3×10^{14} vg/kg) (see Table 3). AAVrh74 was chosen by one group (Sarepta's trial) because of its simian origin, which should decrease the likelihood of pre-existing immunity. Indeed, in a population of DMD patients, AAVrh74 sero-prevalence was shown to be low (measured in fewer than 20% of the patients tested) [223], and the average titres were also amongst the lowest [224]. Quite surprisingly, this seems to be a specific feature of DMD, as higher titres of antibodies are measured in non-DMD children, possibly owing to the small size of the population or to a disease-specific effect on AAV biology [224]. Another study even showed higher levels of antibodies against rAAVrh74 than against other serotypes in a healthy child population, probably because of cross-reactivity with serotypes present in humans [225]. Furthermore, AAVrh74's safety has been demonstrated in a preclinical dose-escalation study in Duchenne's model mice and in NHPs [226,227], as well as in humans in a clinical trial targeting LGMD, though the doses used were 100-fold lower than in the current DMD trial (1×10^{12} or 3×10^{12} vg/kg in the LGMD trial versus 2×10^{14} vg/kg in the DMD trial) [228]. The minimal effective dose was defined as 2×10^{14} vg/kg in *mdx* mice, a DMD model, and safety was confirmed in NHPs at doses reaching up to 6×10^{14} vg/kg [226,227,229]. One year after a single injection of 2×10^{14} vg/kg of AAVrh74-MHCK7-coΔR4-R23ΔCT (SRP-9001) in four patients, the first results are encouraging in terms of safety [177]. No SAEs were reported, and 18 mild or moderate events were deemed treatment-related. As previously observed in haemophilia [230,231] and SMA [44] clinical trials, liver enzymes peaked and diminished with a glucocorticoid course ($n = 3$). No adverse immune responses occurred, and, as expected, a transitory T cell response and the development of stable titres of antibodies against AAVrh74 were observed. The product was highly expressed, as seen in biceps brachii biopsies. Whether the treatment has any beneficial effects remains to be assessed more closely, although a clinically meaningful improvement of 2.2 to 7 points on the NorthStar Ambulatory Assessment score (NSAA) multi-parametric scale (maximum score of 34) suggests motor function improvement. A comparison with a historical cohort of untreated patients and longer time of treatments is needed to draw more definite conclusions. These encouraging results preclude dose escalation, and a new randomized, placebo-controlled clinical trial with a much larger sample size is under way (NCT03769116).

Table 3. Micro-dystrophin clinical trials in DMD patients. NSAA: NorthStar Ambulatory Assessment score, SM: Skeletal muscle.

Trial Promoter/Product Name/Reference	Vector	Promoter	Micro-Dystrophin Domains	Dose vg/kg of Body Weight	Expression in SM	NSAA	Serious Adverse Events
Sarepta SRP-9001-101 [177]	AAVrh74	MHCK7 (SM and cardiac)	coΔR4-R23/ΔCT	2×10^{14}	95.8% of normal	5.5 points increase after 1 year	
Pfizer PF-06939926	AAV9	Human muscle specific	-	1×10^{14} 3×10^{14}	23.6% of normal 29.5% of normal	2 points increase after 1 year	In 1 patient at 3×10^{14} vg/kg: complement activation, acute kidney failure, thrombocytopenia
Solid SGT-001	AAV9	CK8	ΔR2-R15/ΔR18-R22/ΔCT	2×10^{14} 2 doses			Complement activation, acute kidney failure, thrombocytopenia (2 SAEs in 6th patient)

However, at very close doses, two other products composed of an rAAV9, a muscle-specific promoter and a micro-dystrophin transgene led to product-related SAEs. In one trial (product PF-06930026, Pfizer), the six participants included to date have shown a mean of ≈40% dystrophin-positive fibres at 1×10^{14} vg/kg and ≈70% at 3×10^{14} vk/kg in a bicep biopsy taken two months after injection, corresponding altogether to ≈24 to 30% of normal dystrophin expression [187]. The NSAA score increased by 4.5 points after one year in two participants treated with the lowest dose. However, one child treated with 3×10^{14} vg/kg developed a rapid antibody response with complement activation, acute kidney injury, haemolysis and thrombocytopenia. A transient 2-fold elevation of liver serum enzymes was observed, though it was not considered significant enough to indicate hepatic failure. Suspected complement-mediated nephropathy resulted in a protocol-driven pause of enrolment. Haemodialysis together with a course of complement inhibitor solved the problem in fifteen days. In the third trial (product SGT-001, Solid Biosciences), which differs slightly by the construct used (product SGT-001, different promoter and integration of the nNOS-binding domain in the transgene), similar treatment-related toxic events were seen in two patients at doses of 5×10^{13} and 2×10^{14} vg/kg. To date, six patients have been included, three at low and three at high doses. The preliminary results showed weak dystrophin expression in the three patients who received low dosages. The first patient injected at 5×10^{13} vg/kg developed complement activation, kidney failure and platelet count drops without signs of liver damage. The clinical hold [186] was lifted in 2018 after full symptom resorption following treatment with a modified course of steroids and a complement inhibitor and a change in the study design (the inclusion of an intravenous glucocorticoid administration in the first weeks following drug injection). A second patient dosed at 2×10^{14} vg/kg developed the same symptoms together with cardiopulmonary decline, leading to a second FDA hold of the trial. The SAEs fully resolved, but the clinical trial remained on hold on the grounds of remaining questions related to the mode of production of the product [185,232] and was finally allowed to continue in October 2020 [233]. A dose-finding study in a canine model of the pathology did not evidence any safety issue for this product at doses reaching 5×10^{14} vg/kg [221,234]. These SAEs could be related to the AAV9 capsid, though no severe side effects were observed in the SMA trial with this serotype at an equivalent dosage. The genetic background might account for the different effects between the SMA and DMD trials, whether for vector processing or the immune response. In the absence of liver injury, an immune response-mediated platelet drop, complement activation and ensuing nephropathy might be a reasonable hypothetical pathogenic mechanism. This could also be in line with the incidence of the age of the patients, as younger children are included in the only trial without SAEs, and the immune system is immature at a younger age [235]. The information on the three DMD trials is summarised in Table 3.

4.3. XLMTM Trial

XLMTM is a very rare congenital centronuclear myopathy caused by mutations in the *MTM1* gene, affecting 1/50,000 boys [236]. Skeletal and respiratory muscles are deeply affected, and many patients decease before one year of age, mainly from respiratory failure. The *MTM1* gene encodes a lipid phosphatase, myotubularin, involved in PI_3P dephosphorylation and membrane remodelling [237,238]. The myotubularin cDNA, together with the regulatory elements, can be packaged in an AAV, and two very good murine and canine models of the disease recapitulate the main features of the pathology, noticeably, histological defects specific to centronuclear myopathies, generalised muscle hypotrophy and weakness, and lifespan reduction [239–241]. With these tools in hand, Buj-Bello and collaborators established a very convincing proof of concept, first by intramuscular injection with an rAAV2/1-CMV-*mtm1* product [242], and next using the whole-body delivery of an rAAV2/8-Des-*mtm1* product in mouse and canine models [163]. In both models, a single intravenous injection of a dose of $\approx 3 \times 10^{13}$ vg/kg led to an important improvement of muscle and respiratory functions, and survival was largely extended. Importantly, therapeutic effects were also observed, though to a lesser extent, in older mice, showing that pathology reversal, essential in patients presenting the symptoms at birth, could be achievable. In 4-year-old, long-term survivor dogs, gait, respiratory and neurological functions remained comparable to the ones of wild-type, age-matched controls, despite a progressive decline in the vector copy number in muscles, which reached a plateau after three years of age, and a diminution of muscle force [243]. A dose study carried out in the canine model established the dose-dependency of the therapy, with a significant correction achieved from 2×10^{14} vg/kg, a quasi-normalisation of the phenotype at 5×10^{14} vg/kg and no significant side-effects, apart from the expectable humoral immune response towards the vector and a thickening of the heart septal wall without functional consequences [244]. In this protocol, muscle expression defects evidenced by a transcriptomics approach were corrected by the mid-dose of 2×10^{14} vg/kg [245]. Considering that the doses reversing the pathology are in the 10^{14} vg/kg range and challenge vector production, an additional efficacy study was carried out in three infant NHPs [246]. Eight weeks after intravenous injection, a dose of 8×10^{14} vg/kg did not lead to significant treatment-related adverse events and produced MTM1 protein expression at levels 8- to 20-fold higher than endogenous levels in target skeletal muscles [246]. Importantly, despite a high vector copy number in the liver, the myotubularin protein level remained normal, and serum markers of liver damage did not peak significantly. Altogether, these results led to the initiation of a clinical trial in 2017 on XLMTM infants. The ASPIRO phase 1/2 trial aims at treating ventilatory-assisted patients aged less than 5 years with ascending doses (1×10^{14} vg/kg or 3×10^{14} vg/kg) of an rAAV2/8-Des-h*MTM1* vector (AT132 product, Audentes Therapeutics [San Francisco, USA], NCT03199469). Until very recently, the results were strikingly positive. To date, twenty-three patients have been treated, six at 1×10^{14} vg/kg and 17 at 3×10^{14} vg/kg: the CHOP-INTEND (Children's Hospital Of Philadelphia INfant Test of Neuromuscular Disorders) has improved by various levels, the limb and trunk strength have increased, and new developmental skills have been achieved, such as controlling head movement, rolling over or sitting unassisted [247–249]. Respiratory function has improved significantly resulting in patients being weaned off ventilators completely. One SAE possibly related to the product occurred and was resolved by a course of intravenous steroids and supportive care. However, since the 5 May 2020, three patients treated with the highest dose have died. All three patients had progressive liver dysfunction characterised by hyperbilirubinemia starting a few weeks after dosing. Preliminary findings suggest that two children died from sepsis and one from gastrointestinal bleeding. The FDA put the trial on hold on the 29 June [189]. This tragic event remains hard to rationalize, as 14 out of 17 children treated with the high dose have not developed complications to date. The common features of the three deceased children were an older age (the boys were at the higher end of the age cut-off), a heavier weight and a pre-existing hepatobiliary disease of an unknown severity, although one can assume it to be mild, as hepatic disorders were an exclusion criterion. This condition might have facilitated liver toxicity due to the large doses of vectors. This toxicity is reminiscent of the one observed in NHPs [200,201], and the activation of complement through the formation of vector–antibody complexes, which have been implicated in lethal systemic inflammation with an adenovirus vector [250], has been hypothesized [251]. Of note, some children dosed at 1×10^{14} vg/kg also had pre-existing liver disorders and did not develop the complications, despite being years out from treatment.

5. Improvement of the Therapeutic Toolbox

5.1. Towards Safer Next-Generation Muscle and CNS-Restricted AAVs

It is becoming increasingly evident that AAVs should be chosen carefully for every clinical application, considering specificities such as the patient's genetic background, age, disease progression, sex, immunological state and targeted tissues. Capsid engineering is commonly used to develop safer next-generation AAV variants. These methods rely either on rational design in which capsids are tailored by targeted modifications, or on directed evolution, consisting of recovering new capsids from randomly generated high-complexity libraries after selective pressure on a tissue of interest.

For neuromuscular disorders, the improvement of muscle transduction; reduction of off-targeting, especially in the liver; and development of vectors escaping the immune response are major endeavours.

AAV2.5, obtained by replacing five residues in the AAV2 capsid with corresponding orthogonal residues of AAV1 [146] and several other variants generated by variable combinations of 32 capsids' amino acids [252], improved muscle transduction compared with parental serotype 2 or 1 but were not assessed for whole-body distribution. Three AAV2 variants, AAV2i8, a chimeric capsid obtained by replacing a receptor-binding hexapeptide motif in the AAV2 capsid with corresponding residues in the AAV8 capsid [253], and two variants obtained by peptide insertions in a hypervariable loop [254,255] showed equivalent or improved targeting in skeletal muscles, with an important reduction in the liver in comparison with AAV2. AAV2i8 was also shown to be less likely to be serum-neutralized than the parental capsid [253]. The ratio of skeletal muscle/liver transduction was also better than for AAV9 in mice [256] but not in NHPs [257]. An additional insertion of a galactose-binding footprint on AAV2i8 did not improve the ratio further in mice [256].

Three other variants proved even more efficient than AAV9 for improving the muscle/liver transduction ratio: (1) AAV-9.45 is an AAV9 variant obtained by the random integration of amino acids and showing reduced liver expression and identical muscle and heart transduction when compared with AAV9 [258]. (2) AAVpo1 is a natural pig isolate that transduces muscles and the heart to a slightly lower level than AAV9 but presents the advantage of being completely detargeted from the liver [259]. (3) AAV-B1 is a chimeric AAV isolated from a shuffled library consisting of 11 parental serotypes and displaying reduced liver transduction and at least 10-fold higher muscle and CNS tropism than AAV9 [260].

A series of mutations on surface phosphorylable residues of the AAV1 and AAV9 capsid improved vector stability and led to 3 to 10 times lower transduction in the liver than in muscles [261]. AAVM41 was isolated from a chimeric AAV1 and AAV9 capsid's shuffled library and reduced both skeletal muscle and liver targeting while preserving heart transduction compared with AAV9, suggesting that this serotype could be of interest for rescuing cardiac pathologies [262,263]. Tyrosine-specific modifications of the AAV6 capsid can improve vector muscle entry [264].

Several variants demonstrated interesting characteristics regarding immune evasion. Bat AAV serotype 10HB transduced muscle with a higher muscle/liver ratio than primate AAV and showed a reduced sensitivity to antibody neutralisation [265]. A method consisting of purifying new AAV2 variants by rabbit antibody-specific affinity chromatography resulted in the identification of several antibody-resistant clones, though neither the relevance to human sera nor variant biodistribution were assessed [266]. The AAV1 variant CAM130 isolated through multiple rounds of neutralizing-antibody escape from several species evaded neutralizing antibodies, even at high concentrations, in mice, NHPs and human sera, while maintaining the tissue tropism of the parental AAV1, suggesting it could be suitable for clinical trials in large populations, as seropositivity is a common exclusion criterion [267]. Finally, after applying the double selection of variants resistant to human-serum neutralization and selected after local muscle transduction, the AAV mutant MuS12 was isolated and showed immune response escape together with the preservation of muscle tropism, although transduction was largely reduced by the intravenous route in comparison with AAV9 [268]. Considering that this variant transduces muscle differently according to the route of administration, low vascular permeability was hypothesised. Future protocol improvement could aim at selecting new variants after intravascular injection.

Altogether, these new vectors have the potency to improve targeting efficiency and reduce the off-target effects and immune response. Their respective characteristics are summarised in Table 4.

Table 4. Next-generation recombinant AAVs for improved muscle targeting, reduced liver targeting and/or a reduced immune response.

Reference	AAV Name	Parental AAV	Method	Compared with	Receptor	Muscle Transduction	Heart Transduction	Liver Transduction	Immune Response
[253]	AAV2i8	AAV2	Rational design: replacement of receptor-binding hexapeptide with corresponding residue in AAV8	AAV2/AAV8	Not HS	=AAV8 >AAV2	=AAV8 >AAV2	<AAV2 and AAV8 (40-fold lower)	Lower cross reactivity to AAV2 antibody
[257]				AAV9 (rhesus monkey)		<AAV9 (122-fold lower)	<AAV9 (46-fold lower)	<AAV9 (11-fold lower)	ND
[256]	AAV2i8G9	AAV2i8	Rational design: graft galactose-binding footprint of AAV9 in VP3 AAV2i8	AAV2i8/AAV9	Not HS Glycan	>AAV2i8>AAV9	AAV2i8 < AAV2i8G9 < AAV9	AAV2i8 < AAV2i8G9 < AAV9 (5-fold lower)	ND
[146]	AAV2.5	AAV2	Rational design: AAV2 capsid with 5 mutations from AAV1	AAV2	HS	>AAV2 (2- to 5-fold)	NA	NA	-No cellular immune response to capsid. -Lower cross-reaction to AAV2 NAb
[254]	AAV2 587 MTP	AAV2	Rational design: insertion of muscle-targeting peptide in AAV2 capsid	AAV2	Not heparin	≥AAV2 (2-fold)	>AAV2 (7-fold)	<AAV2 (2.5-fold)	ND
[258]	AAV9.45	AAV9	Directed evolution: random mutagenesis of surface-exposed regions of AAV9	AAV9	ND	≈AAV9	≈AAV9	<AAV9 (10- to 25-fold lower)	ND
[259]	AAV po1	NA	Natural pig isolate	AAV9	ND	<AAV9 (≈2–4-fold)	<AAV9 (≈3-fold)	<AAV9 (≈140-fold)	ND
				AAV5	ND	>AAV5 (1.5-fold)	<AAV5 (30-fold)	<AAV5 (≈125-fold)	No pre-existing immunity No cross-neutralisation by antisera against all common AAVs
[261]	AAV9-Y731F AAV1-Y445F/Y731F	AAV9 AAV1	Rational design: tyrosine mutations	Other mutants (no AAV of reference)	ND	Skeletal muscle > heart ≈ liver liver (3–10-fold lower) < skeletal muscle (3–10-fold lower) < heart			ND
[255]	AAV2-VNSTRLP	AAV2	Directed evolution: from AAV2 display peptide library with in vitro selection for heart tropism	AAV2 AAV9	ND	≈AAV2 <AAV9	>AAV2 (>10-fold) <AAV9	<AAV2 (≈10-fold) <AAV9	ND
[263]	AAVM41	AAV1/6/7/8	Directed evolution by shuffling the capsids of AAV1 to AAV9 and in vivo selection on skeletal muscle	AAV9 AAV6	ND	<AAV9 >AAV6	≈AAV9 >AAV6 (up to 13-fold)	<AAV9 <AAV6	Lower cross reactivity
[260]	AAVB1	AAV8 AAVrh43 (mostly)	Directed evolution of DNA shuffled library and selection on brain tissues	AAV9	Not SA Not Galactose	>AAV9 (10- to 26-fold higher)	>AAV9 (14-fold higher)	<AAV9 (3.6-fold lower)	Modestly more resistant to neutralisation than AAV9
[252]	AAVC4 AAVC7 AAVG4	NA	Ancestral reconstruction from NHP and human AAV by combinatorial variation of 32 amino acids and selection on muscle cells	AAV1	Not SA Not galactose Not HS	>AAV1 (10–31-fold higher)	NA	NA	Not resistant to neutralisation with IVIG

Table 4. *Cont.*

Reference	AAV Name	Parental AAV	Method	Compared with	Receptor	Muscle Transduction	Heart Transduction	Liver Transduction	Immune Response
[265]	AAV10HB	NA	Isolation from bat faecal and intestinal tissues	AAV2 AAV8	ND	Ratio muscle/liver = 8.8 Ratio muscle/liver < 1 for AAV2 and AAV8			Reduced neutralisation with IVIG ≈AAV2
[266]	Several variants: r2.4/r2.15	AAV2	Directed evolution: random mutagenesis and selection of heparin binding or neutralising serum binding	AAV2	Not heparin	ND	ND	ND	Reduced neutralisation with serum/AAV2
[268]	Mus12	Capsid shuffled library	Directed evolution: shuffled library selected on patients' sera and amplified in vivo in mouse muscle	AAV1/AAV2/ AAV2.5/AAV6/ AAV8/AAV9	ND	IM ≈ AAV6 ≈ AAV9 IV < AAV9	ND	ND	Immune escape
[267]	CAM130	AAV1	Directed evolution: rational mutagenesis on AAV1 capsid residues in contact with antibodies, library generation and evolution on vascular endothelial cells	AAV1	ND	ND	>AAV1 (2-fold)	=AAV1	Neutralisation escape to murine, NHP and human sera

IVIG: intravenous human IgG; HS: heparan sulfate; SA: sialic acid; NA: not applicable; ND: not determined.

5.2. Enhancing the Repertoire of Muscle and CNS-Restricted Promoters

With the development of in silico analysis technologies, a multistep, genome-wide data-mining strategy was performed to identify conserved skeletal muscle-specific cis-regulatory modules (Sk-CRMs) in highly expressed muscle-specific genes. Sk-CRM4, containing binding sites for the E2A, CEBP, LRF, MyoD and SREBP transcription factors, boosted transgene expression driven by Des or C5.12 promoters in heart and skeletal muscles (up to 400-fold), with a significant improvement of the *mdx* mouse phenotype [269]. Similarly, a 1030 bp modular muscle hybrid (MH) promoter composed of two enhancers (from the *Des* and *Mck* genes, respectively), a proximal promoter and an intron (modified from the *Mck* gene and core promoter) was more efficient than the Des promoter in skeletal and cardiac muscles, with limited expression in non-muscle tissues compared with the CMV promoter, showing a high potential for muscular gene therapy [270].

The presence of large promoters limits the size available for the transgene in the cassette, which proves problematic for several muscle genes. Promoterless cassettes were recently tested for liver expression. In this strategy, the transgene, flanked by homology arms, is brought into the cells by rAAV and integrates by nuclease-free homologous recombination downstream of the native promoter, where it is regulated like the endogenous gene is. Despite promising results in hepatocytes [271,272], the promoterless strategy might be limited for muscle application, as muscle cells are mostly quiescent and homologous recombination is restricted. Nonetheless, it might prove interesting for satellite cell targeting, provided that muscle progenitor targeting can be achieved.

5.3. Detargeting with miRNA-Based Elements

While regulatory elements such as introns, polyA signals or the Woodchuck hepatitis virus post-transcriptional regulatory element (WPRE) can be added to improve global transgene expression, miRNA-based sequences can mitigate tissue-specific transgene expression [273]. MiRNAs are small (approximately 22-nucleotide-long) non-coding RNAs post-transcriptionally silencing gene expression in plants and animals. Once bound to complementary target sites (TS) in mRNA, they either reduce its stability or inhibit its translation, which results altogether in the reduction of protein expression [274]. While the number of identified miRNAs has constantly increased since their discovery in *Caenorhabditis elegans* in 1993 [275], the miRBase database reports 1917 annotated hairpin precursors and 2654 mature sequences in the human genome [276]. Some miRNAs present a tissue-specific pattern of expression, with expression detectable only in a particular tissue or at least 20-fold higher than in other tissues [277]. Amongst tissue-specific miRNAs are found the MyomiRs, a family of miRNAs expressed in both cardiac and skeletal muscles, namely, miR-1, miR-133a, miR-122a, miR-124a, miR-208b, miR-499 and miR-486, with the exception of miR-208a and miR-206, which are specifically expressed in the heart and skeletal muscles, respectively [278].

One strategy to improve the specificity of AAV-mediated gene delivery, overriding the broad tissue tropism of AAV vectors and/or promoter leakage in non-targeted tissues, is based on the miRNA-mediated post-transcriptional regulation of the transgene. Indeed, the insertion of miRNA TS into the 3′UTR of a gene expression cassette limits transgene expression in tissues expressing the corresponding miRNA [279]. Due to the small size of miRNAs, it is therefore feasible to insert different miRNA TS in the 3′UTR of the expression cassette to detarget specific cell types depending on the application. For neuromuscular disorders, this strategy was applied for the reduction of expression in the heart [280], liver [281,282] and APCs [283–287].

The control of heart transgene expression is of utmost importance, because even if no specific cardiac toxicity has been reported to date in clinical trials, preclinical reports have evidenced the danger of transgene cardiac overexpression when the heart is not the primary target [280] or even when it is [288]. The insertion of the cardiac-specific miR-208a TS in the cassette was shown to prevent cardiac transgene expression and rescued the cardiac toxicity resulting from transgene overexpression in this organ, while maintaining the efficient expression of the transgene in skeletal muscles [280].

MiR-122 is highly expressed in the liver. The insertion of miR-122 TS in the 3′UTR of a reporter gene was able to prevent protein expression in the liver after rAAV9 intravenous administration without interfering with cardiac protein expression [281,282]. The level of transgene repression was related to the number of repetitive miRNA TS used. However, the recent paper of Kraszewska et al. challenges this apparently safe approach. Indeed, in some genetic backgrounds, transgene expression was completely repressed not only in the liver, but also in the cardiac muscle, linked with the presence of miR-122 in these animals' hearts. MiR-122 was also shown to be present in the human cardiac tissues of patients with cardiomyopathy and in human iPSC-derived cardiomyocytes. The cardiac expression showed high variability between different mouse strains, sexes and human individuals [289]. This publication challenges the liver-specificity of miR-122 and warns against miRNA inter-individual variability.

As previously mentioned, preventing transgene expression in APCs may avoid undesirable adaptive immune responses directed against the transgene product. A miR-based approach aiming at inhibiting transgene expression in APCs by inserting four targets of the endogenous miR-142-3p (exclusively expressed in the hematopoietic lineage) at the 3′ end of the transgene coding sequence [283] allowed escaping a deleterious adaptive immune response after gene delivery with either lentiviral [284] or AAV vectors [285–287].

Most importantly, it is crucial to verify during preclinical studies that miRNAs are not reduced by the miRNA TS and that their natural targets are not misregulated, as that could induce detrimental side effects. To our knowledge, no clinical trial has used miRNA TS in the cassette to date. If it ever happens, checking beforehand the mean level of the targeted miRNA in the treated population will be essential, as miRNA expression can substantially vary between individuals, sexes or pathologies [289–291].

6. From Preclinical Studies to Clinical Trials … and Back: General Point of View

The ongoing clinical trials summarised herein show spectacular results in terms of efficacy, especially in SMA and XLMTM, two very severe conditions characterised by generalised muscle weakness and respiratory deficiency often leading to infant deaths in the first years of life. The DMD trials need further investigation. However, this very beautiful landscape has lately been obscured by SAEs in two DMD and in the XLMLM trials, leading to three children's deaths in the last case. It is hard to find common features in the two situations, as, apart from the doses, which are very close, the genetic background, the age of the patients and the vector capsids are different. DMD-related adverse events have been proposed to be caused by adverse immune reactions driving acute kidney failure, while XLMTM fatal hepatotoxicity is not associated with obvious immunotoxicity. Future investigations will undoubtedly document these side-effects and help with the design of next-generation products, but for now, with the current state of our knowledge, a lot of effort has to be put into designing the safest therapeutic strategies for future trials, especially as some diseases are not prone to being good candidates in terms of the benefit/risk ratio. Several factors have to be considered in the "ideal" trial design (see Figure 1).

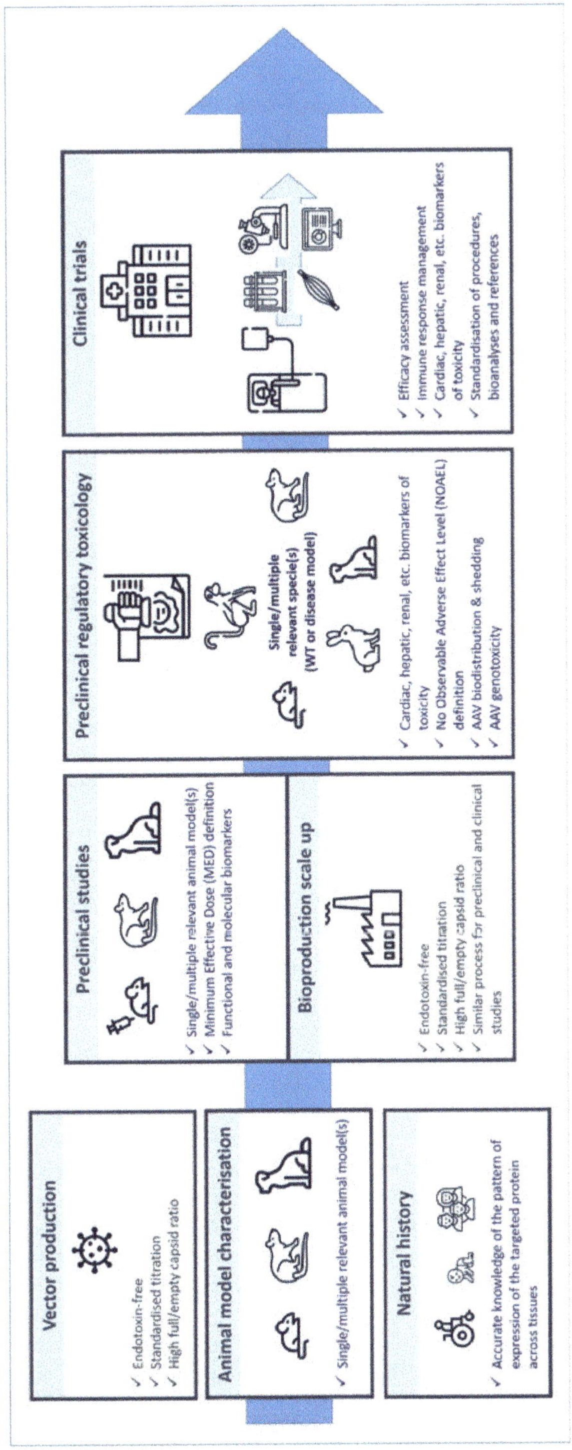

Figure 1. Of rAAV and men: the ideal journey.

6.1. Defining the Product

The transgene definition is obviously central and often evident, except in large proteins, where shorter forms have to be assessed with extreme care in preclinical studies to choose the best product possible. However, improvements can be made even with a full-length transgene. For example, codon optimisation leading to better protein expression has proven beneficial [196,292]. A thoughtful choice of promoter and other regulatory elements, such as WPRE addition, which proved efficient in enhancing transgene expression [293], can also improve the efficacy of the products. As for capsids, the choices are widening deeply, with new attractive vectors improving specific targeting. However, these vectors have to be assessed more closely in terms of safety before they can be considered for clinical usage. All these choices are crucial for specific tissue targeting and the reduction of off-target effects. Fundamentally, a very good knowledge of the levels of the protein to replace in every tissue of the healthy population is necessary to design the best targeted product. The addition of miRNA TS to detarget the liver [281], heart [280] and APCs [294] proved efficacious in reducing ectopic expression, but has not been tried in neuromuscular-deficient patients to date, probably because the underlying risks have to be assessed more closely. It is also worth mentioning that even with perfect targeting, the overexpression of transgenic protein in the targeted organs can also lead to toxic effects [295].

6.2. Manufacturing AAV

Vector production is a critical process for gene therapy success and safety. Importantly, the methods used for vector production and titration are not standardised, complicating the comparability of different clinical trials. Indeed, it was previously shown that the same production of rAAV8 led to significant variations in titres when dosed in 16 different laboratories [296]. In the absence of consensual methods, a common standard used for all clinic-intended rAAV production could help to correct the titres amongst trials.

Moreover, according to the method of production/purification used, various quantities of toxic contaminants can be found in AAV production. Indeed, endotoxins are known to be able to activate the human immune system and lead to SAEs and often contaminate AAV production. It is therefore crucial to reduce their load in the final product [297]. Their safety limit is defined as 5.0 International Units of endotoxin per kilogram of body mass by the FDA and the European Pharmacopeia for intravenous usage in humans [298,299], but could be raised further depending on the patient status (age, disease severity, etc). Of note, other contaminants, not known or tested, could also play a role in the safety and/or efficacy of the product. Notably, the presence of empty capsids in the final product was shown to reduce transduction efficiency and may participate in side effects [300].

6.3. Choosing the Best Preclinical Models

While proof-of-concept and preclinical studies aim at determining the minimal effective dose of the product, toxicology studies evaluate its safety/toxicity. Although using relevant animal models mimicking the human disease in proof-of-concept and preclinical studies seems obvious, the choice of animal models in toxicology studies remains unclear. The assessment of chemical drug toxicity is traditionally performed on wild-type animals such as rats, dogs or monkeys, as they are relevant for phase 1 clinical trials aiming at determining the safety in the general population. However, gene therapy cannot ethically be tested in healthy individuals. The crucial question is the relevance of extrapolating toxicology findings to human clinical trials. Vector entry relies on cell surface receptors and co-receptors, and its internal traffic requires components of the host cells. As these processes are not well deciphered in humans, inter-species variability could preclude toxicity results in some species. In line with this idea, piglets did not show liver failure and haemorrhage when administered the same dose of rAAVhu68 as NHPs [200]. Additionally, NHPs are not necessarily the best species as assumed, since toxicology studies in XLMTM at 8×10^{14} vg/kg did not detect coagulation defects or acute toxicity [246], while three children died at 3×10^{14} vg/kg. This is not completely surprising, as membrane and cytoskeletal remodelling likely alter vector processing in neuromuscular diseases, possibly significantly modifying AAV efficacy. In line with these considerations, the International Council for Harmonisation of Technical Requirements for Pharmaceuticals for Human Use (ICH) defines a relevant species as *"one in which the test material is pharmacologically active"* and states that if *"no relevant species exists, the use of relevant transgenic animals expressing the human receptor or the use of homologous proteins should be considered. [...] In certain cases, studies performed in animal models of disease may be used as an acceptable alternative to toxicity studies in normal animals"* [299].

6.4. Defining the Dose

For intravenous administration, the AVV dose administered is proportional to body weight, regardless of the age, gender, genetic background and disease severity of the individual treated. However, any of these factors could influence AAV efficacy and toxicity. The relative weight of organs is not proportional to body weight during development. For example, the liver/body ratio is higher in children than in adults [301], which could lead to variable levels of transduction and influence vector biodistribution at different ages. The immune systems of young children are not fully mature [235] and could therefore facilitate AAV transfer. Although it is not the only difference, the two DMD trials showing toxicity events potentially linked to immune responses were performed in older patients than the trial without any SAEs. Sex was also shown to impact AAV transfer in hepatic tissue but not in other tissues, with the male liver being more transduced than females' [302]. Finally, the disease itself can modify the structure and function of various organs, with high variability between individuals of the same age. Indeed, in the XLMTM trial, the three deceased patients had pre-existing hepatobiliary diseases and were older than the other infants without SAEs treated at the same dose. Although it is challenging, finding a better and universal normalisation method for dose calculation could probably improve clinical trial standardisation and safety.

6.5. Circumventing Immune Response

As previously mentioned, the adaptive immune response directed against a viral-derived vector restricts the full therapeutic potential of in vivo gene therapy [303]. Thus, in the first clinical trial showing safe and efficacious liver targeting with an rAAV2 vector carrying the human factor IX transgene (under the control of a liver-specific promoter), transgene expression was only transient [109]. A decline in expression starting at four weeks was associated with transient liver transaminases and the detection of CD8+ T cells directed against the AAV2 capsid. This unexpected deleterious cellular immune response was vector-dose-dependent and, in the absence of preclinical animal models, is still poorly understood. Nevertheless, a short prednisolone treatment quickly given in response to liver injury is often sufficient to stabilise transgene expression and has been used since then [230,231]. As a result, anti-AAV neutralizing antibodies are one of the most important remaining barriers, either impairing the efficacy of gene transfer in a set of patients with a cross-reactive pre-existing immunity against wild-type AAV, or precluding the redosing of patients developing a rapid and strong humoral response after the first vector injection. In preclinical models, numerous strategies targeting the host have been proposed such as plasmapheresis [304], direct tissue injection or isolated organ perfusion and immunosuppression combining rituximab (anti-CD20 depleting monoclonal antibody) and others drugs, or synthetic particles encapsulating rapamycin [305,306]. Other strategies target the rAAV vector itself, such as the use of alternative and less-prevalent serotypes, empty decoy capsids [307], exosome-enveloped AAV vectors or the generation of novel AAV capsids with optimised biodistributions and transduction efficacy, as well as the capacity to evade NAbs, as discussed above [47,305]. The most promising approach to including patients non-eligible to date was recently reported with the use of an IgG-cleaving endopeptidase from *Streptococcus pyogenes* (IdeS) [308]. The IdeS enzyme very rapidly (in a few hours) cleaves human IgG into F(ab')2 and Fc fragments, and is safe and efficient in patients with donor-specific antibodies undergoing kidney transplantation [309]. In both mouse and NHPs, Leborgne et al. reported that IdeS treatment was able to decrease pre-existing anti-AAV antibodies to a level sufficient to enable efficient liver gene transfer, even in the setting of vector re-administration [308]. Equivalent properties were demonstrated with IdeZ, a homolog of IdeS [310]. In July 2020, Sarepta Therapeutics announced an agreement with Hansa Biopharma to develop and promote imlifidase (the commercial IdeS) as a pre-treatment for DMD and LGMD gene therapy.

6.6. Assessing Long-Term Efficacy

Since the clinical trials using whole-body delivery in neuromuscular diseases are quite recent, the long-term assessment of efficacy will be made available in the next few years. For CNS-targeted treatments, a relative stability of the treatment is to be expected, as neurons are the longest living cells of the body. However, muscles, while being in a post-mitotic state, are remodelled during growth and following exercise, which could dilute the therapeutic effect. Targeting the treatment to satellite stem cells could hence be useful to help maintain long-term efficacy. Unfortunately, AAV-driven attempts to target satellite cells have failed. While reinjection might prove complicated, it might be worth using lentiviral vectors to target stem cells, as it was demonstrated efficient in transducing satellite cells in vivo [311], or new rAAV-rDNA integrating vectors, which proved their efficacy for directed integration in dividing and quiescent cells [312]. Importantly, it was recently shown that the AAV virus is found in episomal and randomly integrated transcriptionally active forms in human samples of liver tissues, and while it is impossible to know the time of infection, considering the large number of samples, it certainly suggests long-term persistence of the virus [89]. Whether this is also true for rAAV genomes remains to be determined.

6.7. Assessing Long-Term Toxicity

Serious concerns about the long-term safety of rAAV vectors were raised after several genotoxic studies performed in mice. Recombinant AAV2, 8 and 9 vectors were shown to integrate in the *Rian* locus on chromosome 12, irrespective of viral transgene, mouse genotype, sex or genetic background [104,105]. This insertion upregulates non-coding RNA and genes proximal to the *Rian* locus and is associated with an increased rate of hepatocellular carcinoma (HCC). The trans-regulatory elements carried by the vectors influence genotoxicity: sometimes, insertion is seen, but adjacent oncogenes are not overexpressed and HCC does not develop. Capsid-specific properties may also influence genotoxicity. These results were all obtained in neonatal mice, and neither integration in the *Rian* locus nor HCC were ever observed in older animals [106]. Other results obtained with sc vectors evidenced insertion within proto-oncogenes injected in young adult mice [313]. As no liver tumours have been seen to date after rAAV treatment in humans, the risk of insertional mutagenesis is probably very low, if it ever exists. Nonetheless, patients should be followed longitudinally to monitor long-term effects.

Figure 1 summarises the main steps necessary to push forward an AAV-based gene therapy medicinal product from preclinical studies to clinical trials

Without minimising the importance of the tragic toxic events seen in the current clinical trials, it is worth emphasizing that AAV-mediated gene therapy is the only treatment that led to highly significant disease improvement in severely affected human patients. It is now necessary to go back to bench work in order to decipher the pathogenic mechanisms underlying AAV-linked toxicity and design safer next-generation therapeutic cassettes. Indeed, AAV therapy remains the main source of hope for patients affected by neuromuscular disorders, and there are many more diseases to treat. Importantly, our laboratory is planning a new rAAV-based clinical trial using micro-dystrophin transfer in DMD patients, in partnership with Sarepta [314]. The baseline study, aiming at collecting data on the natural disease course in DMD male subjects aged from 5 to 9 years of age, is currently ongoing (GNT-014-MDYF, NCT03882827), and the interventional gene therapy trial should start in early 2021.

Author Contributions: Writing—original draft preparation, L.B., D.-A.G. and N.D.; writing—review and editing, L.B., D.-A.G. and N.D; and funding acquisition: N.D. All authors have read and agreed to the published version of the manuscript.

Funding: This work was funded by Généthon. D.-A.G. was supported by grants from Paris Île-de-France Region (DIM Thérapie Génique) and Génopole d'Evry (ATIGE).

Acknowledgments: We are very grateful to Stephen Lupton, for English proof-reading and to the Flaticon database community for the icons used in Figure 1 and in the graphical abstract.

Conflicts of Interest: The authors declare no conflict of interest. The funders had no role in the design of the study; in the collection, analyses or interpretation of data; in the writing of the manuscript; or in the decision to publish the results.

Abbreviations

NHP: non-human primate; SMA: spinal muscular atrophy; DMD: Duchenne myopathy disease; XLMTM: X-linked myotubular myopathy; CNS: central nervous system; AAV: adeno-associated virus; SMN: survival motor neuron; FDA: Food and Drug Administration; EMA: European Medicines Agency; SAE: serious-adverse event; miR: microRNA; PMO: phosphorodiamidate morpholino oligomer; ITR: inverted-terminal-repeat; LGMD: limb-girdle muscular dystrophy; NAbs: neutralizing antibodies; Sk-CRMs: skeletal muscle-specific cis-regulatory modules; WPRE: woodchuck hepatitis virus post-transcriptional regulatory element; APC: antigen presenting cells; sc: self-complementary.

References

1. Sajer, S.; Guardiero, G.S.; Scicchitano, B.M. Myokines in Home-Based Functional Electrical Stimulation-Induced Recovery of Skeletal Muscle in Elderly and Permanent Denervation. *Eur. J. Transl. Myol.* **2018**, *28*, 7905. [CrossRef] [PubMed]
2. Kern, H.; Carraro, U. Home-Based Functional Electrical Stimulation of Human Permanent Denervated Muscles: A Narrative Review on Diagnostics, Managements, Results and Byproducts Revisited 2020. *Diagnostics* **2020**, *10*, 529. [CrossRef] [PubMed]
3. Carraro, U. Thirty years of translational research in Mobility Medicine: Collection of abstracts of the 2020 Padua Muscle Days. *Eur. J. Transl. Myol.* **2020**, *30*, 8826. [CrossRef] [PubMed]
4. Sarabon, N.; Kozinc, Z.; Lofler, S.; Hofer, C. Resistance Exercise, Electrical Muscle Stimulation, and Whole-Body Vibration in Older Adults: Systematic Review and Meta-Analysis of Randomized Controlled Trials. *J. Clin. Med.* **2020**, *9*, 2902. [CrossRef]
5. Benarroch, L.; Bonne, G.; Rivier, F.; Hamroun, D. The 2020 version of the gene table of neuromuscular disorders (nuclear genome). *Neuromuscul. Disord. NMD* **2019**, *29*, 980–1018. [CrossRef]
6. Kraker, J.; Zivkovic, S.A. Autoimmune neuromuscular disorders. *Curr. Neuropharmacol.* **2011**, *9*, 400–408. [CrossRef]
7. Herbelet, S.; Rodenbach, A.; Paepe, B.; De Bleecker, J.L. Anti-Inflammatory and General Glucocorticoid Physiology in Skeletal Muscles Affected by Duchenne Muscular Dystrophy: Exploration of Steroid-Sparing Agents. *Int. J. Mol. Sci.* **2020**, *21*, 4596. [CrossRef]
8. Goto, M.; Komaki, H.; Takeshita, E.; Abe, Y.; Ishiyama, A.; Sugai, K.; Sasaki, M.; Goto, Y.; Nonaka, I. Long-term outcomes of steroid therapy for Duchenne muscular dystrophy in Japan. *Brain Dev.* **2016**, *38*, 785–791. [CrossRef]
9. Biggar, W.D.; Harris, V.A.; Eliasoph, L.; Alman, B. Long-term benefits of deflazacort treatment for boys with Duchenne muscular dystrophy in their second decade. *Neuromuscul. Disord. NMD* **2006**, *16*, 249–255. [CrossRef]
10. Martiniuk, F.; Mehler, M.; Pellicer, A.; Tzall, S.; La Badie, G.; Hobart, C.; Ellenbogen, A.; Hirschhorn, R. Isolation of a cDNA for human acid alpha-glucosidase and detection of genetic heterogeneity for mRNA in three alpha-glucosidase-deficient patients. *Proc. Natl. Acad. Sci. USA* **1986**, *83*, 9641–9644. [CrossRef]
11. FDA. Myozyme Approval Letter. 2006. Available online: https://www.accessdata.fda.gov/drugsatfda_docs/nda/2006/125141s0000_Myozyme_Approv.pdf (accessed on 25 November 2020).
12. EMA. Myozyme Alglucosidase Alfa. 2014. Available online: https://www.ema.europa.eu/en/documents/overview/myozyme-epar-summary-public_en.pdf (accessed on 25 November 2020).
13. Kishnani, P.S.; Corzo, D.; Nicolino, M.; Byrne, B.; Mandel, H.; Hwu, W.L.; Leslie, N.; Levine, J.; Spencer, C.; McDonald, M.; et al. Recombinant human acid [alpha]-glucosidase: Major clinical benefits in infantile-onset Pompe disease. *Neurology* **2007**, *68*, 99–109. [CrossRef]

14. Nicolino, M.; Byrne, B.; Wraith, J.E.; Leslie, N.; Mandel, H.; Freyer, D.R.; Arnold, G.L.; Pivnick, E.K.; Ottinger, C.J.; Robinson, P.H.; et al. Clinical outcomes after long-term treatment with alglucosidase alfa in infants and children with advanced Pompe disease. *Genet. Med. Off. J. Am. Coll. Med. Genet.* **2009**, *11*, 210–219. [CrossRef]
15. van der Ploeg, A.T.; Clemens, P.R.; Corzo, D.; Escolar, D.M.; Florence, J.; Groeneveld, G.J.; Herson, S.; Kishnani, P.S.; Laforet, P.; Lake, S.L.; et al. A randomized study of alglucosidase alfa in late-onset Pompe's disease. *N. Engl. J. Med.* **2010**, *362*, 1396–1406. [CrossRef] [PubMed]
16. Hahn, S.H.; Kronn, D.; Leslie, N.D.; Pena, L.D.M.; Tanpaiboon, P.; Gambello, M.J.; Gibson, J.B.; Hillman, R.; Stockton, D.W.; Day, J.W.; et al. Efficacy, safety profile, and immunogenicity of alglucosidase alfa produced at the 4,000-liter scale in US children and adolescents with Pompe disease: ADVANCE, a phase IV, open-label, prospective study. *Genet. Med. Off. J. Am. Coll. Med. Genet.* **2018**, *20*, 1284–1294. [CrossRef] [PubMed]
17. FDA. FDA Grants Accelerated Approval to First Drug for Duchenne Muscular Dystrophy. 2016. Available online: https://www.fda.gov/news-events/press-announcements/fda-grants-accelerated-approval-first-drug-duchenne-muscular-dystrophy (accessed on 25 November 2020).
18. Koenig, M.; Hoffman, E.P.; Bertelson, C.J.; Monaco, A.P.; Feener, C.; Kunkel, L.M. Complete cloning of the Duchenne muscular dystrophy (DMD) cDNA and preliminary genomic organization of the DMD gene in normal and affected individuals. *Cell* **1987**, *50*, 509–517. [CrossRef]
19. England, S.B.; Nicholson, L.V.; Johnson, M.A.; Forrest, S.M.; Love, D.R.; Zubrzycka-Gaarn, E.E.; Bulman, D.E.; Harris, J.B.; Davies, K.E. Very mild muscular dystrophy associated with the deletion of 46% of dystrophin. *Nature* **1990**, *343*, 180–182. [CrossRef] [PubMed]
20. Charleston, J.S.; Schnell, F.J.; Dworzak, J.; Donoghue, C.; Lewis, S.; Chen, L.; Young, G.D.; Milici, A.J.; Voss, J.; DeAlwis, U.; et al. Eteplirsen treatment for Duchenne muscular dystrophy: Exon skipping and dystrophin production. *Neurology* **2018**, *90*, e2146–e2154. [CrossRef] [PubMed]
21. Mendell, J.R.; Rodino-Klapac, L.R.; Sahenk, Z.; Roush, K.; Bird, L.; Lowes, L.P.; Alfano, L.; Gomez, A.M.; Lewis, S.; Kota, J.; et al. Eteplirsen for the treatment of Duchenne muscular dystrophy. *Ann. Neurol.* **2013**, *74*, 637–647. [CrossRef] [PubMed]
22. Mendell, J.R.; Goemans, N.; Lowes, L.P.; Alfano, L.N.; Berry, K.; Shao, J.; Kaye, E.M.; Mercuri, E.; Eteplirsen Study, G.; Telethon Foundation, D.M.D.I.N. Longitudinal effect of eteplirsen versus historical control on ambulation in Duchenne muscular dystrophy. *Ann. Neurol.* **2016**, *79*, 257–271. [CrossRef] [PubMed]
23. Alfano, L.N.; Charleston, J.S.; Connolly, A.M.; Cripe, L.; Donoghue, C.; Dracker, R.; Dworzak, J.; Eliopoulos, H.; Frank, D.E.; Lewis, S.; et al. Long-term treatment with eteplirsen in nonambulatory patients with Duchenne muscular dystrophy. *Medicine* **2019**, *98*, e15858. [CrossRef] [PubMed]
24. Kinane, T.B.; Mayer, O.H.; Duda, P.W.; Lowes, L.P.; Moody, S.L.; Mendell, J.R. Long-Term Pulmonary Function in Duchenne Muscular Dystrophy: Comparison of Eteplirsen-Treated Patients to Natural History. *J. Neuromuscul. Dis.* **2018**, *5*, 47–58. [CrossRef] [PubMed]
25. Khan, N.; Eliopoulos, H.; Han, L.; Kinane, T.B.; Lowes, L.P.; Mendell, J.R.; Gordish-Dressman, H.; Henricson, E.K.; McDonald, C.M.; Eteplirsen, I.; et al. Eteplirsen Treatment Attenuates Respiratory Decline in Ambulatory and Non-Ambulatory Patients with Duchenne Muscular Dystrophy. *J. Neuromuscul. Dis.* **2019**, *6*, 213–225. [CrossRef] [PubMed]
26. Bushby, K.; Finkel, R.; Wong, B.; Barohn, R.; Campbell, C.; Comi, G.P.; Connolly, A.M.; Day, J.W.; Flanigan, K.M.; Goemans, N.; et al. Ataluren treatment of patients with nonsense mutation dystrophinopathy. *Muscle Nerve* **2014**, *50*, 477–487. [CrossRef] [PubMed]
27. McDonald, C.M.; Campbell, C.; Torricelli, R.E.; Finkel, R.S.; Flanigan, K.M.; Goemans, N.; Heydemann, P.; Kaminska, A.; Kirschner, J.; Muntoni, F.; et al. Ataluren in patients with nonsense mutation Duchenne muscular dystrophy (ACT DMD): A multicentre, randomised, double-blind, placebo-controlled, phase 3 trial. *Lancet* **2017**, *390*, 1489–1498. [CrossRef]
28. Mercuri, E.; Muntoni, F.; Osorio, A.N.; Tulinius, M.; Buccella, F.; Morgenroth, L.P.; Gordish-Dressman, H.; Jiang, J.; Trifillis, P.; Zhu, J.; et al. Safety and effectiveness of ataluren: Comparison of results from the STRIDE Registry and CINRG DMD Natural History Study. *J. Comp. Eff. Res.* **2020**, *9*, 341–360. [CrossRef]
29. FDA. FDA Approves First Drug for Spinal Muscular Atrophy. 2016. Available online: https://www.fda.gov/news-events/press-announcements/fda-approves-first-drug-spinal-muscular-atrophy (accessed on 25 November 2020).

30. EMA. First Medicine for Spinal Muscular Atrophy. 2017. Available online: https://www.ema.europa.eu/en/news/first-medicine-spinal-muscular-atrophy (accessed on 25 November 2020).
31. Lefebvre, S.; Burglen, L.; Reboullet, S.; Clermont, O.; Burlet, P.; Viollet, L.; Benichou, B.; Cruaud, C.; Millasseau, P.; Zeviani, M.; et al. Identification and characterization of a spinal muscular atrophy-determining gene. *Cell* **1995**, *80*, 155–165. [CrossRef]
32. Mercuri, E.; Darras, B.T.; Chiriboga, C.A.; Day, J.W.; Campbell, C.; Connolly, A.M.; Iannaccone, S.T.; Kirschner, J.; Kuntz, N.L.; Saito, K.; et al. Nusinersen versus Sham Control in Later-Onset Spinal Muscular Atrophy. *N. Engl. J. Med.* **2018**, *378*, 625–635. [CrossRef]
33. De Vivo, D.C.; Bertini, E.; Swoboda, K.J.; Hwu, W.L.; Crawford, T.O.; Finkel, R.S.; Kirschner, J.; Kuntz, N.L.; Parsons, J.A.; Ryan, M.M.; et al. Nusinersen initiated in infants during the presymptomatic stage of spinal muscular atrophy: Interim efficacy and safety results from the Phase 2 NURTURE study. *Neuromuscul. Disord. NMD* **2019**, *29*, 842–856. [CrossRef]
34. FDA. FDA Approves Oral Treatment for Spinal Muscular Atrophy. 2020. Available online: https://www.fda.gov/news-events/press-announcements/fda-approves-oral-treatment-spinal-muscular-atrophy (accessed on 25 November 2020).
35. Poirier, A.; Weetall, M.; Heinig, K.; Bucheli, F.; Schoenlein, K.; Alsenz, J.; Bassett, S.; Ullah, M.; Senn, C.; Ratni, H.; et al. Risdiplam distributes and increases SMN protein in both the central nervous system and peripheral organs. *Pharmacol. Res. Perspect.* **2018**, *6*, e00447. [CrossRef]
36. Ratni, H.; Ebeling, M.; Baird, J.; Bendels, S.; Bylund, J.; Chen, K.S.; Denk, N.; Feng, Z.; Green, L.; Guerard, M.; et al. Discovery of Risdiplam, a Selective Survival of Motor Neuron-2 (SMN2) Gene Splicing Modifier for the Treatment of Spinal Muscular Atrophy (SMA). *J. Med. Chem.* **2018**, *61*, 6501–6517. [CrossRef]
37. Sturm, S.; Gunther, A.; Jaber, B.; Jordan, P.; Al Kotbi, N.; Parkar, N.; Cleary, Y.; Frances, N.; Bergauer, T.; Heinig, K.; et al. A phase 1 healthy male volunteer single escalating dose study of the pharmacokinetics and pharmacodynamics of risdiplam (RG7916, RO7034067), a SMN2 splicing modifier. *Br. J. Clin. Pharmacol.* **2019**, *85*, 181–193. [CrossRef] [PubMed]
38. Gaudet, D.; Methot, J.; Dery, S.; Brisson, D.; Essiembre, C.; Tremblay, G.; Tremblay, K.; de Wal, J.; Twisk, J.; van den Bulk, N.; et al. Efficacy and long-term safety of alipogene tiparvovec (AAV1-LPLS447X) gene therapy for lipoprotein lipase deficiency: An open-label trial. *Gene Ther.* **2013**, *20*, 361–369. [CrossRef] [PubMed]
39. Yla-Herttuala, S. Endgame: Glybera finally recommended for approval as the first gene therapy drug in the European union. *Mol. Ther. J. Am. Soc. Gene Ther.* **2012**, *20*, 1831–1832. [CrossRef] [PubMed]
40. Russell, S.; Bennett, J.; Wellman, J.A.; Chung, D.C.; Yu, Z.F.; Tillman, A.; Wittes, J.; Pappas, J.; Elci, O.; McCague, S.; et al. Efficacy and safety of voretigene neparvovec (AAV2-hRPE65v2) in patients with RPE65-mediated inherited retinal dystrophy: A randomised, controlled, open-label, phase 3 trial. *Lancet* **2017**, *390*, 849–860. [CrossRef]
41. FDA. FDA Approves Novel Gene Therapy to Treat Patients with a Rare Form of Inherited Vision Loss. 2017. Available online: https://www.fda.gov/news-events/press-announcements/fda-approves-novel-gene-therapy-treat-patients-rare-form-inherited-vision-loss (accessed on 25 November 2020).
42. EMA. New Gene Therapy for Rare Inherited Disorder Causing Vision Loss Recommended for Approval. 2018. Available online: https://www.ema.europa.eu/en/documents/press-release/new-gene-therapy-rare-inherited-disorder-causing-vision-loss-recommended-approval_en.pdf (accessed on 25 November 2020).
43. FDA. FDA Approves Innovative Gene Therapy to Treat Pediatric Patients with Spinal Muscular Atrophy, a Rare Disease and Leading Genetic Cause of Infant Mortality. 2019. Available online: https://www.fda.gov/news-events/press-announcements/fda-approves-innovative-gene-therapy-treat-pediatric-patients-spinal-muscular-atrophy-rare-disease (accessed on 25 November 2020).
44. Mendell, J.R.; Al-Zaidy, S.; Shell, R.; Arnold, W.D.; Rodino-Klapac, L.R.; Prior, T.W.; Lowes, L.; Alfano, L.; Berry, K.; Church, K.; et al. Single-Dose Gene-Replacement Therapy for Spinal Muscular Atrophy. *N. Engl. J. Med.* **2017**, *377*, 1713–1722. [CrossRef]
45. EMA. New Gene Therapy to Treat Spinal Muscular Atrophy (Corrected). 2020. Available online: https://www.ema.europa.eu/en/news/new-gene-therapy-treat-spinal-muscular-atrophy-corrected (accessed on 25 November 2020).
46. Sonntag, F.; Schmidt, K.; Kleinschmidt, J.A. A viral assembly factor promotes AAV2 capsid formation in the nucleolus. *Proc. Natl. Acad. Sci. USA* **2010**, *107*, 10220–10225. [CrossRef]

47. Ogden, P.J.; Kelsic, E.D.; Sinai, S.; Church, G.M. Comprehensive AAV capsid fitness landscape reveals a viral gene and enables machine-guided design. *Science* **2019**, *366*, 1139–1143. [CrossRef]
48. Russell, D.W.; Miller, A.D.; Alexander, I.E. Adeno-associated virus vectors preferentially transduce cells in S phase. *Proc. Natl. Acad. Sci. USA* **1994**, *91*, 8915–8919. [CrossRef]
49. Atchison, R.W.; Casto, B.C.; Hammon, W.M. Adenovirus-Associated Defective Virus Particles. *Science* **1965**, *149*, 754–756. [CrossRef]
50. Hoggan, M.D.; Blacklow, N.R.; Rowe, W.P. Studies of small DNA viruses found in various adenovirus preparations: Physical, biological, and immunological characteristics. *Proc. Natl. Acad. Sci. USA* **1966**, *55*, 1467–1474. [CrossRef]
51. Parks, W.P.; Green, M.; Pina, M.; Melnick, J.L. Physicochemical characterization of adeno-associated satellite virus type 4 and its nucleic acid. *J. Virol.* **1967**, *1*, 980–987. [CrossRef] [PubMed]
52. Bantel-Schaal, U.; zur Hausen, H. Characterization of the DNA of a defective human parvovirus isolated from a genital site. *Virology* **1984**, *134*, 52–63. [CrossRef]
53. Rutledge, E.A.; Halbert, C.L.; Russell, D.W. Infectious clones and vectors derived from adeno-associated virus (AAV) serotypes other than AAV type 2. *J. Virol.* **1998**, *72*, 309–319. [CrossRef]
54. Gao, G.P.; Alvira, M.R.; Wang, L.; Calcedo, R.; Johnston, J.; Wilson, J.M. Novel adeno-associated viruses from rhesus monkeys as vectors for human gene therapy. *Proc. Natl. Acad. Sci. USA* **2002**, *99*, 11854–11859. [CrossRef] [PubMed]
55. Gao, G.; Vandenberghe, L.H.; Alvira, M.R.; Lu, Y.; Calcedo, R.; Zhou, X.; Wilson, J.M. Clades of Adeno-associated viruses are widely disseminated in human tissues. *J. Virol.* **2004**, *78*, 6381–6388. [CrossRef] [PubMed]
56. Mori, S.; Wang, L.; Takeuchi, T.; Kanda, T. Two novel adeno-associated viruses from cynomolgus monkey: Pseudotyping characterization of capsid protein. *Virology* **2004**, *330*, 375–383. [CrossRef] [PubMed]
57. Schmidt, M.; Voutetakis, A.; Afione, S.; Zheng, C.; Mandikian, D.; Chiorini, J.A. Adeno-associated virus type 12 (AAV12): A novel AAV serotype with sialic acid- and heparan sulfate proteoglycan-independent transduction activity. *J. Virol.* **2008**, *82*, 1399–1406. [CrossRef] [PubMed]
58. Schmidt, M.; Govindasamy, L.; Afione, S.; Kaludov, N.; Agbandje-McKenna, M.; Chiorini, J.A. Molecular characterization of the heparin-dependent transduction domain on the capsid of a novel adeno-associated virus isolate, AAV(VR-942). *J. Virol.* **2008**, *82*, 8911–8916. [CrossRef]
59. Gao, G.; Vandenberghe, L.H.; Wilson, J.M. New recombinant serotypes of AAV vectors. *Curr. Gene Ther.* **2005**, *5*, 285–297. [CrossRef]
60. Gao, G.; Zhong, L.; Danos, O. Exploiting natural diversity of AAV for the design of vectors with novel properties. *Methods Mol. Biol.* **2011**, *807*, 93–118. [CrossRef]
61. DiMattia, M.A.; Nam, H.J.; Van Vliet, K.; Mitchell, M.; Bennett, A.; Gurda, B.L.; McKenna, R.; Olson, N.H.; Sinkovits, R.S.; Potter, M.; et al. Structural insight into the unique properties of adeno-associated virus serotype 9. *J. Virol.* **2012**, *86*, 6947–6958. [CrossRef] [PubMed]
62. Calcedo, R.; Vandenberghe, L.H.; Gao, G.; Lin, J.; Wilson, J.M. Worldwide epidemiology of neutralizing antibodies to adeno-associated viruses. *J. Infect. Dis.* **2009**, *199*, 381–390. [CrossRef] [PubMed]
63. Calcedo, R.; Wilson, J.M. Humoral Immune Response to AAV. *Front. Immunol.* **2013**, *4*, 341. [CrossRef] [PubMed]
64. Boutin, S.; Monteilhet, V.; Veron, P.; Leborgne, C.; Benveniste, O.; Montus, M.F.; Masurier, C. Prevalence of serum IgG and neutralizing factors against adeno-associated virus (AAV) types 1, 2, 5, 6, 8, and 9 in the healthy population: Implications for gene therapy using AAV vectors. *Hum. Gene Ther.* **2010**, *21*, 704–712. [CrossRef]
65. Summerford, C.; Samulski, R.J. Membrane-associated heparan sulfate proteoglycan is a receptor for adeno-associated virus type 2 virions. *J. Virol.* **1998**, *72*, 1438–1445. [CrossRef]
66. Mietzsch, M.; Broecker, F.; Reinhardt, A.; Seeberger, P.H.; Heilbronn, R. Differential adeno-associated virus serotype-specific interaction patterns with synthetic heparins and other glycans. *J. Virol.* **2014**, *88*, 2991–3003. [CrossRef]
67. Walters, R.W.; Yi, S.M.; Keshavjee, S.; Brown, K.E.; Welsh, M.J.; Chiorini, J.A.; Zabner, J. Binding of adeno-associated virus type 5 to 2,3-linked sialic acid is required for gene transfer. *J. Biol. Chem.* **2001**, *276*, 20610–20616. [CrossRef]

68. Kaludov, N.; Brown, K.E.; Walters, R.W.; Zabner, J.; Chiorini, J.A. Adeno-associated virus serotype 4 (AAV4) and AAV5 both require sialic acid binding for hemagglutination and efficient transduction but differ in sialic acid linkage specificity. *J. Virol.* **2001**, *75*, 6884–6893. [CrossRef]
69. Wu, Z.; Miller, E.; Agbandje-McKenna, M.; Samulski, R.J. Alpha2,3 and alpha2,6 N-linked sialic acids facilitate efficient binding and transduction by adeno-associated virus types 1 and 6. *J. Virol.* **2006**, *80*, 9093–9103. [CrossRef]
70. Shen, S.; Bryant, K.D.; Brown, S.M.; Randell, S.H.; Asokan, A. Terminal N-linked galactose is the primary receptor for adeno-associated virus 9. *J. Biol. Chem.* **2011**, *286*, 13532–13540. [CrossRef]
71. Bell, C.L.; Vandenberghe, L.H.; Bell, P.; Limberis, M.P.; Gao, G.P.; Van Vliet, K.; Agbandje-McKenna, M.; Wilson, J.M. The AAV9 receptor and its modification to improve In Vivo lung gene transfer in mice. *J. Clin. Investig.* **2011**, *121*, 2427–2435. [CrossRef] [PubMed]
72. Bell, C.L.; Gurda, B.L.; Van Vliet, K.; Agbandje-McKenna, M.; Wilson, J.M. Identification of the galactose binding domain of the adeno-associated virus serotype 9 capsid. *J. Virol.* **2012**, *86*, 7326–7333. [CrossRef] [PubMed]
73. Di Pasquale, G.; Davidson, B.L.; Stein, C.S.; Martins, I.; Scudiero, D.; Monks, A.; Chiorini, J.A. Identification of PDGFR as a receptor for AAV-5 transduction. *Nat. Med.* **2003**, *9*, 1306–1312. [CrossRef] [PubMed]
74. Akache, B.; Grimm, D.; Pandey, K.; Yant, S.R.; Xu, H.; Kay, M.A. The 37/67-kilodalton laminin receptor is a receptor for adeno-associated virus serotypes 8, 2, 3, and 9. *J. Virol.* **2006**, *80*, 9831–9836. [CrossRef] [PubMed]
75. Kashiwakura, Y.; Tamayose, K.; Iwabuchi, K.; Hirai, Y.; Shimada, T.; Matsumoto, K.; Nakamura, T.; Watanabe, M.; Oshimi, K.; Daida, H. Hepatocyte growth factor receptor is a coreceptor for adeno-associated virus type 2 infection. *J. Virol.* **2005**, *79*, 609–614. [CrossRef] [PubMed]
76. Summerford, C.; Bartlett, J.S.; Samulski, R.J. AlphaVbeta5 integrin: A co-receptor for adeno-associated virus type 2 infection. *Nat. Med.* **1999**, *5*, 78–82. [CrossRef] [PubMed]
77. Qing, K.; Mah, C.; Hansen, J.; Zhou, S.; Dwarki, V.; Srivastava, A. Human fibroblast growth factor receptor 1 is a co-receptor for infection by adeno-associated virus 2. *Nat. Med.* **1999**, *5*, 71–77. [CrossRef]
78. Pillay, S.; Meyer, N.L.; Puschnik, A.S.; Davulcu, O.; Diep, J.; Ishikawa, Y.; Jae, L.T.; Wosen, J.E.; Nagamine, C.M.; Chapman, M.S.; et al. Corrigendum: An essential receptor for adeno-associated virus infection. *Nature* **2016**, *539*, 456. [CrossRef]
79. Pillay, S.; Meyer, N.L.; Puschnik, A.S.; Davulcu, O.; Diep, J.; Ishikawa, Y.; Jae, L.T.; Wosen, J.E.; Nagamine, C.M.; Chapman, M.S.; et al. An essential receptor for adeno-associated virus infection. *Nature* **2016**, *530*, 108–112. [CrossRef]
80. Xiao, P.J.; Samulski, R.J. Cytoplasmic trafficking, endosomal escape, and perinuclear accumulation of adeno-associated virus type 2 particles are facilitated by microtubule network. *J. Virol.* **2012**, *86*, 10462–10473. [CrossRef]
81. Buller, R.M.; Janik, J.E.; Sebring, E.D.; Rose, J.A. Herpes simplex virus types 1 and 2 completely help adenovirus-associated virus replication. *J. Virol.* **1981**, *40*, 241–247. [CrossRef] [PubMed]
82. Georg-Fries, B.; Biederlack, S.; Wolf, J.; zur Hausen, H. Analysis of proteins, helper dependence, and seroepidemiology of a new human parvovirus. *Virology* **1984**, *134*, 64–71. [CrossRef]
83. Kotin, R.M.; Menninger, J.C.; Ward, D.C.; Berns, K.I. Mapping and direct visualization of a region-specific viral DNA integration site on chromosome 19q13-qter. *Genomics* **1991**, *10*, 831–834. [CrossRef]
84. Kotin, R.M.; Siniscalco, M.; Samulski, R.J.; Zhu, X.D.; Hunter, L.; Laughlin, C.A.; McLaughlin, S.; Muzyczka, N.; Rocchi, M.; Berns, K.I. Site-specific integration by adeno-associated virus. *Proc. Natl. Acad. Sci. USA* **1990**, *87*, 2211–2215. [CrossRef]
85. Samulski, R.J.; Zhu, X.; Xiao, X.; Brook, J.D.; Housman, D.E.; Epstein, N.; Hunter, L.A. Targeted integration of adeno-associated virus (AAV) into human chromosome 19. *EMBO J.* **1991**, *10*, 3941–3950. [CrossRef]
86. Kotin, R.M.; Linden, R.M.; Berns, K.I. Characterization of a preferred site on human chromosome 19q for integration of adeno-associated virus DNA by non-homologous recombination. *EMBO J.* **1992**, *11*, 5071–5078. [CrossRef]
87. Lamartina, S.; Sporeno, E.; Fattori, E.; Toniatti, C. Characteristics of the adeno-associated virus preintegration site in human chromosome 19: Open chromatin conformation and transcription-competent environment. *J. Virol.* **2000**, *74*, 7671–7677. [CrossRef]

88. Huser, D.; Gogol-Doring, A.; Lutter, T.; Weger, S.; Winter, K.; Hammer, E.M.; Cathomen, T.; Reinert, K.; Heilbronn, R. Integration preferences of wildtype AAV-2 for consensus rep-binding sites at numerous loci in the human genome. *PLoS Pathog.* **2010**, *6*, e1000985. [CrossRef]
89. La Bella, T.; Imbeaud, S.; Peneau, C.; Mami, I.; Datta, S.; Bayard, Q.; Caruso, S.; Hirsch, T.Z.; Calderaro, J.; Morcrette, G.; et al. Adeno-associated virus in the liver: Natural history and consequences in tumour development. *Gut* **2020**, *69*, 737–747. [CrossRef]
90. Nault, J.C.; Datta, S.; Imbeaud, S.; Franconi, A.; Mallet, M.; Couchy, G.; Letouze, E.; Pilati, C.; Verret, B.; Blanc, J.F.; et al. Recurrent AAV2-related insertional mutagenesis in human hepatocellular carcinomas. *Nat. Genet.* **2015**, *47*, 1187–1193. [CrossRef]
91. Zhang, H.G.; Wang, Y.M.; Xie, J.F.; Liang, X.; Hsu, H.C.; Zhang, X.; Douglas, J.; Curiel, D.T.; Mountz, J.D. Recombinant adenovirus expressing adeno-associated virus cap and rep proteins supports production of high-titer recombinant adeno-associated virus. *Gene Ther.* **2001**, *8*, 704–712. [CrossRef] [PubMed]
92. Miao, C.H.; Snyder, R.O.; Schowalter, D.B.; Patijn, G.A.; Donahue, B.; Winther, B.; Kay, M.A. The kinetics of rAAV integration in the liver. *Nat. Genet.* **1998**, *19*, 13–15. [CrossRef] [PubMed]
93. Nakai, H.; Storm, T.A.; Kay, M.A. Recruitment of single-stranded recombinant adeno-associated virus vector genomes and intermolecular recombination are responsible for stable transduction of liver In Vivo. *J. Virol.* **2000**, *74*, 9451–9463. [CrossRef] [PubMed]
94. Nakai, H.; Yant, S.R.; Storm, T.A.; Fuess, S.; Meuse, L.; Kay, M.A. Extrachromosomal recombinant adeno-associated virus vector genomes are primarily responsible for stable liver transduction In Vivo. *J. Virol.* **2001**, *75*, 6969–6976. [CrossRef]
95. Fisher, K.J.; Jooss, K.; Alston, J.; Yang, Y.; Haecker, S.E.; High, K.; Pathak, R.; Raper, S.E.; Wilson, J.M. Recombinant adeno-associated virus for muscle directed gene therapy. *Nat. Med.* **1997**, *3*, 306–312. [CrossRef]
96. Rutledge, E.A.; Russell, D.W. Adeno-associated virus vector integration junctions. *J. Virol.* **1997**, *71*, 8429–8436. [CrossRef]
97. Nakai, H.; Iwaki, Y.; Kay, M.A.; Couto, L.B. Isolation of recombinant adeno-associated virus vector-cellular DNA junctions from mouse liver. *J. Virol.* **1999**, *73*, 5438–5447. [CrossRef]
98. Nakai, H.; Montini, E.; Fuess, S.; Storm, T.A.; Grompe, M.; Kay, M.A. AAV serotype 2 vectors preferentially integrate into active genes in mice. *Nat. Genet.* **2003**, *34*, 297–302. [CrossRef]
99. Miller, D.G.; Trobridge, G.D.; Petek, L.M.; Jacobs, M.A.; Kaul, R.; Russell, D.W. Large-scale analysis of adeno-associated virus vector integration sites in normal human cells. *J. Virol.* **2005**, *79*, 11434–11442. [CrossRef]
100. Nakai, H.; Wu, X.; Fuess, S.; Storm, T.A.; Munroe, D.; Montini, E.; Burgess, S.M.; Grompe, M.; Kay, M.A. Large-scale molecular characterization of adeno-associated virus vector integration in mouse liver. *J. Virol.* **2005**, *79*, 3606–3614. [CrossRef]
101. Inagaki, K.; Lewis, S.M.; Wu, X.; Ma, C.; Munroe, D.J.; Fuess, S.; Storm, T.A.; Kay, M.A.; Nakai, H. DNA palindromes with a modest arm length of greater, similar 20 base pairs are a significant target for recombinant adeno-associated virus vector integration in the liver, muscles, and heart in mice. *J. Virol.* **2007**, *81*, 11290–11303. [CrossRef] [PubMed]
102. Miller, D.G.; Rutledge, E.A.; Russell, D.W. Chromosomal effects of adeno-associated virus vector integration. *Nat. Genet.* **2002**, *30*, 147–148. [CrossRef] [PubMed]
103. Yang, C.C.; Xiao, X.; Zhu, X.; Ansardi, D.C.; Epstein, N.D.; Frey, M.R.; Matera, A.G.; Samulski, R.J. Cellular recombination pathways and viral terminal repeat hairpin structures are sufficient for adeno-associated virus integration In Vivo and in vitro. *J. Virol.* **1997**, *71*, 9231–9247. [CrossRef] [PubMed]
104. Chandler, R.J.; LaFave, M.C.; Varshney, G.K.; Trivedi, N.S.; Carrillo-Carrasco, N.; Senac, J.S.; Wu, W.; Hoffmann, V.; Elkahloun, A.G.; Burgess, S.M.; et al. Vector design influences hepatic genotoxicity after adeno-associated virus gene therapy. *J. Clin. Investig.* **2015**, *125*, 870–880. [CrossRef] [PubMed]
105. Donsante, A.; Miller, D.G.; Li, Y.; Vogler, C.; Brunt, E.M.; Russell, D.W.; Sands, M.S. AAV vector integration sites in mouse hepatocellular carcinoma. *Science* **2007**, *317*, 477. [CrossRef]
106. Li, H.; Malani, N.; Hamilton, S.R.; Schlachterman, A.; Bussadori, G.; Edmonson, S.E.; Shah, R.; Arruda, V.R.; Mingozzi, F.; Wright, J.F.; et al. Assessing the potential for AAV vector genotoxicity in a murine model. *Blood* **2011**, *117*, 3311–3319. [CrossRef] [PubMed]

107. Kaplitt, M.G.; Leone, P.; Samulski, R.J.; Xiao, X.; Pfaff, D.W.; O'Malley, K.L.; During, M.J. Long-term gene expression and phenotypic correction using adeno-associated virus vectors in the mammalian brain. *Nat. Genet.* **1994**, *8*, 148–154. [CrossRef]
108. Pruchnic, R.; Cao, B.; Peterson, Z.Q.; Xiao, X.; Li, J.; Samulski, R.J.; Epperly, M.; Huard, J. The use of adeno-associated virus to circumvent the maturation-dependent viral transduction of muscle fibers. *Hum. Gene Ther.* **2000**, *11*, 521–536. [CrossRef]
109. Manno, C.S.; Pierce, G.F.; Arruda, V.R.; Glader, B.; Ragni, M.; Rasko, J.J.; Ozelo, M.C.; Hoots, K.; Blatt, P.; Konkle, B.; et al. Successful transduction of liver in hemophilia by AAV-Factor IX and limitations imposed by the host immune response. *Nat. Med.* **2006**, *12*, 342–347. [CrossRef]
110. Zincarelli, C.; Soltys, S.; Rengo, G.; Rabinowitz, J.E. Analysis of AAV serotypes 1-9 mediated gene expression and tropism in mice after systemic injection. *Mol. Ther. J. Am. Soc. Gene Ther.* **2008**, *16*, 1073–1080. [CrossRef]
111. Sanftner, L.M.; Sommer, J.M.; Suzuki, B.M.; Smith, P.H.; Vijay, S.; Vargas, J.A.; Forsayeth, J.R.; Cunningham, J.; Bankiewicz, K.S.; Kao, H.; et al. AAV2-mediated gene delivery to monkey putamen: Evaluation of an infusion device and delivery parameters. *Exp. Neurol.* **2005**, *194*, 476–483. [CrossRef] [PubMed]
112. Xiao, W.; Chirmule, N.; Berta, S.C.; McCullough, B.; Gao, G.; Wilson, J.M. Gene therapy vectors based on adeno-associated virus type 1. *J. Virol.* **1999**, *73*, 3994–4003. [CrossRef] [PubMed]
113. Louboutin, J.P.; Wang, L.; Wilson, J.M. Gene transfer into skeletal muscle using novel AAV serotypes. *J. Gene Med.* **2005**, *7*, 442–451. [CrossRef] [PubMed]
114. Riaz, M.; Raz, Y.; Moloney, E.B.; van Putten, M.; Krom, Y.D.; van der Maarel, S.M.; Verhaagen, J.; Raz, V. Differential myofiber-type transduction preference of adeno-associated virus serotypes 6 and 9. *Skelet. Muscle* **2015**, *5*, 37. [CrossRef] [PubMed]
115. Ohshima, S.; Shin, J.H.; Yuasa, K.; Nishiyama, A.; Kira, J.; Okada, T.; Takeda, S. Transduction efficiency and immune response associated with the administration of AAV8 vector into dog skeletal muscle. *Mol. Ther. J. Am. Soc. Gene Ther.* **2009**, *17*, 73–80. [CrossRef]
116. Wang, Z.; Zhu, T.; Qiao, C.; Zhou, L.; Wang, B.; Zhang, J.; Chen, C.; Li, J.; Xiao, X. Adeno-associated virus serotype 8 efficiently delivers genes to muscle and heart. *Nat. Biotechnol.* **2005**, *23*, 321–328. [CrossRef]
117. Nakai, H.; Fuess, S.; Storm, T.A.; Muramatsu, S.; Nara, Y.; Kay, M.A. Unrestricted hepatocyte transduction with adeno-associated virus serotype 8 vectors in mice. *J. Virol.* **2005**, *79*, 214–224. [CrossRef]
118. Inagaki, K.; Fuess, S.; Storm, T.A.; Gibson, G.A.; McTiernan, C.F.; Kay, M.A.; Nakai, H. Robust systemic transduction with AAV9 vectors in mice: Efficient global cardiac gene transfer superior to that of AAV8. *Mol. Ther. J. Am. Soc. Gene Ther.* **2006**, *14*, 45–53. [CrossRef]
119. Pacak, C.A.; Mah, C.S.; Thattaliyath, B.D.; Conlon, T.J.; Lewis, M.A.; Cloutier, D.E.; Zolotukhin, I.; Tarantal, A.F.; Byrne, B.J. Recombinant adeno-associated virus serotype 9 leads to preferential cardiac transduction In Vivo. *Circ. Res.* **2006**, *99*, e3–e9. [CrossRef]
120. Bostick, B.; Ghosh, A.; Yue, Y.; Long, C.; Duan, D. Systemic AAV-9 transduction in mice is influenced by animal age but not by the route of administration. *Gene Ther.* **2007**, *14*, 1605–1609. [CrossRef]
121. Pacak, C.A.; Sakai, Y.; Thattaliyath, B.D.; Mah, C.S.; Byrne, B.J. Tissue specific promoters improve specificity of AAV9 mediated transgene expression following intra-vascular gene delivery in neonatal mice. *Genet. Vaccines Ther.* **2008**, *6*, 13. [CrossRef] [PubMed]
122. Pan, X.; Yue, Y.; Zhang, K.; Lostal, W.; Shin, J.H.; Duan, D. Long-term robust myocardial transduction of the dog heart from a peripheral vein by adeno-associated virus serotype-8. *Hum. Gene Ther.* **2013**, *24*, 584–594. [CrossRef] [PubMed]
123. Sarkar, R.; Mucci, M.; Addya, S.; Tetreault, R.; Bellinger, D.A.; Nichols, T.C.; Kazazian, H.H., Jr. Long-term efficacy of adeno-associated virus serotypes 8 and 9 in hemophilia a dogs and mice. *Hum. Gene Ther.* **2006**, *17*, 427–439. [CrossRef] [PubMed]
124. Toromanoff, A.; Cherel, Y.; Guilbaud, M.; Penaud-Budloo, M.; Snyder, R.O.; Haskins, M.E.; Deschamps, J.Y.; Guigand, L.; Podevin, G.; Arruda, V.R.; et al. Safety and Efficacy of Regional Intravenous (RI) Versus Intramuscular (IM) Delivery of rAAV1 and rAAV8 to Nonhuman Primate Skeletal Muscle. *Mol. Ther. J. Am. Soc. Gene Ther.* **2008**, *16*, 1291–1299. [CrossRef]
125. Gregorevic, P.; Blankinship, M.J.; Allen, J.M.; Crawford, R.W.; Meuse, L.; Miller, D.G.; Russell, D.W.; Chamberlain, J.S. Systemic delivery of genes to striated muscles using adeno-associated viral vectors. *Nat. Med.* **2004**, *10*, 828–834. [CrossRef]

126. Thomas, G.D. Functional muscle ischemia in Duchenne and Becker muscular dystrophy. *Front. Physiol.* **2013**, *4*, 381. [CrossRef]
127. Katwal, A.B.; Konkalmatt, P.R.; Piras, B.A.; Hazarika, S.; Li, S.S.; John Lye, R.; Sanders, J.M.; Ferrante, E.A.; Yan, Z.; Annex, B.H.; et al. Adeno-associated virus serotype 9 efficiently targets ischemic skeletal muscle following systemic delivery. *Gene Ther.* **2013**, *20*, 930–938. [CrossRef]
128. Arnett, A.L.; Konieczny, P.; Ramos, J.N.; Hall, J.; Odom, G.; Yablonka-Reuveni, Z.; Chamberlain, J.R.; Chamberlain, J.S. Adeno-associated viral (AAV) vectors do not efficiently target muscle satellite cells. *Mol. Ther. Methods Clin. Dev.* **2014**, *1*. [CrossRef]
129. Jiang, H.; Pierce, G.F.; Ozelo, M.C.; de Paula, E.V.; Vargas, J.A.; Smith, P.; Sommer, J.; Luk, A.; Manno, C.S.; High, K.A.; et al. Evidence of multiyear factor IX expression by AAV-mediated gene transfer to skeletal muscle in an individual with severe hemophilia B. *Mol. Ther. J. Am. Soc. Gene Ther.* **2006**, *14*, 452–455. [CrossRef]
130. Bish, L.T.; Morine, K.; Sleeper, M.M.; Sanmiguel, J.; Wu, D.; Gao, G.; Wilson, J.M.; Sweeney, H.L. Adeno-associated virus (AAV) serotype 9 provides global cardiac gene transfer superior to AAV1, AAV6, AAV7, and AAV8 in the mouse and rat. *Hum. Gene Ther.* **2008**, *19*, 1359–1368. [CrossRef]
131. Vandendriessche, T.; Thorrez, L.; Acosta-Sanchez, A.; Petrus, I.; Wang, L.; Ma, L.; De Waele, L.; Iwasaki, Y.; Gillijns, V.; Wilson, J.M.; et al. Efficacy and safety of adeno-associated viral vectors based on serotype 8 and 9 vs. lentiviral vectors for hemophilia B gene therapy. *J. Thromb. Haemost.* **2007**, *5*, 16–24. [CrossRef] [PubMed]
132. Zincarelli, C.; Soltys, S.; Rengo, G.; Koch, W.J.; Rabinowitz, J.E. Comparative cardiac gene delivery of adeno-associated virus serotypes 1-9 reveals that AAV6 mediates the most efficient transduction in mouse heart. *Clin. Transl. Sci.* **2010**, *3*, 81–89. [CrossRef] [PubMed]
133. Foust, K.D.; Kaspar, B.K. Over the barrier and through the blood: To CNS delivery we go. *Cell Cycle* **2009**, *8*, 4017–4018. [CrossRef] [PubMed]
134. Foust, K.D.; Nurre, E.; Montgomery, C.L.; Hernandez, A.; Chan, C.M.; Kaspar, B.K. Intravascular AAV9 preferentially targets neonatal neurons and adult astrocytes. *Nat. Biotechnol.* **2009**, *27*, 59–65. [CrossRef] [PubMed]
135. Gray, S.J.; Matagne, V.; Bachaboina, L.; Yadav, S.; Ojeda, S.R.; Samulski, R.J. Preclinical differences of intravascular AAV9 delivery to neurons and glia: A comparative study of adult mice and nonhuman primates. *Mol. Ther. J. Am. Soc. Gene Ther.* **2011**, *19*, 1058–1069. [CrossRef] [PubMed]
136. Samaranch, L.; Salegio, E.A.; San Sebastian, W.; Kells, A.P.; Foust, K.D.; Bringas, J.R.; Lamarre, C.; Forsayeth, J.; Kaspar, B.K.; Bankiewicz, K.S. Adeno-associated virus serotype 9 transduction in the central nervous system of nonhuman primates. *Hum. Gene Ther.* **2012**, *23*, 382–389. [CrossRef] [PubMed]
137. Duque, S.; Joussemet, B.; Riviere, C.; Marais, T.; Dubreil, L.; Douar, A.M.; Fyfe, J.; Moullier, P.; Colle, M.A.; Barkats, M. Intravenous administration of self-complementary AAV9 enables transgene delivery to adult motor neurons. *Mol. Ther. J. Am. Soc. Gene Ther.* **2009**, *17*, 1187–1196. [CrossRef]
138. Bevan, A.K.; Duque, S.; Foust, K.D.; Morales, P.R.; Braun, L.; Schmelzer, L.; Chan, C.M.; McCrate, M.; Chicoine, L.G.; Coley, B.D.; et al. Systemic gene delivery in large species for targeting spinal cord, brain, and peripheral tissues for pediatric disorders. *Mol. Ther. J. Am. Soc. Gene Ther.* **2011**, *19*, 1971–1980. [CrossRef]
139. Yue, Y.; Pan, X.; Hakim, C.H.; Kodippili, K.; Zhang, K.; Shin, J.H.; Yang, H.T.; McDonald, T.; Duan, D. Safe and bodywide muscle transduction in young adult Duchenne muscular dystrophy dogs with adeno-associated virus. *Hum. Mol. Genet.* **2015**, *24*, 5880–5890. [CrossRef]
140. Shin, J.H.; Pan, X.; Hakim, C.H.; Yang, H.T.; Yue, Y.; Zhang, K.; Terjung, R.L.; Duan, D. Microdystrophin ameliorates muscular dystrophy in the canine model of duchenne muscular dystrophy. *Mol. Ther. J. Am. Soc. Gene Ther.* **2013**, *21*, 750–757. [CrossRef]
141. Kornegay, J.N.; Li, J.; Bogan, J.R.; Bogan, D.J.; Chen, C.; Zheng, H.; Wang, B.; Qiao, C.; Howard, J.F., Jr.; Xiao, X. Widespread muscle expression of an AAV9 human mini-dystrophin vector after intravenous injection in neonatal dystrophin-deficient dogs. *Mol. Ther. J. Am. Soc. Gene Ther.* **2010**, *18*, 1501–1508. [CrossRef] [PubMed]
142. Yue, Y.; Ghosh, A.; Long, C.; Bostick, B.; Smith, B.F.; Kornegay, J.N.; Duan, D. A single intravenous injection of adeno-associated virus serotype-9 leads to whole body skeletal muscle transduction in dogs. *Mol. Ther. J. Am. Soc. Gene Ther.* **2008**, *16*, 1944–1952. [CrossRef] [PubMed]

143. Mendell, J.R.; Sahenk, Z.; Malik, V.; Gomez, A.M.; Flanigan, K.M.; Lowes, L.P.; Alfano, L.N.; Berry, K.; Meadows, E.; Lewis, S.; et al. A phase 1/2a follistatin gene therapy trial for becker muscular dystrophy. *Mol. Ther. J. Am. Soc. Gene Ther.* **2015**, *23*, 192–201. [CrossRef] [PubMed]

144. Mendell, J.R.; Sahenk, Z.; Al-Zaidy, S.; Rodino-Klapac, L.R.; Lowes, L.P.; Alfano, L.N.; Berry, K.; Miller, N.; Yalvac, M.; Dvorchik, I.; et al. Follistatin Gene Therapy for Sporadic Inclusion Body Myositis Improves Functional Outcomes. *Mol. Ther. J. Am. Soc. Gene Ther.* **2017**, *25*, 870–879. [CrossRef] [PubMed]

145. Mendell, J.R.; Campbell, K.; Rodino-Klapac, L.; Sahenk, Z.; Shilling, C.; Lewis, S.; Bowles, D.; Gray, S.; Li, C.; Galloway, G.; et al. Dystrophin immunity in Duchenne's muscular dystrophy. *N. Engl. J. Med.* **2010**, *363*, 1429–1437. [CrossRef] [PubMed]

146. Bowles, D.E.; McPhee, S.W.; Li, C.; Gray, S.J.; Samulski, J.J.; Camp, A.S.; Li, J.; Wang, B.; Monahan, P.E.; Rabinowitz, J.E.; et al. Phase 1 gene therapy for Duchenne muscular dystrophy using a translational optimized AAV vector. *Mol. Ther. J. Am. Soc. Gene Ther.* **2012**, *20*, 443–455. [CrossRef]

147. Smith, B.K.; Collins, S.W.; Conlon, T.J.; Mah, C.S.; Lawson, L.A.; Martin, A.D.; Fuller, D.D.; Cleaver, B.D.; Clement, N.; Phillips, D.; et al. Phase I/II trial of adeno-associated virus-mediated alpha-glucosidase gene therapy to the diaphragm for chronic respiratory failure in Pompe disease: Initial safety and ventilatory outcomes. *Hum. Gene Ther.* **2013**, *24*, 630–640. [CrossRef]

148. Brooks, A.R.; Harkins, R.N.; Wang, P.; Qian, H.S.; Liu, P.; Rubanyi, G.M. Transcriptional silencing is associated with extensive methylation of the CMV promoter following adenoviral gene delivery to muscle. *J. Gene Med.* **2004**, *6*, 395–404. [CrossRef]

149. Qin, L.; Ding, Y.; Pahud, D.R.; Chang, E.; Imperiale, M.J.; Bromberg, J.S. Promoter attenuation in gene therapy: Interferon-gamma and tumor necrosis factor-alpha inhibit transgene expression. *Hum. Gene Ther.* **1997**, *8*, 2019–2029. [CrossRef]

150. Duan, B.; Cheng, L.; Gao, Y.; Yin, F.X.; Su, G.H.; Shen, Q.Y.; Liu, K.; Hu, X.; Liu, X.; Li, G.P. Silencing of fat-1 transgene expression in sheep may result from hypermethylation of its driven cytomegalovirus (CMV) promoter. *Theriogenology* **2012**, *78*, 793–802. [CrossRef]

151. Harms, J.S.; Splitter, G.A. Interferon-gamma inhibits transgene expression driven by SV40 or CMV promoters but augments expression driven by the mammalian MHC I promoter. *Hum. Gene Ther.* **1995**, *6*, 1291–1297. [CrossRef] [PubMed]

152. Qin, J.Y.; Zhang, L.; Clift, K.L.; Hulur, I.; Xiang, A.P.; Ren, B.Z.; Lahn, B.T. Systematic comparison of constitutive promoters and the doxycycline-inducible promoter. *PLoS ONE* **2010**, *5*, e10611. [CrossRef] [PubMed]

153. Al-Zaidy, S.; Pickard, A.S.; Kotha, K.; Alfano, L.N.; Lowes, L.; Paul, G.; Church, K.; Lehman, K.; Sproule, D.M.; Dabbous, O.; et al. Health outcomes in spinal muscular atrophy type 1 following AVXS-101 gene replacement therapy. *Pediatr. Pulmonol.* **2019**, *54*, 179–185. [CrossRef] [PubMed]

154. Al-Zaidy, S.A.; Kolb, S.J.; Lowes, L.; Alfano, L.N.; Shell, R.; Church, K.R.; Nagendran, S.; Sproule, D.M.; Feltner, D.E.; Wells, C.; et al. AVXS-101 (Onasemnogene Abeparvovec) for SMA1: Comparative Study with a Prospective Natural History Cohort. *J. Neuromuscul. Dis.* **2019**, *6*, 307–317. [CrossRef] [PubMed]

155. Dabbous, O.; Maru, B.; Jansen, J.P.; Lorenzi, M.; Cloutier, M.; Guerin, A.; Pivneva, I.; Wu, E.Q.; Arjunji, R.; Feltner, D.; et al. Survival, Motor Function, and Motor Milestones: Comparison of AVXS-101 Relative to Nusinersen for the Treatment of Infants with Spinal Muscular Atrophy Type 1. *Adv. Ther.* **2019**, *36*, 1164–1176. [CrossRef] [PubMed]

156. Lowes, L.P.; Alfano, L.N.; Arnold, W.D.; Shell, R.; Prior, T.W.; McColly, M.; Lehman, K.J.; Church, K.; Sproule, D.M.; Nagendran, S.; et al. Impact of Age and Motor Function in a Phase 1/2A Study of Infants with SMA Type 1 Receiving Single-Dose Gene Replacement Therapy. *Pediatr. Neurol.* **2019**, *98*, 39–45. [CrossRef]

157. Hartigan-O'Connor, D.; Kirk, C.J.; Crawford, R.; Mule, J.J.; Chamberlain, J.S. Immune evasion by muscle-specific gene expression in dystrophic muscle. *Mol. Ther. J. Am. Soc. Gene Ther.* **2001**, *4*, 525–533. [CrossRef]

158. Yuasa, K.; Sakamoto, M.; Miyagoe-Suzuki, Y.; Tanouchi, A.; Yamamoto, H.; Li, J.; Chamberlain, J.S.; Xiao, X.; Takeda, S. Adeno-associated virus vector-mediated gene transfer into dystrophin-deficient skeletal muscles evokes enhanced immune response against the transgene product. *Gene Ther.* **2002**, *9*, 1576–1588. [CrossRef]

159. Weeratna, R.D.; Wu, T.; Efler, S.M.; Zhang, L.; Davis, H.L. Designing gene therapy vectors: Avoiding immune responses by using tissue-specific promoters. *Gene Ther.* **2001**, *8*, 1872–1878. [CrossRef]

160. Fabre, E.E.; Bigey, P.; Orsini, C.; Scherman, D. Comparison of promoter region constructs for In Vivo intramuscular expression. *J. Gene Med.* **2006**, *8*, 636–645. [CrossRef]
161. Talbot, G.E.; Waddington, S.N.; Bales, O.; Tchen, R.C.; Antoniou, M.N. Desmin-regulated lentiviral vectors for skeletal muscle gene transfer. *Mol. Ther. J. Am. Soc. Gene Ther.* **2010**, *18*, 601–608. [CrossRef] [PubMed]
162. Aikawa, R.; Huggins, G.S.; Snyder, R.O. Cardiomyocyte-specific gene expression following recombinant adeno-associated viral vector transduction. *J. Biol. Chem.* **2002**, *277*, 18979–18985. [CrossRef] [PubMed]
163. Childers, M.K.; Joubert, R.; Poulard, K.; Moal, C.; Grange, R.W.; Doering, J.A.; Lawlor, M.W.; Rider, B.E.; Jamet, T.; Daniele, N.; et al. Gene therapy prolongs survival and restores function in murine and canine models of myotubular myopathy. *Sci. Transl. Med.* **2014**, *6*, 220ra210. [CrossRef] [PubMed]
164. Shieh, P.B.; Kuntz, N.; Smith, B.; Dowling, J.J.; Müller-Felber, W.; Bönnemann, C.G.; Servais, L.; Muntoni, F.; Blaschek, A.; Neuhaus, S.; et al. ASPIRO Gene Therapy Trial In X-Linked Myotubular Myopathy (XLMTM): Update on Preliminary Safety And Efficacy Findings up to 72 Weeks Post-Treatment (1053). *Neurology* **2020**, *94*, 1053.
165. Amacher, S.L.; Buskin, J.N.; Hauschka, S.D. Multiple regulatory elements contribute differentially to muscle creatine kinase enhancer activity in skeletal and cardiac muscle. *Mol. Cell. Biol.* **1993**, *13*, 2753–2764. [CrossRef]
166. Nguyen, Q.G.; Buskin, J.N.; Himeda, C.L.; Fabre-Suver, C.; Hauschka, S.D. Transgenic and tissue culture analyses of the muscle creatine kinase enhancer Trex control element in skeletal and cardiac muscle indicate differences in gene expression between muscle types. *Transgenic Res.* **2003**, *12*, 337–349. [CrossRef]
167. Jaynes, J.B.; Chamberlain, J.S.; Buskin, J.N.; Johnson, J.E.; Hauschka, S.D. Transcriptional regulation of the muscle creatine kinase gene and regulated expression in transfected mouse myoblasts. *Mol. Cell. Biol.* **1986**, *6*, 2855–2864. [CrossRef]
168. Takeshita, F.; Takase, K.; Tozuka, M.; Saha, S.; Okuda, K.; Ishii, N.; Sasaki, S. Muscle creatine kinase/SV40 hybrid promoter for muscle-targeted long-term transgene expression. *Int. J. Mol. Med.* **2007**, *19*, 309–315. [CrossRef]
169. Hauser, M.A.; Robinson, A.; Hartigan-O'Connor, D.; Williams-Gregory, D.A.; Buskin, J.N.; Apone, S.; Kirk, C.J.; Hardy, S.; Hauschka, S.D.; Chamberlain, J.S. Analysis of muscle creatine kinase regulatory elements in recombinant adenoviral vectors. *Mol. Ther. J. Am. Soc. Gene Ther.* **2000**, *2*, 16–25. [CrossRef]
170. Wang, B.; Li, J.; Fu, F.H.; Chen, C.; Zhu, X.; Zhou, L.; Jiang, X.; Xiao, X. Construction and analysis of compact muscle-specific promoters for AAV vectors. *Gene Ther.* **2008**, *15*, 1489–1499. [CrossRef]
171. Sahenk, Z.; Galloway, G.; Clark, K.R.; Malik, V.; Rodino-Klapac, L.R.; Kaspar, B.K.; Chen, L.; Braganza, C.; Montgomery, C.; Mendell, J.R. AAV1.NT-3 gene therapy for charcot-marie-tooth neuropathy. *Mol. Ther. J. Am. Soc. Gene Ther.* **2014**, *22*, 511–521. [CrossRef] [PubMed]
172. Rodino-Klapac, L.R.; Lee, J.S.; Mulligan, R.C.; Clark, K.R.; Mendell, J.R. Lack of toxicity of alpha-sarcoglycan overexpression supports clinical gene transfer trial in LGMD2D. *Neurology* **2008**, *71*, 240–247. [CrossRef] [PubMed]
173. Mendell, J.R.; Rodino-Klapac, L.R.; Rosales, X.Q.; Coley, B.D.; Galloway, G.; Lewis, S.; Malik, V.; Shilling, C.; Byrne, B.J.; Conlon, T.; et al. Sustained alpha-sarcoglycan gene expression after gene transfer in limb-girdle muscular dystrophy, type 2D. *Ann. Neurol.* **2010**, *68*, 629–638. [CrossRef] [PubMed]
174. Mendell, J.R.; Rodino-Klapac, L.R.; Rosales-Quintero, X.; Kota, J.; Coley, B.D.; Galloway, G.; Craenen, J.M.; Lewis, S.; Malik, V.; Shilling, C.; et al. Limb-girdle muscular dystrophy type 2D gene therapy restores alpha-sarcoglycan and associated proteins. *Ann. Neurol.* **2009**, *66*, 290–297. [CrossRef] [PubMed]
175. Salva, M.Z.; Himeda, C.L.; Tai, P.W.; Nishiuchi, E.; Gregorevic, P.; Allen, J.M.; Finn, E.E.; Nguyen, Q.G.; Blankinship, M.J.; Meuse, L.; et al. Design of tissue-specific regulatory cassettes for high-level rAAV-mediated expression in skeletal and cardiac muscle. *Mol. Ther. J. Am. Soc. Gene Ther.* **2007**, *15*, 320–329. [CrossRef] [PubMed]
176. Sun, B.; Young, S.P.; Li, P.; Di, C.; Brown, T.; Salva, M.Z.; Li, S.; Bird, A.; Yan, Z.; Auten, R.; et al. Correction of multiple striated muscles in murine Pompe disease through adeno-associated virus-mediated gene therapy. *Mol. Ther. J. Am. Soc. Gene Ther.* **2008**, *16*, 1366–1371. [CrossRef]
177. Mendell, J.R.; Sahenk, Z.; Lehman, K.; Nease, C.; Lowes, L.P.; Miller, N.F.; Iammarino, M.A.; Alfano, L.N.; Nicholl, A.; Al-Zaidy, S.; et al. Assessment of Systemic Delivery of rAAVrh74.MHCK7.micro-dystrophin in Children With Duchenne Muscular Dystrophy: A Nonrandomized Controlled Trial. *JAMA Neurol.* **2020**. [CrossRef]

178. Li, X.; Eastman, E.M.; Schwartz, R.J.; Draghia-Akli, R. Synthetic muscle promoters: Activities exceeding naturally occurring regulatory sequences. *Nat. Biotechnol.* **1999**, *17*, 241–245. [CrossRef]
179. Cordier, L.; Gao, G.P.; Hack, A.A.; McNally, E.M.; Wilson, J.M.; Chirmule, N.; Sweeney, H.L. Muscle-specific promoters may be necessary for adeno-associated virus-mediated gene transfer in the treatment of muscular dystrophies. *Hum. Gene Ther.* **2001**, *12*, 205–215. [CrossRef]
180. Fougerousse, F.; Bartoli, M.; Poupiot, J.; Arandel, L.; Durand, M.; Guerchet, N.; Gicquel, E.; Danos, O.; Richard, I. Phenotypic correction of alpha-sarcoglycan deficiency by intra-arterial injection of a muscle-specific serotype 1 rAAV vector. *Mol. Ther. J. Am. Soc. Gene Ther.* **2007**, *15*, 53–61. [CrossRef]
181. Ziegler, R.J.; Lonning, S.M.; Armentano, D.; Li, C.; Souza, D.W.; Cherry, M.; Ford, C.; Barbon, C.M.; Desnick, R.J.; Gao, G.; et al. AAV2 vector harboring a liver-restricted promoter facilitates sustained expression of therapeutic levels of alpha-galactosidase A and the induction of immune tolerance in Fabry mice. *Mol. Ther. J. Am. Soc. Gene Ther.* **2004**, *9*, 231–240. [CrossRef] [PubMed]
182. Poupiot, J.; Costa Verdera, H.; Hardet, R.; Colella, P.; Collaud, F.; Bartolo, L.; Davoust, J.; Sanatine, P.; Mingozzi, F.; Richard, I.; et al. Role of Regulatory T Cell and Effector T Cell Exhaustion in Liver-Mediated Transgene Tolerance in Muscle. *Mol. Ther. Methods Clin. Dev.* **2019**, *15*, 83–100. [CrossRef] [PubMed]
183. Bartolo, L.; Li Chung Tong, S.; Chappert, P.; Urbain, D.; Collaud, F.; Colella, P.; Richard, I.; Ronzitti, G.; Demengeot, J.; Gross, D.A.; et al. Dual muscle-liver transduction imposes immune tolerance for muscle transgene engraftment despite preexisting immunity. *JCI Insight* **2019**, *4*. [CrossRef] [PubMed]
184. Colella, P.; Sellier, P.; Costa Verdera, H.; Puzzo, F.; van Wittenberghe, L.; Guerchet, N.; Daniele, N.; Gjata, B.; Marmier, S.; Charles, S.; et al. AAV Gene Transfer with Tandem Promoter Design Prevents Anti-transgene Immunity and Provides Persistent Efficacy in Neonate Pompe Mice. *Mol. Ther. Methods Clin. Dev.* **2019**, *12*, 85–101. [CrossRef] [PubMed]
185. GlobeNewswire. Solid Biosciences Provides Update Regarding SGT-001 Phase I/II Clinical Hold on IGNITE DMD. 2020. Available online: https://www.globenewswire.com/news-release/2020/05/07/2029328/0/en/Solid-Biosciences-Provides-Update-regarding-SGT-001-Phase-I-II-Clinical-Hold-on-IGNITE-DMD.html (accessed on 25 November 2020).
186. GlobeNewswire. Solid Biosciences Announces Clinical Hold On SGT-001 Phase I/II Clinical Trial for Duchenne Muscular Dystrophy. 2018. Available online: https://www.globenewswire.com/news-release/2018/03/14/1422770/0/en/Solid-Biosciences-Announces-Clinical-Hold-On-SGT-001-Phase-I-II-Clinical-Trial-For-Duchenne-Muscular-Dystrophy.html (accessed on 25 November 2020).
187. Binks, M. *Early, Initial Data from C3391001, a First-In-Human Safety Study of PF-06939926, a Mini-Dystrophin Gene Therapy for the Potential Treatment of DMD*; Parent Project Muscular Dystrophy: Hackensack, NJ, USA, 2019.
188. Novartis. AveXis Presents AVXS-101 IT Data Demonstrating Remarkable Increases in HFMSE Scores and a Consistent Clinically Meaningful Response in Older Patients with SMA Type 2. 2020. Available online: https://www.novartis.com/news/media-releases/avexis-presents-avxs-101-it-data-demonstrating-remarkable-increases-hfmse-scores-and-consistent-clinically-meaningful-response-older-patients-sma-type-2 (accessed on 25 November 2020).
189. Pharma, F. Audentes' Gene Therapy AT132 Hit with FDA Clinical Hold after Second Patient Death. 2020. Available online: https://www.firstwordpharma.com/node/1736153 (accessed on 25 November 2020).
190. Rodino-klapac, L.R.; Pozsgai, E.R.; Lewis, S.; Griffin, D.A.; Meadows, A.S.; Lehman, K.; Church, K.; Lowes, L.; Mendel, J.R. Systemic Gene Transfer with AAVrh74.MHCK7.SGCB Increased β-sarcoglycan Expression in Patients with Limb Girdle Muscular Dystrophy Type 2E. *Neuropediatrics* **2019**, *50*, S1–S55.
191. Zhang, Z.; Lotti, F.; Dittmar, K.; Younis, I.; Wan, L.; Kasim, M.; Dreyfuss, G. SMN deficiency causes tissue-specific perturbations in the repertoire of snRNAs and widespread defects in splicing. *Cell* **2008**, *133*, 585–600. [CrossRef] [PubMed]
192. Wijngaarde, C.A.; Blank, A.C.; Stam, M.; Wadman, R.I.; van den Berg, L.H.; van der Pol, W.L. Cardiac pathology in spinal muscular atrophy: A systematic review. *Orphanet J. Rare Dis.* **2017**, *12*, 67. [CrossRef]
193. Meyer, K.; Ferraiuolo, L.; Schmelzer, L.; Braun, L.; McGovern, V.; Likhite, S.; Michels, O.; Govoni, A.; Fitzgerald, J.; Morales, P.; et al. Improving single injection CSF delivery of AAV9-mediated gene therapy for SMA: A dose-response study in mice and nonhuman primates. *Mol. Ther. J. Am. Soc. Gene Ther.* **2015**, *23*, 477–487. [CrossRef]

194. Foust, K.D.; Wang, X.; McGovern, V.L.; Braun, L.; Bevan, A.K.; Haidet, A.M.; Le, T.T.; Morales, P.R.; Rich, M.M.; Burghes, A.H.; et al. Rescue of the spinal muscular atrophy phenotype in a mouse model by early postnatal delivery of SMN. *Nat. Biotechnol.* **2010**, *28*, 271–274. [CrossRef]
195. Bevan, A.K.; Hutchinson, K.R.; Foust, K.D.; Braun, L.; McGovern, V.L.; Schmelzer, L.; Ward, J.G.; Petruska, J.C.; Lucchesi, P.A.; Burghes, A.H.; et al. Early heart failure in the SMNDelta7 model of spinal muscular atrophy and correction by postnatal scAAV9-SMN delivery. *Hum. Mol. Genet.* **2010**, *19*, 3895–3905. [CrossRef]
196. Dominguez, E.; Marais, T.; Chatauret, N.; Benkhelifa-Ziyyat, S.; Duque, S.; Ravassard, P.; Carcenac, R.; Astord, S.; Pereira de Moura, A.; Voit, T.; et al. Intravenous scAAV9 delivery of a codon-optimized SMN1 sequence rescues SMA mice. *Hum. Mol. Genet.* **2011**, *20*, 681–693. [CrossRef] [PubMed]
197. Valori, C.F.; Ning, K.; Wyles, M.; Mead, R.J.; Grierson, A.J.; Shaw, P.J.; Azzouz, M. Systemic delivery of scAAV9 expressing SMN prolongs survival in a model of spinal muscular atrophy. *Sci. Transl. Med.* **2010**, *2*, 35ra42. [CrossRef] [PubMed]
198. McCarty, D.M. Self-complementary AAV vectors; advances and applications. *Mol. Ther. J. Am. Soc. Gene Ther.* **2008**, *16*, 1648–1656. [CrossRef] [PubMed]
199. Finkel, R.S.; Mercuri, E.; Darras, B.T.; Connolly, A.M.; Kuntz, N.L.; Kirschner, J.; Chiriboga, C.A.; Saito, K.; Servais, L.; Tizzano, E.; et al. Nusinersen versus Sham Control in Infantile-Onset Spinal Muscular Atrophy. *N. Engl. J. Med.* **2017**, *377*, 1723–1732. [CrossRef]
200. Hinderer, C.; Katz, N.; Buza, E.L.; Dyer, C.; Goode, T.; Bell, P.; Richman, L.K.; Wilson, J.M. Severe Toxicity in Nonhuman Primates and Piglets Following High-Dose Intravenous Administration of an Adeno-Associated Virus Vector Expressing Human SMN. *Hum. Gene Ther.* **2018**, *29*, 285–298. [CrossRef]
201. Hordeaux, J.; Wang, Q.; Katz, N.; Buza, E.L.; Bell, P.; Wilson, J.M. The Neurotropic Properties of AAV-PHP.B Are Limited to C57BL/6J Mice. *Mol. Ther. J. Am. Soc. Gene Ther.* **2018**, *26*, 664–668. [CrossRef]
202. Assinger, A. Platelets and infection—An emerging role of platelets in viral infection. *Front. Immunol.* **2014**, *5*, 649. [CrossRef]
203. Harper, S.Q.; Hauser, M.A.; DelloRusso, C.; Duan, D.; Crawford, R.W.; Phelps, S.F.; Harper, H.A.; Robinson, A.S.; Engelhardt, J.F.; Brooks, S.V.; et al. Modular flexibility of dystrophin: Implications for gene therapy of Duchenne muscular dystrophy. *Nat. Med.* **2002**, *8*, 253–261. [CrossRef]
204. Le Guiner, C.; Montus, M.; Servais, L.; Cherel, Y.; Francois, V.; Thibaud, J.L.; Wary, C.; Matot, B.; Larcher, T.; Guigand, L.; et al. Forelimb treatment in a large cohort of dystrophic dogs supports delivery of a recombinant AAV for exon skipping in Duchenne patients. *Mol. Ther. J. Am. Soc. Gene Ther.* **2014**, *22*, 1923–1935. [CrossRef]
205. Goyenvalle, A.; Babbs, A.; Wright, J.; Wilkins, V.; Powell, D.; Garcia, L.; Davies, K.E. Rescue of severely affected dystrophin/utrophin-deficient mice through scAAV-U7snRNA-mediated exon skipping. *Hum. Mol. Genet.* **2012**, *21*, 2559–2571. [CrossRef]
206. Goyenvalle, A.; Vulin, A.; Fougerousse, F.; Leturcq, F.; Kaplan, J.C.; Garcia, L.; Danos, O. Rescue of dystrophic muscle through U7 snRNA-mediated exon skipping. *Science* **2004**, *306*, 1796–1799. [CrossRef] [PubMed]
207. Vulin, A.; Barthelemy, I.; Goyenvalle, A.; Thibaud, J.L.; Beley, C.; Griffith, G.; Benchaouir, R.; le Hir, M.; Unterfinger, Y.; Lorain, S.; et al. Muscle function recovery in golden retriever muscular dystrophy after AAV1-U7 exon skipping. *Mol. Ther. J. Am. Soc. Gene Ther.* **2012**, *20*, 2120–2133. [CrossRef] [PubMed]
208. Bish, L.T.; Sleeper, M.M.; Forbes, S.C.; Wang, B.; Reynolds, C.; Singletary, G.E.; Trafny, D.; Morine, K.J.; Sanmiguel, J.; Cecchini, S.; et al. Long-term restoration of cardiac dystrophin expression in golden retriever muscular dystrophy following rAAV6-mediated exon skipping. *Mol. Ther. J. Am. Soc. Gene Ther.* **2012**, *20*, 580–589. [CrossRef] [PubMed]
209. Barbash, I.M.; Cecchini, S.; Faranesh, A.Z.; Virag, T.; Li, L.; Yang, Y.; Hoyt, R.F.; Kornegay, J.N.; Bogan, J.R.; Garcia, L.; et al. MRI roadmap-guided transendocardial delivery of exon-skipping recombinant adeno-associated virus restores dystrophin expression in a canine model of Duchenne muscular dystrophy. *Gene Ther.* **2013**, *20*, 274–282. [CrossRef] [PubMed]
210. Gregorevic, P.; Allen, J.M.; Minami, E.; Blankinship, M.J.; Haraguchi, M.; Meuse, L.; Finn, E.; Adams, M.E.; Froehner, S.C.; Murry, C.E.; et al. rAAV6-microdystrophin preserves muscle function and extends lifespan in severely dystrophic mice. *Nat. Med.* **2006**, *12*, 787–789. [CrossRef] [PubMed]
211. Gregorevic, P.; Blankinship, M.J.; Allen, J.M.; Chamberlain, J.S. Systemic microdystrophin gene delivery improves skeletal muscle structure and function in old dystrophic mdx mice. *Mol. Ther. J. Am. Soc. Gene Ther.* **2008**, *16*, 657–664. [CrossRef] [PubMed]

212. Rodgers, B.D.; Bishaw, Y.; Kagel, D.; Ramos, J.N.; Maricelli, J.W. Micro-dystrophin Gene Therapy Partially Enhances Exercise Capacity in Older Adult mdx Mice. *Mol. Ther. Methods Clin. Dev.* **2020**, *17*, 122–132. [CrossRef] [PubMed]

213. Rodino-Klapac, L.R.; Janssen, P.M.; Montgomery, C.L.; Coley, B.D.; Chicoine, L.G.; Clark, K.R.; Mendell, J.R. A translational approach for limb vascular delivery of the micro-dystrophin gene without high volume or high pressure for treatment of Duchenne muscular dystrophy. *J. Transl. Med.* **2007**, *5*, 45. [CrossRef]

214. Rodino-Klapac, L.R.; Montgomery, C.L.; Bremer, W.G.; Shontz, K.M.; Malik, V.; Davis, N.; Sprinkle, S.; Campbell, K.J.; Sahenk, Z.; Clark, K.R.; et al. Persistent expression of FLAG-tagged micro dystrophin in nonhuman primates following intramuscular and vascular delivery. *Mol. Ther. J. Am. Soc. Gene Ther.* **2010**, *18*, 109–117. [CrossRef]

215. Foster, H.; Sharp, P.S.; Athanasopoulos, T.; Trollet, C.; Graham, I.R.; Foster, K.; Wells, D.J.; Dickson, G. Codon and mRNA sequence optimization of microdystrophin transgenes improves expression and physiological outcome in dystrophic mdx mice following AAV2/8 gene transfer. *Mol. Ther. J. Am. Soc. Gene Ther.* **2008**, *16*, 1825–1832. [CrossRef]

216. Le Guiner, C.; Servais, L.; Montus, M.; Larcher, T.; Fraysse, B.; Moullec, S.; Allais, M.; Francois, V.; Dutilleul, M.; Malerba, A.; et al. Long-term microdystrophin gene therapy is effective in a canine model of Duchenne muscular dystrophy. *Nat. Commun.* **2017**, *8*, 16105. [CrossRef] [PubMed]

217. Bostick, B.; Yue, Y.; Lai, Y.; Long, C.; Li, D.; Duan, D. Adeno-associated virus serotype-9 microdystrophin gene therapy ameliorates electrocardiographic abnormalities in mdx mice. *Hum. Gene Ther.* **2008**, *19*, 851–856. [CrossRef] [PubMed]

218. Hakim, C.H.; Wasala, N.B.; Pan, X.; Kodippili, K.; Yue, Y.; Zhang, K.; Yao, G.; Haffner, B.; Duan, S.X.; Ramos, J.; et al. A Five-Repeat Micro-Dystrophin Gene Ameliorated Dystrophic Phenotype in the Severe DBA/2J-mdx Model of Duchenne Muscular Dystrophy. *Mol. Ther. Methods Clin. Dev.* **2017**, *6*, 216–230. [CrossRef] [PubMed]

219. Bostick, B.; Shin, J.H.; Yue, Y.; Duan, D. AAV-microdystrophin therapy improves cardiac performance in aged female mdx mice. *Mol. Ther. J. Am. Soc. Gene Ther.* **2011**, *19*, 1826–1832. [CrossRef] [PubMed]

220. Shin, J.H.; Nitahara-Kasahara, Y.; Hayashita-Kinoh, H.; Ohshima-Hosoyama, S.; Kinoshita, K.; Chiyo, T.; Okada, H.; Okada, T.; Takeda, S. Improvement of cardiac fibrosis in dystrophic mice by rAAV9-mediated microdystrophin transduction. *Gene Ther.* **2011**, *18*, 910–919. [CrossRef] [PubMed]

221. Hakim, C.H.; Clement, N.; Wasala, L.P.; Yang, H.T.; Yue, Y.; Zhang, K.; Kodippili, K.; Adamson-Small, L.; Pan, X.; Schneider, J.S.; et al. Micro-dystrophin AAV Vectors Made by Transient Transfection and Herpesvirus System Are Equally Potent in Treating mdx Mouse Muscle Disease. *Mol. Ther. Methods Clin. Dev.* **2020**, *18*, 664–678. [CrossRef]

222. Lai, Y.; Thomas, G.D.; Yue, Y.; Yang, H.T.; Li, D.; Long, C.; Judge, L.; Bostick, B.; Chamberlain, J.S.; Terjung, R.L.; et al. Dystrophins carrying spectrin-like repeats 16 and 17 anchor nNOS to the sarcolemma and enhance exercise performance in a mouse model of muscular dystrophy. *J. Clin. Investig.* **2009**, *119*, 624–635. [CrossRef]

223. Griffin, D.A.; Potter, R.A.; Pozsgai, E.R.; Peterson, E.L.; Rodino-Klapac, L.R. Adeno-Associated Virus Serotype rh74 Prevalence in Muscular Dystrophy Population. *Mol. Ther.* **2019**, *27* (Suppl. 1), 342.

224. Zygmunt, D.A.; Crowe, K.E.; Flanigan, K.M.; Martin, P.T. Comparison of Serum rAAV Serotype-Specific Antibodies in Patients with Duchenne Muscular Dystrophy, Becker Muscular Dystrophy, Inclusion Body Myositis, or GNE Myopathy. *Hum. Gene Ther.* **2017**, *28*, 737–746. [CrossRef]

225. Fu, H.; Meadows, A.S.; Pineda, R.J.; Kunkler, K.L.; Truxal, K.V.; McBride, K.L.; Flanigan, K.M.; McCarty, D.M. Differential Prevalence of Antibodies Against Adeno-Associated Virus in Healthy Children and Patients with Mucopolysaccharidosis III: Perspective for AAV-Mediated Gene Therapy. *Hum. Gene Ther. Clin. Dev.* **2017**, *28*, 187–196. [CrossRef]

226. Potter, R.A.; Griffin, D.A.; Heller, K.N.; Peterson, E.L.; Clark, E.K.; Mendell, J.R.; Rodino-Klapac, L.R. Dose-escalation study of systematically delivered rAAVrh74.MHCK7.micro-dystrophin in the mdx mouse model of Duchenne muscular dystrophy. In Proceedings of the American Society of Gene and Cell Therapy Annual Meeting, Chicago, IL, USA, 16–19 May 2018.

227. Potter, R.A.; Griffin, D.A.; Heller, K.N.; Peterson, E.L.; Johnson, R.W.; Clark, E.K.; Mendell, J.R.; Rodino-Klapac, L.R. Dose Escalation Study of Systemically Delivered AAVrh74.MHCK7.Micro-Dystrophin in the Mdx Mouse Model of DMD. *Mol. Ther.* **2018**, *26* (Suppl. 1), 5.

228. Mendell, J.R.; Chicoine, L.G.; Al-Zaidy, S.A.; Sahenk, Z.; Lehman, K.; Lowes, L.; Miller, N.; Alfano, L.; Galliers, B.; Lewis, S.; et al. Gene Delivery for Limb-Girdle Muscular Dystrophy Type 2D by Isolated Limb Infusion. *Hum. Gene Ther.* **2019**, *30*, 794–801. [CrossRef] [PubMed]
229. Asher, D.R.; Thapa, K.; Dharia, S.D.; Khan, N.; Potter, R.A.; Rodino-Klapac, L.R.; Mendell, J.R. Clinical development on the frontier: Gene therapy for duchenne muscular dystrophy. *Expert Opin. Biol. Ther.* **2020**, *20*, 263–274. [CrossRef] [PubMed]
230. Nathwani, A.C.; Tuddenham, E.G.; Rangarajan, S.; Rosales, C.; McIntosh, J.; Linch, D.C.; Chowdary, P.; Riddell, A.; Pie, A.J.; Harrington, C.; et al. Adenovirus-associated virus vector-mediated gene transfer in hemophilia B. *N. Engl. J. Med.* **2011**, *365*, 2357–2365. [CrossRef]
231. George, L.A.; Sullivan, S.K.; Giermasz, A.; Rasko, J.E.J.; Samelson-Jones, B.J.; Ducore, J.; Cuker, A.; Sullivan, L.M.; Majumdar, S.; Teitel, J.; et al. Hemophilia B Gene Therapy with a High-Specific-Activity Factor IX Variant. *N. Engl. J. Med.* **2017**, *377*, 2215–2227. [CrossRef]
232. Solid Biosciences. Letter to the Duchenne Community about the Status of the IGNITE DMD Clinical Trial. 2018. Available online: https://www.solidbio.com/about/media/news/letter-to-the-duchenne-community-about-the-status-of-the-ignite-dmd-clinical-trial (accessed on 25 November 2020).
233. Solid Biosciences. Solid Biosciences Announces FDA Lifts Clinical Hold on IGNITE DMD Clinical Trial. 2020. Available online: https://www.solidbio.com/about/media/press-releases/solid-biosciences-announces-fda-lifts-clinical-hold-on-ignite-dmd-clinical-trial (accessed on 25 November 2020).
234. Hakim, C.H.; Kodippili, K.; Jenkins, G.; Yang, H.; Pan, X.; Lessa, T.; Leach, S.; Emter, C.; Yue, Y.; Zhang, K.; et al. AAV micro-dystrophin therapy ameliorates muscular dystrophy in young adult Duchenne muscular dystrophy dogs for up to thirty months following injection. *Mol. Ther.* **2018**, *26* (Suppl. 1), 5.
235. Simon, A.K.; Hollander, G.A.; McMichael, A. Evolution of the immune system in humans from infancy to old age. *Proc. Biol. Sci.* **2015**, *282*, 20143085. [CrossRef]
236. Laporte, J.; Hu, L.J.; Kretz, C.; Mandel, J.L.; Kioschis, P.; Coy, J.F.; Klauck, S.M.; Poustka, A.; Dahl, N. A gene mutated in X-linked myotubular myopathy defines a new putative tyrosine phosphatase family conserved in yeast. *Nat. Genet.* **1996**, *13*, 175–182. [CrossRef]
237. Blondeau, F.; Laporte, J.; Bodin, S.; Superti-Furga, G.; Payrastre, B.; Mandel, J.L. Myotubularin, a phosphatase deficient in myotubular myopathy, acts on phosphatidylinositol 3-kinase and phosphatidylinositol 3-phosphate pathway. *Hum. Mol. Genet.* **2000**, *9*, 2223–2229. [CrossRef]
238. Taylor, G.S.; Maehama, T.; Dixon, J.E. Myotubularin, a protein tyrosine phosphatase mutated in myotubular myopathy, dephosphorylates the lipid second messenger, phosphatidylinositol 3-phosphate. *Proc. Natl. Acad. Sci. USA* **2000**, *97*, 8910–8915. [CrossRef]
239. Buj-Bello, A.; Laugel, V.; Messaddeq, N.; Zahreddine, H.; Laporte, J.; Pellissier, J.F.; Mandel, J.L. The lipid phosphatase myotubularin is essential for skeletal muscle maintenance but not for myogenesis in mice. *Proc. Natl. Acad. Sci. USA* **2002**, *99*, 15060–15065. [CrossRef] [PubMed]
240. Beggs, A.H.; Bohm, J.; Snead, E.; Kozlowski, M.; Maurer, M.; Minor, K.; Childers, M.K.; Taylor, S.M.; Hitte, C.; Mickelson, J.R.; et al. MTM1 mutation associated with X-linked myotubular myopathy in Labrador Retrievers. *Proc. Natl. Acad. Sci. USA* **2010**, *107*, 14697–14702. [CrossRef] [PubMed]
241. Grange, R.W.; Doering, J.; Mitchell, E.; Holder, M.N.; Guan, X.; Goddard, M.; Tegeler, C.; Beggs, A.H.; Childers, M.K. Muscle function in a canine model of X-linked myotubular myopathy. *Muscle Nerve* **2012**, *46*, 588–591. [CrossRef] [PubMed]
242. Buj-Bello, A.; Fougerousse, F.; Schwab, Y.; Messaddeq, N.; Spehner, D.; Pierson, C.R.; Durand, M.; Kretz, C.; Danos, O.; Douar, A.M.; et al. AAV-mediated intramuscular delivery of myotubularin corrects the myotubular myopathy phenotype in targeted murine muscle and suggests a function in plasma membrane homeostasis. *Hum. Mol. Genet.* **2008**, *17*, 2132–2143. [CrossRef] [PubMed]
243. Elverman, M.; Goddard, M.A.; Mack, D.; Snyder, J.M.; Lawlor, M.W.; Meng, H.; Beggs, A.H.; Buj-Bello, A.; Poulard, K.; Marsh, A.P.; et al. Long-term effects of systemic gene therapy in a canine model of myotubular myopathy. *Muscle Nerve* **2017**, *56*, 943–953. [CrossRef] [PubMed]
244. Mack, D.L.; Poulard, K.; Goddard, M.A.; Latournerie, V.; Snyder, J.M.; Grange, R.W.; Elverman, M.R.; Denard, J.; Veron, P.; Buscara, L.; et al. Systemic AAV8-Mediated Gene Therapy Drives Whole-Body Correction of Myotubular Myopathy in Dogs. *Mol. Ther. J. Am. Soc. Gene Ther.* **2017**, *25*, 839–854. [CrossRef]

245. Dupont, J.B.; Guo, J.; Renaud-Gabardos, E.; Poulard, K.; Latournerie, V.; Lawlor, M.W.; Grange, R.W.; Gray, J.T.; Buj-Bello, A.; Childers, M.K.; et al. AAV-Mediated Gene Transfer Restores a Normal Muscle Transcriptome in a Canine Model of X-Linked Myotubular Myopathy. *Mol. Ther. J. Am. Soc. Gene Ther.* **2020**, *28*, 382–393. [CrossRef]
246. Phillips, A.; Belle, A.; Guo, J.; Ton, J.; Stinchcombe, T.; Buj-Bello, A.; Gray, J.T. Nonhuman Primate Safety and Potency of an AAV Vector for XLMTM Produced by Transient Transfection at 500L. *Mol. Ther.* **2017**, *25* (Suppl. 1), 102–103.
247. Kuntz, N.; Shieh, P.B.; Smith, B.; Bonnemann, C.G.; Dowling, J.J.; Lawlor, M.W.; Müller-Felber, W.; Noursalehi, M.; Rico, S.; Servais, L.; et al. ASPIRO phase 1/2 gene therapy trial In X-linked myotubular myopathy: Preliminary safety and efficacy findings. *Neuromuscul. Disord.* **2018**, *28* (Suppl. 2), S91. [CrossRef]
248. Kuntz, N.; Shieh, P.B.; Smith, B.; Bonnemann, C.G.; Dowling, J.J.; Lawlor, M.W.; Müller-Felber, W.; Noursalehi, M.; Rico, S.; Servais, L.; et al. ASPIRO phase 1/2 gene therapy trail in X-linked myotubular myopathy (XLMTM): Preliminary safety and efficacy findings. *Mol. Ther.* **2018**, *26* (Suppl. 1), 4.
249. Kuntz, N.; Shie, P.B.; Smith, B.; Bonnemann, C.G.; Dowling, J.J.; Lawlor, M.W.; Mavilio, F.; Muller-Felber, W.; Noursalehi, M.; Rico, S.; et al. Gene therapy for X-linked myotubular myopathy with AT132 (rAAV8-Des-hMTM1): Preliminary results from the ASPIRO phase-1/2 study. *Hum. Gene Ther.* **2018**, *27*, A2–A169.
250. Somanathan, S.; Calcedo, R.; Wilson, J.M. Adenovirus-Antibody Complexes Contributed to Lethal Systemic Inflammation in a Gene Therapy Trial. *Mol. Ther. J. Am. Soc. Gene Ther.* **2020**, *28*, 784–793. [CrossRef] [PubMed]
251. Wilson, J.M.; Flotte, T.R. Moving Forward After Two Deaths in a Gene Therapy Trial of Myotubular Myopathy. *Hum. Gene Ther.* **2020**, *31*, 695–696. [CrossRef] [PubMed]
252. Santiago-Ortiz, J.; Ojala, D.S.; Westesson, O.; Weinstein, J.R.; Wong, S.Y.; Steinsapir, A.; Kumar, S.; Holmes, I.; Schaffer, D.V. AAV ancestral reconstruction library enables selection of broadly infectious viral variants. *Gene Ther.* **2015**, *22*, 934–946. [CrossRef]
253. Asokan, A.; Conway, J.C.; Phillips, J.L.; Li, C.; Hegge, J.; Sinnott, R.; Yadav, S.; DiPrimio, N.; Nam, H.J.; Agbandje-McKenna, M.; et al. Reengineering a receptor footprint of adeno-associated virus enables selective and systemic gene transfer to muscle. *Nat. Biotechnol.* **2010**, *28*, 79–82. [CrossRef]
254. Yu, C.Y.; Yuan, Z.; Cao, Z.; Wang, B.; Qiao, C.; Li, J.; Xiao, X. A muscle-targeting peptide displayed on AAV2 improves muscle tropism on systemic delivery. *Gene Ther.* **2009**, *16*, 953–962. [CrossRef]
255. Ying, Y.; Muller, O.J.; Goehringer, C.; Leuchs, B.; Trepel, M.; Katus, H.A.; Kleinschmidt, J.A. Heart-targeted adeno-associated viral vectors selected by In Vivo biopanning of a random viral display peptide library. *Gene Ther.* **2010**, *17*, 980–990. [CrossRef]
256. Shen, S.; Horowitz, E.D.; Troupes, A.N.; Brown, S.M.; Pulicherla, N.; Samulski, R.J.; Agbandje-McKenna, M.; Asokan, A. Engraftment of a galactose receptor footprint onto adeno-associated viral capsids improves transduction efficiency. *J. Biol. Chem.* **2013**, *288*, 28814–28823. [CrossRef]
257. Tarantal, A.F.; Lee, C.C.I.; Martinez, M.L.; Asokan, A.; Samulski, R.J. Systemic and Persistent Muscle Gene Expression in Rhesus Monkeys with a Liver De-Targeted Adeno-Associated Virus Vector. *Hum. Gene Ther.* **2017**, *28*, 385–391. [CrossRef]
258. Pulicherla, N.; Shen, S.; Yadav, S.; Debbink, K.; Govindasamy, L.; Agbandje-McKenna, M.; Asokan, A. Engineering liver-detargeted AAV9 vectors for cardiac and musculoskeletal gene transfer. *Mol. Ther. J. Am. Soc. Gene Ther.* **2011**, *19*, 1070–1078. [CrossRef]
259. Tulalamba, W.; Weinmann, J.; Pham, Q.H.; El Andari, J.; VandenDriessche, T.; Chuah, M.K.; Grimm, D. Distinct transduction of muscle tissue in mice after systemic delivery of AAVpo1 vectors. *Gene Ther.* **2020**, *27*, 170–179. [CrossRef] [PubMed]
260. Choudhury, S.R.; Fitzpatrick, Z.; Harris, A.F.; Maitland, S.A.; Ferreira, J.S.; Zhang, Y.; Ma, S.; Sharma, R.B.; Gray-Edwards, H.L.; Johnson, J.A.; et al. In Vivo Selection Yields AAV-B1 Capsid for Central Nervous System and Muscle Gene Therapy. *Mol. Ther. J. Am. Soc. Gene Ther.* **2016**, *24*, 1247–1257. [CrossRef] [PubMed]
261. Hakim, C.H.; Yue, Y.; Shin, J.H.; Williams, R.R.; Zhang, K.; Smith, B.F.; Duan, D. Systemic gene transfer reveals distinctive muscle transduction profile of tyrosine mutant AAV-1, -6, and -9 in neonatal dogs. *Mol. Ther. Methods Clin. Dev.* **2014**, *1*, 14002. [CrossRef] [PubMed]
262. Yang, L.; Li, J.; Xiao, X. Directed evolution of adeno-associated virus (AAV) as vector for muscle gene therapy. *Methods Mol. Biol.* **2011**, *709*, 127–139. [CrossRef]

263. Yang, L.; Jiang, J.; Drouin, L.M.; Agbandje-McKenna, M.; Chen, C.; Qiao, C.; Pu, D.; Hu, X.; Wang, D.Z.; Li, J.; et al. A myocardium tropic adeno-associated virus (AAV) evolved by DNA shuffling and In Vivo selection. *Proc. Natl. Acad. Sci. USA* **2009**, *106*, 3946–3951. [CrossRef] [PubMed]
264. Qiao, C.; Zhang, W.; Yuan, Z.; Shin, J.H.; Li, J.; Jayandharan, G.R.; Zhong, L.; Srivastava, A.; Xiao, X.; Duan, D. Adeno-associated virus serotype 6 capsid tyrosine-to-phenylalanine mutations improve gene transfer to skeletal muscle. *Hum. Gene Ther.* **2010**, *21*, 1343–1348. [CrossRef] [PubMed]
265. Li, Y.; Li, J.; Liu, Y.; Shi, Z.; Liu, H.; Wei, Y.; Yang, L. Bat adeno-associated viruses as gene therapy vectors with the potential to evade human neutralizing antibodies. *Gene Ther.* **2019**, *26*, 264–276. [CrossRef]
266. Maheshri, N.; Koerber, J.T.; Kaspar, B.K.; Schaffer, D.V. Directed evolution of adeno-associated virus yields enhanced gene delivery vectors. *Nat. Biotechnol.* **2006**, *24*, 198–204. [CrossRef]
267. Tse, L.V.; Klinc, K.A.; Madigan, V.J.; Castellanos Rivera, R.M.; Wells, L.F.; Havlik, L.P.; Smith, J.K.; Agbandje-McKenna, M.; Asokan, A. Structure-guided evolution of antigenically distinct adeno-associated virus variants for immune evasion. *Proc. Natl. Acad. Sci. USA* **2017**, *114*, E4812–E4821. [CrossRef]
268. Li, C.; Wu, S.; Albright, B.; Hirsch, M.; Li, W.; Tseng, Y.S.; Agbandje-McKenna, M.; McPhee, S.; Asokan, A.; Samulski, R.J. Development of Patient-specific AAV Vectors After Neutralizing Antibody Selection for Enhanced Muscle Gene Transfer. *Mol. Ther. J. Am. Soc. Gene Ther.* **2016**, *24*, 53–65. [CrossRef]
269. Sarcar, S.; Tulalamba, W.; Rincon, M.Y.; Tipanee, J.; Pham, H.Q.; Evens, H.; Boon, D.; Samara-Kuko, E.; Keyaerts, M.; Loperfido, M.; et al. Next-generation muscle-directed gene therapy by in silico vector design. *Nat. Commun.* **2019**, *10*, 492. [CrossRef] [PubMed]
270. Piekarowicz, K.; Bertrand, A.T.; Azibani, F.; Beuvin, M.; Julien, L.; Machowska, M.; Bonne, G.; Rzepecki, R. A Muscle Hybrid Promoter as a Novel Tool for Gene Therapy. *Mol. Ther. Methods Clin. Dev.* **2019**, *15*, 157–169. [CrossRef] [PubMed]
271. Porro, F.; Bortolussi, G.; Barzel, A.; De Caneva, A.; Iaconcig, A.; Vodret, S.; Zentilin, L.; Kay, M.A.; Muro, A.F. Promoterless gene targeting without nucleases rescues lethality of a Crigler-Najjar syndrome mouse model. *EMBO Mol. Med.* **2017**, *9*, 1346–1355. [CrossRef] [PubMed]
272. Barzel, A.; Paulk, N.K.; Shi, Y.; Huang, Y.; Chu, K.; Zhang, F.; Valdmanis, P.N.; Spector, L.P.; Porteus, M.H.; Gaensler, K.M.; et al. Promoterless gene targeting without nucleases ameliorates haemophilia B in mice. *Nature* **2015**, *517*, 360–364. [CrossRef]
273. Powell, S.K.; Rivera-Soto, R.; Gray, S.J. Viral expression cassette elements to enhance transgene target specificity and expression in gene therapy. *Discov. Med.* **2015**, *19*, 49–57.
274. Bartel, D.P. MicroRNAs: Genomics, biogenesis, mechanism, and function. *Cell* **2004**, *116*, 281–297. [CrossRef]
275. Lee, R.C.; Feinbaum, R.L.; Ambros, V. The *C. elegans* heterochronic gene lin-4 encodes small RNAs with antisense complementarity to lin-14. *Cell* **1993**, *75*, 843–854. [CrossRef]
276. Kozomara, A.; Birgaoanu, M.; Griffiths-Jones, S. miRBase: From microRNA sequences to function. *Nucleic Acids Res.* **2019**, *47*, D155–D162. [CrossRef]
277. Lee, E.J.; Baek, M.; Gusev, Y.; Brackett, D.J.; Nuovo, G.J.; Schmittgen, T.D. Systematic evaluation of microRNA processing patterns in tissues, cell lines, and tumors. *RNA* **2008**, *14*, 35–42. [CrossRef]
278. Vechetti, I.J., Jr.; Wen, Y.; Chaillou, T.; Murach, K.A.; Alimov, A.P.; Figueiredo, V.C.; Dal-Pai-Silva, M.; McCarthy, J.J. Life-long reduction in myomiR expression does not adversely affect skeletal muscle morphology. *Sci. Rep.* **2019**, *9*, 5483. [CrossRef]
279. Brown, B.D.; Gentner, B.; Cantore, A.; Colleoni, S.; Amendola, M.; Zingale, A.; Baccarini, A.; Lazzari, G.; Galli, C.; Naldini, L. Endogenous microRNA can be broadly exploited to regulate transgene expression according to tissue, lineage and differentiation state. *Nat. Biotechnol.* **2007**, *25*, 1457–1467. [CrossRef] [PubMed]
280. Roudaut, C.; Le Roy, F.; Suel, L.; Poupiot, J.; Charton, K.; Bartoli, M.; Richard, I. Restriction of calpain3 expression to the skeletal muscle prevents cardiac toxicity and corrects pathology in a murine model of limb-girdle muscular dystrophy. *Circulation* **2013**, *128*, 1094–1104. [CrossRef] [PubMed]
281. Geisler, A.; Jungmann, A.; Kurreck, J.; Poller, W.; Katus, H.A.; Vetter, R.; Fechner, H.; Muller, O.J. microRNA122-regulated transgene expression increases specificity of cardiac gene transfer upon intravenous delivery of AAV9 vectors. *Gene Ther.* **2011**, *18*, 199–209. [CrossRef] [PubMed]
282. Qiao, C.; Yuan, Z.; Li, J.; He, B.; Zheng, H.; Mayer, C.; Li, J.; Xiao, X. Liver-specific microRNA-122 target sequences incorporated in AAV vectors efficiently inhibits transgene expression in the liver. *Gene Ther.* **2011**, *18*, 403–410. [CrossRef] [PubMed]

283. Brown, B.D.; Venneri, M.A.; Zingale, A.; Sergi Sergi, L.; Naldini, L. Endogenous microRNA regulation suppresses transgene expression in hematopoietic lineages and enables stable gene transfer. *Nat. Med.* **2006**, *12*, 585–591. [CrossRef] [PubMed]

284. Brown, B.D.; Cantore, A.; Annoni, A.; Sergi, L.S.; Lombardo, A.; Della Valle, P.; D'Angelo, A.; Naldini, L. A microRNA-regulated lentiviral vector mediates stable correction of hemophilia B mice. *Blood* **2007**, *110*, 4144–4152. [CrossRef]

285. Majowicz, A.; Maczuga, P.; Kwikkers, K.L.; van der Marel, S.; van Logtenstein, R.; Petry, H.; van Deventer, S.J.; Konstantinova, P.; Ferreira, V. Mir-142-3p target sequences reduce transgene-directed immunogenicity following intramuscular adeno-associated virus 1 vector-mediated gene delivery. *J. Gene Med.* **2013**, *15*, 219–232. [CrossRef]

286. Carpentier, M.; Lorain, S.; Chappert, P.; Lalfer, M.; Hardet, R.; Urbain, D.; Peccate, C.; Adriouch, S.; Garcia, L.; Davoust, J.; et al. Intrinsic Transgene Immunogenicity Gears CD8(+) T-cell Priming After rAAV-Mediated Muscle Gene Transfer. *Mol. Ther. J. Am. Soc. Gene Ther.* **2015**, *23*, 697–706. [CrossRef]

287. Boisgerault, F.; Gross, D.A.; Ferrand, M.; Poupiot, J.; Darocha, S.; Richard, I.; Galy, A. Prolonged gene expression in muscle is achieved without active immune tolerance using microrRNA 142.3p-regulated rAAV gene transfer. *Hum. Gene Ther.* **2013**, *24*, 393–405. [CrossRef]

288. Belbellaa, B.; Reutenauer, L.; Messaddeq, N.; Monassier, L.; Puccio, H. High levels of frataxin overexpression leads to mitochondrial and cardiac toxicity in mouse model. *Mol. Ther. Methods Clin. Dev.* **2020**, *19*, 120–138. [CrossRef]

289. Kraszewska, I.; Tomczyk, M.; Andrysiak, K.; Biniecka, M.; Geisler, A.; Fechner, H.; Zembala, M.; Stepniewski, J.; Dulak, J.; Jazwa-Kusior, A. Variability in Cardiac miRNA-122 Level Determines Therapeutic Potential of miRNA-Regulated AAV Vectors. *Mol. Ther. Methods Clin. Dev.* **2020**, *17*, 1190–1201. [CrossRef] [PubMed]

290. Guo, L.; Zhang, Q.; Ma, X.; Wang, J.; Liang, T. miRNA and mRNA expression analysis reveals potential sex-biased miRNA expression. *Sci. Rep.* **2017**, *7*, 39812. [CrossRef] [PubMed]

291. Gao, W.; He, H.W.; Wang, Z.M.; Zhao, H.; Lian, X.Q.; Wang, Y.S.; Zhu, J.; Yan, J.J.; Zhang, D.G.; Yang, Z.J.; et al. Plasma levels of lipometabolism-related miR-122 and miR-370 are increased in patients with hyperlipidemia and associated with coronary artery disease. *Lipids Health Dis.* **2012**, *11*, 55. [CrossRef] [PubMed]

292. Puzzo, F.; Colella, P.; Biferi, M.G.; Bali, D.; Paulk, N.K.; Vidal, P.; Collaud, F.; Simon-Sola, M.; Charles, S.; Hardet, R.; et al. Rescue of Pompe disease in mice by AAV-mediated liver delivery of secretable acid alpha-glucosidase. *Sci. Transl. Med.* **2017**, *9*. [CrossRef] [PubMed]

293. Paterna, J.C.; Moccetti, T.; Mura, A.; Feldon, J.; Bueler, H. Influence of promoter and WHV post-transcriptional regulatory element on AAV-mediated transgene expression in the rat brain. *Gene Ther.* **2000**, *7*, 1304–1311. [CrossRef]

294. Brown, B.D.; Naldini, L. Exploiting and antagonizing microRNA regulation for therapeutic and experimental applications. *Nat. Rev. Genet.* **2009**, *10*, 578–585. [CrossRef]

295. Dressman, D.; Araishi, K.; Imamura, M.; Sasaoka, T.; Liu, L.A.; Engvall, E.; Hoffman, E.P. Delivery of alpha- and beta-sarcoglycan by recombinant adeno-associated virus: Efficient rescue of muscle, but differential toxicity. *Hum. Gene Ther.* **2002**, *13*, 1631–1646. [CrossRef]

296. Ayuso, E.; Blouin, V.; Lock, M.; McGorray, S.; Leon, X.; Alvira, M.R.; Auricchio, A.; Bucher, S.; Chtarto, A.; Clark, K.R.; et al. Manufacturing and characterization of a recombinant adeno-associated virus type 8 reference standard material. *Hum. Gene Ther.* **2014**, *25*, 977–987. [CrossRef]

297. Prior, H.; Baldrick, P.; Beken, S.; Booler, H.; Bower, N.; Brooker, P.; Brown, P.; Burlinson, B.; Burns-Naas, L.A.; Casey, W.; et al. Opportunities for use of one species for longer-term toxicology testing during drug development: A cross-industry evaluation. *Regul. Toxicol. Pharmacol. RTP* **2020**, *113*, 104624. [CrossRef]

298. EMA. Guideline on Quality, Non-Clinical and Clinical Requirements for Investigational Advanced Therapy Medicinal Products in Clinical Trials. 2019. Available online: https://www.ema.europa.eu/en/documents/scientific-guideline/draft-guideline-quality-non-clinical-clinical-requirements-investigational-advanced-therapy_en.pdf (accessed on 25 November 2020).

299. EMA. ICH Guideline S6 (R1)–Preclinical Safety Evaluation of Biotechnology-Derived Pharmaceuticals. 2011. Available online: https://www.ema.europa.eu/en/documents/scientific-guideline/ich-s6r1-preclinical-safety-evaluation-biotechnology-derived-pharmaceuticals-step-5_en.pdf (accessed on 25 November 2020).

300. Gao, K.; Li, M.; Zhong, L.; Su, Q.; Li, J.; Li, S.; He, R.; Zhang, Y.; Hendricks, G.; Wang, J.; et al. Empty Virions In AAV8 Vector Preparations Reduce Transduction Efficiency And May Cause Total Viral Particle Dose-Limiting Side-Effects. *Mol. Ther. Methods Clin. Dev.* **2014**, *1*, 20139. [CrossRef]
301. Deland, F.H.; North, W.A. Relationship between liver size and body size. *Radiology* **1968**, *91*, 1195–1198. [CrossRef] [PubMed]
302. Davidoff, A.M.; Ng, C.Y.; Zhou, J.; Spence, Y.; Nathwani, A.C. Sex significantly influences transduction of murine liver by recombinant adeno-associated viral vectors through an androgen-dependent pathway. *Blood* **2003**, *102*, 480–488. [CrossRef] [PubMed]
303. Ronzitti, G.; Gross, D.A.; Mingozzi, F. Human Immune Responses to Adeno-Associated Virus (AAV) Vectors. *Front. Immunol.* **2020**, *11*, 670. [CrossRef] [PubMed]
304. Chicoine, L.G.; Montgomery, C.L.; Bremer, W.G.; Shontz, K.M.; Griffin, D.A.; Heller, K.N.; Lewis, S.; Malik, V.; Grose, W.E.; Shilling, C.J.; et al. Plasmapheresis eliminates the negative impact of AAV antibodies on microdystrophin gene expression following vascular delivery. *Mol. Ther. J. Am. Soc. Gene Ther.* **2014**, *22*, 338–347. [CrossRef]
305. Mingozzi, F.; High, K.A. Overcoming the Host Immune Response to Adeno-Associated Virus Gene Delivery Vectors: The Race between Clearance, Tolerance, Neutralization, and Escape. *Annu. Rev. Virol.* **2017**, *4*, 511–534. [CrossRef]
306. Meliani, A.; Boisgerault, F.; Hardet, R.; Marmier, S.; Collaud, F.; Ronzitti, G.; Leborgne, C.; Costa Verdera, H.; Simon Sola, M.; Charles, S.; et al. Antigen-selective modulation of AAV immunogenicity with tolerogenic rapamycin nanoparticles enables successful vector re-administration. *Nat. Commun.* **2018**, *9*, 4098. [CrossRef]
307. Mingozzi, F.; Anguela, X.M.; Pavani, G.; Chen, Y.; Davidson, R.J.; Hui, D.J.; Yazicioglu, M.; Elkouby, L.; Hinderer, C.J.; Faella, A.; et al. Overcoming preexisting humoral immunity to AAV using capsid decoys. *Sci. Transl. Med.* **2013**, *5*, 194ra192. [CrossRef]
308. Leborgne, C.; Barbon, E.; Alexander, J.M.; Hanby, H.; Delignat, S.; Cohen, D.M.; Collaud, F.; Muraleetharan, S.; Lupo, D.; Silverberg, J.; et al. IgG-cleaving endopeptidase enables In Vivo gene therapy in the presence of anti-AAV neutralizing antibodies. *Nat. Med.* **2020**, *26*, 1096–1101. [CrossRef]
309. Jordan, S.C.; Lorant, T.; Choi, J.; Kjellman, C.; Winstedt, L.; Bengtsson, M.; Zhang, X.; Eich, T.; Toyoda, M.; Eriksson, B.M.; et al. IgG Endopeptidase in Highly Sensitized Patients Undergoing Transplantation. *N. Engl. J. Med.* **2017**, *377*, 442–453. [CrossRef]
310. Elmore, Z.C.; Oh, D.K.; Simon, K.E.; Fanous, M.M.; Asokan, A. Rescuing AAV gene transfer from neutralizing antibodies with an IgG-degrading enzyme. *JCI Insight* **2020**, *5*. [CrossRef]
311. Jonuschies, J.; Antoniou, M.; Waddington, S.; Boldrin, L.; Muntoni, F.; Thrasher, A.; Morgan, J. The human desmin promoter drives robust gene expression for skeletal muscle stem cell-mediated gene therapy. *Curr. Gene Ther.* **2014**, *14*, 276–288. [CrossRef] [PubMed]
312. Lisowski, L.; Lau, A.; Wang, Z.; Zhang, Y.; Zhang, F.; Grompe, M.; Kay, M.A. Ribosomal DNA integrating rAAV-rDNA vectors allow for stable transgene expression. *Mol. Ther. J. Am. Soc. Gene Ther.* **2012**, *20*, 1912–1923. [CrossRef] [PubMed]
313. Rosas, L.E.; Grieves, J.L.; Zaraspe, K.; La Perle, K.M.; Fu, H.; McCarty, D.M. Patterns of scAAV vector insertion associated with oncogenic events in a mouse model for genotoxicity. *Mol. Ther. J. Am. Soc. Gene Ther.* **2012**, *20*, 2098–2110. [CrossRef] [PubMed]
314. GlobeNewswire. Sarepta Therapeutics and Genethon Announce a Gene Therapy Research Collaboration for the Treatment of Duchenne Muscular Dystrophy. 2017. Available online: https://www.globenewswire.com/news-release/2017/06/21/1027114/0/en/Sarepta-Therapeutics-and-Genethon-Announce-a-Gene-Therapy-Research-Collaboration-for-the-Treatment-of-Duchenne-Muscular-Dystrophy.html (accessed on 25 November 2020).

Journal of
Personalized
Medicine

MDPI

Article

Gene Editing Targeting the DUX4 Polyadenylation Signal: A Therapy for FSHD?

Romains Joubert [1], Virginie Mariot [1], Marine Charpentier [2], Jean Paul Concordet [2] and Julie Dumonceaux [1,3,*]

1 NIHR Biomedical Research Centre, University College London, Great Ormond Street Institute of Child Health and Great Ormond Street Hospital NHS Trust, London WC1N 1EH, UK; romain.joubert@kcl.ac.uk (R.J.); virginie.mariot@ucl.ac.uk (V.M.)
2 Genome Structure and Instability laboratory, CNRS UMR 7196, Inserm U1154, National Museum of Natural History, 75005 Paris, France; marine.charpentier@univ-rennes1.fr (M.C.); jean-paul.concordet@mnhn.fr (J.P.C.)
3 Northern Ireland Center for Stratified Medicine, Biomedical Sciences Research Institute, Londonderry BT47 6SB, UK
* Correspondence: j.dumonceaux@ucl.ac.uk; Tel.: +44-(0)207-905-2860

Abstract: Facioscapulohumeral dystrophy (FSHD, OMIM: 158900, 158901) is the most common dystrophy in adults and so far, there is no treatment. Different loci of the disease have been characterized and they all lead to the aberrant expression of the DUX4 protein, which impairs the function of the muscle, ultimately leading to cell death. Here, we used gene editing to try to permanently shut down DUX4 expression by targeting its poly(A) sequence. We used transcription activator-like effector nucleases (TALEN) and CRISPR-Cas9 nucleases in vitro on FSHD myoblasts. More than 150 TOPO clones were sequenced and only indels were observed in 4%. Importantly, in 2 of them, the DUX4 poly(A) signal was eliminated at the genomic level but DUX4 mRNA was still produced thanks to the use of a non-canonical upstream poly(A) signal sequence. These experiments show that targeting DUX4 PAS at the genomic level might not be an appropriate gene editing strategy for FSHD therapy.

Keywords: facioscapulohumeral dystrophy; FSHD; TALEN; CRISPR-Cas9; gene editing; muscle; polyadenylation; D4Z4; DUX4

check for updates

Citation: Joubert, R.; Mariot, V.; Charpentier, M.; Concordet, J.P.; Dumonceaux, J. Gene Editing Targeting the DUX4 Polyadenylation Signal: A Therapy for FSHD? *J. Pers. Med.* **2020**, *11*, 7. https://dx.doi.org/10.3390/jpm11010007

Received: 3 November 2020
Accepted: 18 December 2020
Published: 23 December 2020

Publisher's Note: MDPI stays neutral with regard to jurisdictional claims in published maps and institutional affiliations.

1. Introduction

Facioscapulohumeral dystrophy (FSHD, OMIM: 158900, 158901) is one of the most frequent forms of inherited muscular dystrophy with an estimated prevalence of 1:8000 to 1:20,000 [1]. FSHD patients present a selective and asymmetric atrophy of facial, shoulder, arm and leg muscles. The genetic disease is usually transmitted via an autosomal dominant mode of inheritance. The vast majority of patients (95%, FSHD1) show partial deletion of a 3.3 kb tandemly repeated sequence named D4Z4 located on the sub-telomeric region of chromosome 4 [2]. In the general population, the number of repeats varies from 11 to 150, whereas FSHD1 patients carry between 1 and 10 repeats. The 5% remaining patients (FSHD2) do not display this D4Z4 contraction but present mutations in epigenetic modifier genes including SMCHD1, DNMT3B and LRIF1 [3–5]. Both FSHD1 and 2 patients are phenotypically indistinguishable and share common features. The mutations cause a chromatin relaxation leading to a D4Z4 derepression associated with aberrant expression of the DUX4 transcription, whose ORF is within each D4Z4 repeat [3,6,7]. When expressed in mucles, DUX4 is highly toxic, leading to a transcriptional deregulation cascade, with ultimately myopathic effects (for review see [8]) and the spreading of the pathological phenotype [9,10]. DUX4 is expressed at very low levels but is robustly found in adult and fetal FSHD biopsies and muscle cells [11–13]. There is now a scientific consensus that FSHD is caused by this aberrant expression of DUX4 is muscles.

At least 5 different *DUX4* mRNAs have been characterized leading to either a truncated protein (DUX4-s) or a full-length protein (DUX4-FL) [11]. The *DUX4-s* isoform is non-pathogenic, rarely found in muscle cell or biopsies and results from the use of a cryptic splicing site located within the ORF, leading to a truncated and non-functional protein. The two *DUX4-FL* isoforms are pathogenic, only differ by the presence or the absence of intron 1 in the 3′UTR of the transcript, and lead to the same toxic protein (for review [14]). All these 5 isoforms use the same PAS located outside of the D4Z4 region, in the subtelomeric region (called PLAM) of the chromosome 4 that is only present in the 4qA variant of chromosome 4, [15–18].

Because we and others have previously shown that the use of antisense oligonucleotides targeting the 3′ end elements involved in *DUX4* mRNA processing is an efficient therapeutic strategy for FSHD [19–22], in this study, we performed gene editing using transcription activator-like effector nuclease (TALEN) and CRISPR-Cas9 technology to permanently inhibit DUX4 expression by targeting its poly(A) signal sequence (PAS). Even though TALENs initially proved highly efficient nucleases for gene editing, CRISPR-Cas9 has now superseded TALEN technology due to the simplicity of designing guide RNAs for sequence-specific DNA cleavage by Cas9 proteins, the only constraint being the presence of a PAM motif, specific to the different Cas9 proteins available for gene editing, next to the target sequence. Indeed, gene editing has been successful in correcting mutations responsible for genetic diseases in several proof-of-principle studies including for muscular diseases [23].

Here, we designed five TALEN pairs and seven guide RNAs targeting DNA cleavage at the polyadenylation signal of DUX4. FSHD cells were transfected with the nucleases and an oligonucleotide to favour homology directed repairs (HDR). More than 150 clones were generated and sequenced but only two presented an insertion of the oligonucleotides destroying the DUX4 PAS. However, a redirection of the polyadenylation site was observed in the modified clones, similar to what we previously observed with an antisense strategy targeting the DUX4 pre-mRNA with antisense oligonucleotides [19]. Whereas it is possible to permanently destroy the polyadenylation site of DUX4, this may lead to the use of an alternative PAS, and DUX4 may not be efficiently silenced.

2. Results

2.1. Nuclease Design and Selection

To achieve permanent inhibition of DUX4 by targeting the polyadenylation site of DUX4 (Figure 1A), we designed five TALEN pairs (Figure 1B) and six single guide RNAs (sgRNAs), four for *Streptococcus pyogenes* Cas9 endonuclease recognizing the NGG and NAG PAM motifs (Figure 1C) and two for *Neisseria meningitides* Cas9 endonuclease recognizing the NNNNGATT PAM motif (Figure 1D).

The efficiency of both TALENs and guide RNAs was tested by transfection of the corresponding plasmids into U20S cells. We have verified that U20S cells contain the 4qA allele and can therefore be used for testing nucleases that target the DUX4 PAS specific to 4qA. We used the T7 endonuclease 1 assay to evaluate the nuclease efficiencies. After transfection of the cells with the nuclease plasmids, cells were harvested in bulk and a PCR flanking the PAS region was realized to determine the percentage of indels using the T7 endonuclease 1 that that specifically cleaves DNA hetero- but not homoduplexes. For the TALENs, the highest percentage of indels was 45% and was achieved with pair 406/407 (Figure 2A), whereas with CRISPR/Cas9, the highest percentage was achieved with SPCas9/sgRNA24 with 22% (Figure 2B,C). The TALEN pair 406/407 was therefore chosen for further experiments.

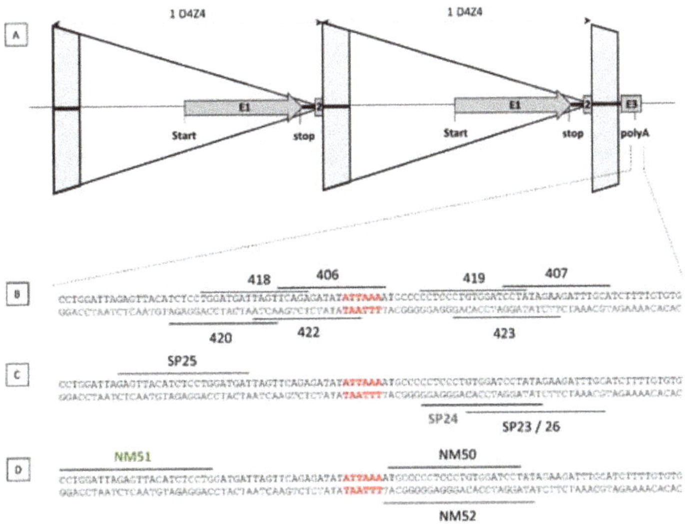

Figure 1. TALEN and gRNA design. TALENs and guide RNAs were designed to target the D4Z4 region (**A**). 5 TALEN pairs were designed, and their positions are indicated (**B**). Six single guide RNAs (sgRNAs), among them, 4 are complexed to *Streptococcus pyogenes* Cas9 endonuclease (**C**) and 2 to *Neisseria meningitides* Cas9 (**D**). The PAS is indicated in red.

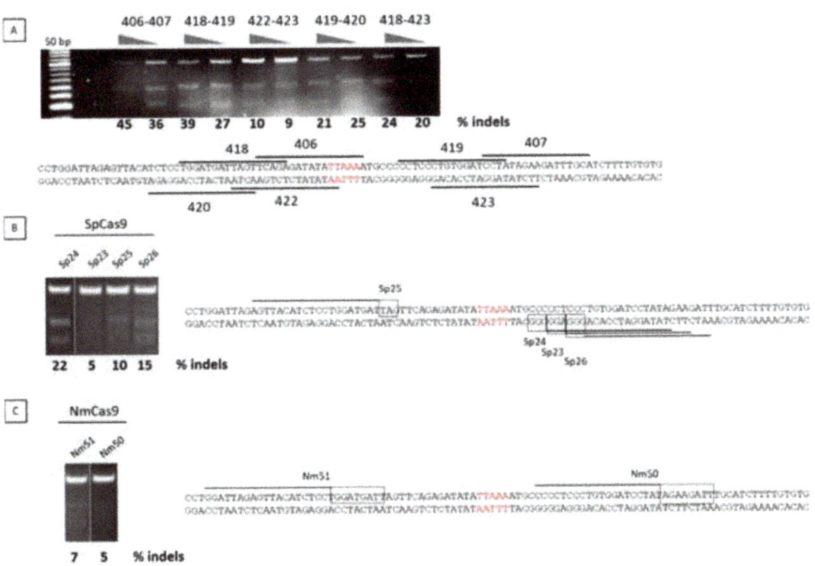

Figure 2. Nuclease selection. U2OS cells were transfected with expression plasmids for TALENs (**A**) or for guide RNAs and cognate *Streptococcus pyogenes* Cas9 (SPCas9 in **B**) or *Neisseria meningitides* Cas9 (NmCas9 in **C**) proteins. Two days after transfection, DNA was extracted and the frequency of indels in the targeted DUX4 PAS was analysed by the T7E1 assay. The target region was amplified by PCR, then PCR products were denatured, followed by re-annealing, leading to a population of double strand fragments, some of which containing mismatches that are detected and cleaved by the T7 endonuclease. The cleavage products are visible on an agarose gel. PCR product is 500 bp long and after T7E1 digestion, expected bands are roughly 200 and 300 bp long depending on the position of mutations induced by the nuclease tested. The proportion of indels was estimated from quantification of the different bands corresponding to non-cleaved and cleaved PCR products.

2.2. Transfection of the TALEN Expression Plasmids into FSHD Cells

The protocol we followed is described in Figure 3A. Briefly, immortalized FSHD cells were transfected with the TALEN pair 406/407, single-stranded DNA donor oligonucleotides (including a miR1) and a CD4-expressing plasmid for the selection of the transfected cells using magnetic beads coupled to anti-CD4 antibodies (MACS column). After enrichment, 3 to 17% of the transfected cells were harvested and genomic DNA (gDNA) of the population was sequenced after cloning of the PCR products into TOPO-TA plasmids. More than 150 TOPO clones were sequenced and small nucleotide insertions or deletions (indels) were observed in 4% of the TOPO clones. We next performed clonal expansion, and isolated and sequenced 277 cellular clones. Importantly, because chromosomes 4 and 10 share 98% homology, a PCR differentiating these 2 chromosomes was absolutely required. We designed two pairs specifically recognizing each chromosome. Both pairs give rise to PCR products similar in length (Figure 3B). We confirmed the specificity of the PCR products after digestion by MseI and XmaI restriction enzymes (Figure 3C). Indeed, MseI restriction site is present only in chromosome 4, whereas the XmaI site is present only on chromosome 10. PCR products were also sequenced to definitely assess the specificity of the primers.

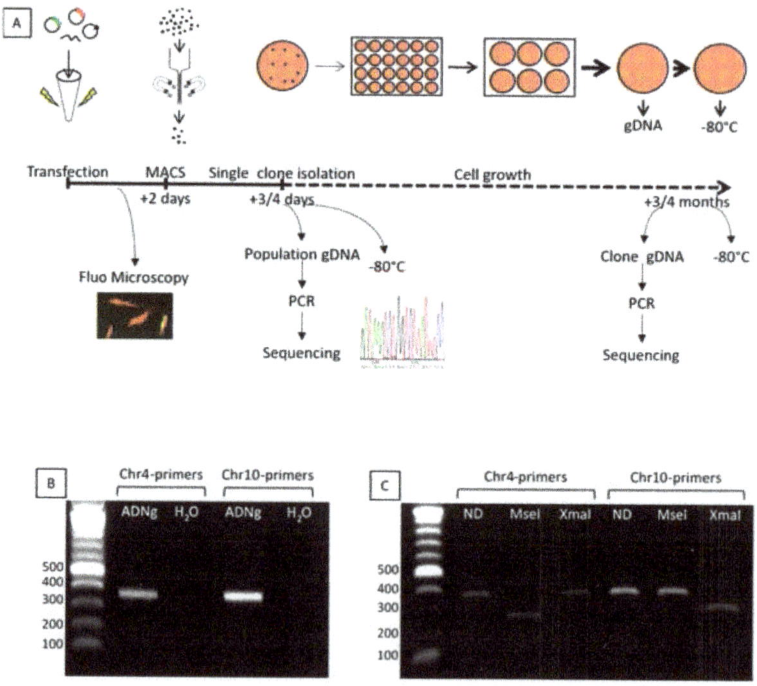

Figure 3. Flow chart and primers validation. The flow chart is described (**A**). Genomic modifications are assessed by sequencing, Primers were designed to specifically recognize chromosomes 4 or 10. The PCR products are similar in length (**B**) but digestion by MseI (present only on chromosome 4) or XmaI (present only on chromosome 10) confirmed the PCR specificity (**C**).

2.3. Selection of the Genetically Modified Clones

Only two clones showed targeted modifications of the DUX4 PAS. The clone#1 presented the insertion of the mir1 sequence into the DUX4 PAS as a result of DNA repair using ssDNA oligonucleotide miR-1 co-transfected with the TALEN pair 406/407 (Figure 4A). We chose to incorporate the target sequence of miR-1 into the PAS site because this miRNA is specifically expressed during differentiation in parallel to DUX4 mRNA [24] and the addition of target sequences for miR-1 may be sufficient to inhibit the translation of residual DUX4 mRNA.

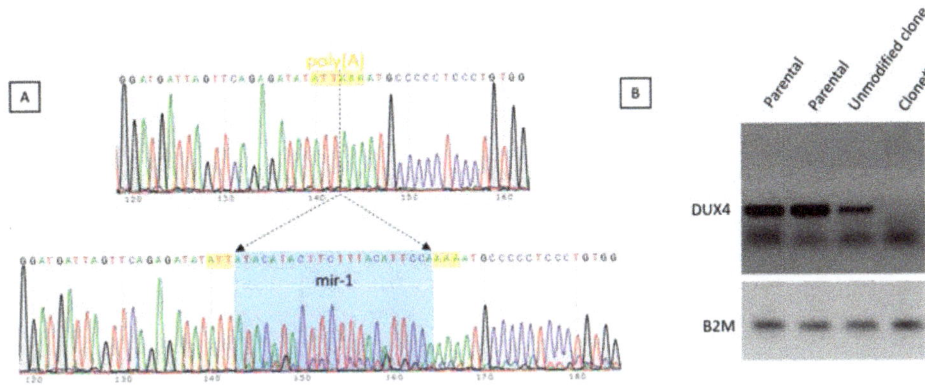

Figure 4. Characteristics of clone#1. This clone was selected after transfection of TALEN pairs 406-407 with a miR-1 ssDNA oligonucleotide. (**A**): The miR-1 sequence (in blue) is inserted within the PAS (in yellow). (**B**): Myoblasts from clone#1 were treated with 0.4 mM chaetocin and harvested 24 h later. DUX4 expression was assessed after RT-PCR Parental FSHD cells and one unmodified clone obtained in parallel to clone#1 were used as controls.

The capability of the clone to express DUX4 was assessed on myoblasts treated with chaetocin, an inhibitor of the SUV39H1 methyltransferase responsible for establishing the chromatin repressive mark H3K9me3 and leading to a decreased H3K9me3 at D4Z4 and to the expression of *DUX4-FL* [25,26]. We observed that after disruption of the PAS by insertion of the miR-1 sequence, no DUX4 mRNA was detected (Figure 4B). It was not possible to further investigate DUX4 expression in this clone because we noted a decrease in its proliferative capacity and it stopped dividing.

The second clone presented a complex genotype with a recombination between the ss-DNA oligonucleotides miR-random and miR-upstream and the genomic DNA (Figure 5A).

In order to determine if DUX4 expression was maintained in this clone, we performed a RT-PCR using primers flanking the two introns of DUX4 which can amplify both the DUX4-FL isoforms (DUX4-FL1: 368 bp and DUX4-FL2: 504 bp). In the parental cells, DUX4-FL2 only was amplified (Figure 5B). In the clone, it was not possible to amplify DUX4 mRNA most of the time, but a band corresponding to DUX4-FL2 was also detected (Figure 5B). A redirection of the polyA and/or cleavage site was thus investigated by 3′RACE nested PCR. A switch was observed in the mutated clone (Figure 5C) and the sequence of this lower band revealed that the cleavage occurred ~40 nt upstream of the canonical one (Figure 5D), as we previously observed with the use of some antisense oligonucleotides targeting the polyadenylation site of DUX4 pre-mRNA [19]. This result indicates that while it is technically possible to eliminate the DUX4 poly(A) signal sequence, it does not necessarily result in DUX4 silencing as another alternative poly(A) site may be used.

We also measured the expression of three well characterized DUX4 downstream genes: TRIM43, MBD3L2 and LEUTX (Figure 5E). In wells A and C, we observed a 95% ($p < 0.0001$) decrease of MBD3L2 and LEUTX, and a 85–90% ($p < 0.0001$) decrease of TRIM43. In well B, expression of these genes was also reduced but to a lower extent, a 78% ± 7 decrease for MBD3L2 ($p < 0.0001$), a 75% ± 9 decrease for LEUTX ($p < 0.0001$) while the apparent decrease was TRIM43 was not statistically significant. When wells A and C are compared to well B, statistical differences were observed for TRIM43 expression ($p < 0.0001$). For MBD3L2 and LEUTX, a 68% decrease or more was observed between well B and wells A/C (Figure 5E). The use of the alternative PAS in well B, leads to a lower activation of the genes downstream of DUX4 when compared to the parental cells but to a higher expression when compared to wells A/C.

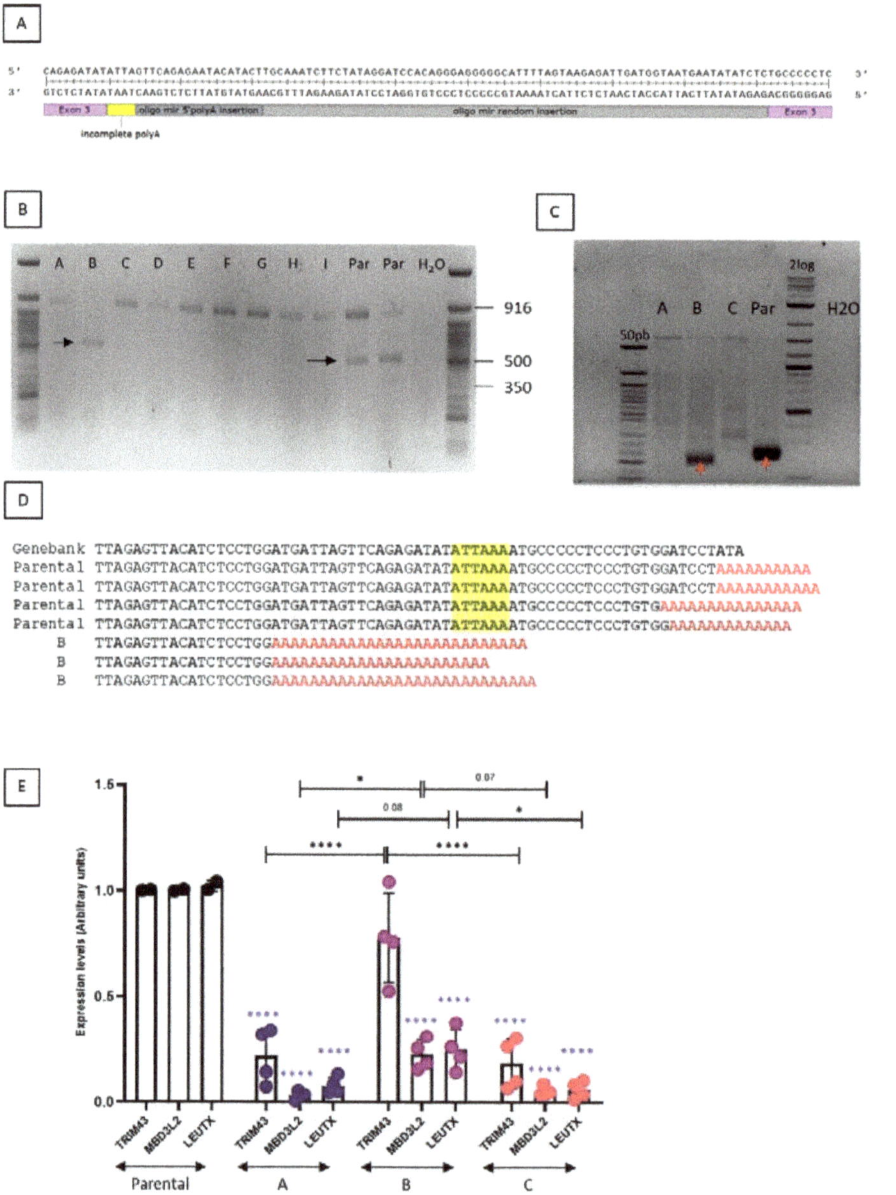

Figure 5. Characteristics of clone#2. (**A**): This clone was selected after transfection of TALEN pair 406-407 with both the miR-5′ and miR-random. The chromosomal modifications are shown. (**B**): DUX4 mRNA expression was assessed by RT-PCR in different wells with cells of the clone carrying the modification and in the parental unmodified FSHD cells (Par). (**C**): 3′RACE nested PCR using forward primers located in Exon 3 shows the redirection of the cleavage site in the modified clone (B) compared to the parental cells (Par), as indicated by the red arrows. (**D**): The sequence shows the redirection of the cleavage site. The PAS is highlighted in yellow. (**E**): Expression levels of 3 genes downstream of DUX4 (TRIM43, MBD3l2 and LEUTX) were measured in parental cells and wells A, B and C. Data are presented as means ± SD; ****: $p < 0.0001$; *: $p < 0.05$ by one-way ANOVA with Tukey's post hoc test. When the parental cells are compared to the clone, statistical analysis are presented in blue. Other analyses are indicated on the graph. Par: parental.

3. Discussion

During the past 10 years, several strategies aiming at inhibiting DUX4 expression have been developed (for review see [27]) and the poly(A) signal was already successfully targeted to inhibit DUX4 expression [19,21]. Targeting the key elements of DUX4 mRNA's 3′UTR is attractive for several reasons: (i) correct polyadenylation of mRNAs is required for their stability, nuclear export and efficient translation, and targeting the polyadenylation signal can result in decreased gene expression (for review see [20]). (ii) DUX4 expression is common in FSHD1 and FSHD2 patients and therefore targeting DUX4 will be beneficial to all FSHD patients. (iii) Contraction of the D4Z4 array on chromosome 10, which carries a single point mutation in the polyadenylation signal sequence found on chromosome 4, does not lead to DUX4 expression [18]. (iv) DUX4 expression has been described in keratinocytes [28] and in EBV-immortalized FSHD lymphoblastoid cell lines [29]. Moreover, DUX4 expression [28,29] is normally suppressed in post-mitotic tissues with the exception of testis where alternative 3′exons are used with a polyadenylation signal in exon 7 that is approximately 6.5 kb further telomeric to the one used in muscles [11]. Moreover, DUX4 might act as a transcription factor to activate cleavage-stage-specific gene expression and may be activated in the primary spermatocytes during spermatogonia and it may therefore be important to target DUX4 in muscles preferentially. Targeting the polyadenylation signal located in exon 3 may lead to the inactivation of the pathogenic DUX4 only, the one that is expressed in the skeletal muscle, whereas the DUX4 mRNA expressed in testis may be spared.

We decided to not incorporate a resistance gene in our constructs in order to be as close as possible to the context of a clinical application. After transfection, we detected two clones with genetic modifications. This might be due the absence of selection and also to the difficulty to transfect myoblasts. In order to increase our chance to inhibit DUX4 expression, we co-transfected a double strand oligonucleotide carrying the miR1 target sequence: in the case of the use of an alternative polyadenylation (APA) signal located downstream of the regular one, DUX4 translation would have been strongly impaired by its presence in the DUX4 mRNA. Indeed, miR1 is a well described myomir, which promotes myoblast-to-myotube differentiation and is up-regulated during this process [30].

It was technically possible to destroy the DUX4 PAS at genomic level, but unfortunately, a redirection of the polyadenylation site, leading to a full length DUX4 mRNA was observed, leading to a lower activation of several genes downstream DUX4. In the literature, only two DUX4 APA sites have been described: one in exon 7 [11], which may be used in testis only and one in exon 3, located ~40 nt upstream of the canonical one, and sometimes observed after treatment of FSHD cells with antisense oligonucelotides targeting the DUX4 PAS [19]. Interestingly, here we report the same polyadenylation site redirection as with the antisense oligonucleotides, suggesting that a steric congestion at the polyadenylation signal and cleavage site provided by the antisense oligonucleotide may not be responsible for the redirection. This rather suggests the presence of an unknown, non-canonical poly(A) signal upstream of the ATTAAA, which may be used in the absence of the canonical poly(A) signal sequence. The use of this proximal APA has been described for other mRNAs in the literature (For review see [20]). It leads to a shorter mRNA but to the same protein since exon 3 is a non-coding exon. Our results suggest that targeting the DUX4 poly(A) signal using TALENs to destroy the PAS is possible but may not lead to an efficient extinction of DUX4 and may not be a suitable strategy for FSHD. A strategy targeting both the canonical and the alternative PAS detected here could give a more robust extinction of DUX4 expression. This would be possible by inducing the corresponding deletion with programmable nucleases, for example using two guide RNAs flanking the PAS and APA sequences with Cas9.

4. Material and Methods

4.1. TALEN and Guide RNA Plasmid Design and Validation

TALEN and guide RNA expression plasmids were constructed using standard molecular biology cloning methods as detailed in [31]. Expression plasmids were transfected into U2OS (established in 1967 from a moderately differentiated sarcoma of the tibia of a 15-year-old girl [32,33]) cells using Amaxa-mediated electroporation (Lonza, Colmar, France) and DNA cleavage activity was assayed indirectly by the T7E1 assay. Primers were hFSHDcr12 fw (5′ gtctgtctttgcccgcttcc 3′) and hFSHDcr12 rev (5′ tgcctacactctgcctacagga 3′). PCR product is 500 bp long and after T7E1 digestion, expected bands are roughly 200 and 300 bp long depending on the position of mutations induced by the nucleases tested.

4.2. Cell Culture

The immortalized myoblasts (54-2) were grown as previously described (Krom et al. 2012). Briefly, cells were grown in 64% DMEM, 16% 199 medium, 20% FBS, supplemented with fetuin 25 μg/mL, β-FGF 0.5 ng/mL, hEGF 5 ng/mL, Insulin 5 μg/mL, Dexametasone 0.2 μg/mL, Gentamycin 50 μg/mL. The transfection was performed using Neon Transfection System (Life technologies, Courtaboeuf, France) on 10^6 cells with 0.21 μg of both TALEN plasmids, 0.2 μg of CD4 expression plasmid, and 0.38 μg of single-stranded donor oligonucleotide (oligonucleotide miR-1 5′GATTAGAGTTACATCTCCTGGATGATTAGTTCAGAGATATATTatacatacttctttacattccaAAAATGCCCCCTCCCTGTGGATCCTATAGAAGATTTGCATCTTTTGTGTG3′; oligonucleotide miR-random 5′GATTAGAGTTACATCTCCTGGATGATTAGTTCAGAGATATATTcattaccatcaatctcttactAAAATGCCCCCTCCCTGTGGATCCTATAGAAGATTTGCATCTTTTGTGTG3′; oligonucleotide miR-1-upstream 5′ACCTGGATTAGAGTTACATCTCCTGGATGATTAGTTCAGAGAatacatacttctttacattccaTATATTAAAATGCCCCCTCCCTGTGGATCCTATAGAAGATTTGCATCTTTT3′). After 48 h of incubation, cells were enriched on MACS columns with a CD4 antibody (Miltenyi Biotec, Paris, France) following manufacturer's recommendation. Monoclonal cell population was performed by limiting dilutions, 3 or 4 days later according to the confluency. Briefly, cells were diluted to seed around 1 cells/well in 96-well plate. After checking under microscope, and scanning the entire plate, wells with only one cell colony were selected and, generally after 1–2 weeks, expanded when confluent in 24-well plate, 6-well plate, and 10 cm dish. For chaetocin treatment, when myoblasts reaching 80% confluency were treated with 0.4 mM chaetocin (Sigma-Aldrich C9492, Saint Quentin Fallavier, France).

4.3. Genomic DNA Analysis

Genomic DNA was extracted from myoblast culture using EZNA tissues DNA kit (VWR) according to the manufacturer's instruction. PCR to amplify DUX4 genomic sequence were performed using GoTaq Green Master Mix (Promega, Madison, WI, USA) and primers specific of chromosome 4 (5′-AGCTGCCAGCGCGGAGCT-3′ and 5′-TGATCACCGAAGTTATGTAAAC-3′), or chromosome 10 (5′-AGGTGCCAGCACGGAGCG-3′ and 5′-TGATCACCGAAGTTATGTAAAT-3′). The size of the resulted amplicon is 362bp and after digestion by MseI and XmaI, the chromosomic origin can be identified because MseI is specific of chromosome 4 and XmaI is specific of chromosome 10. The different amplicons were eventually cloned using Topo-TA cloning kit (Life technologies) and sequenced with M13 primer (5′-GTAAAACGACGGCCAG-3′).

4.4. RT-PCR and RT-qPCR

RT-PCR has been described previously [10,13]. 3′RACE PCR was published in [19]. The primers used for the expression levels of the genes downstream of DUX4 were: B2M (Fwd 5′-CTCTCTTTCTGGCCTGGAGG-3′; Rev 5′-TGCTGGATGACGTGAGTAAACC-3′; amplicon size: 67 bp), TRIM43 (Fwd 5′-ACCCATCACTGGACTGGTGT-3′; Rev 5′-CACATCCTCAAAGAGCCTGA-3′; amplicon size: 100 bp), MBD3L2 (Fwd 5′-CGTTCACCTCTTTTCCAAGC-3′; Rev 5′-AGTCTCATGGGGAGAGCAGA-3′; amplicon size: 142 bp) and LEUTX (Fwd 5′-TGGCTACAATGGGGAAACTG-3′; Rev 5′-CTGCTGCCTCTTCCATTT G-3′; amplicon size: 98 bp).

4.5. Statistical Analysis

GraphPad Prism software was used for statistical analyses. Differences between groups were evaluated by one-analysis of variance (ANOVA) followed by Tukey's post hoc tests. ****: $p < 0.0001$; ***: $p < 0.001$; **: $p < 0.01$; *: $p < 0.05$.

Author Contributions: Conceptualization: J.D. and J.P.C.; methodology: J.D. and J.P.C.; performed experiments V.M., R.J., M.C.; writing—original draft preparation: J.D.; funding acquisition: J.D. and J.P.C. All authors have read and agreed to the published version of the manuscript.

Funding: This research was funded by the FSH Society Grant FSHS-82013-02 and ANR grant ANR-II-INSB-0014.

Acknowledgments: TALEN repeat plasmids were kindly provided by the Zhang lab (University of Beijing, China) and we thank the Church and Joung labs for guide RNA and Sp Cas9 plasmids obtained from Addgene. All research at Great Ormond Street Hospital NHS Foundation Trust and UCL Great Ormond Street Institute of Child Health is made possible by the NIHR Great Ormond Street Hospital Biomedical Research Centre. The views expressed are those of the author(s) and not necessarily those of the NHS, the NIHR or the Department of Health.

Conflicts of Interest: The authors declare no conflict of interest.

References

1. Deenen, J.C.; Arnts, H.; van der Maarel, S.M.; Padberg, G.W.; Verschuuren, J.J.; Bakker, E.; Weinreich, S.S.; Verbeek, A.L.; van Engelen, B.G. Population-based incidence and prevalence of facioscapulohumeral dystrophy. *Neurology* **2014**, *83*, 1056–1059. [CrossRef] [PubMed]
2. Van Deutekom, J.C.; Wijmenga, C.; van Tienhoven, E.A.; Gruter, A.M.; Hewitt, J.E.; Padberg, G.W.; van Ommen, G.J.; Hofker, M.H.; Frants, R.R. FSHD associated DNA rearrangements are due to deletions of integral copies of a 3.2 kb tandemly repeated unit. *Hum. Mol. Genet.* **1993**, *2*, 2037–2042. [CrossRef] [PubMed]
3. Lemmers, R.J.; Tawil, R.; Petek, L.M.; Balog, J.; Block, G.J.; Santen, G.W.; Amell, A.M.; van der Vliet, P.J.; Almomani, R.; Straasheijm, K.R.; et al. Digenic inheritance of an SMCHD1 mutation and an FSHD-permissive D4Z4 allele causes facioscapulohumeral muscular dystrophy type 2. *Nat. Genet.* **2012**, *44*, 1370–1374. [CrossRef] [PubMed]
4. Van den Boogaard, M.L.; Lemmers, R.; Balog, J.; Wohlgemuth, M.; Auranen, M.; Mitsuhashi, S.; van der Vliet, P.J.; Straasheijm, K.R.; van den Akker, R.F.P.; Kriek, M.; et al. Mutations in DNMT3B Modify Epigenetic Repression of the D4Z4 Repeat and the Penetrance of Facioscapulohumeral Dystrophy. *Am. J. Hum. Genet.* **2016**, *98*, 1020–1029. [CrossRef]
5. Hamanaka, K.; Sikrova, D.; Mitsuhashi, S.; Masuda, H.; Sekiguchi, Y.; Sugiyama, A.; Shibuya, K.; Lemmers, R.; Goossens, R.; Ogawa, M.; et al. Homozygous nonsense variant in LRIF1 associated with facioscapulohumeral muscular dystrophy. *Neurology* **2020**. [CrossRef]
6. Gabriels, J.; Beckers, M.C.; Ding, H.; De Vriese, A.; Plaisance, S.; van der Maarel, S.M.; Padberg, G.W.; Frants, R.R.; Hewitt, J.E.; Collen, D.; et al. Nucleotide sequence of the partially deleted D4Z4 locus in a patient with FSHD identifies a putative gene within each 3.3 kb element. *Gene* **1999**, *236*, 25–32. [CrossRef]
7. Gaillard, M.C.; Roche, S.; Dion, C.; Tasmadjian, A.; Bouget, G.; Salort-Campana, E.; Vovan, C.; Chaix, C.; Broucqsault, N.; Morere, J.; et al. Differential DNA methylation of the D4Z4 repeat in patients with FSHD and asymptomatic carriers. *Neurology* **2014**, *19*, 733–742. [CrossRef]
8. Lim, K.R.Q.; Nguyen, Q.; Yokota, T. DUX4 Signalling in the Pathogenesis of Facioscapulohumeral Muscular Dystrophy. *Int. J. Mol. Sci.* **2020**, *21*, 729. [CrossRef]
9. Tassin, A.; Laoudj-Chenivesse, D.; Vanderplanck, C.; Barro, M.; Charron, S.; Ansseau, E.; Chen, Y.W.; Mercier, J.; Coppee, F.; Belayew, A. DUX4 expression in FSHD muscle cells: How could such a rare protein cause a myopathy? *J. Cell Mol. Med.* **2012**, *17*, 76–89. [CrossRef]
10. Ferreboeuf, M.; Mariot, V.; Furling, D.; Butler-Browne, G.; Mouly, V.; Dumonceaux, J. Nuclear protein spreading: Implication for pathophysiology of neuromuscular diseases. *Hum. Mol. Genet.* **2014**, *23*, 4125–4133. [CrossRef]
11. Snider, L.; Geng, L.N.; Lemmers, R.J.; Kyba, M.; Ware, C.B.; Nelson, A.M.; Tawil, R.; Filippova, G.N.; van der Maarel, S.M.; Tapscott, S.J.; et al. Facioscapulohumeral dystrophy: Incomplete suppression of a retrotransposed gene. *PLoS Genet.* **2010**, *6*, e1001181. [CrossRef] [PubMed]
12. Broucqsault, N.; Morere, J.; Gaillard, M.C.; Dumonceaux, J.; Torrents, J.; Salort-Campana, E.; Maues de Paula, A.; Bartoli, M.; Fernandez, C.; Chesnais, A.L.; et al. Dysregulation of 4q35- and muscle-specific genes in fetuses with a short D4Z4 array linked to Facio-Scapulo-Humeral Dystrophy. *Hum. Mol. Genet.* **2013**, *22*, 4206–4214. [CrossRef] [PubMed]

13. Ferreboeuf, M.; Mariot, V.; Bessieres, B.; Vasiljevic, A.; Attie-Bitach, T.; Collardeau, S.; Morere, J.; Roche, S.; Magdinier, F.; Robin-Ducellier, J.; et al. DUX4 and DUX4 downstream target genes are expressed in fetal FSHD muscles. *Hum. Mol. Genet.* **2014**, *23*, 171–181. [CrossRef] [PubMed]
14. Sidlauskaite, E.; Le Gall, L.; Mariot, V.; Dumonceaux, J. DUX4 Expression in FSHD Muscles: Focus on Its mRNA Regulation. *J. Pers. Med.* **2020**, *10*, 73. [CrossRef] [PubMed]
15. Dixit, M.; Ansseau, E.n.; Tassin, A.; Winokur, S.; Shi, R.; Qian, H.; Sauvage, S.; Mattéotti, C.; van Acker, A.M.; Leo, O.; et al. DUX4, a candidate gene of facioscapulohumeral muscular dystrophy, encodes a transcriptional activator of PITX1. *Proc. Natl. Acad. Sci. USA* **2007**, *104*, 18157–18162. [CrossRef]
16. Lemmers, R.J.; de Kievit, P.; Sandkuijl, L.; Padberg, G.W.; van Ommen, G.J.; Frants, R.R.; van der Maarel, S.M. Facioscapulohumeral muscular dystrophy is uniquely associated with one of the two variants of the 4q subtelomere. *Nat. Genet.* **2002**, *32*, 235–236. [CrossRef]
17. Thomas, N.S.T.; Wiseman, K.; Spurlock, G.; MacDonald, M.; Ustek, D.; Upadhyaya, M. A large patient study confirming that facioscapulohumeral muscular dystrophy (FSHD) disease expression is almost exclusively associated with an FSHD locus located on a 4qA-defined 4qter subtelomere. *J. Med. Genet.* **2007**, *44*, 215–218. [CrossRef]
18. Lemmers, R.J.; van der Vliet, P.J.; Klooster, R.; Sacconi, S.; Camano, P.; Dauwerse, J.G.; Snider, L.; Straasheijm, K.R.; van Ommen, G.J.; Padberg, G.W.; et al. A unifying genetic model for facioscapulohumeral muscular dystrophy. *Science* **2010**, *329*, 1650–1653. [CrossRef]
19. Marsollier, A.C.; Ciszewski, L.; Mariot, V.; Popplewell, L.; Voit, T.; Dickson, G.; Dumonceaux, J. Antisense targeting of 3′ end elements involved in DUX4 mRNA processing is an efficient therapeutic strategy for facioscapulohumeral dystrophy: A new gene-silencing approach. *Hum. Mol. Genet.* **2016**, *25*, 1468–1478. [CrossRef]
20. Marsollier, A.C.; Joubert, R.; Mariot, V.; Dumonceaux, J. Targeting the Polyadenylation Signal of Pre-mRNA: A New Gene Silencing Approach for Facioscapulohumeral Dystrophy. *Int. J. Mol. Sci.* **2018**, *19*, 1347. [CrossRef]
21. Chen, J.C.; King, O.D.; Zhang, Y.; Clayton, N.P.; Spencer, C.; Wentworth, B.M.; Emerson, C.P., Jr.; Wagner, K.R. Morpholino-mediated Knockdown of DUX4 Toward Facioscapulohumeral Muscular Dystrophy Therapeutics. *Mol. Ther.* **2016**, *24*, 1405–1411. [CrossRef] [PubMed]
22. Lim, K.R.Q.; Maruyama, R.; Echigoya, Y.; Nguyen, Q.; Zhang, A.; Khawaja, H.; Sen Chandra, S.; Jones, T.; Jones, P.; Chen, Y.W.; et al. Inhibition of DUX4 expression with antisense LNA gapmers as a therapy for facioscapulohumeral muscular dystrophy. *Proc. Natl. Acad. Sci. USA* **2020**. [CrossRef]
23. Chemello, F.; Bassel-Duby, R.; Olson, E.N. Correction of muscular dystrophies by CRISPR gene editing. *J. Clin. Investig.* **2020**, *130*, 2766–2776. [CrossRef] [PubMed]
24. Guller, I.; Russell, A.P. MicroRNAs in skeletal muscle: Their role and regulation in development, disease and function. *J. Physiol.* **2010**, *588*, 4075–4087. [CrossRef]
25. Zeng, W.; Chen, Y.Y.; Newkirk, D.A.; Wu, B.; Balog, J.; Kong, X.; Ball, A.R., Jr.; Zanotti, S.; Tawil, R.; Hashimoto, N.; et al. Genetic and Epigenetic Characteristics of FSHD-Associated 4q and 10q D4Z4 that are Distinct from Non-4q/10q D4Z4 Homologs. *Hum. Mutat.* **2014**, *35*, 998–1010. [CrossRef]
26. Jones, T.I.; King, O.D.; Himeda, C.L.; Homma, S.; Chen, J.C.; Beermann, M.L.; Yan, C.; Emerson, C.P., Jr.; Miller, J.B.; Wagner, K.R.; et al. Individual epigenetic status of the pathogenic D4Z4 macrosatellite correlates with disease in facioscapulohumeral muscular dystrophy. *Clin. Epigenetics* **2015**, *7*, 37. [CrossRef]
27. Le Gall, L.; Sidlauskaite, E.; Mariot, V.; Dumonceaux, J. Therapeutic Strategies Targeting DUX4 in FSHD. *J. Clin. Med.* **2020**, *9*, 2886. [CrossRef]
28. Gannon, O.M.; Merida de Long, L.; Saunders, N.A. DUX4 Is Derepressed in Late-Differentiating Keratinocytes in Conjunction with Loss of H3K9me3 Epigenetic Repression. *J. Investig. Derm.* **2016**, *136*, 1299–1302. [CrossRef]
29. Banerji, C.R.S.; Panamarova, M.; Zammit, P.S. DUX4 expressing immortalized FSHD lymphoblastoid cells express genes elevated in FSHD muscle biopsies, correlating with the early stages of inflammation. *Hum. Mol. Genet.* **2020**, *29*, 2285–2299. [CrossRef]
30. Townley-Tilson, W.H.; Callis, T.E.; Wang, D. MicroRNAs 1, 133, and 206: Critical factors of skeletal and cardiac muscle development, function, and disease. *Int. J. Biochem. Cell Biol.* **2010**, *42*, 1252–1255. [CrossRef]
31. Renaud, J.B.; Boix, C.; Charpentier, M.; De Cian, A.; Cochennec, J.; Duvernois-Berthet, E.; Perrouault, L.; Tesson, L.; Edouard, J.; Thinard, R.; et al. Improved Genome Editing Efficiency and Flexibility Using Modified Oligonucleotides with TALEN and CRISPR-Cas9 Nucleases. *Cell Rep.* **2016**, *14*, 2263–2272. [CrossRef] [PubMed]
32. Ponten, J.; Saksela, E. Two established in vitro cell lines from human mesenchymal tumours. *Int. J. Cancer* **1967**, *2*, 434–447. [CrossRef] [PubMed]
33. Raile, K.; Hoflich, A.; Kessler, U.; Yang, Y.; Pfuender, M.; Blum, W.F.; Kolb, H.; Schwarz, H.P.; Kiess, W. Human osteosarcoma (U-2 OS) cells express both insulin-like growth factor-I (IGF-I) receptors and insulin-like growth factor-II/mannose-6-phosphate (IGF-II/M6P) receptors and synthesize IGF-II: Autocrine growth stimulation by IGF-II via the IGF-I receptor. *J. Cell Physiol.* **1994**, *159*, 531–541. [CrossRef] [PubMed]

Journal of *Personalized Medicine*

MDPI

Article

Genotype–Phenotype Correlations in Duchenne and Becker Muscular Dystrophy Patients from the Canadian Neuromuscular Disease Registry

Kenji Rowel Q. Lim [1,†], Quynh Nguyen [1,†] and Toshifumi Yokota [1,2,*]

1 Department of Medical Genetics, Faculty of Medicine and Dentistry, University of Alberta, Edmonton, AB T6G2H7, Canada; kenjirow@ualberta.ca (K.R.Q.L.); nguyenth@ualberta.ca (Q.N.)
2 The Friends of Garrett Cumming Research & Muscular Dystrophy Canada, HM Toupin Neurological Science Research Chair, Edmonton, AB T6G2H7, Canada
* Correspondence: toshifumi.yokota@ualberta.ca; Tel.: +1-780-492-1102
† These authors contributed equally to this work.

Received: 29 October 2020; Accepted: 21 November 2020; Published: 23 November 2020

Abstract: Duchenne muscular dystrophy (DMD) is a fatal neuromuscular disorder generally caused by out-of-frame mutations in the *DMD* gene. In contrast, in-frame mutations usually give rise to the milder Becker muscular dystrophy (BMD). However, this reading frame rule does not always hold true. Therefore, an understanding of the relationships between genotype and phenotype is important for informing diagnosis and disease management, as well as the development of genetic therapies. Here, we evaluated genotype–phenotype correlations in DMD and BMD patients enrolled in the Canadian Neuromuscular Disease Registry from 2012 to 2019. Data from 342 DMD and 60 BMD patients with genetic test results were analyzed. The majority of patients had deletions (71%), followed by small mutations (17%) and duplications (10%); 2% had negative results. Two deletion hotspots were identified, exons 3–20 and exons 45–55, harboring 86% of deletions. Exceptions to the reading frame rule were found in 13% of patients with deletions. Surprisingly, C-terminal domain mutations were associated with decreased wheelchair use and increased forced vital capacity. Dp116 and Dp71 mutations were also linked with decreased wheelchair use, while Dp140 mutations significantly predicted cardiomyopathy. Finally, we found that 12.3% and 7% of DMD patients in the registry could be treated with FDA-approved exon 51- and 53-skipping therapies, respectively.

Keywords: Duchenne muscular dystrophy; Becker muscular dystrophy; dystrophinopathy; genotype-phenotype correlations; Canadian Neuromuscular Disease Registry; reading frame rule; dystrophin; multiple logistic regression analysis; exon skipping therapy

1. Introduction

Duchenne muscular dystrophy (DMD) is the most common inherited neuromuscular disorder worldwide, affecting approximately 20 per 100,000 male births (1:5000) [1,2]. DMD is an X-linked recessive disorder that is characterized by progressive body-wide muscle degeneration, with proximal muscle weakness starting at 3–5 years and loss of ambulation during the early teens [3,4]. Cardiac and respiratory symptoms often appear during the third decade of life, which eventually lead to death. DMD is primarily caused by mutations in the *DMD* gene that lead to an absence of dystrophin. Dystrophin is a protein responsible for stabilizing muscle cell membranes during contraction–relaxation cycles; its loss increases the susceptibility of muscles to tear during use [5–7]. There is a milder form of the disease called Becker muscular dystrophy (BMD), which is caused by mutations in the same gene. However, mutations in BMD patients generally only reduce the amount or functionality of the dystrophin produced, as opposed to the complete absence of dystrophin seen in DMD [8–10].

DMD and BMD are part of a group of disorders called the dystrophinopathies, which are all characterized by mutations in the *DMD* gene. Stark differences between the fatal DMD and mild BMD prompt us to understand how differences in genotype (i.e., mutation) impact phenotype (i.e., clinical outcome). This is especially important since there is no cure for DMD at present. To study these genotype–phenotype correlations, among other purposes, dystrophinopathy patient registries were formed by local, national, and international initiatives to collect information on patient clinical outcomes and *DMD* mutations. Perhaps the most extensive of these would be the TREAT-NMD DMD Global Registry [11] and the Leiden Open Variation Database (LOVD) [12,13], each having data from more than 7000 dystrophinopathy patients across the world. Canada in particular has the Canadian Neuromuscular Disease Registry (CNDR), a national patient registry established in 2011 that also contributes to the TREAT-NMD database [14,15]. As of 1 December 2019, with 4310 registrants, dystrophinopathy patients make up the second-largest disease group in the CNDR at 13.3% [15]. Amyotrophic lateral sclerosis has the most number of registered patients at 36.1%; myotonic dystrophy, limb–girdle muscular dystrophy, and spinal muscular atrophy patients make up 10.5%, 5.9%, and 5.3% of CNDR registrants, respectively.

Here, we aimed to evaluate genotype–phenotype correlations specifically in the Canadian DMD/BMD population, using the information on 402 patients from the CNDR. Similar studies have been conducted previously [11,16–22]; however, most of these investigated a limited number of clinical phenotypes. There may also be correlations unique to the Canadian population that would otherwise not be observed from a global database. We particularly examined the relationships between patient genotype and clinical diagnosis (DMD/BMD), as well as between patient genotype and clinical outcomes (e.g., wheelchair use and cardiomyopathy status). We also determined the applicability of recent U.S. Food and Drug Administration (FDA)-approved exon skipping DMD therapeutics to the CNDR DMD patient population, given the increasing entry of this class of therapies into the clinic. Finally, this work provides the most recent characterization of the *DMD* mutation landscape in Canada.

2. Materials and Methods

2.1. Study Population and Design

This study was approved by the University of Alberta Health Research Ethics Board—Health Panel (reference Pro00092569). Participants in the CNDR provided informed consent and agreed to have their data shared for research purposes. For this study, the following information was used from CNDR patient records, which were provided directly from the clinic by neuromuscular specialists in the CNDR network: weight, height, clinical diagnosis, genetic data (test information, mutation type, mutation location), neuromuscular data (motor function, therapies received), cardiac history (presence of cardiomyopathy, left ventricle ejection fraction (LVEF), cardiac medications received), respiratory data (use of non-invasive/invasive ventilation, forced vital capacity (FVC)), and gastrointestinal data (feeding tube use, major nutritional route). Clinical diagnosis (DMD/BMD) was at the discretion of the neuromuscular specialist attending to the patient on the basis of clinical and genetic characteristics. All genetic data were derived from accredited testing laboratories across Canada as part of standard clinical practice. If a patient had information in the registry from more than one visit, data from the most recent visit was considered for analysis. All patient data were de-identified before provision to the study team.

The initial study population consisted of 508 dystrophinopathy patients in the CNDR from 1 January 2012 to 3 July 2019. This included 414 DMD patients, 78 BMD patients, 13 female *DMD* mutation carriers, 2 intermediate muscular dystrophy (IMD) patients, and 1 with an unknown diagnosis (Figure 1). We filtered out patients who did not have genetic testing data or a definite DMD/BMD diagnosis, leaving us with 420 patients (350 DMD patients, 61 BMD patients, 9 female carriers). Data from these patients were used for comparisons of clinical outcomes across groups. For correlational analysis between genotype and clinical diagnosis as phenotype, we focused only on the 342 DMD and 60 BMD patients with non-negative genetic test results. On the other hand, for the analysis between genotype and clinical outcomes (wheelchair use, presence of cardiomyopathy, LVEF, FVC), we restricted our analysis to include only the 342 DMD patients.

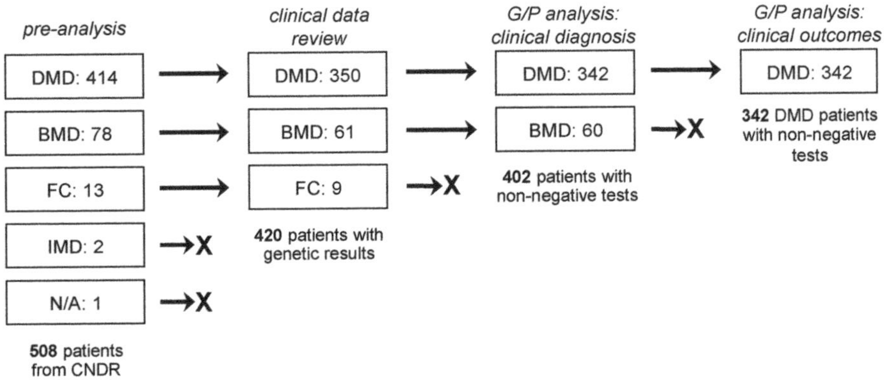

Figure 1. Study population and design. Patient data from the Canadian Neuromuscular Disease Registry between January 2012 and July 2019 were used for this study. The number and groups of patients evaluated for the various analyses performed are shown. DMD, Duchenne muscular dystrophy; BMD, Becker muscular dystrophy; FC, female carrier; IMD, intermediate muscular dystrophy; G/P, genotype–phenotype.

2.2. Statistical Analysis

All statistical analyses and plotting were performed using GraphPad Prism version 8.4.3 (GraphPad Software, San Diego, CA, USA). A two-sided Fisher's exact test was done to determine statistically significant differences between groups of categorical variables, while a two-tailed, unpaired Student's t-test was done for continuous variables. A multiple logistic or linear (least squares) regression analysis was used to construct inferential models studying the relationships between genotypes and clinical outcomes, with the latter serving as dependent variables. Patients with missing information were excluded from the multiple regression analyses by the software. A p-value of less than 0.05 was considered statistically significant.

3. Results

3.1. Clinical Characteristics

Table 1 summarizes the clinical characteristics of the three subgroups in our study population: DMD, BMD, and female carriers. The female carriers all appear to be healthy, at least based on the parameters reviewed. However, the low number of carriers in our cohort ($N = 9$) makes it difficult to accurately compare with other subgroups. Thus, we decided to perform a comparative analysis of clinical characteristics only between DMD and BMD patients.

Table 1. Summary of clinical characteristics for patients with genetic data in our study population.

Characteristic	DMD [1]	BMD [1]	FC [1]	p-Value [2]
Number (N)	350 (83)	61 (15)	9 (2)	-
Age at visit (yr)	10.5 (6.8–14.6)	17.9 (14.1–24.9)	13.0 (11.9–15.0)	<0.0001
Body mass index	18.1 (16.2–22.8)	21.3 (17.6–26.6)	17.0 (15.4–25.0)	0.0093
Neuromuscular parameters				
Wheelchair use				0.0023
>Permanent	39 (11)	2 (3)	0 (0)	
>Intermittent	98 (28)	10 (16)	0 (0)	
>Never	156 (45)	43 (70)	8 (89)	
>Unknown	57 (16)	6 (10)	1 (11)	
Can walk without support	189 (62)	43 (78)	9 (100)	0.0175
Can sit without support	244 (81)	52 (95)	9 (100)	0.013
Uses steroids				0.0218 [3]
>Deflazacort	231 (91)	6 (67)	1 (100)	
>Prednisone	19 (7)	3 (33)	0 (0)	
>Vamorolone	1 (0)	0 (0)	0 (0)	
>Testosterone	3 (1)	0 (0)	0 (0)	
Cardiac parameters				
Cardiomyopathy	37 (11)	10 (17)	0 (0)	0.1928
Age of CM onset (yr)	13.0 (11.0–14.3)	23.0 (16.0–33.0)	-	0.0059
Left ventricle ejection fraction (%)	63.0 (58.0–68.0)	60.0 (50.0–65.0)	68.5 (62.0–70.5)	0.0325
Uses cardiac medication				0.0197 [4]
>ACEi/ARB	69 (70)	9 (43)	-	
>β-blocker	18 (18)	5 (24)	-	
>Digoxin	5 (5)	0 (0)	-	
>Statin	0 (0)	2 (10)	-	
>Antiplatelet	1 (1)	1 (5)	-	
>Anticoagulant	0 (0)	3 (14)	-	
>MRA	3 (3)	0 (0)	-	
Respiratory parameters				
Uses ventilation assistance				>0.9999
>Non-invasive	30 (9)	2 (3)	0 (0)	
>Invasive	1 (0)	0 (0)	0 (0)	
Forced vital capacity (%)	76.0 (55.0–93.0)	88.0 (80.0–100.0)	74.0 (59.0–85.3)	0.0018
Sleep apnea	10 (38)	0 (0)	0 (0)	0.3703
Gastrointestinal parameters				
Uses feeding tube	2 (1)	0 (0)	0 (0)	>0.9999
Major nutritional route				>0.9999
>Oral	132 (99)	16 (100)	2 (100)	
>Enteral	1 (1)	0 (0)	0 (0)	

[1] count data: frequency (%), continuous data: median (interquartile range), [2] DMD versus BMD, [3] vamorolone and testosterone counted as one group, [4] digoxin up to MRA counted as one group. DMD, Duchenne muscular dystrophy; BMD, Becker muscular dystrophy; FC, female carrier.

The DMD patients in our population were significantly younger by 7 years ($p < 0.0001$; mean ages of 10.5 versus 17.9 years old, respectively) and had lower body mass indices (BMIs) by 3 points ($p < 0.005$; mean BMIs of 18.1 versus 21.3, respectively) than the BMD patients. As expected, DMD patients used the wheelchair significantly more than BMD patients ($p < 0.005$), required more support for walking ($p < 0.05$) or sitting ($p < 0.05$), and were mostly on deflazacort therapy ($p < 0.05$). In terms of cardiac outcomes, no significant differences in cardiomyopathy status between DMD and BMD patients were observed in our population. However, the age of cardiomyopathy onset was significantly earlier for DMD at an average of 13.0 years than BMD at an average of 23.0 years ($p < 0.05$). Despite LVEF values being significantly lower in BMD than DMD patients ($p < 0.05$), both subgroups were well within the healthy LVEF range at >50%. These LVEF results likely reflect how patients from both groups also received standard cardiac medications in the form of angiotensin-converting enzyme inhibitors, angiotensin II-receptor blockers, and β-blockers, among others. FVC values were significantly reduced in DMD than in BMD patients ($p < 0.005$; 76.0% versus 88.0% on average, respectively). Perhaps due to scarcity in the available data, no significant differences in other respiratory or gastrointestinal parameters were found between the two patient subgroups.

3.2. Genetic Characteristics

Genetic testing data was available for 350 of 414 DMD patients (85%) and 61 of 78 BMD patients (78%) (Figure 1). The majority of mutations were deletions of at least one exon in the *DMD* gene in 69% (241/350) of DMD patients and 80% (49/61) of BMD patients, or 71% (290/411) of patients in total (Figure 2a). This was followed by small mutations, i.e., point mutations and insertions/deletions within exons or splice sites, in 17% (71/411) of patients, and duplications of at least one exon in 10% (41/411) of patients. Negative results were found for 2% of patients, i.e., these patients were clinically diagnosed as having DMD/BMD, but genetic testing failed to identify a variant. However, as these patients were also not tested via gene sequence analysis, it remains possible that they could have deep intronic mutations in the *DMD* gene that were missed.

Mapping out all large deletions (>1 exon) revealed two mutation hotspots, one from exons 3 to 20 and another from exons 45 to 55 (Figure 2b). More than half of all patients with deletions at ~65% had mutations in the distal hotspot, whereas only ~21% were in the proximal hotspot. Moreover, most deletions in the proximal hotspot were represented by only one patient. The most common deletion was a deletion of exon 45, which was in 18 out of 290 patients (6%) with large deletion mutations (Figure 2c). Out of the 18 most common large deletion mutations, 17 were in the distal exons 45–55 mutation hotspot. Conversely, mapping out all large duplications (>1 exon) in our DMD and BMD patients revealed one hotspot from exons 3–10 (Figure 2d). However, note that most exon duplication patterns were represented by only one patient. The most common duplications were an exon 2 duplication and an extensive exons 5–65 duplication, which were each found in 3 out of 41 patients (7%) with large *DMD* duplication mutations (Figure 2e).

Small mutations were spread out across the entire gene, ultimately affecting all four major dystrophin protein regions: the N-terminal actin-binding domain (exons 2–8), the central rod domain (exons 8–61), the cysteine-rich domain (exons 63–69), and the C-terminal domain (exons 70–79) (Figure 3a,b). Exons were assigned to protein domains following information from the Leiden Muscular Dystrophy dystrophin page (https://www.dmd.nl/). Exon 18 harbored the greatest number of small mutations in our combined DMD and BMD population (Figure 3b). More than half (51%) of all identified small mutations were nonsense point mutations, followed by 27% being small insertions/deletions, 13% being splice site mutations, and 4% being missense mutations (Figure 3c). Interestingly, two DMD patients each carried two different small mutations—one with c.8729A>T and c.8734A>G (both missense mutations; reported in the LOVD to frequently co-segregate with each other and are classified as benign), and one with c.10127T > C (a missense mutation) and c.10133dup (a frameshifting insertion mutation). There was also one DMD patient who had both a duplication of exon 61 and a nonsense c.9100C > T point mutation; for purposes of this study, this patient was grouped with other duplication mutation carriers. A survey of nonsense point mutations in our population showed that 47% (17/36) involved a C-to-T transition (Figure 3d).

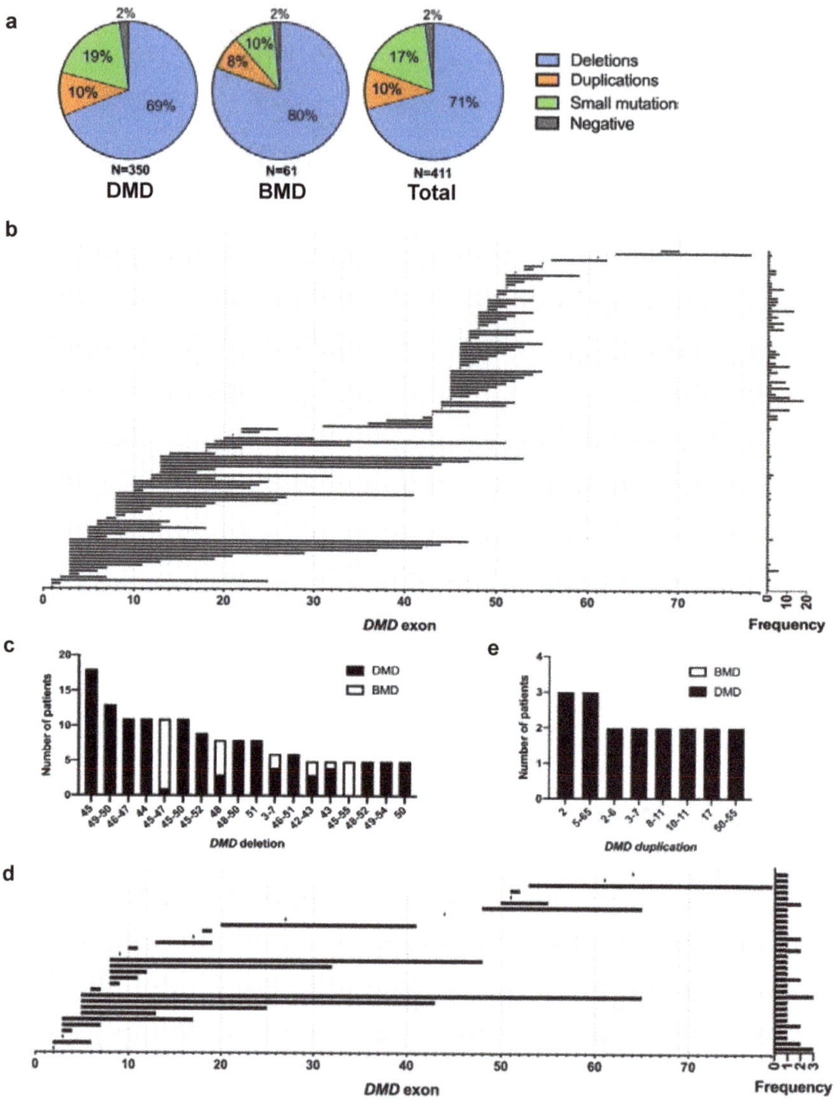

Figure 2. Overview of genetic characteristics in the study population. (**a**) *DMD* mutations in patients grouped according to type (deletions, duplications, small mutations); (**b**) Map of large *DMD* deletions (>1 exon) in Duchenne and Becker muscular dystrophy patients (DMD, BMD), with their frequencies (# patients) on the right (N = 290); (**c**) Top 18 most common large *DMD* deletions in DMD and BMD patients; (**d**) Corresponding map of large *DMD* duplications (>1 exon) in DMD and BMD patients (N = 41); (**e**) Top 8 most common large *DMD* duplications in DMD and BMD patients.

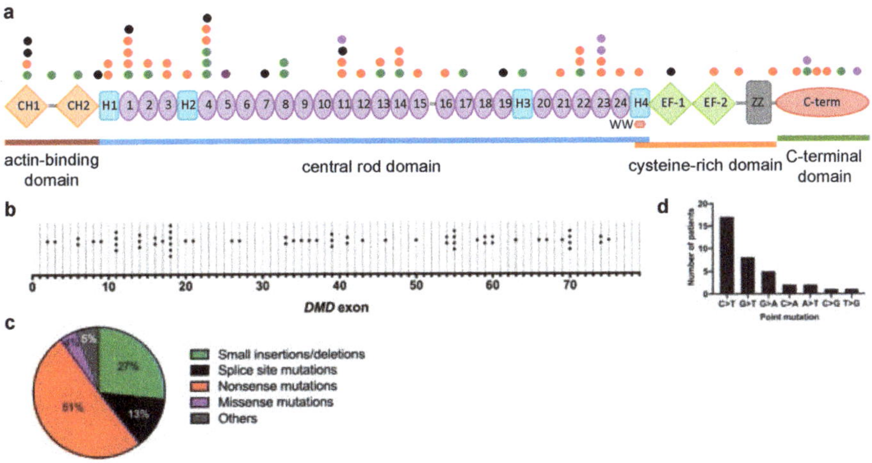

Figure 3. Overview of small mutations in the study population. (**a**) The positions of small mutations identified in Duchenne and Becker muscular dystrophy (DMD, BMD) patients are shown according to the domain/region of the dystrophin protein they affect, with each dot representing a unique mutation. The color of the dots correspond to the legend in (**c**); (**b**) The positions of small mutations, shown according to the *DMD* exon they are located in; (**c**) Distribution of *DMD* small mutations according to type (small insertions/deletions, splice site mutations, nonsense mutations, missense mutations, others); (**d**) Frequency of point mutation types in DMD and BMD patients. ($N = 71$).

3.3. Relationships between Genotype and DMD/BMD Diagnosis as Phenotype

The reading frame rule predicts at least 90% of the time [11,22] if a given *DMD* mutation will lead to a DMD or BMD phenotype. Most out-of-frame mutations give rise to DMD, while most in-frame mutations give rise to BMD [8]. To determine how well this rule holds in our population, we examined the frequency of out-of-frame and in-frame deletions in our DMD and BMD patients from the CNDR (Figure S1a–c). Of the 238 DMD patients in our cohort with deletion mutations not involving either exon 1 or 79, 87% (208/238) had out-of-frame mutations and 13% (30/238) had in-frame mutations (Figure 4a). On the other hand, of the 49 BMD patients with corresponding deletions, 16% (8/49) had out-of-frame mutations and 84% (41/49) had in-frame mutations.

Considering the deletions themselves, 96% (208/216) of observed out-of-frame deletions led to DMD, with only 4% (8/216) leading to BMD (Figure 4b). The in-frame deletions displayed a less skewed behavior—with 42% (30/71) giving rise to DMD and 58% (41/71) to BMD. Since the in-frame deletions did not predominantly favor one phenotype over the other to the same extent as out-of-frame deletions, we decided to map them out across the *DMD* exons. This will allow us to see if the location of the in-frame deletion is a key determinant of whether a patient develops DMD or BMD. The majority of in-frame deletions leading to DMD were found to start within the N-terminal exons 3–20 hotspot (Figure 4 and Figure S1b). In particular, of the 19 in-frame deletions solely associated with DMD, 14 or 74% of them started in this region. DMD-associated N-terminal in-frame deletions also tended to partially or completely remove more functional domains on the resulting dystrophin protein than their BMD-associated counterparts (Table S1). On the other hand, 67% (10/15) of in-frame deletions located at the distal half of the gene past exon 43 led to a BMD phenotype or to a mix of either a DMD or BMD phenotype (Figure 4c).

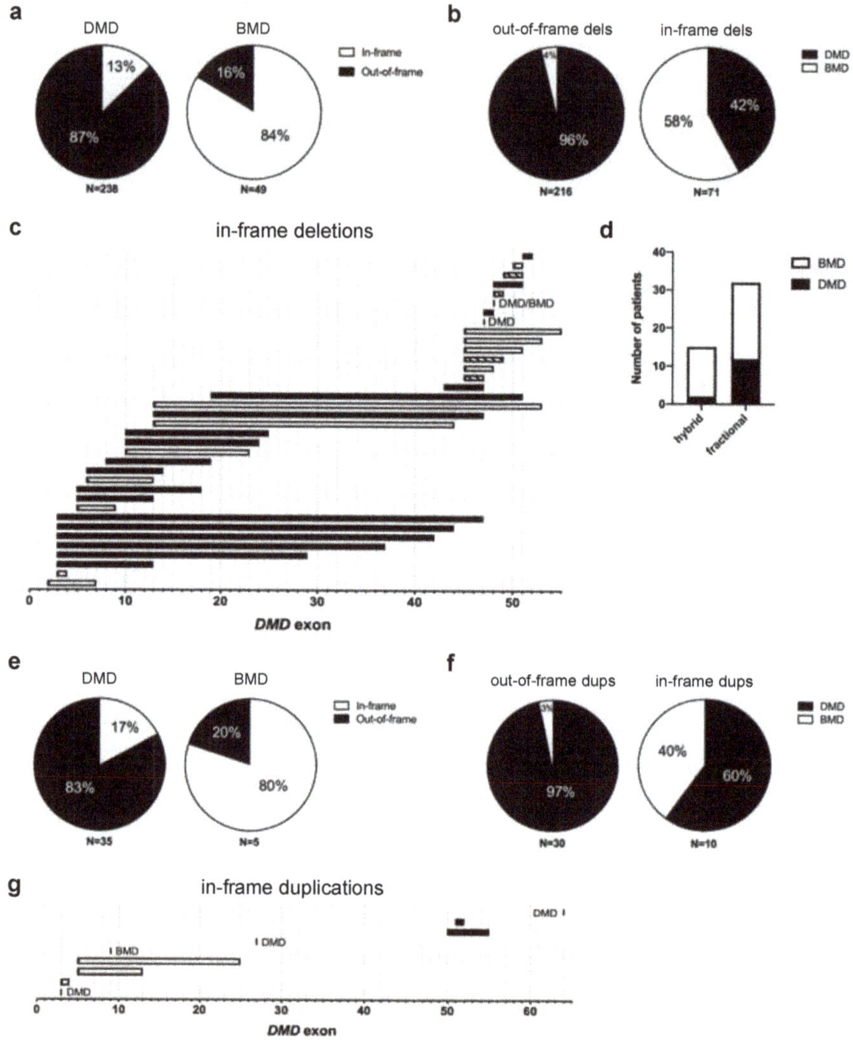

Figure 4. Analysis of large deletions and duplications, and their effect on the *DMD* reading frame. (**a**) Distribution of in-frame and out-of-frame deletions in Duchenne and Becker muscular dystrophy (DMD, BMD) patients; (**b**) Distribution of phenotypes associated with in-frame and out-of-frame deletions; (**c**) Map of in-frame *DMD* deletions in DMD and BMD patients, black: DMD, white: BMD, striped: both; (**d**) Frequencies of DMD and BMD patients with hybrid or fractional repeat-forming in-frame deletions; (**e**–**g**) Corresponding plots of (**a**–**c**) for duplications in DMD and BMD patients.

As these distal in-frame deletions all occur within the central rod domain of the dystrophin protein, one could model in silico how well these preserve the filamentous, helical structure of the region. Depending on where the exon breakpoints are, an in-frame deletion can give rise to either a hybrid or a fractional repeat unit in the rod domain. Hybrid repeats maintain the filamentous structure of the rod domain, whereas fractional repeats disrupt it [23–25]. Using the eDystrophin database (http://edystrophin.genouest.org/) [25], we obtained modeling predictions for the repeat structures formed by the various distal in-frame deletions (Table 2). Although hybrid repeat-forming deletions were found in more BMD than DMD patients, no significant association was found between clinical phenotype (DMD/BMD) and the predicted repeat structure formed by an in-frame deletion in the exons 45–55 hotspot region (Figure 4d). Interestingly, despite giving rise to a predicted fractional repeat unit, the in-frame deletion of exons 45–47 led to BMD 91% of the time (10/11 patients) rather than DMD (Table 2).

Table 2. Repeat structure modeling of in-frame *DMD* deletions within the exons 45–55 hotspot.

In-Frame Deletion	Frequency DMD (%)	Frequency BMD (%)	Predicted Repeat Structure [1]
45–47	1 (9.1)	10 (90.9)	Fractional
45–48	0 (0)	4 (100)	Hybrid
45–49	1 (50)	1 (50)	Fractional
45–51	0 (0)	2 (100)	Hybrid
45–53	0 (0)	1 (100)	Hybrid
45–55	0 (0)	5 (100)	Hybrid
47	2 (100)	0 (0)	Fractional
47–48	1 (100)	0 (0)	Hybrid
48	3 (37.5)	5 (62.5)	Fractional
48–49	1 (33.3)	2 (66.7)	Fractional
48–51	2 (100)	0 (0)	Fractional
49–51	1 (50)	1 (50)	Hybrid
50–51	0 (0)	2 (100)	Fractional
51–52	2 (100)	0 (0)	Fractional

[1] Information obtained from the online eDystrophin database.

We next examined the frequency of out-of-frame and in-frame duplications in our DMD and BMD patient population (Figure S2a,b). Of the 35 DMD patients in our cohort with duplication mutations, 83% (29/35) had out-of-frame mutations and 6% (6/35) had in-frame mutations (Figure 4e). Meanwhile, we only had five BMD patients with duplication mutations, one of which had an out-of-frame mutation, with the remaining four having in-frame mutations. In terms of the duplications themselves, out-of-frame duplications led to DMD 97% (29/30) of the time and to BMD 3% (1/30) of the time; in-frame duplications led to DMD in 60% (6/10) of cases and to BMD in 40% (4/10) of cases (Figure 4f). Similarly, as we did with the deletions, we mapped out all in-frame duplication patterns across the *DMD* exons (Figure 4g). Only nine unique in-frame duplications were found in our population, with those at the proximal end of the gene mostly associated with BMD and those at the distal end all associated with DMD.

Notably, less than 10% of small mutations (6/71) were associated with BMD in our study population. Due to the low representation of this mutation type among BMD patients, an analysis of genotype–phenotype correlations may be premature and therefore was not performed.

3.4. Relationships between Genotype and Clinical Outcome as Phenotype

We then proceeded to perform a series of multiple regression analyses to determine any relationships between patient genotypes and clinical outcomes, focusing on data from DMD patients (Figure 1). For genotype, we considered the location of the mutation according to which dystrophin protein domain/s or dystrophin isoform/s they affect. Exons were once again assigned to protein domains following information from the Leiden Muscular Dystrophy dystrophin page (https://www.dmd.nl/). For clinical outcomes, we looked at wheelchair use (combined permanent and intermittent use), cardiomyopathy status (presence or absence), LVEF, and FVC. In constructing these models, we also took into account the effect of other parameters such as age, BMI, steroid use (past or present), and use of cardiac medications, as appropriate. The results of these analyses are summarized in Table S2 and Table S3.

Multiple logistic regression analysis revealed that there is a 6.136 times increase in odds (95% confidence interval (CI): 1.44, 33.99; $p < 0.05$) that a DMD patient will require wheelchair use when they have mutations affecting the dystrophin rod domain (Table S2). Mutations affecting the C-terminal domain yielded an odds ratio of 0.0281 (95% CI: 0.001, 0.30; $p < 0.005$), indicating that their presence was associated with decreased wheelchair use in our DMD patient population. A similar relationship was found for mutations affecting the Dp116 and Dp71 isoforms (both $p < 0.005$). Across all models with wheelchair use as the selected outcome, age had an odds ratio greater than 1.75 ($p < 0.0005$), and BMI as well as steroid use were not significant predictors. All area under the receiving operator curve (AUC) values were at least 0.93. When cardiomyopathy status was used as an outcome, only mutations affecting the Dp140 isoform showed a significant relationship, with an odds ratio of 0.3662 (95% CI: 0.14, 0.92; $p < 0.05$) (Table S2). Age gave an odds ratio of at least 1.31 ($p < 0.0005$), with BMI and steroid use not being significant predictors of cardiomyopathy status; AUC values were at least 0.83. Unfortunately, models could not be generated for the other genotype categories, as these groups did not have any patients with cardiomyopathy.

Multiple linear regression analysis with LVEF as the outcome yielded no genotypes as significant predictors (Table S3). Age, steroid use, and use of cardiac medications all yielded significant estimates (β) in the produced regression models (individual $R^2 > 0.3$). Age and use of cardiac medications gave negative estimates ($p < 0.0005$ and $p < 0.005$, respectively), while steroid use gave positive estimates ($p < 0.05$). On the other hand, when FVC was used as an outcome, mutations in the C-terminal domain gave a significant β in the model at −19.24 (95% CI: −36.56, −1.91; $p < 0.05$). No other genotype categories yielded significant β values. Age and steroid use had significant estimates in all produced models for FVC (individual R^2 values >0.4), with age having negative β values ($p < 0.0005$) and steroid use having positive β values ($p < 0.005$).

3.5. Applicability of Exon Skipping Therapy to DMD Patients in Canada

A particularly promising approach to treat DMD is exon skipping using small single-stranded nucleic acid analogues called antisense oligonucleotides (AOs). In this strategy, AOs are designed to bind specific splicing enhancer sequences in out-of-frame *DMD* exons by base pairing. This results in the exclusion of targeted exons from the final mRNA transcript, restoring the reading frame and thereby allowing for the synthesis of shorter, partially functional dystrophin proteins [26,27]. With the increasing number of exon skipping therapies entering the clinic and receiving FDA approval, we sought to determine their applicability to DMD patients in Canada. We evaluated the applicability of the top 10 single exon skipping strategies that can treat the most number of patients registered in the global TREAT-NMD DMD database [11], and we also evaluated two multiple exon skipping strategies that target exons within the *DMD* mutation hotspots [18]. Exon 51 skipping treated the most number of DMD patients with deletions at 17%, as well as the most number of DMD patients overall (with deletions, duplications, and small mutations) at 12.3% in our cohort, which was similar to the trend observed worldwide in a previous TREAT-NMD study [11] (Table 3). This was followed by exon 45 skipping at 15.8% of DMD patients with deletions or 11.1% of all DMD patients and then by exon 44 skipping at 12.9% of DMD patients with deletions or 9.4% of all DMD patients. Exon 53 skipping is only the fourth most applicable single exon skipping therapy in our cohort, as opposed to being ranked second among TREAT-NMD DMD patients [11]. For the multiple exon skipping strategies, exons 45–55 skipping was applicable to 66.8% of DMD patients with deletions or 50.9% of all DMD patients in Canada (Table 3). Exons 3–9 skipping was less applicable, at 7.9% of all DMD patients with deletions or 9.1% of all DMD patients.

Table 3. Applicability of single and multiple exon skipping strategies to DMD patients in Canada.

Exon/s to Skip	% of DMD Patients with Deletions	% of all DMD Patients	Rank in TREAT-NMD [1]	Rank in Our Population
51	17.0	12.3	1	1
53	9.5	7.0	2	4
45	15.8	11.1	3	2
44	12.9	9.4	4	3
43	5.0	4.1	5	7
46	7.9	5.8	6	5
50	5.0	3.8	7	8
52	3.7	2.6	8	10
55	5.0	4.7	9	6
8	2.9	3.5	10	9
45–55	66.8	50.9	n/a	n/a
3–9	7.9	9.1	n/a	n/a

[1] Rank information obtained from Bladen et al. (2015) [11].

4. Discussion

We characterized *DMD* mutation data from DMD/BMD patients registered in the CNDR between 2012 and 2019, with a subsequent analysis of genotype–phenotype correlations. This study partly builds on previous work done by the Canadian Pediatric Neuromuscular Group (CPNG) in 2011, who studied the spectrum of *DMD* mutations in 773 patients across Canada from 2000 to 2009 [16]. We observed a similar abundance of mutation types across patients as the CPNG, with deletions forming the largest group (71% here compared to 64% from the CPNG study), followed by small mutations and duplications (Figure 2a). We found similar *DMD* mutation hotspots, with the exception that the CPNG observed a more extensive duplication hotspot from exons 2–20. In terms of overall genetic characteristics, our findings were largely consistent with those from global database studies (TREAT-NMD, LOVD) [11,18], indicating underlying commonalities in *DMD* gene mutability between patients in Canada and the rest of the world.

Perhaps the most well-known genotype–phenotype correlation in the field concerns the reading frame rule [8]. As in other studies (e.g., [11,16,18,19]) we found exceptions to this rule, with only 87% of DMD patients in our population having out-of-frame deletions and 84% of BMD patients having in-frame deletions (Figure 4a), for a total exception rate of 13%, which was higher than what was observed in the TREAT-NMD and LOVD databases [11,18]. Examining the 36 in-frame deletion patterns in our cohort revealed that deletion location and size matter, particularly if it affects dystrophin protein-binding domains mostly concentrated at the N-terminal end of the protein (Figure 4c, Table S1). In-frame deletions within the rod domain-coding region past exon 45, which do not code for any known protein-binding domains, were mostly associated with BMD. However, the number of impacted binding domains does not completely predict the disease phenotype of in-frame deletions. Consider our in-frame deletions that start on exon 13: exons 13–44 and 13–53 deletions lead to BMD, while the sandwiched exons 13–47 deletion leads to DMD. All three affect the same dystrophin protein-binding domains (Table S1) and yet have varying clinical consequences.

It is possible that regions other than the currently known protein-binding domains may be more critical for dystrophin function. For instance, a previous study looked at 97 patients from the Universal Mutation Database (UMD)-DMD registry with in-frame deletions before exon 35 and suggested that certain protein-binding domains may be dispensable to dystrophin function [28]. Characterizing these other potential critical regions in the *DMD* gene would be essential to understanding patients with mutations not governed by the reading frame rule. These regions can be identified through a combination of extensive patient database study and in vitro validation with patient-derived cells or induced pluripotent stem cell-derived models [29] of patient mutations. The identification of such regions will also benefit the development of gene replacement or correction therapies for DMD [24] to ensure that the dystrophin protein variants used or produced by these approaches are as functionally close as possible to the full-length version.

One concern for in-frame deletions affecting the central rod domain is also whether or not they can preserve its repeating, filamentous structure. Intuitively, in-frame mutations that can maintain this structure would be more likely to lead to BMD. While we observed this to be somewhat true for hybrid repeat-forming deletions, the same surprisingly cannot be said for fractional repeat-forming deletions (Figure 4d). In fact, one study of LOVD patients with in-frame mutations between exons 42 and 57 even found that fractional repeat-forming deletions were more commonly associated with BMD (72% of the time) than DMD [24]. The same study showed that the position of in-frame mutations relative to hinge 3 (exons 50–51) better determines phenotype than the predicted repeat structure formed by the deletions, which is a finding corroborated by another report [30]. This suggests that other parameters should be considered when evaluating the consequences of in-frame mutations on dystrophin structure, such as effects on overall protein flexibility or intra-protein interactions between residues. However, it is important to point out that knowing this information would still not be sufficient to explain certain cases, such as why the same in-frame deletion leads to a mix of DMD and BMD patients (e.g., deletions of exons 45–47, 45–49, 48, 48–49, and 49–51; Figure 4c and Figure S1b). In these cases, genetic modifiers [31,32] or spontaneous exon skipping events (as discussed in the next paragraph) may play a role in determining patient phenotypes.

We also saw a few out-of-frame deletion patients in our cohort to be exceptions to the reading frame rule, particularly those with deletions in exons 3–6, 3–7, 3–21, 7–8, 42–43, and 43 (Figure S1b). Two mechanisms have been proposed to explain such exceptions. The first is the use of alternative translational start sites further downstream in the *DMD* transcript [33–35]. For instance, a series of immunofluorescence experiments performed on skeletal muscle biopsies from exons 3 to 7 deletion patients suggested that there was a potential alternative initiation codon in exon 8 [34]. Dystrophin was not detectable when antibodies recognizing the 5′ end of exon 8 in the protein were used; however, dystrophin was detected using antibodies recognizing the 3′ end of exon 8. This may explain why a deletion of exons 3–7 is typically associated with BMD or with milder DMD phenotypes [18,35,36]. The second mechanism is the occurrence of spontaneous exon skipping events that convert out-of-frame into in-frame mutations. A well-documented example is the spontaneous skipping of exon 44 that occurs when the exons flanking it are deleted [37,38]. In fact, exon 44-skippable deletions are usually associated with a higher number of dystrophin-revertant fibers and milder DMD phenotypes such as prolonged ambulation [36,39–42]. In addition, of the six out-of-frame deletions that we have listed as exceptions, five of them can be converted into in-frame deletions with the skipping of just one exon adjacent to the deletion. This spontaneous exon skipping may be tied to how the junction sequences formed by a deletion influences splicing, i.e., if it creates or destroys exon splicing silencer/enhancer sequences [37]. Further study into this phenomenon may also provide hints regarding the formation of dystrophin-revertant fibers.

As for correlations between genotypes and clinical outcomes, it is important to emphasize that the regression analysis performed here produces an inferential model, i.e., a model that best describes the study population at its current state. There were a number of limitations with the study population as it is now that may have affected the analysis, mostly concerning low sample sizes for each mutation pattern observed and incomplete availability of clinical outcome data for all patients. The majority of DMD patients analyzed were within the younger range as well (Table 1), and so there may be some bias in the observed phenotypes. For practical reasons, we also limited our analysis to genotypes classified according to the protein domain or the dystrophin isoform affected by the respective patient mutations. We acknowledge that use of other stratification procedures may lead to differing conclusions.

With these in mind, we saw an increased likelihood of wheelchair use associated with mutations affecting the rod domain and, conversely, a decreased likelihood with mutations affecting the C-terminal domain and Dp116/71 isoforms in our DMD patient population (Table S2). It is interesting that a positive association with rod domain mutations was observed. Previous reports have shown that certain rod domain-coding mutations are associated with prolonged ambulation in DMD patients, e.g., exon 44-skippable deletions [36,40–42]. Once a sufficient number of patients are available, it may be worthwhile to further stratify rod domain mutations to pinpoint the importance of specific sub-regions. The finding regarding the C-terminal domain is striking, since one would expect it to be critical in localizing dystrophin to the muscle membrane [7]; note that C-terminal domain mutations were also significantly, positively correlated with FVC in our DMD patient cohort (Table S3). Interestingly, there has been a previous case of an 8-year-old boy reported to be asymptomatic despite having a nonsense mutation truncating the C-terminal domain [43]. Microdystrophins lacking most or all of the C-terminal domain have also been promising in *mdx* mice with improvements in skeletal and cardiac muscle phenotypes [44–46]. Our results complement such findings, inviting closer investigation into the importance of the C-terminal domain for dystrophin function in muscle. However, it is also important to note that our result is based on a small number of patients with C-terminal domain mutations ($n = 10$), and so further validation by conducting a regression analysis with a larger sample size is recommended.

The association of Dp116 and Dp71 with motor function was likewise unexpected, as these isoforms are not normally expressed in differentiated skeletal muscle. Dp116 is exclusively expressed in Schwann cells [47], and Dp71 displays mostly ubiquitous expression but is difficult to detect in differentiated skeletal muscle [48,49]. While some reports are now claiming otherwise [50,51], i.e., that these isoforms are in fact expressed in muscle (one study is described in the next section for Dp116), their functional significance in muscle remains unknown. As for other factors included in the model for wheelchair use, it was surprising that steroid use did not have a significant impact, contrary to a previous TREAT-NMD DMD registry report [17]. However, this observation may be restricted to the particular demographic of the population under study.

Mutations affecting Dp140 was the only genotype group determined to be a significant predictor of cardiomyopathy (Table S2); no significant genotypes were found as predictors for LVEF (Table S3). Dp140 is a non-muscle dystrophin isoform typically expressed in the central nervous system and the kidneys [52]; its expression in the heart (or skeletal muscles) has not yet been demonstrated. Based on our analysis, Dp140 mutations are apparently associated with the lack of cardiomyopathy. One group previously studied the relationship between cardiac dysfunction (LVEF <53%)-free survival and dystrophin isoform mutations, but they did not find any significant association with respect to Dp140 [51]. Instead, the authors observed that Dp116 mutations were significantly linked to better rates of cardiac dysfunction-free survival, which we did not see in our analysis. Note that Dp116 was thought to be a non muscle dystrophin isoform; however, this study demonstrated that Dp116 mRNA expression was detectable in both human cardiac and skeletal muscle samples. Therefore, it remains possible that Dp140 may have a role in the heart, but this will have to be supported first by in vivo validation of cardiac Dp140 expression similar to what was done for Dp116 in the study above, and then by further confirmation of our result in other patient registries. Considering other factors in our model, steroid use was not a significant predictor for cardiomyopathy, but it was significantly, positively correlated with LVEF. This may be explained in part by the fact that our DMD patient cohort is relatively young and not well-suited for observing cardiac symptoms that manifest relatively late in the disease. Cardiac medications were significantly, negatively correlated with LVEF, but they may reflect the bias that patients with reduced LVEF are typically the ones receiving such treatments—the factor was included more as a control for other predictors.

There are other reports of genotype–phenotype correlations with respect to cardiac outcomes in the literature, with proximal/N-terminal mutations generally associated with worse cardiac symptoms than distal/C-terminal mutations [21,53–55]. Still, some studies demonstrated a lack of correlation altogether [21,56,57]. This issue of non-agreement across genotype–phenotype correlation studies is not only true for cardiac outcomes but also for skeletal muscle phenotypes. This clearly indicates the need for further work in this area, starting perhaps by standardizing data collection procedures to maximize comparability across patient registries as well as the amount of information obtained from each patient.

Within the last five years, we have seen the approval of three exon skipping AOs for DMD therapy by the FDA: eteplirsen (brand name Exondys 51, Sarepta) for skipping exon 51 in 2016 [26], and golodirsen (Vyondys 53, Sarepta) in 2019 [58] as well as viltolarsen (Viltepso, NS Pharma) in 2020 [59] for skipping exon 53; another AO, the exon 45-skipping casimersen (SRP-4045, Sarepta) is currently under FDA review. These FDA-approved therapies can treat a combined 26.5% of DMD patients with deletions or 19.3% of all DMD patients in Canada (Table 3), which is incredibly encouraging. Notably, the applicability of single exon skipping strategies was different for patients in Canada compared to global estimates from the TREAT-NMD DMD database [11], suggesting potential implications for future clinical trials. These findings highlight one of the major limitations associated with personalized therapies such as exon skipping, i.e., low patient applicability. One way to overcome this would be to develop multi-exon skipping strategies such as exons 45–55 skipping, which could treat more than half of all DMD patients (Table 3). Our data and those from other patient registries [18,60] also show that exons 45–55 deletions are commonly associated with mild BMD or asymptomatic phenotypes (Figure 4c and Figure S1b), confirming the viability of the approach as a treatment for DMD.

This last point raises a concern for other exon skipping strategies, i.e., if the in-frame-skipped dystrophin proteins they produce are indeed functional or associated with mild phenotypes. We have seen how some deletions lead to a DMD phenotype despite being in-frame, e.g., in our population, 42% of in-frame deletions were in DMD patients (Figure 4b). Encouragingly, the majority of patients with deletions equivalent to exon 51-skipped transcripts showed mild phenotypes [61], bearing well for eteplirsen. Therefore, consulting patient registries such as the CNDR when designing exon skipping strategies is recommended. Finally, despite the promise of exon skipping therapy, it cannot correct all mutations, and there remain concerns regarding its efficacy in patients. The continued development of other therapeutic approaches such as gene replacement with mini/microdystrophins or gene correction with genome editing strategies, as informed by genotype–phenotype correlation studies from patient registries, remains critically important.

Supplementary Materials: The following are available online at http://www.mdpi.com/2075-4426/10/4/241/s1, Figure S1: Summary of large *DMD* gene deletions, Figure S2: Summary of large *DMD* gene duplications, Table S1: In-frame deletions and their effects on dystrophin protein-binding domains, Table S2: Multiple logistic regression analysis for wheelchair use and cardiomyopathy status, Table S3: Multiple linear regression analysis for left ventricle ejection fraction (LVEF) and forced vital capacity (FVC).

Author Contributions: Conceptualization, K.R.Q.L. and Q.N.; methodology, K.R.Q.L., Q.N., and T.Y.; investigation, K.R.Q.L. and Q.N.; writing—original draft preparation, K.R.Q.L. and Q.N.; writing—review and editing, K.R.Q.L., Q.N., and T.Y.; supervision, T.Y.; project administration, K.R.Q.L., T.Y.; funding acquisition, T.Y. All authors have read and agreed to the published version of the manuscript.

Funding: This research was funded by the Friends of Garrett Cumming Research & Muscular Dystrophy Canada HM Toupin Neurological Science Research Chair, Canadian Institutes of Health Research (CIHR) FDN 143251, 169193, and the Women and Children's Health Research Institute (WCHRI) IG 2874.

Acknowledgments: We would like to thank the CNDR Investigator Network for collecting, compiling, and providing the patient data used in this study. We would also like to thank Matthew Pietrosanu, a statistical consultant at the Training and Consulting Centre (TCC) in the Department of Mathematical and Statistical Sciences, University of Alberta, for his advice on the statistical analysis performed in this study.

Conflicts of Interest: The authors declare no conflict of interest.

References

1. Crisafulli, S.; Sultana, J.; Fontana, A.; Salvo, F.; Messina, S.; Trifirò, G. Global epidemiology of Duchenne muscular dystrophy: An updated systematic review and meta-analysis. *Orphanet J. Rare Dis.* **2020**, *15*, 141.
2. Mendell, J.R.; Shilling, C.; Leslie, N.D.; Flanigan, K.M.; Al-Dahhak, R.; Gastier-Foster, J.; Kneile, K.; Dunn, D.M.; Duval, B.; Aoyagi, A.; et al. Evidence-based path to newborn screening for Duchenne muscular dystrophy. *Ann. Neurol.* **2012**, *71*, 304–313. [PubMed]
3. Yiu, E.M.; Kornberg, A.J. Duchenne muscular dystrophy. *J. Paediatr. Child Health* **2015**, *51*, 759–764. [PubMed]

4. Manzur, A.; Kinali, M.; Muntoni, F. Update on the management of Duchenne muscular dystrophy. *Arch. Dis. Child.* **2008**, *93*, 986–990. [PubMed]
5. Petrof, B.J.; Shrager, J.B.; Stedman, H.H.; Kelly, A.M.; Sweeney, H.L. Dystrophin protects the sarcolemma from stresses developed during muscle contraction. *Proc. Natl. Acad. Sci. USA* **1993**, *90*, 3710–3714.
6. Ervasti, J.M.; Campbell, K.P. A role for the dystrophin-glycoprotein complex as a transmembrane linker between laminin and actin. *J. Cell Biol.* **1993**, *122*, 809–823.
7. Ervasti, J.M. Dystrophin, its interactions with other proteins, and implications for muscular dystrophy. *Biochim. Biophys. Acta* **2007**, *1772*, 108–117.
8. Monaco, A.P.; Bertelson, C.J.; Liechti-Gallati, S.; Moser, H.; Kunkel, L.M. An explanation for the phenotypic differences between patients bearing partial deletions of the DMD locus. *Genomics* **1988**, *2*, 90–95.
9. Anthony, K.; Cirak, S.; Torelli, S.; Tasca, G.; Feng, L.; Arechavala-Gomeza, V.; Armaroli, A.; Guglieri, M.; Straathof, C.S.; Verschuuren, J.J.; et al. Dystrophin quantification and clinical correlations in Becker muscular dystrophy: Implications for clinical trials. *Brain* **2011**, *134*, 3547–3559.
10. Nicolas, A.; Raguénès-Nicol, C.; Ben Yaou, R.; Ameziane-Le Hir, S.; Chéron, A.; Vié, V.; Claustres, M.; Leturcq, F.; Delalande, O.; Hubert, J.-F.; et al. Becker muscular dystrophy severity is linked to the structure of dystrophin. *Hum. Mol. Genet.* **2015**, *24*, 1267–1279.
11. Bladen, C.L.; Salgado, D.; Monges, S.; Foncuberta, M.E.; Kekou, K.; Kosma, K.; Dawkins, H.; Lamont, L.; Roy, A.J.; Chamova, T.; et al. The TREAT-NMD DMD Global Database: Analysis of more than 7,000 Duchenne muscular dystrophy mutations. *Hum. Mutat.* **2015**, *36*, 395–402.
12. Aartsma-Rus, A.; Van Deutekom, J.C.T.; Fokkema, I.F.; Van Ommen, G.-J.B.; Den Dunnen, J.T. Entries in the Leiden Duchenne muscular dystrophy mutation database: An overview of mutation types and paradoxical cases that confirm the reading-frame rule. *Muscle Nerve* **2006**, *34*, 135–144. [PubMed]
13. White, S.J.; den Dunnen, J.T. Copy number variation in the genome; the human DMD gene as an example. *Cytogenet. Genome Res.* **2006**, *115*, 240–246. [PubMed]
14. Wei, Y.; McCormick, A.; MacKenzie, A.; O'Ferrall, E.; Venance, S.; Mah, J.K.; Selby, K.; McMillan, H.J.; Smith, G.; Oskoui, M.; et al. The Canadian Neuromuscular Disease Registry: Connecting patients to national and international research opportunities. *Paediatr. Child Health* **2018**, *23*, 20–26. [PubMed]
15. Hodgkinson, V.; Lounsberry, J.; M'Dahoma, S.; Russell, A.; Benstead, T.; Brais, B.; Campbell, C.; Johnston, W.; Lochmüller, H.; McCormick, A.; et al. The Canadian Neuromuscular Disease Registry 2010–2019: A Decade of Facilitating Clinical Research Through a Nationwide, Pan-Neuromuscular Disease Registry. *J. Neuromuscul. Dis.* **2020**, 1–9. [CrossRef]
16. Mah, J.K.; Selby, K.; Campbell, C.; Nadeau, A.; Tarnopolsky, M.; McCormick, A.; Dooley, J.M.; Kolski, H.; Skalsky, A.J.; Smith, R.G.; et al. A Population-Based Study of Dystrophin Mutations in Canada. *Can. J. Neurol. Sci. / J. Can. des Sci. Neurol.* **2011**, *38*, 465–474.
17. Koeks, Z.; Bladen, C.L.; Salgado, D.; van Zwet, E.; Pogoryelova, O.; McMacken, G.; Monges, S.; Foncuberta, M.E.; Kekou, K.; Kosma, K.; et al. Clinical Outcomes in Duchenne Muscular Dystrophy: A Study of 5345 Patients from the TREAT-NMD DMD Global Database. *J. Neuromuscul. Dis.* **2017**, *4*, 293–306. [PubMed]
18. Echigoya, Y.; Lim, K.R.Q.; Nakamura, A.; Yokota, T. Multiple Exon Skipping in the Duchenne Muscular Dystrophy Hot Spots: Prospects and Challenges. *J. Pers. Med.* **2018**, *8*, 41.
19. Tuffery-Giraud, S.; Béroud, C.; Leturcq, F.; Yaou, R.B.; Hamroun, D.; Michel-Calemard, L.; Moizard, M.P.; Bernard, R.; Cossée, M.; Boisseau, P.; et al. Genotype-phenotype analysis in 2,405 patients with a dystrophinopathy using the UMD-DMD database: A model of nationwide knowledgebase. *Hum. Mutat.* **2009**, *30*, 934–945.
20. Juan-Mateu, J.; Gonzalez-Quereda, L.; Rodriguez, M.J.; Baena, M.; Verdura, E.; Nascimento, A.; Ortez, C.; Baiget, M.; Gallano, P. DMD Mutations in 576 Dystrophinopathy Families: A Step Forward in Genotype-Phenotype Correlations. *PLoS ONE* **2015**, *10*, e0135189.
21. Magri, F.; Govoni, A.; D'Angelo, M.G.; Del Bo, R.; Ghezzi, S.; Sandra, G.; Turconi, A.C.; Sciacco, M.; Ciscato, P.; Bordoni, A.; et al. Genotype and phenotype characterization in a large dystrophinopathic cohort with extended follow-up. *J. Neurol.* **2011**, *258*, 1610–1623. [PubMed]

22. Vengalil, S.; Preethish-Kumar, V.; Polavarapu, K.; Mahadevappa, M.; Sekar, D.; Purushottam, M.; Thomas, P.T.; Nashi, S.; Nalini, A. Duchenne Muscular Dystrophy and Becker Muscular Dystrophy Confirmed by Multiplex Ligation-Dependent Probe Amplification: Genotype-Phenotype Correlation in a Large Cohort. *J. Clin. Neurol.* **2017**, *13*, 91. [PubMed]
23. Menhart, N. Hybrid spectrin type repeats produced by exon-skipping in dystrophin. *Biochim. Biophys. Acta Proteins Proteom.* **2006**, *1764*, 993–999.
24. Yokota, T.; Duddy, W.; Partridge, T. Optimizing exon skipping therapies for DMD. *Acta Myol.* **2007**, *26*, 179–184. [PubMed]
25. Nicolas, A.; Lucchetti-Miganeh, C.; Yaou, R.; Kaplan, J.-C.; Chelly, J.; Leturcq, F.; Barloy-Hubler, F.; Le Rumeur, E. Assessment of the structural and functional impact of in-frame mutations of the DMD gene, using the tools included in the eDystrophin online database. *Orphanet J. Rare Dis.* **2012**, *7*, 45. [PubMed]
26. Lim, K.R.Q.; Maruyama, R.; Yokota, T. Eteplirsen in the treatment of Duchenne muscular dystrophy. *Drug Des. Devel. Ther.* **2017**, *11*, 533–545.
27. Lim, K.R.Q.; Yokota, T. Invention and Early History of Exon Skipping and Splice Modulation. In *Exon Skipping and Inclusion Therapies: Methods and Protocols*; Yokota, T., Maruyama, R., Eds.; Springer: New York, NY, USA, 2018; pp. 3–30.
28. Gibbs, E.M.; Barthélémy, F.; Douine, E.D.; Hardiman, N.C.; Shieh, P.B.; Khanlou, N.; Crosbie, R.H.; Nelson, S.F.; Miceli, M.C. Large in-frame 5′ deletions in DMD associated with mild Duchenne muscular dystrophy: Two case reports and a review of the literature. *Neuromuscul. Disord.* **2019**, *29*, 863–873.
29. Long, C.; Li, H.; Tiburcy, M.; Rodriguez-Caycedo, C.; Kyrychenko, V.; Zhou, H.; Zhang, Y.; Min, Y.-L.; Shelton, J.M.; Mammen, P.P.A.; et al. Correction of diverse muscular dystrophy mutations in human engineered heart muscle by single-site genome editing. *Sci. Adv.* **2018**, *4*, eaap9004.
30. Carsana, A.; Frisso, G.; Tremolaterra, M.R.; Lanzillo, R.; Vitale, D.F.; Santoro, L.; Salvatore, F. Analysis of Dystrophin Gene Deletions Indicates that the Hinge III Region of the Protein Correlates with Disease Severity. *Ann. Hum. Genet.* **2005**, *69*, 253–259.
31. Heydemann, A.; Doherty, K.R.; McNally, E.M. Genetic modifiers of muscular dystrophy: Implications for therapy. *Biochim. Biophys. Acta Mol. Basis Dis.* **2007**, *1772*, 216–228.
32. Vo, A.H.; McNally, E.M. Modifier genes and their effect on Duchenne muscular dystrophy. *Curr. Opin. Neurol.* **2015**, *28*, 528–534. [PubMed]
33. Gurvich, O.L.; Maiti, B.; Weiss, R.B.; Aggarwal, G.; Howard, M.T.; Flanigan, K.M. DMD exon 1 truncating point mutations: Amelioration of phenotype by alternative translation initiation in exon 6. *Hum. Mutat.* **2009**, *30*, 633–640. [PubMed]
34. Winnard, A.V.; Mendell, J.R.; Prior, T.W.; Florence, J.; Burghes, A.H. Frameshift deletions of exons 3-7 and revertant fibers in Duchenne muscular dystrophy: Mechanisms of dystrophin production. *Am. J. Hum. Genet.* **1995**, *56*, 158–166. [PubMed]
35. Muntoni, F.; Gobbi, P.; Sewry, C.; Sherratt, T.; Taylor, J.; Sandhu, S.K.; Abbs, S.; Roberts, R.; Hodgson, S.V.; Bobrow, M. Deletions in the 5′ region of dystrophin and resulting phenotypes. *J. Med. Genet.* **1994**, *31*, 843–847. [PubMed]
36. Wang, R.T.; Barthelemy, F.; Martin, A.S.; Douine, E.D.; Eskin, A.; Lucas, A.; Lavigne, J.; Peay, H.; Khanlou, N.; Sweeney, L.; et al. DMD genotype correlations from the Duchenne Registry: Endogenous exon skipping is a factor in prolonged ambulation for individuals with a defined mutation subtype. *Hum. Mutat.* **2018**, *39*, 1193–1202.
37. Dwianingsih, E.K.; Malueka, R.G.; Nishida, A.; Itoh, K.; Lee, T.; Yagi, M.; Iijima, K.; Takeshima, Y.; Matsuo, M. A novel splicing silencer generated by DMD exon 45 deletion junction could explain upstream exon 44 skipping that modifies dystrophinopathy. *J. Hum. Genet.* **2014**, *59*, 423–429.
38. Aartsma-Rus, A.; Muntoni, F. 194th ENMC international workshop. 3rd ENMC workshop on exon skipping: Towards clinical application of antisense-mediated exon skipping for Duchenne muscular dystrophy 8–10 December 2012, Naarden, The Netherlands. *Neuromuscul. Disord.* **2013**, *23*, 934–944.
39. Anthony, K.; Arechavala-Gomeza, V.; Ricotti, V.; Torelli, S.; Feng, L.; Janghra, N.; Tasca, G.; Guglieri, M.; Barresi, R.; Armaroli, A.; et al. Biochemical Characterization of Patients With In-Frame or Out-of-Frame DMD Deletions Pertinent to Exon 44 or 45 Skipping. *JAMA Neurol.* **2014**, *71*, 32.
40. Bello, L.; Morgenroth, L.P.; Gordish-Dressman, H.; Hoffman, E.P.; McDonald, C.M.; Cirak, S. DMD genotypes and loss of ambulation in the CINRG Duchenne Natural History Study. *Neurology* **2016**, *87*, 401–409.

41. Brogna, C.; Coratti, G.; Pane, M.; Ricotti, V.; Messina, S.; D'Amico, A.; Bruno, C.; Vita, G.; Berardinelli, A.; Mazzone, E.; et al. Long-term natural history data in Duchenne muscular dystrophy ambulant patients with mutations amenable to skip exons 44, 45, 51 and 53. *PLoS ONE* **2019**, *14*, e0218683.
42. van den Bergen, J.C.; Ginjaar, H.B.; Niks, E.H.; Aartsma-Rus, A.; Verschuuren, J.J.G.M. Prolonged Ambulation in Duchenne Patients with a Mutation Amenable to Exon 44 Skipping. *J. Neuromuscul. Dis.* **2014**, *1*, 91–94. [PubMed]
43. Suminaga, R.; Takeshima, Y.; Wada, H.; Yagi, M.; Matsuo, M. C-Terminal Truncated Dystrophin Identified in Skeletal Muscle of an Asymptomatic Boy with a Novel Nonsense Mutation of the Dystrophin Gene. *Pediatr. Res.* **2004**, *56*, 739–743. [PubMed]
44. Yue, Y.; Liu, M.; Duan, D. C-Terminal-Truncated Microdystrophin Recruits Dystrobrevin and Syntrophin to the Dystrophin-Associated Glycoprotein Complex and Reduces Muscular Dystrophy in Symptomatic Utrophin/Dystrophin Double-Knockout Mice. *Mol. Ther.* **2006**, *14*, 79–87. [PubMed]
45. Yue, Y.; Li, Z.; Harper, S.Q.; Davisson, R.L.; Chamberlain, J.S.; Duan, D. Microdystrophin Gene Therapy of Cardiomyopathy Restores Dystrophin-Glycoprotein Complex and Improves Sarcolemma Integrity in the Mdx Mouse Heart. *Circulation* **2003**, *108*, 1626–1632. [PubMed]
46. Sweeney, H.L.; Barton, E.R. The dystrophin-associated glycoprotein complex: What parts can you do without? *Proc. Natl. Acad. Sci. USA* **2000**, *97*, 13464–13466.
47. Matsuo, M.; Awano, H.; Matsumoto, M.; Nagai, M.; Kawaguchi, T.; Zhang, Z.; Nishio, H. Dystrophin Dp116: A yet to Be Investigated Product of the Duchenne Muscular Dystrophy Gene. *Genes* **2017**, *8*, 251.
48. Richard, C.A.; Howard, P.L.; D'Souza, V.N.; Klamut, H.J.; Ray, P.N. Cloning and characterization of alternatively spliced isoforms of Dp71. *Hum. Mol. Genet.* **1995**, *4*, 1475–1483.
49. de León, M.B.; Montañez, C.; Gómez, P.; Morales-Lázaro, S.L.; Tapia-Ramírez, V.; Valadez-Graham, V.; Recillas-Targa, F.; Yaffe, D.; Nudel, U.; Cisneros, B. Dystrophin Dp71 Expression Is Down-regulated during Myogenesis. *J. Biol. Chem.* **2005**, *280*, 5290–5299.
50. Kawaguchi, T.; Niba, E.; Rani, A.; Onishi, Y.; Koizumi, M.; Awano, H.; Matsumoto, M.; Nagai, M.; Yoshida, S.; Sakakibara, S.; et al. Detection of Dystrophin Dp71 in Human Skeletal Muscle Using an Automated Capillary Western Assay System. *Int. J. Mol. Sci.* **2018**, *19*, 1546.
51. Yamamoto, T.; Awano, H.; Zhang, Z.; Sakuma, M.; Kitaaki, S.; Matsumoto, M.; Nagai, M.; Sato, I.; Imanishi, T.; Hayashi, N.; et al. Cardiac Dysfunction in Duchenne Muscular Dystrophy Is Less Frequent in Patients With Mutations in the Dystrophin Dp116 Coding Region Than in Other Regions. *Circ. Genomic Precis. Med.* **2018**, *11*, e001782.
52. Lidov, H.G.W.; Selig, S.; Kunkel, L.M. Dp140: A novel 140 kDa CNS transcript from the dystrophin locus. *Hum. Mol. Genet.* **1995**, *4*, 329–335. [PubMed]
53. Jefferies, J.L.; Eidem, B.W.; Belmont, J.W.; Craigen, W.J.; Ware, S.M.; Fernbach, S.D.; Neish, S.R.; Smith, E.O.B.; Towbin, J.A. Genetic predictors and remodeling of dilated cardiomyopathy in muscular dystrophy. *Circulation* **2005**, *112*, 2799–2804. [PubMed]
54. Tandon, A.; Jefferies, J.L.; Villa, C.R.; Hor, K.N.; Wong, B.L.; Ware, S.M.; Gao, Z.; Towbin, J.A.; Mazur, W.; Fleck, R.J.; et al. Dystrophin Genotype–Cardiac Phenotype Correlations in Duchenne and Becker Muscular Dystrophies Using Cardiac Magnetic Resonance Imaging. *Am. J. Cardiol.* **2015**, *115*, 967–971. [PubMed]
55. Kaspar, R.W.; Allen, H.D.; Ray, W.C.; Alvarez, C.E.; Kissel, J.T.; Pestronk, A.; Weiss, R.B.; Flanigan, K.M.; Mendell, J.R.; Montanaro, F. Analysis of Dystrophin Deletion Mutations Predicts Age of Cardiomyopathy Onset in Becker Muscular Dystrophy. *Circ. Cardiovasc. Genet.* **2009**, *2*, 544–551.
56. Ashwath, M.L.; Jacobs, I.B.; Crowe, C.A.; Ashwath, R.C.; Super, D.M.; Bahler, R.C. Left ventricular dysfunction in duchenne muscular dystrophy and genotype. *Am. J. Cardiol.* **2014**, *114*, 284–289.
57. Nguyen, Q.; Yokota, T. Antisense oligonucleotides for the treatment of cardiomyopathy in Duchenne muscular dystrophy. *Am. J. Transl. Res.* **2019**, *11*, 1202–1218.
58. Anwar, S.; Yokota, T. Golodirsen for Duchenne muscular dystrophy. *Drugs Today* **2020**, *56*, 491–504.
59. Roshmi, R.R.; Yokota, T. Viltolarsen for the treatment of Duchenne muscular dystrophy. *Drugs Today* **2019**, *55*, 627–639.
60. Béroud, C.; Tuffery-Giraud, S.; Matsuo, M.; Hamroun, D.; Humbertclaude, V.; Monnier, N.; Moizard, M.-P.; Voelckel, M.-A.; Calemard, L.M.; Boisseau, P.; et al. Multiexon skipping leading to an artificial DMD protein lacking amino acids from exons 45 through 55 could rescue up to 63% of patients with Duchenne muscular dystrophy. *Hum. Mutat.* **2007**, *28*, 196–202.

61. Waldrop, M.A.; Ben Yaou, R.; Lucas, K.K.; Martin, A.S.; O'Rourke, E.; Ferlini, A.; Muntoni, F.; Leturcq, F.; Tuffery-Giraud, S.; Weiss, R.B.; et al. Clinical Phenotypes of DMD Exon 51 Skip Equivalent Deletions: A Systematic Review. *J. Neuromuscul. Dis.* **2020**, *7*, 217–229.

Publisher's Note: MDPI stays neutral with regard to jurisdictional claims in published maps and institutional affiliations.

Journal of
Personalized
Medicine

Review

Advances in Genetic Characterization and Genotype–Phenotype Correlation of Duchenne and Becker Muscular Dystrophy in the Personalized Medicine Era

Omar Sheikh [1] and Toshifumi Yokota [1,2,*]

1 Department of Medical Genetics, University of Alberta Faculty of Medicine and Dentistry, Edmonton, AB T6G 2H7, Canada; osheikh1@ualberta.ca

2 The Friends of Garrett Cumming Research & Muscular Dystrophy Canada HM Toupin Neurological Science Research Chair, Edmonton, AB T6G 2H7, Canada

* Correspondence: toshifumi.yokota@ualberta.ca

Received: 10 August 2020; Accepted: 1 September 2020; Published: 3 September 2020

Abstract: Currently, Duchenne muscular dystrophy (DMD) and the related condition Becker muscular dystrophy (BMD) can be usually diagnosed using physical examination and genetic testing. While BMD features partially functional dystrophin protein due to in-frame mutations, DMD largely features no dystrophin production because of out-of-frame mutations. However, BMD can feature a range of phenotypes from mild to borderline DMD, indicating a complex genotype–phenotype relationship. Despite two mutational hot spots in dystrophin, mutations can arise across the gene. The use of multiplex ligation amplification (MLPA) can easily assess the copy number of all exons, while next-generation sequencing (NGS) can uncover novel or confirm hard-to-detect mutations. Exon-skipping therapy, which targets specific regions of the dystrophin gene based on a patient's mutation, is an especially prominent example of personalized medicine for DMD. To maximize the benefit of exon-skipping therapies, accurate genetic diagnosis and characterization including genotype–phenotype correlation studies are becoming increasingly important. In this article, we present the recent progress in the collection of mutational data and optimization of exon-skipping therapy for DMD/BMD.

Keywords: Duchenne muscular dystrophy (DMD); exon-skipping therapies; next-generation sequencing (NGS); Sanger sequencing; multiplex ligation probe amplification (MLPA); multiplex polymerase chain reaction (PCR); comparative genomic hybridization array (CGH); viltolarsen; eteplirsen; golodirsen

1. Introduction

Duchenne muscular dystrophy (DMD), a severe neuromuscular disorder, affects the skeletal and cardiac muscle of 1 in 5000 newborn boys [1], with very few treatment options available [2]. Frame-shifting mutations in the dystrophin gene [3] cause DMD by removing production of the 427 kDa protein dystrophin [4]. Without dystrophin, progressive muscle wasting occurs [5]. By contrast, in-frame dystrophin deletion mutations lead to the related condition Becker muscular dystrophy (BMD), which ranges in phenotype from subclinical to borderline DMD [6]. For this reason, the term DBMD is used to indicate the range of conditions that arise from dystrophin mutations. Though most mutations reported fit the dystrophin reading frame rule stated above, there are a significant number of exceptions, highlighting an intricate genotype–phenotype relationship [7]. Given the range of mutations underlying DBMD, precise genetic diagnosis and genotype–phenotype correlation analysis are crucial to design mutation-specific therapeutics like exon skipping [8,9]. Box 1 describes the keywords used in this article. A better understanding of genotype–phenotype relationships in DBMD patients may lead to better design of exon-skipping therapies [9]. In this review, we describe the recent advances in molecular diagnostic approaches for DMD/BMD and discuss how exon-skipping therapy can be optimized.

Box 1. Definitions of keywords used in this article.

Genotype: genes that encode physical characteristics of an organism.
Phenotype: the observed characteristics resulting from the expression of those genes.
Intron: non-coding region of DNA that is removed by splicing prior to translation.
Exon: coding region of gene that appears in the mature RNA transcript.
In-frame mutation: a mutation that does not disrupt the reading frame of a gene during the transcription, likely not interfering with protein production.
Out-of-frame mutation (also known as frameshift mutation): a mutation that disrupts the reading frame, likely destroying protein production.

2. Sequencing and Genetic Diagnosis Methodologies Relevant to DBMD

Sequencing and mutation detection strategies are intertwined in DBMD. The key strategies used to study this disorder are listed below in Table 1.

Table 1. Sequencing and genetic diagnosis methodologies relevant to Duchenne/Becker muscular dystrophy (DBMD).

Methodology	Brief Description
Sanger sequencing	Low throughput, conventional strategy with lower cost than more advanced sequencing [10]. Allows for sequencing the dystrophin gene.
Next-generation sequencing (NGS)	Class of more advanced sequencing strategies with high throughput [11]. Can examine whole single genes, panels of multiples genes, all protein-coding genes, or entire genomes [12]. Single gene sequencing is especially powerful in DMD [11].
Quantitative Southern blot	Originally the only reliable method for detecting duplication and identifying carriers [13]; however, this method requires several hybridization steps [14].
Multiplex polymerase chain reaction (PCR)	A strategy that can detect the vast majority of DBMD gene deletions. An improved multiplex PCR assay can detect deletions and duplications in all 79 exons of the DMD gene [15].
Multiplex ligation-dependent probe amplification (MLPA)	A prominent first-pass tool for assessing the genetics of DBMD [11]. MLPA can screen all 79 dystrophin gene exons for deletions and duplications in DBMD patients and carriers but cannot detect most small mutations [12].
Comparative genome hybridization array (CGH)	This tool probes dystrophin exons and introns and can pinpoint the location of intronic breakpoints. CGH is a compelling alternative to MLPA [16].

Sequencing methodologies provide precise genetic testing that can clarify the mutations seen in patients. Sanger sequencing is performed using nucleotides that lack a 3′-hydroxyl group, preventing the DNA polymerase from continuing the DNA chain at that position [17]. Though low throughput, it can complete partial sequencing cheaply.

Southern blot was originally used to examine DMD mutations before other techniques replaced it. Southern blot analysis using cDNA probes, which were established earlier [18], has been used to detect deletions and duplications of the dystrophin gene [19–21]. Southern blotting, however, is no longer commonly employed for DMD since it is time-consuming and requires several hybridization steps [11].

The use of multiplex polymerase chain reaction (PCR) for mutation detection has played a more prominent role in DMD genetic diagnosis [13]. Multiplex PCR, which allows for rapid detection of mutations using small or suboptimal samples of genomic DNA, is more efficient than Southern blotting [13]. One study indicated that the majority of the deletions detected by use of cDNA probes and Southern blot in the study could have been also characterized by multiplex PCR [20]. In 2006, Stockley et al. established the use of quantitative multiplex PCR to screen all 79 exons for deletions and duplications [15], strengthening the technique's applicability in DMD.

Multiplex ligation-dependent probe amplification (MLPA) acts well as a first-pass assessment of DMD due to its speed and cheap cost [22]. This technique detects exon deletions and duplications. An MLPA probe consists of two probe oligonucleotides that hybridize to adjacent sites of the target sequence, followed by probe ligation. Probes hybridized are amplified by PCR and quantified, providing amplification products of unique size. The MLPA approach then provides the relative copy number of target sequences [22], which can detect most of the deletions and duplications in the DMD gene.

One rising alternative to MLPA is comparative genomic hybridization (CGH) array [16]. Since 2004, this approach has marked a new milestone for genetic diagnosis [23]. CGH is performed by using probes covering dystrophin exons and introns conjugated to a glass slide. Control and patient DNA is fragmented and hybridized to the probes, allowing for the detection of the relative abundance of each exon. However, unlike MLPA, it can pinpoint the location of breakpoints within introns [16]. This method can be applied to screen the genome both at the whole-gene level and the individual exon level for many disease genes including DMD [24]. The CGH platform can detect precise intron breakpoints in high resolution and sensitivity [25] while also being completely scalable [26]. Through the use of CGH, the ability to capture intronic mutations is notably improved [27]. Due to the high resolution of CGH [28], this technique has been used to probe intronic mutations in dystrophin using patient data [29,30]. Therefore, CGH is also a recommended technique used first to look at DBMD genetics.

Next-generation sequencing (NGS), which refers to sequencing strategies featuring a much greater sequencing volume than Sanger sequencing [10], is another prominent strategy relevant to DBMD [11] and can be used alongside other strategies such as MLPA to provide a reliable genetic diagnosis. Overall, targeted NGS can bolster a more precise understanding of ambiguous mutations [12] in contrast to MLPA which cannot identify some dystrophin mutations [11].

NGS features several potential diagnostic uses. For instance, NGS can accurately identify pathogenic small mutations in DBMD patients without a large deletion/duplication, especially in non-coding regions [31]. Therefore, this technique has great potential to improve the molecular diagnosis of DBMD. Lastly, whole-exome sequencing, which solely concentrates on the coding exon regions of the genome, is useful for the quick examination of exonic mutations [32]. Though this technique is not widely used, it has been used to identify small mutations giving rise to DBMD [33–35]. The broad range of NGS methodologies available supports precise genetic diagnoses [12].

3. Exon-Skipping Therapies for DMD

Exon-skipping therapy is based on the observation that not all of the 79 dystrophin exons are essential for functional protein [36]. Patients with in-frame deletions typically feature a milder BMD phenotype, despite not having all exons, which forms the basis for the approach of exon skipping [37]. Synthetic antisense oligonucleotides (AONs), which are engineered to resist nuclease degradation, are typically used to target mRNA of the dystrophin gene for removal, thereby restoring the reading frame and promoting the production of partially functional protein [36]. This truncated protein then compensates for the function of the full-length protein. Currently, many exon-skipping therapies are in clinical testing [38]. Thus far, exon skipping has shown effectiveness in delaying DMD progression [36]. Eteplirsen, which is designed to skip exon 51 [39], and golodirsen, which is designed to skip exon 53 [40], gained conditional approval in the US in 2016 and 2019, respectively. A newly approved AON, viltolarsen, has been especially promising. Based on compelling evidence of efficacy, viltolarsen received approval in Japan for the skipping of exon 53 [41] and was conditionally approved by the U.S. Food and Drug Administration (FDA) in August 2020 [42]. A Phase II trial of viltolarsen demonstrated that the low dose group (40 mg/kg) rose from an average dystrophin production baseline of 0.3% to 5.7% of normal while the high dose group rose from an average dystrophin production baseline of 0.6% to 5.9% of normal in Western blots [43]. In parallel, the trial strengthened the evidence that viltolarsen can stabilize or improve muscle strength and functionality based on timed tests. Of the three approved therapies, which are compared in Table 2, viltolarsen has produced the highest observed increases in dystrophin production.

Table 2. Comparison of FDA-approved exon-skipping therapies for Duchenne muscular dystrophy (DMD). Mean dystrophin protein production (as a percentage), relative to healthy controls, is presented based on Western blot data. Baseline values are included for reference.

Therapy	Baseline (% of Normal)	Dystrophin Production (% of Normal)	Side Effects
Eteplirsen [44]	0.08	0.93	No severe or moderate adverse events 8 mild events considered related to treatment [45]
Golodirsen [46]	0.095	1.019	2 moderate adverse events (infection and pyrexia) 8 mild events considered related to treatment
Viltolarsen [43]	0.3 (dose of 40 mg/kg) 0.6 (dose of 80 mg/kg)	5.7 (dose of 40 mg/kg) 5.9 (dose of 80 mg/kg)	No severe or moderate adverse events No mild events considered related to treatment

For this therapy to effectively treat patients, it must produce a stable dystrophin protein. In one study, researchers examined the stability of edited in-frame dystrophins lacking exons 45–53, exons 46–54, and exon 47–55, respectively; the edited protein lacking exons 46–54 featured the greatest stability [47]. Though this study provides biochemical and computational prediction of exon-skipping therapies, it does not demonstrate these results in vivo [9]. Nevertheless, exon-skipping schemes can cause a myriad of consequences at the protein structure level, which could influence therapeutic effectiveness.

In DMD, exon skipping is still challenged by its mutation-specific nature. Such therapies could be spread too thinly across many different mutations even though it can potentially treat many patients in total. For example, though 47% and 90% of nonsense mutations could be treated using single and double exon-skipping, respectively, this therapy development could necessitate targeting 68 of dystrophin's 79 exons [48]. Although technically more challenging, double exon skipping substantially raising the applicability of exon-skipping therapies compared to single exon skipping highlights the power of skipping more than one exon. In a dystrophic dog model, double exon skipping of DMD exons six and eight induced by cocktail AONs resulted in the systemic correction of the reading frame and truncated dystrophin expression in skeletal muscles accompanied by improved running speed [49]. The potential of multi-exon skipping is supported by the milder BMD phenotypes observed with the absence of exons 45–55 [50]. In particular, these patients largely featured no mortality and delayed loss of ambulation [51]. Multi-exon skipping of exons 45–55 is expected to benefit 47% of DMD patients [51]. In a DMD mouse model with a deletion mutation in exon 52, exons 45–55 skipping was induced by cocktail AONs, leading to systemic dystrophin expression and functional rescue [52]. Overall, successful development of multi-exon skipping will significantly expand the applicability and optimize the function and stability of truncated dystrophin.

4. Patient Registries and the Personalization of Exon Skipping

To better understand which patients are amenable to mutation-specific therapies, including exon-skipping, patient data must be collected broadly through studies and registries. In a foundational study, Baumbach et al. observed that 56% of DMD patients have detectable deletions, 29% of which mapped to a region proximal to the 5′ end of the gene whereas 69% mapped to a region located centrally [53]. The Leiden patient registry reflects one major collection of data on the genetics of DBMD [7]. A large-scale study on the UMD-DMD registry from 2008 was performed on 2405 French patients with DBMD [54]. DMD patients featured 61% large deletions and 13% duplications whereas BMD patients featured 81% large deletions and 6% duplications. Comparatively, this indicates a similar deletion rate to Baumbach et al. Furthermore, this database study indicated that 24% of mutations are de novo events, reinforcing the relatively frequent occurrence of mutations in the dystrophin gene. Finally, this large-scale approach to genotype–phenotype in analysis coincides with the development of other international DMD patients' registries [54].

TREAT-NMD, an EU-funded multinational network, aims to establish comprehensive information on the natural history of DMD by acquiring data from a large number of patients from a variety of countries not limited to Europe [55]. Currently, the TREAT-NMD database contains a lot of mutational data [56], though as of 2015 15% and 57% of mutations submitted to the registry were from the Americas and Europe, respectively. In parallel, researchers across many countries are collecting mutational data on DBMD patients across the world. These efforts supplement consolidation of patient data into a global registry like TREAT-NMD [57–67].

In 2015, TREAT-NMD's global database was used to assess more than 7000 dystrophin mutations [56]. Among large mutations, which comprise 80% of total mutations, 86% are deletions and 14% are duplications. This study, beyond providing an overview of mutations observed in a global group of DMD patients, also concludes that the skipping of exons 51 (14% of patients), 45 (9% of patients), 53 (8.1% of patients), and 44 (7.6% of patients) could apply to significant minorities of the registry's patients.

Inspired by the TREAT-NMD global registry, Japan established its own registry called Remudy. In a 2013 study examining 688 DBMD patients, the deletion of exons was most frequent followed by point mutations and duplications [68,69]. The most recent published analysis of Remudy concluded, based on a set of 1197 Japanese DMD patients, that 107 patients could benefit from exon 51 skipping while 111 could benefit from exon 53 skipping [70].

5. Genotype–Phenotype Correlation Studies to Predict the Likely Outcomes of Exon-Skipping Therapies

Through documenting the genotype–phenotype relationship, researchers may better design mutation-specific therapies such as exon-skipping. A greater understanding of genotype–phenotype relationships has been supported by data from clinical studies. Although the reading-frame rule holds in approximately 90% of DBMD cases [7], there are important exceptions. A 2007 review pooled DBMD patient data, based on MLPA, Southern blotting, or PCR analysis, concluded that in-frame deletion patterns result in a mixture of DMD and BMD phenotypes [71]. The deletion of exons 45–47, for instance, featured a 15% occurrence of DMD whereas the deletion of exons 45–51 featured a 48% occurrence of DMD (13 out of 27 patients).

Assessing the genotype–phenotype relationship in a subset of DMD patients might more directly indicate the merits of potential exon-skipping therapies. The 5′ region of the gene, which includes exons 3–9, may be associated with complex genotype–phenotype correlations [72]. In one case study, a patient with an in-frame deletion of exon five featured a more severe than expected BMD phenotype despite the continued recognition of exon six [73]. By contrast, an in-frame deletion of exons 3–9, according to one study, mostly leads to a BMD phenotype [74,75]. The two closely examined patients featured especially mild BMD with only mild heart impairment. In addition, Nakamura et al. reported a patient with this deletion showing only a slight decrease in cardiac function but without muscle involvement at the age of 27 years. By examining this deletion in vivo, the researchers concluded that the removal of exons 3–9 via multi-exon skipping likely generates a mild BMD phenotype. Based on these observations, removal of exons 3–9 is a promising treatment for DMD patients with mutations in this region.

Findlay et al. examined 41 patients enrolled in the United Dystrophinopathy Project focusing on in-frame deletions around exon 45 [8]. All patients with Δ45–46 deletions (n = 4) carried a diagnosis of DMD whereas most patients with Δ45–47 deletions (n = 17) and Δ45–48 deletions (n = 19) were diagnosed with BMD. Based on these findings, the skipping of exon 46 for patients missing exon 45 may not rescue the DMD phenotype. Instead, the study illustrates how the skipping of exons 46–47 or 46–48 for these patients has a greater likelihood of producing a BMD phenotype. As a result of this cohort study, a clinical case can be made for multi-exon skipping, which remains in preclinical testing [76]. From this example, we can see how genotype–phenotype correlations can support the design of exon-skipping therapies, improving their personalization.

A systematic review of dystrophinopathy data from the published literature and unpublished databases examined 135 DBMD patients with in-frame deletions equivalent to the skipping of exon 51 [77]. Of these patients, the majority (n = 81) had BMD whereas 16 patients had more severe phenotypes and 6 had no definitive phenotype. The authors conclude that exon 51 skipping therapy, overall, is likely to produce milder BMD phenotypes in many patients.

To understand the genotype–phenotype relationships of in-frame deletions within the exons 45–55 mutational hot spot, 43 patients with DBMD patients were examined using MLPA, Southern blotting, and multiplex PCR [51]. The deletions examined are as follows: Δ45–55 (n = 7), Δ45–51 (n = 6), Δ45–48 (n = 5), Δ45–57 (n = 3). Researchers subdivided these groups into two groups based on truncated dystrophin conformation: hybrid type (Δ45–55, Δ45–58, Δ45–51) and fractional type (Δ45–57 and Δ45–49). Hybrid type conformation (n = 18) at large features a lower proportion of wheelchair-bound patients than the fractional type conformation (n = 6). Log-rank tests revealed a statistically significant difference between the hybrid and fractional groups ($p < 0.05$) of the age at which patients became wheelchair-bound. In other words, the fractional type appears to more consistently lead to an earlier loss of ambulation. This study provides another manner of predicting the viability of dystrophin protein produced by exon-skipping.

Larger studies of in-frame deletions can more strongly guide exon-skipping development [71]. Looking at in-frame deletions within the hotspot region, researchers determined that some mutations were unexpectedly severe, leading to a DMD phenotype rather than the expected BMD phenotype. For example, in-frame deletions starting from exon 49 and exon 50 featured 92% DMD and 90% DMD proportions, respectively, reinforcing the fact that not all potential exon-skipping strategies will resolve a severe phenotype.

Genotype–phenotype correlations of in-frame deletions also support multi-exon-skipping therapies, especially removing exons 45–55. In three patients each featuring in-frame deletion of the region, two developed heart failure while featuring no overt skeletal pathology whereas a third patient featured muscle atrophy and weakness [78]. The condition of all remained stable with treatment. A separate study examined nine patients with the same mutation and indicated that all nine patients had quadriceps and calf hypertrophy and no respiratory involvement. Meanwhile, two patients featured dilated cardiomyopathy [79]. These results suggest, like with the previous study, that the deletion of exons 45–55 is associated with a milder condition compared to smaller in-frame deletions in this region. A multi-exon-skipping strategy can recapture this phenotype by removing several exons, rather than simply skipping every exon in the region individually, and potentially treat over 65% of DMD patients featuring deletions [80].

6. Conclusions

Through comprehensive registries of patient data such as TREAT-NMD with the support of newly available genetic diagnosis tools, DBMD patients can be classified based on mutations, which will further help optimize therapy design while offering higher power for clinical trials [55]. Concurrently, the emergence of multi-exon skipping raises the overall applicability of this treatment strategy, although it is technically more challenging. For exon-skipping therapies to be as effective as possible, cohort studies of genotype–phenotype relationships in DBMD patients with the same resulting mutation would support their design [9]. Because BMD can feature a plethora of truncated dystrophins, exon skipping resulting in truncated dystrophins linked to a milder BMD phenotype might be more beneficial. However, caution should be taken in interpreting these data as other factors, such as the variability of exon skipping efficacy among different exons, also need to be taken into account. Nevertheless, larger cohort studies utilizing patient registry data on genotype–phenotype correlation would greatly contribute to the rational design of mutation-specific therapies including exon skipping in the personalized medicine era.

Author Contributions: Literature review and writing—original draft preparation, O.S.; writing—review and editing, O.S., and T.Y.; supervision and funding acquisition, T.Y. All authors have read and agreed to the published version of the manuscript.

Funding: This research was funded by the Friends of Garrett Cumming Research & Muscular Dystrophy Canada HM Toupin Neurological Science Research Chair, Canadian Institutes of Health Research (CIHR) FDN 143251, 169193, Fulbright Canada, and the Women and Children's Health Research Institute (WCHRI) IG 2874.

Conflicts of Interest: The authors declare no conflict of interest.

References

1. Moat, S.J.; Bradley, D.M.; Salmon, R.; Clarke, A.; Hartley, L. Newborn bloodspot screening for Duchenne Muscular Dystrophy: 21 years experience in Wales (UK). *Eur. J. Hum. Genet.* **2013**, *21*, 1049–1053. [CrossRef] [PubMed]
2. Duchenne, G.B. The Pathology of paralysis with muscular degeneration (paralysie myosclerotique), or paralysis with apparent hypertrophy. *Br. Med. J.* **1867**, *2*, 541–542. [CrossRef] [PubMed]
3. Worton, R.G. Duchenne muscular dystrophy: Gene and gene product; mechanism of mutation in the gene. *J. Inherit. Metab. Dis.* **1992**, *15*, 539–550. [CrossRef] [PubMed]
4. Dubowitz, V. The Duchenne Dystrophy Story: From Phenotype to Gene and Potential Treatment. *J. Child Neurol.* **1989**, *4*, 240–250. [CrossRef] [PubMed]
5. Hoffman, E.P.; Brown, R.H.; Kunkel, L.M. Dystrophin: The protein product of the Duchene muscular dystrophy locus. 1987. *Biotechnology* **1987**, *51*, 919–928.
6. Koenig, M.; Beggs, A.H.; Moyer, M.; Scherpf, S.; Heindrich, K.; Bettecken, T.; Meng, G.; Müller, C.R.; Lindlöf, M.; Kaariainen, H.; et al. The molecular basis for duchenne versus becker muscular dystrophy: Correlation of severity with type of deletion. *Am. J. Hum. Genet.* **1989**, *45*, 498–506.
7. Aartsma-Rus, A.; Van Deutekom, J.C.T.; Fokkema, I.F.; Van Ommen, G.J.B.; Den Dunnen, J.T. Entries in the Leiden Duchenne muscular dystrophy mutation database: An overview of mutation types and paradoxical cases that confirm the reading-frame rule. *Muscle Nerve* **2006**, *34*, 135–144. [CrossRef]
8. Findlay, A.R.; Wein, N.; Kaminoh, Y.; Taylor, L.E.; Dunn, D.M.; Mendell, J.R.; King, W.M.; Pestronk, A.; Florence, J.M.; Matthews, K.D.; et al. Clinical phenotypes as predictors of the outcome of skipping around DMD exon 45. *Ann. Neurol.* **2015**, *4*, 668–674. [CrossRef]
9. Nakamura, A. Moving towards successful exon-skipping therapy for Duchenne muscular dystrophy. *J. Hum. Genet.* **2017**, *62*, 871–876. [CrossRef]
10. Slatko, B.E.; Gardner, A.F.; Ausubel, F.M. Overview of Next Generation Sequencing technologies (and bioinformatics) in cancer. *Mol. Biol.* **2018**, *122*, 1–15.
11. Volk, A.E.; Kubisch, C. The rapid evolution of molecular genetic diagnostics in neuromuscular diseases. *Curr. Opin. Neurol.* **2017**, *30*, 523–528. [CrossRef] [PubMed]
12. Zhang, K.; Yang, X.; Lin, G.; Han, Y.; Li, J. Molecular genetic testing and diagnosis strategies for dystrophinopathies in the era of next generation sequencing. *Clin. Chim. Acta* **2019**, *491*, 66–73. [CrossRef] [PubMed]
13. Beggs, A.H.; Koenig, M.; Boyce, F.M.; Kunkel, L.M. Detection of 98% of DMD/BMD gene deletions by polymerase chain reaction. *Hum. Genet.* **1990**, *86*, 45–48. [CrossRef] [PubMed]
14. Schwartz, M.; Dunø, M. Improved molecular diagnosis of dystrophin gene mutations using the multiplex ligation-dependent probe amplification method. *Genet. Test.* **2004**, *8*, 361–367. [CrossRef]
15. Stockley, T.L.; Akber, S.; Bulgin, N.; Ray, P.N. Strategy for Comprehensive Molecular Testing for Duchenne and Becker Muscular Dystrophies. *Genet. Test.* **2006**, *10*, 229–243. [CrossRef]

16. Aartsma-Rus, A.; Ginjaar, I.B.; Bushby, K. The importance of genetic diagnosis for Duchenne muscular dystrophy. *J. Med. Genet.* **2016**, *53*, 145–151. [CrossRef]
17. Sanger, F.; Nicklen, S.; Coulson, A.R. DNA sequencing with chain-terminating inhibitors. *Proc. Natl. Acad. Sci. USA* **1977**, *74*, 5463–5467. [CrossRef]
18. Koenig, M.; Hoffman, E.P.; Bertelson, C.J.; Monaco, A.P.; Feener, C.; Kunkel, L.M. Complete cloning of the duchenne muscular dystrophy (DMD) cDNA and preliminary genomic organization of the DMD gene in normal and affected individuals. *Cell* **1987**, *50*, 509–517. [CrossRef]
19. Hu, X.; Burghes, A.H.M.; Ray, P.N.; Thompson, M.W.; Murphy, E.G.; Worton, R.G. Partial gene duplication in Duchenne and Becker muscular dystrophies. *J. Med. Genet.* **1988**, *25*, 369–376. [CrossRef]
20. Gillard, E.F.; Chamberlain, J.S.; Murphy, E.G.; Duff, C.L.; Smith, B.; Burghes, A.H.M.; Thompson, M.W.; Sutherland, J.; Oss, I.; Bodrug, S.E.; et al. Molecular and phenotypic analysis of patients with deletions within the deletion-rich region of the Duchenne muscular dystrophy (DMD) gene. *Am. J. Hum. Genet.* **1989**, *45*, 507–520.
21. Hiraishi, Y.; Kato, S.; Ishihara, T.; Takano, T. Quantitative Southern blot analysis in the dystrophin gene of Japanese patients with Duchenne or Becker muscular dystrophy: A high frequency of duplications. *J. Med. Genet.* **1992**, *29*, 897–901. [CrossRef] [PubMed]
22. Schouten, J.P.; McElgunn, C.J.; Waaijer, R.; Zwijnenburg, D.; Diepvens, F.; Pals, G. Relative quantification of 40 nucleic acid sequences by multiplex ligation-dependent probe amplification. *Nucleic Acids Res.* **2002**, *30*, e57. [CrossRef] [PubMed]
23. Cheung, S.W.; Bi, W. Novel applications of array comparative genomic hybridization in molecular diagnostics. *Expert Rev. Mol. Diagn.* **2018**, *18*, 531–542. [CrossRef]
24. Dhami, P.; Coffey, A.J.; Abbs, S.; Vermeesch, J.R.; Dumanski, J.P.; Woodward, K.J.; Andrews, R.M.; Langford, C.; Vetrie, D. Exon array CGH: Detection of copy-number changes at the resolution of individual exons in the human genome. *Am. J. Hum. Genet.* **2005**, *76*, 750–762. [CrossRef] [PubMed]
25. Del Gaudio, D.; Yang, Y.; Boggs, B.A.; Schmitt, E.S.; Lee, J.A.; Sahoo, T.; Pham, H.T.; Wiszniewska, J.; Chinault, A.C.; Beaudet, A.L.; et al. Molecular diagnosis of Duchenne/Becker muscular dystrophy: Enhanced detection of dystrophin gene rearrangements by oligonucleotide array-comparative genomic hybridization. *Hum. Mutat.* **2008**, *29*, 1100–1107. [CrossRef] [PubMed]
26. Saillour, Y.; Cossée, M.; Leturcq, F.; Vasson, A.; Beugnet, C.; Poirier, K.; Commere, V.; Sublemontier, S.; Viel, M.; Letourneur, F.; et al. Detection of exonic copy-number changes using a highly efficient oligonucleotide-based comparative genomic hybridization-array method. *Hum. Mutat.* **2008**, *29*, 1083–1090. [CrossRef]
27. Bovolenta, M.; Neri, M.; Fini, S.; Fabris, M.; Trabanelli, C.; Venturoli, A.; Martoni, E.; Bassi, E.; Spitali, P.; Brioschi, S.; et al. A novel custom high density-comparative genomic hybridization array detects common rearrangements as well as deep intronic mutations in dystrophinopathies. *BMC Genom.* **2008**, *9*, 1–12. [CrossRef]
28. Baskin, B.; Stavropoulos, D.J.; Rebeiro, P.A.; Orr, J.; Li, M.; Steele, L.; Marshall, C.R.; Lemire, E.G.; Boycott, K.M.; Gibson, W.; et al. Complex genomic rearrangements in the dystrophin gene due to replication-based mechanisms. *Mol. Genet. Genom. Med.* **2014**, *2*, 539–547. [CrossRef]
29. Ishmukhametova, A.; Van Kien, P.K.; Méchin, D.; Thorel, D.; Vincent, M.C.; Rivier, F.; Coubes, C.; Humbertclaude, V.; Claustres, M.; Tuffery-Giraud, S. Comprehensive oligonucleotide array-comparative genomic hybridization analysis: New insights into the molecular pathology of the DMD gene. *Eur. J. Hum. Genet.* **2012**, *20*, 1096–1100. [CrossRef]
30. Oshima, J.; Magner, D.B.; Lee, J.A.; Breman, A.M.; Schmitt, E.S.; White, L.D.; Crowe, C.A.; Merrill, M.; Jayakar, P.; Rajadhyaksha, A.; et al. Regional genomic instability predisposes to complex dystrophin gene rearrangements. *Hum. Genet.* **2009**, *126*, 411–423. [CrossRef]
31. Lim, B.C.; Lee, S.; Shin, J.Y.; Kim, J.I.; Hwang, H.; Kim, K.J.; Hwang, Y.S.; Seo, J.S.; Chae, J.H. Genetic diagnosis of duchenne and becker muscular dystrophy using next-generation sequencing technology: Comprehensive mutational search in a single platform. *J. Med. Genet.* **2011**, *48*, 731–736. [CrossRef] [PubMed]
32. Kuperberg, M.; Lev, D.; Blumkin, L.; Zerem, A.; Ginsberg, M.; Linder, I.; Carmi, N.; Kivity, S.; Lerman-Sagie, T.; Leshinsky-Silver, E. Utility of Whole Exome Sequencing for Genetic Diagnosis of Previously Undiagnosed Pediatric Neurology Patients. *J. Child Neurol.* **2016**, *31*, 1534–1539. [CrossRef] [PubMed]

33. Luce, L.N.; Carcione, M.; Mazzanti, C.; Ferrer, M.; Szijan, I.; Giliberto, F. Small mutation screening in the DMD gene by whole exome sequencing of an argentine Duchenne/Becker muscular dystrophies cohort. *Neuromuscul. Disord.* **2018**, *28*, 986–995. [CrossRef] [PubMed]
34. Zhang, Y.; Yang, W.; Wen, G.; Wu, Y.; Jing, Z.; Li, D.; Tang, M.; Liu, G.; Wei, X.; Zhong, Y.; et al. Application whole exome sequencing for the clinical molecular diagnosis of patients with Duchenne muscular dystrophy; identification of four novel nonsense mutations in four unrelated Chinese DMD patients. *Mol. Genet. Genom. Med.* **2019**, *7*, 1–8. [CrossRef] [PubMed]
35. Reddy, H.M.; Cho, K.A.; Lek, M.; Estrella, E.; Valkanas, E.; Jones, M.D.; Mitsuhashi, S.; Darras, B.T.; Amato, A.A.; Lidov, H.G.; et al. The sensitivity of exome sequencing in identifying pathogenic mutations for LGMD in the United States. *J. Hum. Genet.* **2017**, *62*, 243–252. [CrossRef] [PubMed]
36. Li, D.; Mastaglia, F.L.; Fletcher, S.; Wilton, S.D. Precision Medicine through Antisense Oligonucleotide-Mediated Exon Skipping. *Trends Pharmacol. Sci.* **2018**, *39*, 982–994. [CrossRef]
37. Wang, R.T.; Barthelemy, F.; Martin, A.S.; Douine, E.D.; Eskin, A.; Lucas, A.; Lavigne, J.; Peay, H.; Khanlou, N.; Sweeney, L.; et al. DMD genotype correlations from the Duchenne Registry: Endogenous exon skipping is a factor in prolonged ambulation for individuals with a defined mutation subtype. *Hum. Mutat.* **2018**, *39*, 1193–1202. [CrossRef]
38. Hoffman, E.P. *Pharmacotherapy of Duchenne Muscular Dystrophy*; Springer-Natur Switzerland AG: Cham, Switzerland, 2019.
39. Aartsma-Rus, A.; Goemans, N. A Sequel to the Eteplirsen Saga: Eteplirsen Is Approved in the United States but Was Not Approved in Europe. *Nucleic Acid Ther.* **2019**, *29*, 13–15. [CrossRef]
40. Sarepta Therapeutics Inc. *Sarepta Therapeutics Announces FDA Approval of VYONDYS 53 (golodirsen) Injection for the Treatment of Duchenne Muscular Dystrophy (DMD) in Patients Amenable to Skipping Exon 53*; Sarepta Therapeutics, Inc.: Cambridge, MA, USA, 2019.
41. Dhillon, S. Viltolarsen: First Approval. *Drugs* **2020**, *80*, 1027–1031. [CrossRef]
42. FDA. FDA Approves Targeted Treatment for Rare Duchenne Muscular Dystrophy Mutation. 2020. Available online: https://www.fda.gov/news-events/press-announcements/fda-approves-targeted-treatment-rare-duchenne-muscular-dystrophy-mutation (accessed on 29 August 2020).
43. Clemens, P.R.; Rao, V.K.; Connolly, A.M.; Harper, A.D.; Mah, J.K.; Smith, E.C.; McDonald, C.M.; Zaidman, C.M.; Morgenroth, L.P.; Osaki, H.; et al. Safety, Tolerability, and Efficacy of Viltolarsen in Boys with Duchenne Muscular Dystrophy Amenable to Exon 53 Skipping: A Phase 2 Randomized Clinical Trial. *JAMA Neurol.* **2020**, *15261*, 1–10. [CrossRef]
44. Charleston, J.S.; Schnell, F.J.; Dworzak, J.; Donoghue, C.; Lewis, S.; Chen, L.; David Young, G.; Milici, A.J.; Voss, J.; Dealwis, U.; et al. Eteplirsen treatment for Duchenne muscular dystrophy. *Neurology* **2018**, *90*, e2135–e2145. [CrossRef] [PubMed]
45. Mendell, J.R.; Goemans, N.; Lowes, L.P.; Alfano, L.N.; Berry, K.; Shao, J.; Kaye, E.M.; Mercuri, E. Longitudinal effect of eteplirsen versus historical control on ambulation in Duchenne muscular dystrophy. *Ann. Neurol.* **2016**, *79*, 257–271. [CrossRef] [PubMed]
46. Frank, D.E.; Schnell, F.J.; Akana, C.; El-Husayni, S.H.; Desjardins, C.A.; Morgan, J.; Charleston, J.S.; Sardone, V.; Domingos, J.; Dickson, G.; et al. Increased dystrophin production with golodirsen in patients with Duchenne muscular dystrophy. *Neurology* **2020**, *94*, e2270–e2282. [CrossRef] [PubMed]
47. Ma, K.M.; Thomas, E.S.; Wereszczynski, J.; Menhart, N. Empirical and Computational Comparison of Alternative Therapeutic Exon Skip Repairs for Duchenne Muscular Dystrophy. *Biochemistry* **2019**, *58*, 2061–2076. [CrossRef]
48. Yokota, T.; Duddy, W.; Echigoya, Y.; Kolski, H. Exon skipping for nonsense mutations in Duchenne muscular dystrophy: Too many mutations, too few patients? *Expert Opin. Biol. Ther.* **2012**, *12*, 1141–1152. [CrossRef]
49. Yokota, T.; Lu, Q.L.; Partridge, T.; Kobayashi, M.; Nakamura, A.; Takeda, S.; Hoffman, E. Efficacy of systemic morpholino exon-skipping in duchenne dystrophy dogs. *Ann. Neurol.* **2009**, *65*, 667–676. [CrossRef]
50. Echigoya, Y.; Lim, K.R.Q.; Nakamura, A.; Yokota, T. Multiple exon skipping in the duchenne muscular dystrophy hot spots: Prospects and challenges. *J. Pers. Med.* **2018**, *8*, 41. [CrossRef]
51. Nakamura, A.; Shiba, N.; Miyazaki, D.; Nishizawa, H.; Inaba, Y.; Fueki, N.; Maruyama, R.; Echigoya, Y.; Yokota, T. Comparison of the phenotypes of patients harboring in-frame deletions starting at exon 45 in the Duchenne muscular dystrophy gene indicates potential for the development of exon skipping therapy. *J. Hum. Genet.* **2017**, *62*, 459–463. [CrossRef]

52. Aoki, Y.; Yokota, T.; Nagata, T.; Nakamura, A.; Tanihata, J.; Saito, T.; Duguez, S.M.R.; Nagaraju, K.; Hoffman, E.P.; Partridge, T.; et al. Bodywide skipping of exons 45–55 in dystrophic mdx52 mice by systemic antisense delivery. *Proc. Natl. Acad. Sci. USA* **2012**, *109*, 13763–13768. [CrossRef]
53. Baumbach, L.; Chamberlain, J.; Ward, P.A.; Farwell, N.; Caskey, C. Molecular and clinical correlations of deletions leading to Duchenne and Becker muscular dystrophy. *Neurology* **1989**, *39*, 465–474. [CrossRef]
54. Tuffery-Giraud, S.; Béroud, C.; Leturcq, F.; Yaou, R.B.; Hamroun, D.; Michel-Calemard, L.; Moizard, M.P.; Bernard, R.; Cossée, M.; Boisseau, P.; et al. Genotype-phenotype analysis in 2405 patients with a dystrophinopathy using the UMD-DMD database: A model of nationwide knowledgebase. *Hum. Mutat.* **2009**, *30*, 934–945. [CrossRef] [PubMed]
55. Koeks, Z.; Bladen, C.L.; Salgado, D.; Van Zwet, E.; Pogoryelova, O.; McMacken, G.; Monges, S.; Foncuberta, M.E.; Kekou, K.; Kosma, K.; et al. Clinical Outcomes in Duchenne Muscular Dystrophy: A Study of 5345 Patients from the TREAT-NMD DMD Global Database. *J. Neuromuscul. Dis.* **2017**, *4*, 293–306. [CrossRef] [PubMed]
56. Bladen, C.L.; Salgado, D.; Monges, S.; Foncuberta, M.E.; Kekou, K.; Kosma, K.; Dawkins, H.; Lamont, L.; Roy, A.J.; Chamova, T.; et al. The TREAT-NMD DMD global database: Analysis of more than 7000 duchenne muscular dystrophy mutations. *Hum. Mutat.* **2015**, *36*, 395–402. [CrossRef] [PubMed]
57. Kong, X.; Zhong, X.; Liu, L.; Cui, S.; Yang, Y.; Kong, L. Genetic analysis of 1051 Chinese families with Duchenne/Becker Muscular Dystrophy. *BMC Med. Genet.* **2019**, *20*, 1–7. [CrossRef] [PubMed]
58. Neri, M.; Rossi, R.; Trabanelli, C.; Mauro, A.; Selvatici, R.; Falzarano, M.S.; Spedicato, N.; Margutti, A.; Rimessi, P.; Fortunato, F.; et al. The Genetic Landscape of Dystrophin Mutations in Italy: A Nationwide Study. *Front. Genet.* **2020**, *11*, 1–15. [CrossRef]
59. Tomar, S.; Moorthy, V.; Sethi, R.; Chai, J.; Low, P.S.; Hong, S.T.K.; Lai, P.S. Mutational spectrum of dystrophinopathies in Singapore: Insights for genetic diagnosis and precision therapy. *Am. J. Med. Genet. Part C Semin. Med. Genet.* **2019**, *181*, 230–244. [CrossRef]
60. Kohli, S.; Saxena, R.; Thomas, E.; Singh, K.; Mahay, S.B.; Puri, R.D. Mutation Spectrum of Dystrophinopathies in India: Implications for Therapy. *Indian J. Pedatrics* **2020**, *87*, 495–504. [CrossRef]
61. Ansar, Z.; Nasir, A.; Moatter, T.; Khan, S.; Kirmani, S.; Ibrahim, S.; Imam, K.; Ather, A.; Samreen, A.; Hasan, Z. MLPA Analyses Reveal a Spectrum of Dystrophin Gene Deletions/Duplications in Pakistani Patients Suspected of Having Duchenne/Becker Muscular Dystrophy: A Retrospective Study. *Genet. Test. Mol. Biomark.* **2019**, *23*, 468–472. [CrossRef]
62. Tran, V.K.; Ta, V.T.; Vu, D.C.; Nguyen, S.T.B.; Do, H.N.; Ta, M.H.; Tran, T.H.; Matsuo, M. Exon deletion patterns of the dystrophin gene in 82 Vietnamese Duchenne/Becker muscular dystrophy patients. *J. Neurogenet.* **2013**, *27*, 170–175. [CrossRef]
63. Cho, A.; Seong, M.W.; Lim, B.C.; Lee, H.J.; Byeon, J.H.; Kim, S.S.; Kim, S.Y.; Choi, S.A.; Wong, A.L.; Lee, J.; et al. Consecutive analysis of mutation spectrum in the dystrophin gene of 507 Korean boys with Duchenne/Becker muscular dystrophy in a single center. *Muscle Nerve* **2017**, *55*, 727–734. [CrossRef]
64. Vieitez, I.; Gallano, P.; González-Quereda, L.; Borrego, S.; Marcos, I.; Millán, J.M.; Jairo, T.; Prior, C.; Molano, J.; Trujillo-Tiebas, M.J.; et al. Mutational spectrum of Duchenne muscular dystrophy in Spain: Study of 284 cases. *Neurología* **2017**, *32*, 377–385. [CrossRef]
65. Todorova, A.; Todorov, T.; Georgieva, B.; Lukova, M.; Guergueltcheva, V.; Kremensky, I.; Mitev, V. MLPA analysis/complete sequencing of the DMD gene in a group of Bulgarian Duchenne/Becker muscular dystrophy patients. *Neuromuscul. Disord.* **2008**, *18*, 667–670. [CrossRef] [PubMed]
66. Ebrahimzadeh-Vesal, R.; Teymoori, A.; Aziminezhad, M.; Hosseini, F.S. Next Generation Sequencing approach to molecular diagnosis of Duchenne muscular dystrophy; identification of a novel mutation. *Gene* **2018**, *644*, 1–3. [CrossRef] [PubMed]
67. Iskandar, K.; Dwianingsih, E.K.; Pratiwi, L.; Kalim, A.S.; Mardhiah, H.; Putranti, A.H.; Nurputra, D.K.; Triono, A.; Herini, E.S.; Malueka, R.G.; et al. The analysis of DMD gene deletions by multiplex PCR in Indonesian DMD/BMD patients: The era of personalized medicine. *BMC Res. Notes* **2019**, *12*, 1–7. [CrossRef]
68. Nakamura, H.; Kimura, E.; Mori-Yoshimura, M.; Komaki, H.; Matsuda, Y.; Goto, K.; Hayashi, Y.K.; Nishino, I.; Takeda, S.; Kawai, M. Characteristics of Japanese Duchenne and Becker muscular dystrophy patients in a novel Japanese national registry of muscular dystrophy (Remudy). *Orphanet J. Rare Dis.* **2013**, *8*, 1–7. [CrossRef] [PubMed]

69. Mori-Yoshimura, M.; Mitsuhashi, S.; Nakamura, H.; Komaki, H.; Goto, K.; Yonemoto, N.; Takeuchi, F.; Hayashi, Y.K.; Murata, M.; Takahashi, Y.; et al. Characteristics of Japanese patients with becker muscular dystrophy and intermediate muscular dystrophy in a Japanese national registry of muscular dystrophy (Remudy): Heterogeneity and clinical variation. *J. Neuromuscul. Dis.* **2018**, *5*, 193–203. [CrossRef]
70. Okubo, M.; Goto, K.; Komaki, H.; Nakamura, H.; Mori-Yoshimura, M.; Hayashi, Y.K.; Mitsuhashi, S.; Noguchi, S.; Kimura, E.; Nishino, I. Comprehensive analysis for genetic diagnosis of Dystrophinopathies in Japan. *Orphanet J. Rare Dis.* **2017**, *12*, 1–8. [CrossRef]
71. Yokota, T.; Duddy, W.; Partridge, T. Optimizing exon skipping therapies for DMD. *Acta Myol.* **2007**, *26*, 179–184.
72. Muntoni, F.; Gobbi, P.; Sewry, C.; Sherratt, T.; Taylor, J.; Sandhu, S.K.; Abbs, S.; Roberts, R.; Hodgson, S.V.; Bobrow, M.; et al. Deletions in the 5′ region of dystrophin and resulting phenotypes. *J. Med. Genet.* **1994**, *31*, 843–847. [CrossRef]
73. Toh, Z.Y.C.; Aung-Htut, M.T.; Pinniger, G.; Adams, A.M.; Krishnaswarmy, S.; Wong, B.L.; Fletcher, S.; Wilton, S.D. Deletion of dystrophin in-frame exon 5 leads to a severe phenotype: Guidance for exon skipping strategies. *PLoS ONE* **2016**, *11*, e0145620. [CrossRef]
74. Nakamura, A.; Fueki, N.; Shiba, N.; Motoki, H.; Miyazaki, D.; Nishizawa, H.; Echigoya, Y.; Yokota, T.; Aoki, Y.; Takeda, S. Deletion of exons 3-9 encompassing a mutational hot spot in the DMD gene presents an asymptomatic phenotype, indicating a target region for multiexon skipping therapy. *J. Hum. Genet.* **2016**, *61*, 663–667. [CrossRef] [PubMed]
75. Heald, A.; Anderson, L.V.; Bushby, K.M.; Shaw, P.J. Becker muscular dystrophy with onset after 60 years. *Neurology* **1994**, *44*, 2388–2390. [CrossRef] [PubMed]
76. Lim, K.R.Q.; Echigoya, Y.; Nagata, T.; Kuraoka, M.; Kobayashi, M.; Aoki, Y.; Partridge, T.; Maruyama, R.; Takeda, S.; Yokota, T. Efficacy of Multi-exon Skipping Treatment in Duchenne Muscular Dystrophy Dog Model Neonates. *Mol. Ther.* **2019**, *27*, 76–86. [CrossRef]
77. Waldrop, M.A.; Ben Yaou, R.; Lucas, K.K.; Martin, A.S.; O'Rourke, E.; Ferlini, A.; Muntoni, F.; Leturcq, F.; Tuffery-Giraud, S.; Weiss, R.B.; et al. Clinical Phenotypes of DMD Exon 51 Skip Equivalent Deletions: A Systematic Review. *J. Neuromuscul. Dis.* **2020**, *7*, 217–229. [CrossRef] [PubMed]
78. Nakamura, A.; Yoshida, K.; Fukushima, K.; Ueda, H.; Urasawa, N.; Koyama, J.; Yazaki, Y.; Yazaki, M.; Sakai, T.; Haruta, S.; et al. Follow-up of three patients with a large in-frame deletion of exons 45–55 in the Duchenne muscular dystrophy (DMD) gene. *J. Clin. Neurosci. Off. J. Neurosurg. Soc. Australas.* **2008**, *15*, 757–763. [CrossRef]
79. Taglia, A.; Petillo, R.; D'Ambrosio, P.; Picillo, E.; Torella, A.; Orsini, C.; Ergoli, M.; Scutifero, M.; Passamano, L.; Palladino, A.; et al. Clinical features of patients with dystrophinopathy sharing the 45–55 exon deletion of DMD gene. *Acta Myol.* **2015**, *34*, 9–13.
80. Echigoya, Y.; Lim, K.R.Q.; Melo, D.; Bao, B.; Trieu, N.; Mizobe, Y.; Maruyama, R.; Mamchaoui, K.; Tanihata, J.; Aoki, Y.; et al. Exons 45–55 Skipping Using Mutation-Tailored Cocktails of Antisense Morpholinos in the DMD Gene. *Mol. Ther.* **2019**, *27*, 2005–2017. [CrossRef]

Journal of
Personalized Medicine

MDPI

Review

A Systematic Review of Genotype–Phenotype Correlation across Cohorts Having Causal Mutations of Different Genes in ALS

Owen Connolly [1,†], Laura Le Gall [1,†], Gavin McCluskey [1,2], Colette G Donaghy [2,3], William J Duddy [1] and Stephanie Duguez [1,*]

1 Northern Ireland Center for Stratified/Personalised Medicine, Biomedical Sciences Research Institute, Ulster University, Londonderry BT47 6SB, Northern Ireland, UK; Connolly-O4@ulster.ac.uk (O.C.); Le_Gall-L@ulster.ac.uk (L.L.G.); gmccluskey05@qub.ac.uk (G.M.); w.duddy@ulster.ac.uk (W.J.D.)

2 Department of Neurology, Altnagelvin Hospital, WHSCT, Londonderry BT47 6SB, Northern Ireland, UK; ColetteG.Donaghy@westerntrust.hscni.net

3 Motor Neurone Disease Care Centre, Royal Victoria Hospital, Belfast BT12 6BA, Northern Ireland, UK

* Correspondence: s.duguez@ulster.ac.uk

† Co-first authors.

Received: 18 April 2020; Accepted: 15 June 2020; Published: 29 June 2020

Abstract: Amyotrophic lateral sclerosis is a rare and fatal neurodegenerative disease characterised by progressive deterioration of upper and lower motor neurons that eventually culminates in severe muscle atrophy, respiratory failure and death. There is a concerning lack of understanding regarding the mechanisms that lead to the onset of ALS and as a result there are no reliable biomarkers that aid in the early detection of the disease nor is there an effective treatment. This review first considers the clinical phenotypes associated with ALS, and discusses the broad categorisation of ALS and ALS-mimic diseases into upper and lower motor neuron diseases, before focusing on the genetic aetiology of ALS and considering the potential relationship of mutations of different genes to variations in phenotype. For this purpose, a systematic review is conducted collating data from 107 original published clinical studies on monogenic forms of the disease, surveying the age and site of onset, disease duration and motor neuron involvement. The collected data highlight the complexity of the disease's genotype–phenotype relationship, and thus the need for a nuanced approach to the development of clinical assays and therapeutics.

Keywords: ALS; MND; ALS variants; genotype–phenotype; ALS genes

---

1. Introduction

Amyotrophic lateral sclerosis, or ALS, is characterised by a progressive and fatal degeneration of upper and/or lower motor neurons (UMN and LMN, respectively) resulting in muscle weakness and wasting. Classical ALS is the most common form of motor neuron disease (MND) [1] and is defined by the selective deterioration of both UMN and LMN [2]. The global incidence of ALS varies between 1 and 2.6 cases per 100,000 people per year [3], with the average age of onset ranging from 54 to 67 years old [4]. The prevalence of ALS increases with age, reaching 1/5000 among people aged 70–79 years old [5]. Consequently, as the population ages, it is expected that the world's total number of cases will reach more than 375,000 by 2040 [6]. Owing to the lack of a reliable diagnostic test, absence of validated biomarkers, and phenotypes that are easily confounded with other MNDs, including primary lateral sclerosis (PLS) and progressive muscular atrophy (PMA), there is a delay of approximately 11–12 months in reaching a definite diagnosis [7]. Currently, diagnosis is based on a set of clinical criteria (El Escorial [8] and revisions [9], and Awaji-Shima criteria [10]) that can be used to stratify patients according to the area of initial onset and the progression of symptoms.

ALS phenotypes vary between patients who can present with different sites of onset and symptom severity (Figure 1). Concomitant impairments in cognitive ability are sometimes associated with the ALS phenotype. A recent finding from Chiò et al. suggested that 20.5% of ALS patients had frontotemporal dementia (FTD), and a further 31.3% had a behavioural, cognitive or non-executive impairment [11].

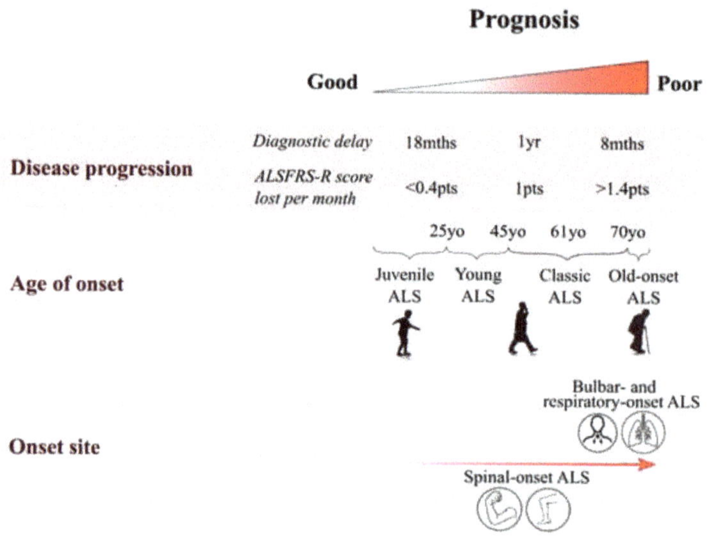

Figure 1. Clinical features of amyotrophic lateral sclerosis (ALS) and their role in prognosis. Diagram summarising the heterogeneity of clinical features in ALS. Multiple features have been associated with a poor prognosis, with an elderly onset being associated with a rapid progression of symptoms and a poor prognosis, especially among elderly females presenting with bulbar-onset phenotype [12]. Disease progression can be assessed either by diagnostic delay or by the ALS functional rating score (ALSFRS: amyotrophic lateral sclerosis functional rating scale). Poor prognosis is associated with patients whose ALS diagnosis has been given less than 8 months after symptom onset, or among those patients losing more than 1.4 points/month on the ALSFRS scale [13].

In the past 30 years, there have been a large number of studies investigating the genetic underpinnings of ALS. To date, over 30 genes have been related to the disease; yet it is important to note that mutations in these genes explain only ~20% of total ALS cases [14] whilst the majority of cases remain unexplained and present no family history. ALS is therefore considered to be a mainly sporadic disease (sALS), with ~80% of cases having no known genetic basis [3], although twin studies have estimated heritability at 40–45% [15] or 61% [16]. Known gene mutations explain some 70% of familial cases (fALS) [17,18], and they have also been identified in 10% of sporadic cases [18]. In European cohorts, the hexanucleotide repeat expansion in the *C9orf72* gene is the most common genetic cause of fALS (33.7%) and sALS (5.1%), followed by *SOD1* (14.8% in fALS and 1.2% in sALS cases), *TARDBP/TDP-43* (4.2% in fALS and 0.8% in sALS), and *FUS* (2.8% in fALS and 0.3% in sALS) [19].

To understand the molecular mechanisms underlying ALS, it is useful to study genotype–phenotype relationships, to determine whether certain gene mutations are associated with specific clinical features or outcomes. Genotype–phenotype relationships have previously been examined for certain gene mutations, and several informatics resources exist to collect genotype–phenotype data [20–24], but a systematic understanding across different gene mutations has not been established. As a step towards this, the present review gathers together the clinical summary statistics from previously studied cohorts across 22 of the more commonly associated genes. Each of the genes are considered in the order in which they were discovered and, where available, a summary of the reported phenotypes associated with each gene is later provided.

2. Pathological Definition of ALS: Clinical Features and Phenotype Variability

2.1. Age of Onset Variation

ALS occurs primarily in patients in their sixth decade, though peak onset is later in sporadic cases (58–63 years) than in familial cases (47–52 years) [25] (Table 1 and Figure 1). Four periods of onset can be defined: juvenile (<25 years old); young (25–45 years); mid–late adulthood (45–70 years); and elderly (>70 years). Juvenile ALS is extremely rare (<1/1,000,000 cases) [26], and is usually associated with slower symptom progression, hence a longer survival time and better prognosis [27]. Some mutations are now described to be associated with juvenile ALS, such as specific mutations of *FUS*, *ALS2* and *SETX* genes [26]. UMN rather than LMN dysfunction is predominant among juvenile ALS cases. Young-onset ALS also shows mainly UMN dysfunction, which is predominant in 60% of those patients [26]. Bulbar-onset ALS is rare in young patients and represents ~15% of cases [27]. In addition, young-onset ALS affects a relatively high proportion of males, with a male:female ratio of 3:1 [26]. These young-onset cases are also associated with a better prognosis than older ALS patients. Elderly-onset patients are more likely to present with bulbar symptoms and are represented by a greater proportion of female patients (M:F ratio 1–1.6) [12,26]. Symptom onset after 80 years is associated with a more aggressive phenotype and poor prognosis, with mean survival times of less than 20 months [12].

Table 1. ALS age of onset variability and their clinical features. Summary of clinical features for ALS in different age periods from Chio et al. [28], Forbes et al. [12], Swinnen et al. [27], Turner at al [26], Sabetelli et al. [29], and Kiernan et al. [25]. In addition to the classical ALS phenotype with age of onset ranging from 45 to 70 years old (mean age ~ 61 years old), three additional age of onset periods (columns) have been observed. Male to female ratios, genetic characteristics, site of onset, estimated survival time, and clinical features are shown where applicable. sALS: sporadic ALS. fALS: familial ALS. MN: motor neurons. UMN: upper motor neurons. LMN: lower motor neurons. -: no data.

	Juvenile	Young	Mid–Late Adulthood	Elderly
Age of onset	≤25 years old	25 to 45 years old	45 to 70 years old	>70 years old
M:F ratio	–	3–3.6:1	1.3–1.56:1	1:1.25
Genetics	Mostly familial cases (*FUS*, *SETX*, *ALS2* mutations)	Mostly familial	~90% sALS ~10% fALS	–
Site of onset: Limb onset Bulbar onset Respiratory/ cognitive onset	 – – –	 – ~16% –	 ~70% ~25% ~5%	 ~40% ~50% (M:F = 1:1.6) –
Survival (from symptoms onset)	Generally longer survival > 10 years		Variable: 50%: <30 months 5–10%: 5–10 years	~20 months
MN involvement: UMN+LMN UMN-predominant LMN-predominant	 – Predominant –	 ~40% ~60% –	 ~80% ~17% –	 ~72% – ~19%

2.2. Site of Onset Variability

The majority of ALS cases (~70%) have spinal onset, usually presenting with focal limb weakness [30] such as foot drop or a weak hand [7]. The disease then tends to spread in a contiguous manner, initiating at distinct focal regions of the body and then propagating from the primarily affected area to adjuvant secondary sites of the body [31].

In 25% of ALS cases, symptoms develop initially in the bulbar-innervated muscles [30,32]. Bulbar-onset ALS is more common in women [7], especially after 70 years (M:F ratio 1:1.6 [12]). Dysarthria almost always predates dysphagia and cognitive impairment is often present [32].

Approximately 3% to 5% of patients [33] present with respiratory or cognitive onset [25]. Thoracic spinal-onset ALS can present as truncal weakness or respiratory impairment and is associated with poor prognosis, with a mean survival time of just 1.4 years [27,34].

Cognitive-onset ALS patients usually present symptoms characteristic of frontotemporal dementia (FTD), such as changes in behaviour, personality and cognition which are all suggestive of frontal impairments [35].

In summary, initial site of symptom onset varies among ALS patients from classic limb-onset to rare cognitive-onset phenotypes, and a poor prognosis is often associated with bulbar and respiratory onset [25].

2.3. *Motor Neuron Involvement in ALS Variants*

ALS patients can present with either a LMN or UMN predominant phenotype (Figure 2). Signs of pure LMN dysfunction are considered as progressive muscular atrophy (PMA), whereas predominant UMN signs are associated with primary lateral sclerosis (PLS) [30]. PMA and PLS are both rare diseases and represent 5% of MND patients [27].

2.3.1. UMN-Dominant ALS Variants

Patients can present predominant UMN dysfunction as in primary lateral sclerosis (PLS) or pseudobulbar palsy. The UMN predominant phenotype can then progress to ALS, which is observed in 40% of PLS cases [36]. Patients diagnosed with PLS for not meeting the diagnostic criteria for ALS can still slowly develop signs of LMN dysfunction and therefore present both UMN and LMN signs [27]. However, LMN involvement and limb atrophy in PLS is exceptionally rare [37] and the prognosis for PLS patients is better than that for patients diagnosed with ALS as symptom progression is relatively slow.

2.3.2. LMN-Dominant ALS Variants

On the contrary, some patients can develop a LMN-dominant phenotype which can be defined as progressive muscular atrophy (PMA), and flail-arm or flail-leg syndrome variants. PMA patients are similar to classic ALS patients without obvious signs of UMN dysfunction. However 50% to 60% of PMA patients develop degeneration of upper motor neurons during the progression of the disease [38], and post-mortem histopathology has demonstrated that some PMA patients show UMN involvement which could not be detected upon clinical examination [39,40]. In patients with flail-arm or flail-leg syndromes, a LMN pattern of weakness and atrophy is observed in the upper limbs or lower limbs, respectively. Similar to PMA, flail-arm and flail-leg syndrome have been described as a LMN variant but can show UMN involvement in the later stages of disease [41]. Involvement of secondary sites should not occur within 12 months of initial onset [42] and prognosis for flail-arm and flail-leg syndrome is better than that seen in ALS, with median survival times of 5 to 6 years [41,43].

2.4. *Non-Motor Involvement in ALS and Overlap with FTD*

For many years, ALS was described as a neurodegenerative disorder with no extra-motor involvement. However, non-motor involvement is now accepted in the ALS phenotype [44], with neuroimaging demonstrating reduced grey matter in motor and non-motor brain regions of ALS patients [45], and histopathology suggesting widespread neuronal and glial TDP-43 pathology in the CNS [46]. In regards to symptomology, a low proportion of ALS patients experience non-motor impairment as a first indication of pathology (3% of sporadic cases and 15% of familial cases) [47]. It has been estimated that approximatively 35% of ALS patients present behavioural and/or cognitive changes (with 15% meeting the Neary criteria [48] for FTD diagnosis (ALS-FTD) [47]). The reported percentage seems to be much lower in most gene-specific studies and varies considerably between them, but it should be noted that the number of patients and studies for which these clinical parameters are reported is relatively small (Table S2). ALS and FTD are sometimes described as part of one continuum, with pure ALS patients (without any non-motor involvement) and pure FTD cases (for whom no motor dysfunction has been described) representing opposite ends of the spectrum.

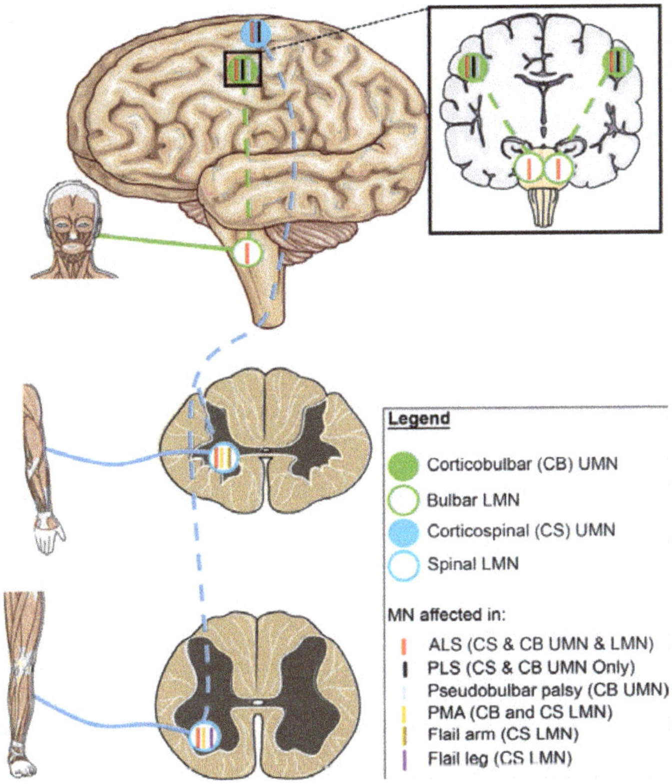

Figure 2. The role of upper and lower motor neurons in different ALS variants. ALS is a disease with high variability in clinical phenotype. "Classic ALS" patients will present with signs of both UMN and LMN degeneration. However, patients with progressive muscular atrophy (PMA) and primary lateral sclerosis (PLS) present with LMN-predominant or UMN-predominant signs, respectively. LMN-predominant patients also include flail-arm syndrome and flail-leg syndrome ALS variants where LMN signs are present in upper or lower limbs, respectively. ALS patients might present symptoms in bulbar-innervated muscles, if UMN signs are predominant, patients are diagnosed with pseudobulbar-palsy. Blue colour circles indicate motor neurons of the corticospinal tract. Green colour circles indicate motor neurons of the corticobulbar tract. Solid circles indicate UMNs and open circles indicate LMNs. Colour of ticks corresponds to colour of variant label and tick location indicates the motor neuron populations affected. ALS: amyotrophic lateral sclerosis. PLS: primary lateral sclerosis. PMA: progressive muscular atrophy. CS: corticospinal. CB: corticobulbar.

ALS patients having FTD usually meet the criteria for behavioural variant FTD characterised by defects in cognitive functions, personality traits and behavioural collapse. Among ALS cases experiencing non-motor dysfunction, language (particularly deficits in verbal fluency) and cognition are the most affected categories [49], and apathy is the most frequently encountered personality impairment [47].

2.4.1. Dementia in ALS Patients—ALS-FTD Variants

ALS-FTD diagnosis is made upon the presence of an ALS phenotype associated with behavioural or cognitive defects that fulfil FTD diagnostic criteria: (1) progressive impairment of behavioural/cognitive functions and observation of at least three behavioural symptoms defined by Rascvosky et al. [50]; or (2) loss of insight and/or presence of psychotic features associated with at least two Rascvosky et al. [50] symptoms; or (3) language impairment combined with semantic dementia (defined in [48]).

2.4.2. Cognitive Changes in Non-Demented ALS Patients—ALSci and ALSbi Variants

Non-demented ALS patients presenting with behavioural impairment are classified as ALSbi-variant, while ALS patients experiencing cognitive impairment including language defects are considered to be ALSci variant [47]. Based on the revised diagnostic criteria from Strong et al. [51], ALS patients can be diagnosed as ALSci variant if either executive impairment (social cognition), or language dysfunction, or a combination of the two features are evident during diagnosis. Diagnostic criteria for ALSbi variant require apathy with or without other behavioural symptoms, or two or more behavioural changes, such as disinhibition, loss of sympathy/empathy, perseverative/stereotypic/compulsive behaviour, hyper orality/dietary change, loss of insight and psychotic symptoms.

3. Genetics of ALS

Superoxide dismutase 1 (SOD1) was the first gene demonstrated to be associated with ALS in 1993 [52]. *SOD1* is ubiquitously expressed in human cells and serves to protect them from harmful reactive oxygen species (ROS). Mutated forms of *SOD1* are believed to result in a toxic gain of function, provoking the presence of misfolded protein aggregates, increased endoplasmic reticulum (ER) stress, and oxidative stress and ultimately accelerating motor neuron degeneration [17].

In 2001, mutations in *ALSIN2 (ALS2)* were shown to be implicated in juvenile forms of ALS [53–55] and PLS [56]. The ALS2 protein has been found to act as a guanine nucleotide exchange factor for the GTPase, Rab5, which is in involved in endosome trafficking [57]. Mutations in *ALS2* have been shown to inhibit activation of Rab5 and its translocation to mitochondria, leaving *ALS2* mutated motor neurons more susceptible to oxidative stress [58]. However, in murine studies, genetic ablation of *ALS2* has failed to recapitulate the pathological features seen in ALS [59,60] although primary motor neurons from these mice did show greater sensitivity to oxidative stress and aberrant morphology, suggesting that *ALS2* mutations may indeed play a role in motor neuron susceptibility in ALS.

Genetic mutations were next reported in 2004 for the *senataxin (SETX)*, *angiogenin (ANG)*, and *vesicle-associated membrane protein-associated protein B (VAPB)* genes. *SETX* plays a role in numerous cellular functions including RNA metabolism and has been shown to regulate RNA polymerase II transcription termination [61] and its yeast homolog, SEN1, has been linked with processing of non-coding RNA [62]. *SETX* mutations are strongly associated with juvenile-onset ALS [63] and associations have been confirmed in American, Italian and Dutch cohorts [63–65]. ANG is highly expressed in the human central nervous system [66] and has been reported to show neuroprotective properties [67]. Indeed, expression of ALS-associated *ANG* variants has been shown to cause motor neuron death in cell culture models [67]. *ANG* has also been reported to play a role in the transcription of ribosomal RNA [68] and many ALS-associated variants are believed to elicit a loss of function in ANG, thus eliminating any neuroprotective functionality [69]. VAPB is a protein closely associated with the endoplasmic reticulum and is thought to be involved in the induction of the unfolded protein response (UPR) [70], as well as cellular processes including lipid transport [71], protein secretion [72], and calcium homeostasis [73]. The P56S mutation in *VAPB* has been implicated in an early-onset and slow-progressing form of fALS [74] and follow-up studies have highlighted how this mutation can result in nuclear envelope defects [75], and provoke VAPB ER aggregates [72]. However, murine models expressing the P56S mutation show widespread VAPB aggregates but demonstrate no motor neuron pathology or ALS phenotypes [76].

The next genetic mutation associated with ALS did not arrive until 2008, when mutations in *TAR DNA-binding protein (TARDBP)*, encoding TDP-43, were reported in patients [77]. TDP-43 is a RNA/DNA-binding protein that plays important roles in several RNA metabolism processes [78]. Ubiquitinated TDP-43 was first shown to be present in CNS inclusions of ALS patients in 2006 [79] and subsequent studies have confirmed TDP-43 as the major protein component of pathological inclusions present in approximately 90% of ALS patients [80]. However, TDP-43 pathology is not unique to ALS and has been reported in numerous neurodegenerative conditions including FTD [79], Parkinson's disease [81], Huntington's disease [82], Alzheimer's disease [83], and dementia with Lewy bodies [84].

Then, in 2009, multiple mutations in the nuclear RNA-binding protein, *Fused in Sarcoma (FUS)* and *FIG4 phosphoinositide 5-phosphatase (FIG4)*, were associated with ALS [85,86]. FUS is another RNA/DNA-binding protein involved in mechanisms of RNA splicing and DNA repair [87] and is implicated in both ALS and FTD [88]. Mutations in *FUS*, particularly those near the nuclear localisation signal (NLS) domain, cause cytoplasmic protein mislocalisation and are associated with a severe phenotype in murine models [89]. *FIG4* is involved in vesicle trafficking due to its role in the regulation of the membrane bound phosphoinositide, PI(3,5)P2 [90]. Mutations in *FIG4* were initially shown to cause neurodegeneration in Charcot–Marie–Tooth (CMT) neuropathy [91]. However, others have questioned the role of *FIG4* in ALS pathology after failing to find pathogenic mutations in their Taiwanese [92] and Italian [93] cohorts.

In 2010, mutations in *Optineurin (OPTN)*, *Spatacsin paraplegia 11 (SPG11)*, *Valosin-containing protein (VCP)*, and *Ataxin-2 (ATXN2)* were all implicated in ALS. Three different *OPTN* mutations were identified in ALS patients [94] and researchers were able to demonstrate the increased immunoreactivity of OPTN in both TDP-43 and SOD1 inclusions found in the spinal cord of sALS patients, suggesting a role for *OPTN* in general ALS pathogenesis.

The link between *SPG11* and ALS was established when mutations were found to be associated with autosomal recessive juvenile ALS [95]. Mutations to *SPG11* are the most common cause of autosomal recessive hereditary spastic paraplegia [96] and loss of function mutations have been shown to elicit lysosomal dysfunction and UMN + LMN degeneration in mice [97]. *ATXN2* encoding the ataxin-2 polyglutamine (polyQ) protein was associated with ALS when researchers identified the presence of intermediate length polyQ expansions (27-33 Qs) in 4.7% of their North-American ALS cohort [98]. Ataxin-2 protein has been shown to regulate mRNA stability and translation [99,100] and upregulation of the fly homolog of Ataxin-2 was found to enhance neurodegeneration in *Drosophila* via its interaction with wild-type and mutated forms of TDP-43 [98]. Involvement of *Ataxin-2* in ALS pathogenesis has since been confirmed in European and Chinese patient cohorts [101,102]. VCP is an ATP-driven chaperone protein that plays a role in ubiquitin-regulated protein degradation [103], autophagy [104], and mRNA processing [105,106]. *VCP* mutations were shown to be present in 1–2% of familial ALS patients in an Italian cohort [107] and mice expressing ALS-associated *VCP* mutations have been shown to develop a slow-progressing ALS phenotype [108].

In 2011, mutations in *ubiquilin-2 (UBQLN2)*, *sequestosome-1 (SQSTM1)*, and *chromosome 9 open reading frame 72 (C9orf72)* were discovered. Ubiquilin-positive inclusions have been implicated in both sALS and fALS [109], whilst mutations in *SQSTM1* have been observed in rare ALS and FTD cases [110] and can be shown to lead to p62 protein inclusions in motor neurons of both patient groups [111]. The G4C2 hexanucleotide repeat expansion mutation (HREM) within *C9orf72* [112,113] is perhaps the most significant genetic mutation associated with ALS thus far, and is estimated to be present in 34% of familial cases, and 5% of sporadic cases in Europe [19,114]. In healthy subjects, the G4C2 repeat length ranges from 2 to 23 units [112], whilst intermediate expansions ranging from 24 to 30 [115] and large expansions ranging from 30 to many hundreds of units have been observed in ALS patients [112,116]. Although rare, *C9orf72* expansions have been implicated in other neurodegenerative and psychiatric diseases including PD [117] and Schizophrenia [118], suggesting a wider role for *C9orf72* in neuropathology and perhaps offering some insight towards the heterogeneous phenotype seen in *C9orf72* ALS.

In 2012, *Profilin 1 (PFN1)* was implicated in familial and sporadic cases of ALS [119]. Mutant *PFN1* has been shown to cause motor neuron degeneration through the formation of insoluble aggregates and disrupted cytoskeleton dynamics in mice [120] and co-aggregation of PFN1 and TDP-43 has been reported in cell lines expressing mutant *PFN1* [119].

Then, in 2013, *heterogeneous nuclear ribonucleoprotein A1 (hnRNPA1)* was reported to be involved in ALS after researchers identified three *hnRNPA1* variants—two of which were associated with familial ALS and the other of which was associated with a sporadic case [121]. hnRNPA1 is known to colocalise with TDP-43 [121] and post-mortem studies have shown that motor neurons of ALS patients display marked reductions in hnRNPA1 alongside concomitant TDP-43 inclusions [122].

In 2014, mutations in *Tubulin alpha-4A (TUBA4A)* and *Matrin-3 (MATR3)* were implicated in ALS. Mutations in *TUBA4A* were first identified in a European and American cohort [123] and then validated in Belgian and Chinese cohorts in 2017 and 2018 [124,125]. *TUBA4A* mutations have been shown to cause cytoskeletal defects in primary motor neurons [123] and are recognised as a rare cause of ALS and FTD [125].

MATR3 was first associated with ALS after exome sequencing identified mutations in Italian, UK and US kindreds, alongside increased levels of MATR3 protein in spinal cord sections of ALS patients relative to controls [126]. MATR3 has been found to interact with TDP-43 and both proteins were shown to co-aggregate in skeletal muscle tissue of ALS patients [126]. *MATR3* is known to play various roles in RNA metabolism and alternative splicing [127,128] and recent evidence suggests ALS-associated *MATR3* mutations play a role in defective nuclear export of FUS and TDP-43 mRNA [129]

In 2015, *NIMA-related kinase 1 (NEK1)* was recognised as an ALS-risk gene [130] and was shown to interact with two other ALS genes, *ALS2* and *VAPB*—both of which are involved in endosomal trafficking. Subsequent studies provided further evidence for the pathogenic role of *NEK1* in ALS [131,132] and pathway analyses have shown NEK1 to interact with C21orf72—both of which are involved in DNA repair mechanisms [133]. Mutations in *Tank-binding kinase 1 (TBK1)* were also associated with ALS in 2015 after exome sequencing identified eight loss of function mutations in 13 fALS pedigrees [134].

Cyclin F (CCNF) was implicated in ALS in 2016 with variants identified in both familial and sporadic cases [135]. In the same study, researchers were able to demonstrate how mutant *CCNF* led to aberrant ubiquitination and aggregation of proteins including TDP-43. More recently, CCNF was shown to be a binding partner of another ALS protein, VCP. Binding of mutated CCNF to VCP increased VCP ATPase activity, which in turn led to increased TDP-43 aggregation in U20S cells [136].

Then, in 2018, the most recent genetic mutations implicated in ALS were discovered when research demonstrated the pathological involvement of *Kinesin family member 5A (KIF5A)* [137]. KIF5A is a protein expressed specifically in neurons and is involved in regulating neuronal microtubule dynamics [138,139]. *KIF5A* is also associated with spastic paraplegia and Charcot–Marie–Tooth neuropathy [140] and mutations have been reported in ALS patients in Chinese [141], European [142,143], and US cohorts [137].

4. Correlation of Genotype/Phenotype: Methods, Results and Discussion

To evaluate whether there is a correlation between associated genes and phenotype in ALS, a systematic search of original papers was performed using key words summarised in Table S1, while adhering to PRISMA guidelines (see checklist in Supplementary Materials).

4.1. Protocol

A systematic search was performed in PubMed using the key words: ALS, genotype phenotype, patient, and onset. To make sure that clinical data would also be obtained for rare genes involved in ALS and listed in Vijayakumar et al. [14], the following search terms were added: ALS, phenotype, patient and the gene name such as *TBK1, VCP, SQSTM1, CCNF, NEK1, OPTN, FIG4, PFN1, ATXN2, VAPB, ANG, ALS2, SPG11, UBQLN2, KIF5A*, and *MATR3*. There were no language, type of study, or publication date restrictions.

4.2. Eligibility Criteria

The search combining the different key words resulted in 355 articles. Reviews and duplicated papers were excluded. To avoid redundancy, papers re-using previously published clinical data were excluded. All studies used in the systematic review were peer-reviewed, written in English, and published original clinical data related to patients affected by monogenic forms of ALS. At least one of the following parameters had to be described in the paper: age of onset, site of onset, motor neuron population being affected (UMN, LMN, UMN+LMN), disease duration, number of patients with FTD, and number of patients with cognitive impairment. A total of 107 papers were then eligible for the analysis (see PRISMA flow chart in Figure 3).

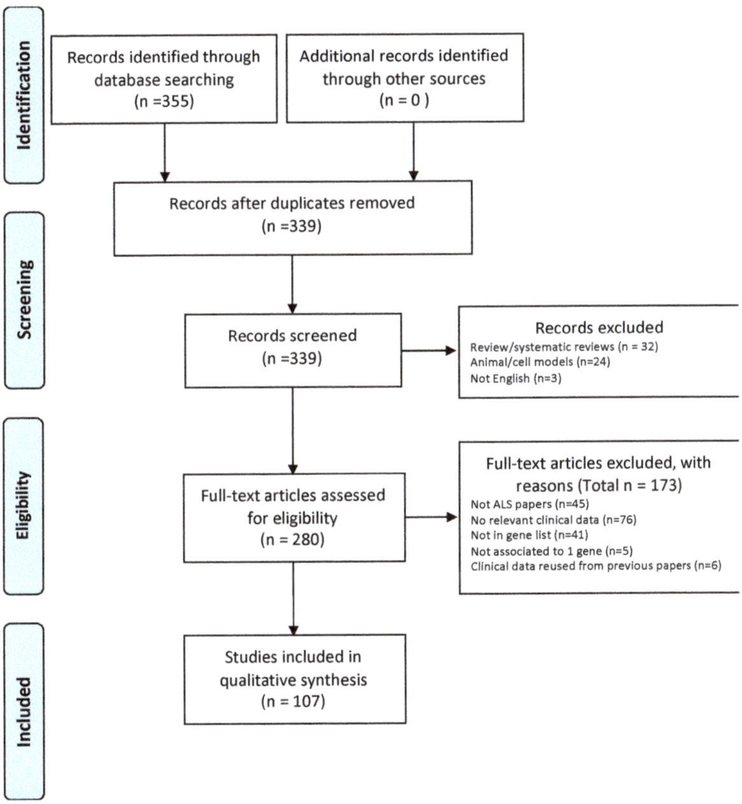

Figure 3. PRISMA flow chart showing how studies have been selected.

4.3. Data Extractions and Synthesis

The papers were thoroughly reviewed by OC, LLG, VM and SD. Key information was extracted from each study, and grouped into cohort characteristics (ethnicity/ country of the study, number of patients), age of onset (distribution and mean and standard deviation (SD)), site of onset (spinal, bulbar, respiratory, other/unknown), motor neurons being affected (UMN, LMN, UMN+LMN), disease duration (mean and SD), percentage of patients with FTD, and percentage of patients with cognitive impairment. All data are collated per gene in Table S2.

For the summary Table (Figure 4), the age of onset and disease duration are presented as the weighted mean ± SD, and the site of onset, motor neuron impairments, and FTD comorbidity are presented as weighted percentages, in all cases taking into account the number of patients studied as described below:

Mean = mean of the parameter of interest given in the referenced study;
n = number of patients studied for the corresponding parameter in the given study;
$Sx^2 = SD^2(n-1) + ((Sx)^2/n)$
Weighted mean $= \frac{\sum Sx}{\sum n}$
Weighted SD $= \sqrt[2]{\frac{(\sum(Sx^2)-(\sum Sx)^2)/\sum n}{(\sum n)-1}}$

4.4. Characteristics of Studies

A total of 1630 ALS patients were included in the systematic review. The total number of reported patients for each gene is shown in Figure 4 column 4. As not all studies reported all clinical parameters, the total number of patients studied for each parameter is reported in the first subcolumn for each parameter. On average, 59% of the population was male, with considerable variation between genes (See Table S2). Most of the studies were conducted in Europe, North America and Asia.

4.5. Overall Findings and Discussion

For most genetic forms of ALS reported in Table S2 and in Figure 4, the age of onset ranges between 50 and 70 years old. Exceptions to this include cases of juvenile ALS, which are observed with mutations in *SPG11* [95,144], *FUS* [145,146] and *ALS2* [53,55] (Table S2). Whilst *FUS* patients are known to show considerable variation in phenotype, with some showing early onset and fast progression, others show a later age of onset and a slower-progressing phenotype [147]. This variation in the *FUS* phenotype has been hypothesised to arise due to the different effects exerted by missense and truncating mutations [148]. Interestingly, the studies reviewed here suggest that *FUS* mutations are indeed associated with a relatively early age of onset (41.8 ± 14.5 years) and a fast-progressing phenotype, with average disease duration lasting 30.6 months (Figure 4). Another gene sometimes associated with early-onset ALS is *SETX*. Patients with *SETX* mutations have been reported to display a slow-progressing phenotype in which bulbar and respiratory muscles seem largely unaffected [149]. However, in one reported case, a patient did go on to experience bulbar symptoms 3 years after onset [150]. Moreover, from the studies retrieved in this review, *SETX* patients do not show an early age of onset nor a particularly slow phenotype. For instance, the average age of onset for *SETX* patients was 59.5 ± 24.7 years with an average disease duration of 43.8 ± 37.5 months.

Many ALS-associated genes show variation in site of onset. Among the 22 genes included in Figure 4, cases of spinal onset are predominant in 19. This is in line with previous findings that suggest spinal onset accounts for approximately two-thirds of ALS cases [32]. For example, *SOD1*, *hnRNAP1*, *TUBA4A*, and *ALS2* show a high percentage of patients with spinal onset (>80%), while spinal onset in *VCP*, *NEK1*, and *TBK1* cases accounted for 50%, 50% and 55% of cases, respectively. Some other ALS-associated gene mutations were associated with a lower proportion of spinal onset, e.g., 33% of *C9orf72* cases, and 40% of *UBQLN2* cases. However, previous research suggests that *C9orf72* ALS demonstrates frequent occurrence of both spinal [151] and bulbar onset [152]. Moreover, it has been reported that site of onset in *C9orf72* ALS can be used to predict disease duration. For instance, the average age of onset in patients with spinal onset was 59.3 years, increasing to 62.3 years in patients with bulbar onset, and male patients with spinal onset seem to display a faster-progressing phenotype [153].

A striking 95% of *SOD1* cases were classified as spinal onset. Indeed, animal studies have provided support for the notion that *SOD1* pathology begins at the periphery and proceeds in a retrograde manner [154,155]. Recently, a homozygous mutation that eliminates the enzymatic activity of *SOD1* was found to result in a severe LMN phenotype and mild cerebellar atrophy in a young child [156] and the presence of a *SOD1* p.D12Y variant was shown to result in a LMN-predominant phenotype [157]. Similarly, seven studies reported a non-negligible percentage of patients with pure LMN signs (Table S2,

Figure 4, 47.6% pure LMN vs. 45.2% UMN+LMN, [158–161]). Overall, these studies seem to suggest that *SOD1* mutations exert profound effects at the distal nerve. In addition, the observation that both overexpression, and absence of *SOD1* activity lead to pathology should be an important consideration in the development of therapeutics that aim to alter *SOD1* levels as a novel treatment in ALS [162].

Figure 4 was sorted in descending order for the percentage of patients showing LMN signs. Not all studies reported UMN and/or LMN signs, and thus the percentage given in this table only represents a small proportion of the studies (see Table S2 for more details). However, it is interesting to see that the majority of the gene mutations do indeed elicit a phenotype that is characterised by both UMN+LMN signs, consistent with the classical clinical definition of ALS. *FUS*, *C9orf72* and *TARDBP* all demonstrated increased presence of both UMN and LMN signs with both neuronal populations affected in 66.7%, 72.7% and 44.4%, respectively. Surprisingly, only 33% of *FIG4*, *PFN1*, *MATR3* and *NEK1* cases showed both UMN and LMN signs, although it should be noted that 4 of the 14 studies reviewed in relation to these genes did not provide details regarding the pattern of motor neuron involvement. Some ALS-associated genes demonstrated >20% of patients with pure LMN signs (*SOD1*, *FUS*, *PFN1*, *ATXN2*, *TARDBP*, *TBK1*, and *hnRNPA1*), while pure UMN signs had >20% preponderance in several genes (*ANG*, *TBK1*, *FIG4*, *MATR3*, *NEK1*, *hnRNPA1*).

Finally, the current review also aimed to collect information regarding the prevalence of cognitive impairments and FTD in ALS. FTD was most frequent in cases with mutations in either *C9orf72*, *SQSTM1*, or *TBK1* (36%, 67%, and 43%, respectively, Figure 4). Indeed, *C9orf72* [163], *SQSTM1* [164], and *TBK1* [165] have all previously been linked with FTD onset. However, in a large screen of 121 patients with FTD, genetic mutations were successfully identified only in *C9orf72* and *SQSTM1*, whilst no *TBK1* variants were identified [166]. It is also worth noting that despite the frequent association between *TARDBP* and FTD, only 12% of cases reviewed here were found to have concomitant FTD symptoms. In relation to general cognitive functioning, reports of impairment were observed across 10 ALS-associated genes, although the number of patients studied for this parameter were often quite low (Table S2), rendering it difficult to form conclusions.

Gene	Frequency in caucasian population (Volk et al. and Chia et al.)	Cohort		Age at onset (years)		Onset phenotype					UMN - LMN- UMN+LMN				Disease Duration			FTD	
		Country	Nb of patients studied	Mean	SD	Nb of patients studied	Spinal onset	Bulbar onset	Respiratory /Cognitive onset	Other/ Unknown	Nb of patients studied	UMN	LMN	UMN+ LMN	Nb of patients studied	Mean (Months)	SD	Nb of patients studied	%
hnRNPA1	<1%	Asia	4	33	2.1	4	100.0%	0.0%	0.0%	0.0%	4	25.0%	75.0%	–	4	360.0	0	–	–
SETX	<1%	Europe, Asia, North America	38	59.5	24.7	38	71.1%	21.1%	2.6%	5.3%	8	0.0%	62.5%	37.5%	18	43.8	37.5	7	14.3%
SOD1	~3%	Europe, Asia	154	52.0	11.6	119	95.0%	1.7%	0.0%	0.8%	42	7.1%	47.6%	45.2%	106	71.7	56.5	3	0.0%
TBK1	~1%	Europe	29	59.5	10.1	29	55.2%	24.1%	0.0%	20.7%	4	25.0%	25.0%	50.0%	24	47.3	38.7	24	42.5%
FUS	~0.9-1.2%	Europe, Asia, North America	142	41.8	14.5	114	74.2%	19.6%	0.0%	6.2%	21	9.5%	19.0%	66.7%	74	30.6	24.7	39	7.9%
OPTN	<1%	Europe, Asia	34	53.2	12.2	31	77.4%	22.6%	0.0%	0.0%	11	0.0%	18.2%	81.8%	16	40.4	46.9	3	0.0%
TUBA4A	<!%	Europe, Asia, North America	15	59.6	8.3	15	80.0%	13.3%	0.0%	6.7%	9	0.0%	11.1%	88.9%	15	53.3	49.7	8	25.0%
TARDBP	~1.2%	Europe, North America, Asia	222	57.3	8.7	205.00	44.0%	23.9%	0.0%	32.2%	27	3.7%	18.5%	44.4%	88.00	63.7	31.9	89	12.4%
ATXN2	<1%	Europe, Asia, South & North America	89	54.2	12.7	52	84.6%	13.5%	0.0%	1.9%	53	3.8%	18.8%	77.4%	34	36.4	17.5	–	–
C9orf72	~8%	Europe, Asia, North America	695	57.5	9.2	520	33.1%	39.0%	0.0%	26.7%	26	17.7%	13.5%	72.7%	459	35.0	20.2	503	35.6%
MATR3	<1%	Europe, Asia, North America	12	55.3	9.6	11	45.6%	45.1%	9.3%	0.0%	6	50.0%	17.0%	33.0%	10	88.2	22.8	7	14.6%
FIG4	<1%	Europe, North America	16	61.3	15.3	16	67.2%	32.8%	0.0%	0.0%	6	50.0%	16.7%	33.3%	12	59.2	41.2	1	100.0%
ANG	<1%	Europe, North America	14	54.9	12.5	14	92.9%	7.1%	0.0%	0.0%	7	57.1%	0.0%	42.9%	13	33.6	23.1	–	–
NEK1	~2%	Europe	16	56.6	9.8	16	50.0%	12.5%	0.0%	37.4%	3	66.0%	0.0%	33.0%	13	42.0	18.0	–	–
KIF5A	<1%	Europe, Asia	23	47.7	9.1	21	76.1%	23.9%	0.0%	0.0%	6	0.0%	0.0%	83.0%	12	31.6	15.6	6	17.0%
CCNF	~2.2%	Austalia, Europe, North America	12	55.3	7.5	12	58.3%	16.7%	0.0%	25.0%	12	8.0%	–	92.0%	12	30.20	18.70	12	8.0%
PFN1	<1%	Europe, North America, Asia	11	58.7	13.2	11	81.6%	9.1%	0.0%	9.3%	3	–	–	33.0%	6	35.6	19.8	–	–
VAPB	<1%	Europe, North America, Asia	23	56.0	12.5	23	82.9%	17.4%	0.0%	0.0%	–	–	–	–	16	38.4	24.4	–	–
UBQLN2	<1%	Europe	10	45.2	17.4	6	40.0%	40.0%	0.0%	20.0%	–	–	–	–	10	30.5	19.6	–	–
SQSTM1	<1%	Europe, Asia, North America	18	56.7	14.7	18	66.7%	22.2%	0.0%	11.1%	–	–	–	–	10	40.9	28.1	3	66.7%
VCP	~1%	Europe, North America	20	11.6	24.8	4	50.0%	25.0%	0.0%	25.0%	–	–	–	–	1	137.0	–	1	100.0%
SPG11	<1%	North Amercia, Asia	26	51.9	13.3	26	76.9%	23.1%	0.0%	0.0%	–	–	–	–	2	276.0	0.0	–	–
ALS2	<1%	Europe	7	37.4	28.1	7.00	85.7%	14.3%	0.0%	0.0%	–	–	–	–	–	–	–	–	–

Figure 4. Table summarising the phenotypes observed in ALS patients with different mutations. A detailed version of this table is accessible as supplemental data (Table S2). PubMed was searched to identify published studies reporting genotype–phenotype data for 23 genes. Column 2 indicates the frequency of genes observed in

Caucasian populations described in previous reviews (Volk et al. [18] and Chia et al. [167]). The ethnicity/origin of cohorts reported across studies is specified in Table S2 and summarised in column 3. For each parameter reported in this table, the number of patients is given in the first subcolumn for each category. For age of onset, motor neuron impairments, disease duration, and FTD, the weighted mean ± SD, and weighted percentages are given, taking into account the numbers of patients studied. Data for each gene were collected from the following reference studies, then summarized: hnRNPA1: [168]; SETX: [169–173]; SOD1: [24,92,160,161,171,172,174–188]; TBK1: [189–192]; FUS: [24,92,145,146,148,170–172,179,180,187,188,193–196]; OPTN: [171,180,187,197–203]; TUBA4A: [123,188,204,205]; TARDBP: [171,172,179,180,187,188,192,206–211]; ATXN2: [101,102,171,180,188,212–214]; C9orf72: [114,170,171,175,180,188,206,215–226]; MATR3: [227–231]; FIG4: [171,192,232]; ANG: [171,180,187,233–235]; NEK1: [131,236]; KIF5A: [137,141,143,237]; CCNF: [135]; PFN1: [238–241]; VAPB: [171,242,243]; UBQLN2: [244,245]; SQSTM1: [171,175,192,246]; VCP: [170,171,175,192,247]; SPG11: [179,248]; ALS2: [55,170]. Results from each separate study are shown in Table S2. Gradient colour for age of onset from dark blue to dark red: dark blue, 16 years; dark red, 60 years. Gradient colour for site of onset and for motor neuron impairment distributions: white, 0%; green, 100%.

5. Conclusions

Over 150 years have passed since ALS was first reported by Charcot and still the aetiology of the disease remains elusive. Although research is progressing and genetic studies continue to identify novel gene associations [14,249–251], many questions remain surrounding the pathological mechanisms associated with already established mutations, their role in the ALS phenotype, and the as yet undiscovered mechanisms that underlie sporadic onset of disease. Here, we have performed a systematic review in an attempt to highlight genotype–phenotype correlations for 23 of the more commonly reported mutated genes in ALS. This has proven to be challenging as many genetic studies do not capture or report a complete summary of clinical data. Whilst it is understandable that such data are difficult to acquire, we hope to illustrate that there is a need for improved and more widely available clinical and informatics resources that would enable genotype–phenotype associations to be easily visualised in ALS.

Whilst we have illustrated the relationship between commonly reported mutated genes and various clinical measures including age and site of onset, disease duration and motor neuron involvement, a limitation of the current review is that we do not consider variation among phenotypes of patients having different mutations of the same gene. For many genes involved in ALS, including *FUS*, *SOD1*, and *TARDBP*, the phenotype may be different depending upon the specific genetic mutation in question. In *SOD1* patients, for instance, the A4V mutation results in a much more aggressive phenotype (death occurring ~1.2 years after onset [252]) than the H46R mutation, for which patients show a relatively mild phenotype (duration of ~17 years [253]). It could be of value in future work to comprehensively review variations in genotype–phenotype correlations among the different mutations reported by single-gene studies, which in turn could contribute towards a comprehensive database of ALS genotype–phenotype correlation. Such a resource could ultimately improve our mechanistic understanding of ALS by enabling a more robust assessment of how the ALS phenotype responds to different variants across multiple genes.

Additional limitations include that many of the studies surveyed are relatively small, involving low numbers of patients, and that, as well as only a subset of studies reporting clinical breakdown of phenotype, ethnic breakdown is also not always reported and some ethnicities have minimal representation.

Despite these limitations, the collected data reveal a landscape of highly variable phenotypic associations, underlining the complexity of the disease, and the need for nuanced approaches to the development of clinical assays and therapeutics.

Supplementary Materials: The following are available online at http://www.mdpi.com/2075-4426/10/3/58/s1, **Table S1**: Key words used in PubMed literature search. **Table S2:** Collated data from clinical studies on monogenic forms of ALS. **Table S3:** PRISMA checklist for systematic review.

Author Contributions: O.C., L.L.G., and S.D. collated the data from the literature. O.C., L.L.G. and S.D. organised the data and wrote the paper. O.C., L.L.G., G.M., C.G.D., W.J.D. and S.D. wrote, discussed, and edited the paper. All authors have read and agreed to the published version of the manuscript.

Funding: This work was financed by the European Union Regional Development Fund (ERDF) EU Sustainable Competitiveness Programme for Northern Ireland, Northern Ireland Public Health Agency (HSC R&D) and Ulster University (PI: A Bjourson). L.L.G. was a recipient of an ArSLA PhD fellowship, O.C. was a recipient of a PhD DELL fellowship and G.M. was a recipient of an IICN fellowship.

Acknowledgments: We would like to thank Vanessa Milla for her input in the systematic search.

Conflicts of Interest: The authors declare no conflict of interest.

References

1. Logroscino, G.; Piccininni, M.; Marin, B.; Nichols, E.; Abd-Allah, F.; Abdelalim, A.; Alahdab, F.; Asgedom, S.W.; Awasthi, A.; Chaiah, Y.; et al. Global, regional, and national burden of motor neuron diseases 1990–2016: A systematic analysis for the Global Burden of Disease Study 2016. *Lancet Neurol.* **2018**, *17*, 1083–1097. [CrossRef]
2. Gordon, P.H. Amyotrophic Lateral Sclerosis: An update for 2013 Clinical Features, Pathophysiology, Management and Therapeutic Trials. *Aging Dis.* **2013**, *4*, 295–310. [CrossRef]

3. Talbott, E.O.; Malek, A.M.; Lacomis, D. The epidemiology of amyotrophic lateral sclerosis. In *Handbook of Clinical Neurology*; Elsevier: Amsterdam, The Netherlands, 2016; Volume 138, pp. 225–238.
4. Chiò, A.; Logroscino, G.; Traynor, B.J.; Collins, J.; Simeone, J.C.; Goldstein, L.A.; White, L.A. Global Epidemiology of Amyotrophic Lateral Sclerosis: A Systematic Review of the Published Literature. *Neuroepidemiology* **2013**, *41*, 118–130. [CrossRef]
5. Mehta, P.; Kaye, W.; Raymond, J.; Wu, R.; Larson, T.; Punjani, R.; Heller, D.; Cohen, J.; Peters, T.; Muravov, O.; et al. Prevalence of Amyotrophic Lateral Sclerosis—United States. *Morb. Mortal. Wkly. Rep.* **2014**, *67*, 216. [CrossRef]
6. Arthur, K.C.; Calvo, A.; Price, T.R.; Geiger, J.T.; Chiò, A.; Traynor, B.J. Projected increase in amyotrophic lateral sclerosis from 2015 to 2040. *Nat. Commun.* **2016**, *7*, 12408. [CrossRef]
7. Salameh, J.; Brown, R.; Berry, J. Amyotrophic Lateral Sclerosis: Review. *Semin. Neurol.* **2015**, *35*, 469–476. [CrossRef] [PubMed]
8. Brooks, B.R.; Antel, J.; Bradley, W.; Cardy, P.; Carpenter, S.; Chou, S.; Conradi, S.; Daube, J.; Denys, E.H.; Festoff, B.; et al. El Escorial World Federation of Neurology criteria for the diagnosis of amyotrophic lateral sclerosis. *J. Neurol. Sci.* **1994**, *124*, 96–107. [CrossRef]
9. Brooks, B.R.; Miller, R.G.; Swash, M.; Munsat, T.L. El Escorial revisited: Revised criteria for the diagnosis of amyotrophic lateral sclerosis. *Amyotroph. Lateral Scler. Other Motor Neuron Disord.* **2000**, *1*, 293–299. [CrossRef] [PubMed]
10. de Carvalho, M.; Dengler, R.; Eisen, A.; England, J.D.; Kaji, R.; Kimura, J.; Mills, K.; Mitsumoto, H.; Nodera, H.; Shefner, J.; et al. Electrodiagnostic criteria for diagnosis of ALS. *Clin. Neurophysiol.* **2008**, *119*, 497–503. [CrossRef] [PubMed]
11. Chiò, A.; Moglia, C.; Canosa, A.; Manera, U.; Vasta, R.; Brunetti, M.; Barberis, M.; Corrado, L.; D'Alfonso, S.; Bersano, E.; et al. Cognitive impairment across ALS clinical stages in a population-based cohort. *Neurology* **2019**, *93*, e984–e994. [CrossRef]
12. Forbes, R.B.; Colville, S.; Swingler, R.J.; Scottish ALS/MND Register. The epidemiology of amyotrophic lateral sclerosis (ALS/MND) in people aged 80 or over. *Age Ageing* **2004**, *33*, 131–134. [CrossRef] [PubMed]
13. Al-Chalabi, A.; Hardiman, O.; Kiernan, M.C.; Chiò, A.; Rix-Brooks, B.; van den Berg, L.H. Amyotrophic lateral sclerosis: Moving towards a new classification system. *Lancet Neurol.* **2016**, *15*, 1182–1194. [CrossRef]
14. Vijayakumar, U.G.; Milla, V.; Cynthia Stafford, M.Y.; Bjourson, A.J.; Duddy, W.; Duguez, S.M.-R. A Systematic Review of Suggested Molecular Strata, Biomarkers and Their Tissue Sources in ALS. *Front. Neurol.* **2019**, *10*, 400. [CrossRef] [PubMed]
15. Wingo, T.S.; Cutler, D.J.; Yarab, N.; Kelly, C.M.; Glass, J.D. The Heritability of Amyotrophic Lateral Sclerosis in a Clinically Ascertained United States Research Registry. *PLoS ONE* **2011**, *6*, e27985. [CrossRef] [PubMed]
16. Al-Chalabi, A.; Fang, F.; Hanby, M.F.; Leigh, P.N.; Shaw, C.E.; Ye, W.; Rijsdijk, F. An estimate of amyotrophic lateral sclerosis heritability using twin data. *J. Neurol. Neurosurg. Psychiatry* **2010**, *81*, 1324–1326. [CrossRef] [PubMed]
17. Mathis, S.; Goizet, C.; Soulages, A.; Vallat, J.-M.; Masson, G. Le Genetics of amyotrophic lateral sclerosis: A review. *J. Neurol. Sci.* **2019**, *399*, 217–226. [CrossRef]
18. Volk, A.E.; Weishaupt, J.H.; Andersen, P.M.; Ludolph, A.C.; Kubisch, C. Current knowledge and recent insights into the genetic basis of amyotrophic lateral sclerosis. *medizinische Genet.* **2018**, *30*, 252–258. [CrossRef]
19. Zou, Z.-Y.; Zhou, Z.-R.; Che, C.-H.; Liu, C.-Y.; He, R.-L.; Huang, H.-P. Genetic epidemiology of amyotrophic lateral sclerosis: A systematic review and meta-analysis. *J. Neurol. Neurosurg. Psychiatry* **2017**, *88*, 540–549. [CrossRef]
20. Yoshida, M.; Takahashi, Y.; Koike, A.; Fukuda, Y.; Goto, J.; Tsuji, S. A mutation database for amyotrophic lateral sclerosis. *Hum. Mutat.* **2010**, *31*, 1003–1010. [CrossRef]
21. McCann, E.P.; Williams, K.L.; Fifita, J.A.; Tarr, I.S.; O'Connor, J.; Rowe, D.B.; Nicholson, G.A.; Blair, I.P. The genotype-phenotype landscape of familial amyotrophic lateral sclerosis in Australia. *Clin. Genet.* **2017**, *92*, 259–266. [CrossRef]
22. Li, H.-F.; Wu, Z.-Y. Genotype-phenotype correlations of amyotrophic lateral sclerosis. *Transl. Neurodegener.* **2016**, *5*, 3. [CrossRef] [PubMed]
23. Sabatelli, M.; Conte, A.; Zollino, M. Clinical and genetic heterogeneity of amyotrophic lateral sclerosis. *Clin. Genet.* **2013**, *83*, 408–416. [CrossRef] [PubMed]

24. Millecamps, S.; Salachas, F.; Cazeneuve, C.; Gordon, P.; Bricka, B.; Camuzat, A.; Guillot-Noël, L.; Russaouen, O.; Bruneteau, G.; Pradat, P.-F.; et al. SOD1, ANG, VAPB, TARDBP, and FUS mutations in familial amyotrophic lateral sclerosis: Genotype-phenotype correlations. *J. Med. Genet.* **2010**, *47*, 554–560. [CrossRef] [PubMed]
25. Kiernan, M.C.; Vucic, S.; Cheah, B.C.; Turner, M.R.; Eisen, A.; Hardiman, O.; Burrell, J.R.; Zoing, M.C. Amyotrophic lateral sclerosis. *Lancet* **2011**, *377*, 942–955. [CrossRef]
26. Turner, M.R.; Barnwell, J.; Al-Chalabi, A.; Eisen, A. Young-onset amyotrophic lateral sclerosis: Historical and other observations. *Brain* **2012**, *135*, 2883–2891. [CrossRef]
27. Swinnen, B.; Robberecht, W. The phenotypic variability of amyotrophic lateral sclerosis. *Nat. Rev. Neurol.* **2014**, *10*, 661–670. [CrossRef]
28. Chiò, A.; Logroscino, G.; Hardiman, O.; Swingler, R.; Mitchell, D.; Beghi, E.; Traynor, B.G. On Behalf of the Eurals Consortium Prognostic factors in ALS: A critical review. *Amyotroph. Lateral Scler.* **2009**, *10*, 310–323. [CrossRef]
29. Sabatelli, M.; Madia, F.; Conte, A.; Luigetti, M.; Zollino, M.; Mancuso, I.; Lo Monaco, M.; Lippi, G.; Tonali, P. Natural history of young-adult amyotrophic lateral sclerosis. *Neurology* **2008**, *71*, 876–881. [CrossRef]
30. Tard, C.; Defebvre, L.; Moreau, C.; Devos, D.; Danel-Brunaud, V. Clinical features of amyotrophic lateral sclerosis and their prognostic value. *Rev. Neurol.* **2017**, *173*, 263–272. [CrossRef]
31. Garden, G.A.; La Spada, A.R. Intercellular (mis)communication in neurodegenerative disease. *Neuron* **2012**, *73*, 886–901. [CrossRef]
32. Hardiman, O.; Al-Chalabi, A.; Chio, A.; Corr, E.M.; Logroscino, G.; Robberecht, W.; Shaw, P.J.; Simmons, Z.; van den Berg, L.H. Amyotrophic lateral sclerosis. *Nat. Rev. Dis. Prim.* **2017**, *3*, 17071. [CrossRef] [PubMed]
33. Bäumer, D.; Talbot, K.; Turner, M.R. Advances in motor neurone disease. *J. R. Soc. Med.* **2014**, *107*, 14–21. [CrossRef] [PubMed]
34. van Es, M.A.; Hardiman, O.; Chio, A.; Al-Chalabi, A.; Pasterkamp, R.J.; Veldink, J.H.; van den Berg, L.H. Amyotrophic lateral sclerosis. *Lancet* **2017**, *390*, 2084–2098. [CrossRef]
35. Lillo, P.; Hodges, J.R. Frontotemporal dementia and motor neurone disease: Overlapping clinic-pathological disorders. *J. Clin. Neurosci.* **2009**, *16*, 1131–1135. [CrossRef]
36. D'Amico, E.; Pasmantier, M.; Lee, Y.-W.; Weimer, L.; Mitsumoto, H. Clinical evolution of pure upper motor neuron disease/dysfunction (PUMMD). *Muscle Nerve* **2013**, *47*, 28–32. [CrossRef]
37. Statland, J.M.; Barohn, R.J.; McVey, A.L.; Katz, J.S.; Dimachkie, M.M. Patterns of Weakness, Classification of Motor Neuron Disease, and Clinical Diagnosis of Sporadic Amyotrophic Lateral Sclerosis. *Neurol. Clin.* **2015**, *33*, 735–748. [CrossRef] [PubMed]
38. Rowland, L.P. Progressive muscular atrophy and other lower motor neuron syndromes of adults. *Muscle Nerve* **2010**, *41*, 161–165. [CrossRef]
39. Riku, Y.; Atsuta, N.; Yoshida, M.; Tatsumi, S.; Iwasaki, Y.; Mimuro, M.; Watanabe, H.; Ito, M.; Senda, J.; Nakamura, R.; et al. Differential motor neuron involvement in progressive muscular atrophy: A comparative study with amyotrophic lateral sclerosis. *BMJ Open* **2014**, *4*, e005213. [CrossRef]
40. Liewluck, T.; Saperstein, D.S. Progressive Muscular Atrophy. *Neurol. Clin.* **2015**, *33*, 761–773. [CrossRef]
41. Wijesekera, L.C.; Mathers, S.; Talman, P.; Galtrey, C.; Parkinson, M.H.; Ganesalingam, J.; Willey, E.; Ampong, M.A.; Ellis, C.M.; Shaw, C.E.; et al. Natural history and clinical features of the flail arm and flail leg ALS variants. *Neurology* **2009**, *72*, 1087–1094. [CrossRef]
42. Zou, Z.-Y.; Chen, S.-D.; Feng, S.-Y.; Liu, C.-Y.; Cui, M.; Chen, S.; Feng, S.-M.; Dong, Q.; Huang, H.; Yu, J.-T. Familial flail leg ALS caused by PFN1 mutation. *J. Neurol. Neurosurg. Psychiatry* **2020**, *91*, 223–224. [CrossRef] [PubMed]
43. Garg, N.; Park, S.B.; Vucic, S.; Yiannikas, C.; Spies, J.; Howells, J.; Huynh, W.; Matamala, J.M.; Krishnan, A.V.; Pollard, J.D.; et al. Differentiating lower motor neuron syndromes. *J. Neurol. Neurosurg. Psychiatry* **2017**, *88*, 474–483. [CrossRef] [PubMed]
44. Goldstein, L.H.; Abrahams, S. Changes in cognition and behaviour in amyotrophic lateral sclerosis: Nature of impairment and implications for assessment. *Lancet Neurol.* **2013**, *12*, 368–380. [CrossRef]
45. Shen, D.; Cui, L.; Fang, J.; Cui, B.; Li, D.; Tai, H. Voxel-Wise Meta-Analysis of Gray Matter Changes in Amyotrophic Lateral Sclerosis. *Front. Aging Neurosci.* **2016**, *8*. [CrossRef]
46. Geser, F.; Brandmeir, N.J.; Kwong, L.K.; Martinez-Lage, M.; Elman, L.; McCluskey, L.; Xie, S.X.; Lee, V.M.-Y.; Trojanowski, J.Q. Evidence of Multisystem Disorder in Whole-Brain Map of Pathological TDP-43 in Amyotrophic Lateral Sclerosis. *Arch. Neurol.* **2008**, *65*. [CrossRef]

47. Crockford, C.; Newton, J.; Lonergan, K.; Chiwera, T.; Booth, T.; Chandran, S.; Colville, S.; Heverin, M.; Mays, I.; Pal, S.; et al. ALS-specific cognitive and behavior changes associated with advancing disease stage in ALS. *Neurology* **2018**, *91*, e1370–e1380. [CrossRef]
48. Neary, D.; Snowden, J.S.; Gustafson, L.; Passant, U.; Stuss, D.; Black, S.; Freedman, M.; Kertesz, A.; Robert, P.H.; Albert, M.; et al. Frontotemporal lobar degeneration: A consensus on clinical diagnostic criteria. *Neurology* **1998**, *51*, 1546–1554. [CrossRef]
49. Bak, T.H.; Chandran, S. What wires together dies together: Verbs, actions and neurodegeneration in motor neuron disease. *Cortex* **2012**, *48*, 936–944. [CrossRef]
50. Rascovsky, K.; Hodges, J.R.; Knopman, D.; Mendez, M.F.; Kramer, J.H.; Neuhaus, J.; van Swieten, J.C.; Seelaar, H.; Dopper, E.G.P.; Onyike, C.U.; et al. Sensitivity of revised diagnostic criteria for the behavioural variant of frontotemporal dementia. *Brain* **2011**, *134*, 2456–2477. [CrossRef]
51. Strong, M.J.; Abrahams, S.; Goldstein, L.H.; Woolley, S.; Mclaughlin, P.; Snowden, J.; Mioshi, E.; Roberts-South, A.; Benatar, M.; HortobáGyi, T.; et al. Amyotrophic lateral sclerosis—Frontotemporal spectrum disorder (ALS-FTSD): Revised diagnostic criteria. *Amyotroph. Lateral Scler. Front. Degener.* **2017**, *18*, 153–174. [CrossRef]
52. Rosen, D.R.; Siddique, T.; Patterson, D.; Figlewicz, D.A.; Sapp, P.; Hentati, A.; Donaldson, D.; Goto, J.; O'Regan, J.P.; Deng, H.X. Mutations in Cu/Zn superoxide dismutase gene are associated with familial amyotrophic lateral sclerosis. *Nature* **1993**, *362*, 59–62. [CrossRef]
53. Yang, Y.; Hentati, A.; Deng, H.-X.; Dabbagh, O.; Sasaki, T.; Hirano, M.; Hung, W.-Y.; Ouahchi, K.; Yan, J.; Azim, A.C.; et al. The gene encoding alsin, a protein with three guanine-nucleotide exchange factor domains, is mutated in a form of recessive amyotrophic lateral sclerosis. *Nat. Genet.* **2001**, *29*, 160–165. [CrossRef] [PubMed]
54. Hadano, S.; Hand, C.K.; Osuga, H.; Yanagisawa, Y.; Otomo, A.; Devon, R.S.; Miyamoto, N.; Showguchi-Miyata, J.; Okada, Y.; Singaraja, R.; et al. A gene encoding a putative GTPase regulator is mutated in familial amyotrophic lateral sclerosis 2. *Nat. Genet.* **2001**, *29*, 166–173. [CrossRef] [PubMed]
55. Sheerin, U.-M.; Schneider, S.A.; Carr, L.; Deuschl, G.; Hopfner, F.; Stamelou, M.; Wood, N.W.; Bhatia, K.P. ALS2 mutations: Juvenile amyotrophic lateral sclerosis and generalized dystonia. *Neurology* **2014**, *82*, 1065–1067. [CrossRef] [PubMed]
56. Eymard-Pierre, E.; Lesca, G.; Dollet, S.; Santorelli, F.M.; di Capua, M.; Bertini, E.; Boespflug-Tanguy, O. Infantile-Onset Ascending Hereditary Spastic Paralysis Is Associated with Mutations in the Alsin Gene. *Am. J. Hum. Genet.* **2002**, *71*, 518–527. [CrossRef] [PubMed]
57. Otomo, A. ALS2, a novel guanine nucleotide exchange factor for the small GTPase Rab5, is implicated in endosomal dynamics. *Hum. Mol. Genet.* **2003**, *12*, 1671–1687. [CrossRef] [PubMed]
58. Hsu, F.; Spannl, S.; Ferguson, C.; Hyman, A.A.; Parton, R.G.; Zerial, M. Rab5 and Alsin regulate stress-activated cytoprotective signaling on mitochondria. *Elife* **2018**, *7*. [CrossRef]
59. Devon, R.S.; Orban, P.C.; Gerrow, K.; Barbieri, M.A.; Schwab, C.; Cao, L.P.; Helm, J.R.; Bissada, N.; Cruz-Aguado, R.; Davidson, T.-L.; et al. Als2-deficient mice exhibit disturbances in endosome trafficking associated with motor behavioral abnormalities. *Proc. Natl. Acad. Sci. USA* **2006**, *103*, 9595–9600. [CrossRef]
60. Cai, H. Loss of ALS2 Function Is Insufficient to Trigger Motor Neuron Degeneration in Knock-Out Mice But Predisposes Neurons to Oxidative Stress. *J. Neurosci.* **2005**, *25*, 7567–7574. [CrossRef]
61. Suraweera, A.; Lim, Y.; Woods, R.; Birrell, G.W.; Nasim, T.; Becherel, O.J.; Lavin, M.F. Functional role for senataxin, defective in ataxia oculomotor apraxia type 2, in transcriptional regulation. *Hum. Mol. Genet.* **2009**, *18*, 3384–3396. [CrossRef]
62. Steinmetz, E.J.; Warren, C.L.; Kuehner, J.N.; Panbehi, B.; Ansari, A.Z.; Brow, D.A. Genome-Wide Distribution of Yeast RNA Polymerase II and Its Control by Sen1 Helicase. *Mol. Cell* **2006**, *24*, 735–746. [CrossRef] [PubMed]
63. Chen, Y.-Z.; Bennett, C.L.; Huynh, H.M.; Blair, I.P.; Puls, I.; Irobi, J.; Dierick, I.; Abel, A.; Kennerson, M.L.; Rabin, B.A.; et al. DNA/RNA Helicase Gene Mutations in a Form of Juvenile Amyotrophic Lateral Sclerosis (ALS4). *Am. J. Hum. Genet.* **2004**, *74*, 1128–1135. [CrossRef] [PubMed]
64. Avemaria, F.; Lunetta, C.; Tarlarini, C.; Mosca, L.; Maestri, E.; Marocchi, A.; Melazzini, M.; Penco, S.; Corbo, M. Mutation in the senataxin gene found in a patient affected by familial ALS with juvenile onset and slow progression. *Amyotroph. Lateral Scler.* **2011**, *12*, 228–230. [CrossRef] [PubMed]

65. Rudnik-Schöneborn, S.; Arning, L.; Epplen, J.T.; Zerres, K. SETX gene mutation in a family diagnosed autosomal dominant proximal spinal muscular atrophy. *Neuromuscul. Disord.* **2012**, *22*, 258–262. [CrossRef] [PubMed]
66. Wu, D.; Yu, W.; Kishikawa, H.; Folkerth, R.D.; Iafrate, A.J.; Shen, Y.; Xin, W.; Sims, K.; Hu, G. Angiogenin loss-of-function mutations in amyotrophic lateral sclerosis. *Ann. Neurol.* **2007**, *62*, 609–617. [CrossRef] [PubMed]
67. Subramanian, V.; Crabtree, B.; Acharya, K.R. Human angiogenin is a neuroprotective factor and amyotrophic lateral sclerosis associated angiogenin variants affect neurite extension/pathfinding and survival of motor neurons. *Hum. Mol. Genet.* **2007**, *17*, 130–149. [CrossRef]
68. Tsuji, T.; Sun, Y.; Kishimoto, K.; Olson, K.A.; Liu, S.; Hirukawa, S.; Hu, G. Angiogenin Is Translocated to the Nucleus of HeLa Cells and Is Involved in Ribosomal RNA Transcription and Cell Proliferation. *Cancer Res.* **2005**, *65*, 1352–1360. [CrossRef]
69. Thiyagarajan, N.; Ferguson, R.; Subramanian, V.; Acharya, K.R. Structural and molecular insights into the mechanism of action of human angiogenin-ALS variants in neurons. *Nat. Commun.* **2012**, *3*, 1121. [CrossRef]
70. Kanekura, K.; Nishimoto, I.; Aiso, S.; Matsuoka, M. Characterization of Amyotrophic Lateral Sclerosis-linked P56S Mutation of Vesicle-associated Membrane Protein-associated Protein B (VAPB/ALS8). *J. Biol. Chem.* **2006**, *281*, 30223–30233. [CrossRef]
71. Peretti, D.; Dahan, N.; Shimoni, E.; Hirschberg, K.; Lev, S. Coordinated Lipid Transfer between the Endoplasmic Reticulum and the Golgi Complex Requires the VAP Proteins and Is Essential for Golgi-mediated Transport. *Mol. Biol. Cell* **2008**, *19*, 3871–3884. [CrossRef]
72. Tsuda, H.; Han, S.M.; Yang, Y.; Tong, C.; Lin, Y.Q.; Mohan, K.; Haueter, C.; Zoghbi, A.; Harati, Y.; Kwan, J.; et al. The Amyotrophic Lateral Sclerosis 8 Protein VAPB Is Cleaved, Secreted, and Acts as a Ligand for Eph Receptors. *Cell* **2008**, *133*, 963–977. [CrossRef] [PubMed]
73. Morotz, G.M.; De Vos, K.J.; Vagnoni, A.; Ackerley, S.; Shaw, C.E.; Miller, C.C.J. Amyotrophic lateral sclerosis-associated mutant VAPBP56S perturbs calcium homeostasis to disrupt axonal transport of mitochondria. *Hum. Mol. Genet.* **2012**, *21*, 1979–1988. [CrossRef] [PubMed]
74. Nishimura, A.L.; Mitne-Neto, M.; Silva, H.C.A.; Richieri-Costa, A.; Middleton, S.; Cascio, D.; Kok, F.; Oliveira, J.R.M.; Gillingwater, T.; Webb, J.; et al. A Mutation in the Vesicle-Trafficking Protein VAPB Causes Late-Onset Spinal Muscular Atrophy and Amyotrophic Lateral Sclerosis. *Am. J. Hum. Genet.* **2004**, *75*, 822–831. [CrossRef] [PubMed]
75. Tran, D.; Chalhoub, A.; Schooley, A.; Zhang, W.; Ngsee, J.K. A mutation in VAPB that causes amyotrophic lateral sclerosis also causes a nuclear envelope defect. *J. Cell Sci.* **2012**, *125*, 2831–2836. [CrossRef]
76. Qiu, L.; Qiao, T.; Beers, M.; Tan, W.; Wang, H.; Yang, B.; Xu, Z. Widespread aggregation of mutant VAPB associated with ALS does not cause motor neuron degeneration or modulate mutant SOD1 aggregation and toxicity in mice. *Mol. Neurodegener.* **2013**, *8*, 1. [CrossRef]
77. Van Deerlin, V.M.; Leverenz, J.B.; Bekris, L.M.; Bird, T.D.; Yuan, W.; Elman, L.B.; Clay, D.; Wood, E.M.; Chen-Plotkin, A.S.; Martinez-Lage, M.; et al. TARDBP mutations in amyotrophic lateral sclerosis with TDP-43 neuropathology: A genetic and histopathological analysis. *Lancet Neurol.* **2008**, *7*, 409–416. [CrossRef]
78. Prasad, A.; Bharathi, V.; Sivalingam, V.; Girdhar, A.; Patel, B.K. Molecular Mechanisms of TDP-43 Misfolding and Pathology in Amyotrophic Lateral Sclerosis. *Front. Mol. Neurosci.* **2019**, *12*. [CrossRef]
79. Neumann, M.; Sampathu, D.M.; Kwong, L.K.; Truax, A.C.; Micsenyi, M.C.; Chou, T.T.; Bruce, J.; Schuck, T.; Grossman, M.; Clark, C.M.; et al. Ubiquitinated TDP-43 in Frontotemporal Lobar Degeneration and Amyotrophic Lateral Sclerosis. *Science* **2006**, *314*, 130–133. [CrossRef]
80. Ling, S.-C.; Polymenidou, M.; Cleveland, D.W. Converging Mechanisms in ALS and FTD: Disrupted RNA and Protein Homeostasis. *Neuron* **2013**, *79*, 416–438. [CrossRef]
81. Rayaprolu, S.; Fujioka, S.; Traynor, S.; Soto-Ortolaza, A.I.; Petrucelli, L.; Dickson, D.W.; Rademakers, R.; Boylan, K.B.; Graff-Radford, N.R.; Uitti, R.J.; et al. TARDBP mutations in Parkinson's disease. *Parkinsonism Relat. Disord.* **2013**, *19*, 312–315. [CrossRef]
82. Schwab, C.; Arai, T.; Hasegawa, M.; Yu, S.; McGeer, P.L. Colocalization of Transactivation-Responsive DNA-Binding Protein 43 and Huntingtin in Inclusions of Huntington Disease. *J. Neuropathol. Exp. Neurol.* **2008**, *67*, 1159–1165. [CrossRef] [PubMed]

83. King, A.; Sweeney, F.; Bodi, I.; Troakes, C.; Maekawa, S.; Al-Sarraj, S. Abnormal TDP-43 expression is identified in the neocortex in cases of dementia pugilistica, but is mainly confined to the limbic system when identified in high and moderate stages of Alzheimer's disease. *Neuropathology* **2010**, *30*, 408–419. [CrossRef] [PubMed]
84. Nakashima-Yasuda, H.; Uryu, K.; Robinson, J.; Xie, S.X.; Hurtig, H.; Duda, J.E.; Arnold, S.E.; Siderowf, A.; Grossman, M.; Leverenz, J.B.; et al. Co-morbidity of TDP-43 proteinopathy in Lewy body related diseases. *Acta Neuropathol.* **2007**, *114*, 221–229. [CrossRef] [PubMed]
85. Kwiatkowski, T.J.; Bosco, D.A.; Leclerc, A.L.; Tamrazian, E.; Vanderburg, C.R.; Russ, C.; Davis, A.; Gilchrist, J.; Kasarskis, E.J.; Munsat, T.; et al. Mutations in the FUS/TLS gene on chromosome 16 cause familial amyotrophic lateral sclerosis. *Science* **2009**, *323*, 1205–1208. [CrossRef]
86. Chow, C.Y.; Landers, J.E.; Bergren, S.K.; Sapp, P.C.; Grant, A.E.; Jones, J.M.; Everett, L.; Lenk, G.M.; McKenna-Yasek, D.M.; Weisman, L.S.; et al. Deleterious Variants of FIG4, a Phosphoinositide Phosphatase, in Patients with ALS. *Am. J. Hum. Genet.* **2009**, *84*, 85–88. [CrossRef]
87. Lagier-Tourenne, C.; Polymenidou, M.; Cleveland, D.W. TDP-43 and FUS/TLS: Emerging roles in RNA processing and neurodegeneration. *Hum. Mol. Genet.* **2010**, *19*, R46–R64. [CrossRef]
88. Vandoorne, T.; Veys, K.; Guo, W.; Sicart, A.; Vints, K.; Swijsen, A.; Moisse, M.; Eelen, G.; Gounko, N.V.; Fumagalli, L.; et al. Differentiation but not ALS mutations in FUS rewires motor neuron metabolism. *Nat. Commun.* **2019**, *10*, 4147. [CrossRef]
89. Shelkovnikova, T.A.; Peters, O.M.; Deykin, A.V.; Connor-Robson, N.; Robinson, H.; Ustyugov, A.A.; Bachurin, S.O.; Ermolkevich, T.G.; Goldman, I.L.; Sadchikova, E.R.; et al. Fused in Sarcoma (FUS) Protein Lacking Nuclear Localization Signal (NLS) and Major RNA Binding Motifs Triggers Proteinopathy and Severe Motor Phenotype in Transgenic Mice. *J. Biol. Chem.* **2013**, *288*, 25266–25274. [CrossRef]
90. Gentile, F.; Scarlino, S.; Falzone, Y.M.; Lunetta, C.; Tremolizzo, L.; Quattrini, A.; Riva, N. The Peripheral Nervous System in Amyotrophic Lateral Sclerosis: Opportunities for Translational Research. *Front. Neurosci.* **2019**, *13*. [CrossRef]
91. Chow, C.Y.; Zhang, Y.; Dowling, J.J.; Jin, N.; Adamska, M.; Shiga, K.; Szigeti, K.; Shy, M.E.; Li, J.; Zhang, X.; et al. Mutation of FIG4 causes neurodegeneration in the pale tremor mouse and patients with CMT4J. *Nature* **2007**, *448*, 68–72. [CrossRef]
92. Tsai, C.-P.; Soong, B.-W.; Lin, K.-P.; Tu, P.-H.; Lin, J.-L.; Lee, Y.-C. FUS, TARDBP, and SOD1 mutations in a Taiwanese cohort with familial ALS. *Neurobiol. Aging* **2011**, *32*, 553.e13–553.e21. [CrossRef] [PubMed]
93. Verdiani, S.; Origone, P.; Geroldi, A.; Bandettini Di Poggio, M.; Mantero, V.; Bellone, E.; Mancardi, G.; Caponnetto, C.; Mandich, P. The FIG4 gene does not play a major role in causing ALS in Italian patients. *Amyotroph. Lateral Scler. Front. Degener.* **2013**, *14*, 228–229. [CrossRef] [PubMed]
94. Maruyama, H.; Morino, H.; Ito, H.; Izumi, Y.; Kato, H.; Watanabe, Y.; Kinoshita, Y.; Kamada, M.; Nodera, H.; Suzuki, H.; et al. Mutations of optineurin in amyotrophic lateral sclerosis. *Nature* **2010**, *465*, 223–226. [CrossRef] [PubMed]
95. Orlacchio, A.; Babalini, C.; Borreca, A.; Patrono, C.; Massa, R.; Basaran, S.; Munhoz, R.P.; Rogaeva, E.A.; St George-Hyslop, P.H.; Bernardi, G.; et al. SPATACSIN mutations cause autosomal recessive juvenile amyotrophic lateral sclerosis. *Brain* **2010**, *133*, 591–598. [CrossRef] [PubMed]
96. Liao, S.; Shen, L.; Du, J.; Zhao, G.; Wang, X.; Yang, Y.; Xiao, Z.; Yuan, Y.; Jiang, H.; Li, N.; et al. Novel mutations of the SPG11 gene in hereditary spastic paraplegia with thin corpus callosum. *J. Neurol. Sci.* **2008**, *275*, 92–99. [CrossRef]
97. Branchu, J.; Boutry, M.; Sourd, L.; Depp, M.; Leone, C.; Corriger, A.; Vallucci, M.; Esteves, T.; Matusiak, R.; Dumont, M.; et al. Loss of spatacsin function alters lysosomal lipid clearance leading to upper and lower motor neuron degeneration. *Neurobiol. Dis.* **2017**, *102*, 21–37. [CrossRef]
98. Elden, A.C.; Kim, H.-J.; Hart, M.P.; Chen-Plotkin, A.S.; Johnson, B.S.; Fang, X.; Armakola, M.; Geser, F.; Greene, R.; Lu, M.M.; et al. Ataxin-2 intermediate-length polyglutamine expansions are associated with increased risk for ALS. *Nature* **2010**, *466*, 1069–1075. [CrossRef]
99. Fittschen, M.; Lastres-Becker, I.; Halbach, M.V.; Damrath, E.; Gispert, S.; Azizov, M.; Walter, M.; Müller, S.; Auburger, G. Genetic ablation of ataxin-2 increases several global translation factors in their transcript abundance but decreases translation rate. *Neurogenetics* **2015**, *16*, 181–192. [CrossRef]
100. Satterfield, T.F.; Pallanck, L.J. Ataxin-2 and its Drosophila homolog, ATX2, physically assemble with polyribosomes. *Hum. Mol. Genet.* **2006**, *15*, 2523–2532. [CrossRef]

101. Van Damme, P.; Veldink, J.H.; van Blitterswijk, M.; Corveleyn, A.; van Vught, P.W.J.; Thijs, V.; Dubois, B.; Matthijs, G.; van den Berg, L.H.; Robberecht, W. Expanded ATXN2 CAG repeat size in ALS identifies genetic overlap between ALS and SCA2. *Neurology* **2011**, *76*, 2066–2072. [CrossRef]
102. Liu, X.; Lu, M.; Tang, L.; Zhang, N.; Chui, D.; Fan, D. ATXN2 CAG repeat expansions increase the risk for Chinese patients with amyotrophic lateral sclerosis. *Neurobiol. Aging* **2013**, *34*, 2236.e5–2236.e8. [CrossRef] [PubMed]
103. Meyer, H.; Bug, M.; Bremer, S. Emerging functions of the VCP/p97 AAA-ATPase in the ubiquitin system. *Nat. Cell Biol.* **2012**, *14*, 117–123. [CrossRef] [PubMed]
104. Ju, J.-S.; Fuentealba, R.A.; Miller, S.E.; Jackson, E.; Piwnica-Worms, D.; Baloh, R.H.; Weihl, C.C. Valosin-containing protein (VCP) is required for autophagy and is disrupted in VCP disease. *J. Cell Biol.* **2009**, *187*, 875–888. [CrossRef] [PubMed]
105. Verma, R.; Oania, R.S.; Kolawa, N.J.; Deshaies, R.J. Cdc48/p97 promotes degradation of aberrant nascent polypeptides bound to the ribosome. *Elife* **2013**, *2*. [CrossRef]
106. Rumpf, S.; Bagley, J.A.; Thompson-Peer, K.L.; Zhu, S.; Gorczyca, D.; Beckstead, R.B.; Jan, L.Y.; Jan, Y.N. Drosophila Valosin-Containing Protein is required for dendrite pruning through a regulatory role in mRNA metabolism. *Proc. Natl. Acad. Sci. USA* **2014**, *111*, 7331–7336. [CrossRef]
107. Johnson, J.O.; Mandrioli, J.; Benatar, M.; Abramzon, Y.; Van Deerlin, V.M.; Trojanowski, J.Q.; Gibbs, J.R.; Brunetti, M.; Gronka, S.; Wuu, J.; et al. Exome Sequencing Reveals VCP Mutations as a Cause of Familial ALS. *Neuron* **2010**, *68*, 857–864. [CrossRef]
108. Yin, H.Z.; Nalbandian, A.; Hsu, C.-I.; Li, S.; Llewellyn, K.J.; Mozaffar, T.; Kimonis, V.E.; Weiss, J.H. Slow development of ALS-like spinal cord pathology in mutant valosin-containing protein gene knock-in mice. *Cell Death Dis.* **2012**, *3*, e374. [CrossRef]
109. Williams, K.L.; Warraich, S.T.; Yang, S.; Solski, J.A.; Fernando, R.; Rouleau, G.A.; Nicholson, G.A.; Blair, I.P. UBQLN2/ubiquilin 2 mutation and pathology in familial amyotrophic lateral sclerosis. *Neurobiol. Aging* **2012**, *33*, 2527.e3–2527.e10. [CrossRef]
110. Goutman, S.A.; Chen, K.S.; Paez-Colasante, X.; Feldman, E.L. Emerging understanding of the genotype–phenotype relationship in amyotrophic lateral sclerosis. *Handb. Clin. Neurol.* **2018**, *148*, 603–623.
111. Mizuno, Y.; Amari, M.; Takatama, M.; Aizawa, H.; Mihara, B.; Okamoto, K. Immunoreactivities of p62, an ubiqutin-binding protein, in the spinal anterior horn cells of patients with amyotrophic lateral sclerosis. *J. Neurol. Sci.* **2006**, *249*, 13–18. [CrossRef]
112. DeJesus-Hernandez, M.; Mackenzie, I.R.R.; Boeve, B.F.F.; Boxer, A.L.L.; Baker, M.; Rutherford, N.J.J.; Nicholson, A.M.M.; Finch, N.A.A.; Flynn, H.; Adamson, J.; et al. Expanded GGGGCC hexanucleotide repeat in noncoding region of C9ORF72 causes chromosome 9p-linked FTD and ALS. *Neuron* **2011**, *72*, 245–256. [CrossRef] [PubMed]
113. Renton, A.E.; Majounie, E.; Waite, A.; Simón-Sánchez, J.; Rollinson, S.; Gibbs, J.R.; Schymick, J.C.; Laaksovirta, H.; van Swieten, J.C.; Myllykangas, L.; et al. A Hexanucleotide Repeat Expansion in C9ORF72 Is the Cause of Chromosome 9p21-Linked ALS-FTD. *Neuron* **2011**, *72*, 257–268. [CrossRef] [PubMed]
114. Umoh, M.E.; Fournier, C.; Li, Y.; Polak, M.; Shaw, L.; Landers, J.E.; Hu, W.; Gearing, M.; Glass, J.D. Comparative analysis of C9orf72 and sporadic disease in an ALS clinic population. *Neurology* **2016**, *87*, 1024–1030. [CrossRef] [PubMed]
115. Iacoangeli, A.; Al Khleifat, A.; Jones, A.R.; Sproviero, W.; Shatunov, A.; Opie-Martin, S.; Morrison, K.E.; Shaw, P.J.; Shaw, C.E.; Fogh, I.; et al. C9orf72 intermediate expansions of 24–30 repeats are associated with ALS. *Acta Neuropathol. Commun.* **2019**, *7*, 115. [CrossRef] [PubMed]
116. Balendra, R.; Isaacs, A.M. C9orf72-mediated ALS and FTD: Multiple pathways to disease. *Nat. Rev. Neurol.* **2018**, *14*, 544–558. [CrossRef]
117. Cooper-Knock, J.; Frolov, A.; Highley, J.R.; Charlesworth, G.; Kirby, J.; Milano, A.; Hartley, J.; Ince, P.G.; McDermott, C.J.; Lashley, T.; et al. C9ORF72 expansions, parkinsonism, and Parkinson disease: A clinicopathologic study. *Neurology* **2013**, *81*, 808–811. [CrossRef]
118. Devenney, E.M.; Ahmed, R.M.; Halliday, G.; Piguet, O.; Kiernan, M.C.; Hodges, J.R. Psychiatric disorders in C9orf72 kindreds. *Neurology* **2018**, *91*, e1498–e1507. [CrossRef]
119. Wu, C.-H.; Fallini, C.; Ticozzi, N.; Keagle, P.J.; Sapp, P.C.; Piotrowska, K.; Lowe, P.; Koppers, M.; McKenna-Yasek, D.; Baron, D.M.; et al. Mutations in the profilin 1 gene cause familial amyotrophic lateral sclerosis. *Nature* **2012**, *488*, 499–503. [CrossRef]

120. Yang, C.; Danielson, E.W.; Qiao, T.; Metterville, J.; Brown, R.H.; Landers, J.E.; Xu, Z. Mutant PFN1 causes ALS phenotypes and progressive motor neuron degeneration in mice by a gain of toxicity. *Proc. Natl. Acad. Sci. USA* **2016**, *113*, E6209–E6218. [CrossRef]
121. Kim, H.J.; Kim, N.C.; Wang, Y.-D.; Scarborough, E.A.; Moore, J.; Diaz, Z.; MacLea, K.S.; Freibaum, B.; Li, S.; Molliex, A.; et al. Mutations in prion-like domains in hnRNPA2B1 and hnRNPA1 cause multisystem proteinopathy and ALS. *Nature* **2013**, *495*, 467–473. [CrossRef]
122. Honda, H.; Hamasaki, H.; Wakamiya, T.; Koyama, S.; Suzuki, S.O.; Fujii, N.; Iwaki, T. Loss of hnRNPA1 in ALS spinal cord motor neurons with TDP-43-positive inclusions. *Neuropathology* **2015**, *35*, 37–43. [CrossRef] [PubMed]
123. Smith, B.N.; Ticozzi, N.; Fallini, C.; Gkazi, A.S.; Topp, S.; Kenna, K.P.; Scotter, E.L.; Kost, J.; Keagle, P.; Miller, J.W.; et al. Exome-wide Rare Variant Analysis Identifies TUBA4A Mutations Associated with Familial ALS. *Neuron* **2014**, *84*, 324–331. [CrossRef] [PubMed]
124. Perrone, F.; Nguyen, H.P.; Van Mossevelde, S.; Moisse, M.; Sieben, A.; Santens, P.; De Bleecker, J.; Vandenbulcke, M.; Engelborghs, S.; Baets, J.; et al. Investigating the role of ALS genes CHCHD10 and TUBA4A in Belgian FTD-ALS spectrum patients. *Neurobiol. Aging* **2017**, *51*, 177.e9–177.e16. [CrossRef] [PubMed]
125. Li, J.; He, J.; Tang, L.; Chen, L.; Ma, Y.; Fan, D. Screening for TUBA4A mutations in a large Chinese cohort of patients with ALS: Re-evaluating the pathogenesis of TUBA4A in ALS. *J. Neurol. Neurosurg. Psychiatry* **2018**, *89*, 1350–1352. [CrossRef] [PubMed]
126. Johnson, J.O.; Pioro, E.P.; Boehringer, A.; Chia, R.; Feit, H.; Renton, A.E.; Pliner, H.A.; Abramzon, Y.; Marangi, G.; Winborn, B.J.; et al. Mutations in the Matrin 3 gene cause familial amyotrophic lateral sclerosis. *Nat. Neurosci.* **2014**, *17*, 664–666. [CrossRef]
127. Salton, M.; Elkon, R.; Borodina, T.; Davydov, A.; Yaspo, M.-L.; Halperin, E.; Shiloh, Y. Matrin 3 Binds and Stabilizes mRNA. *PLoS ONE* **2011**, *6*, e23882. [CrossRef] [PubMed]
128. Coelho, M.B.; Attig, J.; Bellora, N.; König, J.; Hallegger, M.; Kayikci, M.; Eyras, E.; Ule, J.; Smith, C.W. Nuclear matrix protein Matrin3 regulates alternative splicing and forms overlapping regulatory networks with PTB. *EMBO J.* **2015**, *34*, 653–668. [CrossRef]
129. Boehringer, A.; Garcia-Mansfield, K.; Singh, G.; Bakkar, N.; Pirrotte, P.; Bowser, R. ALS Associated Mutations in Matrin 3 Alter Protein-Protein Interactions and Impede mRNA Nuclear Export. *Sci. Rep.* **2017**, *7*, 14529. [CrossRef]
130. Cirulli, E.T.; Lasseigne, B.N.; Petrovski, S.; Sapp, P.C.; Dion, P.A.; Leblond, C.S.; Couthouis, J.; Lu, Y.-F.; Wang, Q.; Krueger, B.J.; et al. Exome sequencing in amyotrophic lateral sclerosis identifies risk genes and pathways. *Science* **2015**, *347*, 1436–1441. [CrossRef]
131. Brenner, D.; Müller, K.; Wieland, T.; Weydt, P.; Böhm, S.; Lulé, D.; Hübers, A.; Neuwirth, C.; Weber, M.; Borck, G.; et al. NEK1 mutations in familial amyotrophic lateral sclerosis. *Brain* **2016**, *139*, e28. [CrossRef]
132. Kenna, K.P.; van Doormaal, P.T.C.; Dekker, A.M.; Ticozzi, N.; Kenna, B.J.; Diekstra, F.P.; van Rheenen, W.; van Eijk, K.R.; Jones, A.R.; Keagle, P.; et al. NEK1 variants confer susceptibility to amyotrophic lateral sclerosis. *Nat. Genet.* **2016**, *48*, 1037–1042. [CrossRef] [PubMed]
133. van Rheenen, W.; Shatunov, A.; Dekker, A.M.; McLaughlin, R.L.; Diekstra, F.P.; Pulit, S.L.; van der Spek, R.A.A.; Võsa, U.; de Jong, S.; Robinson, M.R.; et al. Genome-wide association analyses identify new risk variants and the genetic architecture of amyotrophic lateral sclerosis. *Nat. Genet.* **2016**, *48*, 1043–1048. [CrossRef] [PubMed]
134. Freischmidt, A.; Wieland, T.; Richter, B.; Ruf, W.; Schaeffer, V.; Müller, K.; Marroquin, N.; Nordin, F.; Hübers, A.; Weydt, P.; et al. Haploinsufficiency of TBK1 causes familial ALS and fronto-temporal dementia. *Nat. Neurosci.* **2015**, *18*, 631–636. [CrossRef] [PubMed]
135. Williams, K.L.; Topp, S.; Yang, S.; Smith, B.; Fifita, J.A.; Warraich, S.T.; Zhang, K.Y.; Farrawell, N.; Vance, C.; Hu, X.; et al. CCNF mutations in amyotrophic lateral sclerosis and frontotemporal dementia. *Nat. Commun.* **2016**, *7*, 11253. [CrossRef]
136. Yu, Y.; Nakagawa, T.; Morohoshi, A.; Nakagawa, M.; Ishida, N.; Suzuki, N.; Aoki, M.; Nakayama, K. Pathogenic mutations in the ALS gene CCNF cause cytoplasmic mislocalization of Cyclin F and elevated VCP ATPase activity. *Hum. Mol. Genet.* **2019**, *28*, 3486–3497. [CrossRef]

137. Nicolas, A.; Kenna, K.P.; Renton, A.E.; Ticozzi, N.; Faghri, F.; Chia, R.; Dominov, J.A.; Kenna, B.J.; Nalls, M.A.; Keagle, P.; et al. Genome-wide Analyses Identify KIF5A as a Novel ALS Gene. *Neuron* **2018**, *97*, 1268–1283.e6. [CrossRef]
138. Campbell, P.D.; Shen, K.; Sapio, M.R.; Glenn, T.D.; Talbot, W.S.; Marlow, F.L. Unique Function of Kinesin Kif5A in Localization of Mitochondria in Axons. *J. Neurosci.* **2014**, *34*, 14717–14732. [CrossRef]
139. Hancock, W.O.; Howard, J. Processivity of the Motor Protein Kinesin Requires Two Heads. *J. Cell Biol.* **1998**, *140*, 1395–1405. [CrossRef]
140. Liu, Y.-T.; Laura, M.; Hersheson, J.; Horga, A.; Jaunmuktane, Z.; Brandner, S.; Pittman, A.; Hughes, D.; Polke, J.M.; Sweeney, M.G.; et al. Extended phenotypic spectrum of KIF5A mutations: From spastic paraplegia to axonal neuropathy. *Neurology* **2014**, *83*, 612–619. [CrossRef]
141. Gu, X.; Li, C.; Chen, Y.; Wei, Q.; Cao, B.; Ou, R.; Yuan, X.; Hou, Y.; Zhang, L.; Liu, H.; et al. Mutation screening of the KIF5A gene in Chinese patients with amyotrophic lateral sclerosis. *J. Neurol. Neurosurg. Psychiatry* **2019**, *90*, 245–246. [CrossRef]
142. Filosto, M.; Piccinelli, S.; Palmieri, I.; Necchini, N.; Valente, M.; Zanella, I.; Biasiotto, G.; Lorenzo, D.; Cereda, C.; Padovani, A. A Novel Mutation in the Stalk Domain of KIF5A Causes a Slowly Progressive Atypical Motor Syndrome. *J. Clin. Med.* **2018**, *8*, 17. [CrossRef] [PubMed]
143. Brenner, D.; Yilmaz, R.; Müller, K.; Grehl, T.; Petri, S.; Meyer, T.; Grosskreutz, J.; Weydt, P.; Ruf, W.; Neuwirth, C.; et al. Hot-spot KIF5A mutations cause familial ALS. *Brain* **2018**, *141*, 688–697. [CrossRef] [PubMed]
144. Faber, I.; Martinez, A.R.M.; de Rezende, T.J.R.; Martins, C.R.; Martins, M.P.; Lourenço, C.M.; Marques, W.; Montecchiani, C.; Orlacchio, A.; Pedroso, J.L.; et al. SPG11 mutations cause widespread white matter and basal ganglia abnormalities, but restricted cortical damage. *NeuroImage Clin.* **2018**, *19*, 848–857. [CrossRef] [PubMed]
145. Liu, Z.-J.; Lin, H.-X.; Liu, G.-L.; Tao, Q.-Q.; Ni, W.; Xiao, B.-G.; Wu, Z.-Y. The investigation of genetic and clinical features in Chinese patients with juvenile amyotrophic lateral sclerosis. *Clin. Genet.* **2017**, *92*, 267–273. [CrossRef] [PubMed]
146. Zou, Z.-Y.; Cui, L.-Y.; Sun, Q.; Li, X.-G.; Liu, M.-S.; Xu, Y.; Zhou, Y.; Yang, X.-Z. De novo FUS gene mutations are associated with juvenile-onset sporadic amyotrophic lateral sclerosis in China. *Neurobiol. Aging* **2013**, *34*, 1312.e1–1312.e8. [CrossRef]
147. Mackenzie, I.R.A.; Ansorge, O.; Strong, M.; Bilbao, J.; Zinman, L.; Ang, L.-C.; Baker, M.; Stewart, H.; Eisen, A.; Rademakers, R.; et al. Pathological heterogeneity in amyotrophic lateral sclerosis with FUS mutations: Two distinct patterns correlating with disease severity and mutation. *Acta Neuropathol.* **2011**, *122*, 87–98. [CrossRef]
148. Waibel, S.; Neumann, M.; Rosenbohm, A.; Birve, A.; Volk, A.E.; Weishaupt, J.H.; Meyer, T.; Müller, U.; Andersen, P.M.; Ludolph, A.C. Truncating mutations in FUS/TLS give rise to a more aggressive ALS-phenotype than missense mutations: A clinico-genetic study in Germany. *Eur. J. Neurol.* **2013**, *20*, 540–546. [CrossRef]
149. Orban, P.; Devon, R.S.; Hayden, M.R.; Leavitt, B.R. Chapter 15 Juvenile amyotrophic lateral sclerosis. In *Handbook of Clinical Neurology*; Elsevier: Amsterdam, The Netherlands, 2007; pp. 301–312.
150. Zhao, Z.; Chen, W.; Wu, Z.; Wang, N.; Zhao, G.; Chen, W.; Murong, S. A novel mutation in the senataxin gene identified in a Chinese patient with sporadic amyotrophic lateral sclerosis. *Amyotroph. Lateral Scler.* **2009**, *10*, 118–122. [CrossRef]
151. Murphy, N.A.; Arthur, K.C.; Tienari, P.J.; Houlden, H.; Chiò, A.; Traynor, B.J. Age-related penetrance of the C9orf72 repeat expansion. *Sci. Rep.* **2017**, *7*, 2116. [CrossRef]
152. Trojsi, F.; Siciliano, M.; Femiano, C.; Santangelo, G.; Lunetta, C.; Calvo, A.; Moglia, C.; Marinou, K.; Ticozzi, N.; Ferro, C.; et al. Comparative Analysis of C9orf72 and Sporadic Disease in a Large Multicenter ALS Population: The Effect of Male Sex on Survival of C9orf72 Positive Patients. *Front. Neurosci.* **2019**, *13*. [CrossRef]
153. Rooney, J.; Fogh, I.; Westeneng, H.-J.; Vajda, A.; McLaughlin, R.; Heverin, M.; Jones, A.; van Eijk, R.; Calvo, A.; Mazzini, L.; et al. C9orf72 expansion differentially affects males with spinal onset amyotrophic lateral sclerosis. *J. Neurol. Neurosurg. Psychiatry* **2017**, *88*, 281. [CrossRef] [PubMed]
154. Liu, K.X.; Edwards, B.; Lee, S.; Finelli, M.J.; Davies, B.; Davies, K.E.; Oliver, P.L. Neuron-specific antioxidant OXR1 extends survival of a mouse model of amyotrophic lateral sclerosis. *Brain* **2015**, *138*, 1167–1181. [CrossRef]

155. Fischer, L.R.; Culver, D.G.; Tennant, P.; Davis, A.A.; Wang, M.; Castellano-Sanchez, A.; Khan, J.; Polak, M.A.; Glass, J.D. Amyotrophic lateral sclerosis is a distal axonopathy: Evidence in mice and man. *Exp. Neurol.* **2004**, *185*, 232–240. [CrossRef] [PubMed]
156. Park, J.H.; Elpers, C.; Reunert, J.; McCormick, M.L.; Mohr, J.; Biskup, S.; Schwartz, O.; Rust, S.; Grüneberg, M.; Seelhöfer, A.; et al. SOD1 deficiency: A novel syndrome distinct from amyotrophic lateral sclerosis. *Brain* **2019**, *142*, 2230–2237. [CrossRef] [PubMed]
157. Tasca, G.; Lattante, S.; Marangi, G.; Conte, A.; Bernardo, D.; Bisogni, G.; Mandich, P.; Zollino, M.; Ragozzino, E.; Udd, B.; et al. SOD1 p.D12Y variant is associated with ALS/distal myopathy spectrum. *Eur. J. Neurol.* **2020**. [CrossRef] [PubMed]
158. Stewart, H.; Rutherford, N.J.; Briemberg, H.; Krieger, C.; Cashman, N.; Fabros, M.; Baker, M.; Fok, A.; DeJesus-Hernandez, M.; Eisen, A.; et al. Clinical and pathological features of amyotrophic lateral sclerosis caused by mutation in the C9ORF72 gene on chromosome 9p. *Acta Neuropathol.* **2012**, *123*, 409–417. [CrossRef] [PubMed]
159. Canosa, A.; Calvo, A.; Moglia, C.; Barberis, M.; Brunetti, M.; Cammarosano, S.; Manera, U.; Ilardi, A.; Restagno, G.; Chiò, A. A novel p.E121G heterozygous missense mutation of SOD1 in an apparently sporadic ALS case with a 14-year course. *Amyotroph. Lateral Scler. Front. Degener.* **2015**, *16*, 127–128. [CrossRef]
160. Corcia, P.; Vourc'h, P.; Blasco, H.; Couratier, P.; Dangoumau, A.; Bellance, R.; Desnuelle, C.; Viader, F.; Pautot, V.; Millecamps, S.; et al. Phenotypic and genotypic studies of ALS cases in ALS-SMA families. *Amyotroph. Lateral Scler. Front. Degener.* **2018**, *19*, 432–437. [CrossRef]
161. Battistini, S.; Ricci, C.; Giannini, F.; Calzavara, S.; Greco, G.; Del Corona, A.; Mancuso, M.; Battistini, N.; Siciliano, G.; Carrera, P. G41S SOD1 mutation: A common ancestor for six ALS Italian families with an aggressive phenotype. *Amyotroph. Lateral Scler.* **2010**, *11*, 210–215. [CrossRef]
162. Nizzardo, M.; Simone, C.; Rizzo, F.; Ulzi, G.; Ramirez, A.; Rizzuti, M.; Bordoni, A.; Bucchia, M.; Gatti, S.; Bresolin, N.; et al. Morpholino-mediated SOD1 reduction ameliorates an amyotrophic lateral sclerosis disease phenotype. *Sci. Rep.* **2016**, *6*, 21301. [CrossRef]
163. Rohrer, J.D.; Isaacs, A.M.; Mizielinska, S.; Mead, S.; Lashley, T.; Wray, S.; Sidle, K.; Fratta, P.; Orrell, R.W.; Hardy, J.; et al. C9orf72 expansions in frontotemporal dementia and amyotrophic lateral sclerosis. *Lancet Neurol.* **2015**, *14*, 291–301. [CrossRef]
164. Le Ber, I. SQSTM1 Mutations in French Patients With Frontotemporal Dementia or Frontotemporal Dementia With Amyotrophic Lateral Sclerosis. *JAMA Neurol.* **2013**. [CrossRef] [PubMed]
165. Lamb, R.; Rohrer, J.D.; Real, R.; Lubbe, S.J.; Waite, A.J.; Blake, D.J.; Walters, R.J.; Lashley, T.; Revesz, T.; Holton, J.L.; et al. A novel TBK1 mutation in a family with diverse frontotemporal dementia spectrum disorders. *Mol. Case Stud.* **2019**, *5*, a003913. [CrossRef] [PubMed]
166. Blauwendraat, C.; Wilke, C.; Simón-Sánchez, J.; Jansen, I.E.; Reifschneider, A.; Capell, A.; Haass, C.; Castillo-Lizardo, M.; Biskup, S.; Maetzler, W.; et al. The wide genetic landscape of clinical frontotemporal dementia: Systematic combined sequencing of 121 consecutive subjects. *Genet. Med.* **2018**, *20*, 240–249. [CrossRef]
167. Chia, R.; Chiò, A.; Traynor, B.J. Novel genes associated with amyotrophic lateral sclerosis: Diagnostic and clinical implications. *Lancet Neurol.* **2018**, *17*, 94–102. [CrossRef]
168. Liu, Q.; Shu, S.; Wang, R.R.; Liu, F.; Cui, B.; Guo, X.N.; Lu, C.X.; Li, X.G.; Liu, M.S.; Peng, B.; et al. Whole-exome sequencing identifies a missense mutation in hnRNPA1 in a family with flail arm ALS. *Neurology* **2016**, *87*, 1763–1769. [CrossRef]
169. Tripolszki, K.; Török, D.; Goudenège, D.; Farkas, K.; Sulák, A.; Török, N.; Engelhardt, J.I.; Klivényi, P.; Procaccio, V.; Nagy, N.; et al. High-throughput sequencing revealed a novel SETX mutation in a Hungarian patient with amyotrophic lateral sclerosis. *Brain Behav.* **2017**, *7*, e00669. [CrossRef]
170. Kenna, K.P.; McLaughlin, R.L.; Byrne, S.; Elamin, M.; Heverin, M.; Kenny, E.M.; Cormican, P.; Morris, D.W.; Donaghy, C.G.; Bradley, D.G.; et al. Delineating the genetic heterogeneity of ALS using targeted high-throughput sequencing. *J. Med. Genet.* **2013**, *50*, 776–783. [CrossRef]
171. Cady, J.; Allred, P.; Bali, T.; Pestronk, A.; Goate, A.; Miller, T.M.; Mitra, R.D.; Ravits, J.; Harms, M.B.; Baloh, R.H. Amyotrophic lateral sclerosis onset is influenced by the burden of rare variants in known amyotrophic lateral sclerosis genes. *Ann. Neurol.* **2015**, *77*, 100–113. [CrossRef]

172. Liu, Z.-J.; Lin, H.-X.; Wei, Q.; Zhang, Q.-J.; Chen, C.-X.; Tao, Q.-Q.; Liu, G.-L.; Ni, W.; Gitler, A.D.; Li, H.-F.; et al. Genetic Spectrum and Variability in Chinese Patients with Amyotrophic Lateral Sclerosis. *Aging Dis.* **2019**, *10*, 1199. [CrossRef]
173. Tsai, Y.; Lin, K.; Jih, K.; Tsai, P.; Liao, Y.; Lee, Y. Hand-onset weakness is a common feature of ALS patients with a NEK1 loss-of-function variant. *Ann. Clin. Transl. Neurol.* **2020**, acn3.51064. [CrossRef] [PubMed]
174. Ricci, C.; Giannini, F.; Intini, E.; Battistini, S. Genotype–phenotype correlation and evidence for a common ancestor in two Italian ALS patients with the D124G SOD1 mutation. *Amyotroph. Lateral Scler. Front. Degener.* **2019**, *20*, 611–614. [CrossRef] [PubMed]
175. Dalla Bella, E.; Lombardi, R.; Porretta-Serapiglia, C.; Ciano, C.; Gellera, C.; Pensato, V.; Cazzato, D.; Lauria, G. Amyotrophic lateral sclerosis causes small fiber pathology. *Eur. J. Neurol.* **2016**, *23*, 416–420. [CrossRef]
176. Khani, M.; Alavi, A.; Nafissi, S.; Elahi, E. Observation of c.260A > G mutation in superoxide dismutase 1 that causes p.Asn86Ser in Iranian amyotrophic lateral sclerosis patient and absence of genotype/phenotype correlation. *Iran. J. Neurol.* **2015**, *14*, 152–157. [PubMed]
177. Chiò, A.; Mora, G.; Sabatelli, M.; Caponnetto, C.; Lunetta, C.; Traynor, B.J.; Johnson, J.O.; Nalls, M.A.; Calvo, A.; Moglia, C.; et al. HFE p.H63D polymorphism does not influence ALS phenotype and survival. *Neurobiol. Aging* **2015**, *36*, 2906.e7–2906.e11.
178. Kim, M.-J.; Bae, J.-H.; Kim, J.-M.; Kim, H.R.; Yoon, B.-N.; Sung, J.-J.; Ahn, S.-W. Rapid Progression of Sporadic ALS in a Patient Carrying SOD1 p.Gly13Arg Mutation. *Exp. Neurobiol.* **2016**, *25*, 347–350. [CrossRef]
179. Chen, W.; Xie, Y.; Zheng, M.; Lin, J.; Huang, P.; Pei, Z.; Yao, X. Clinical and genetic features of patients with amyotrophic lateral sclerosis in southern China. *Eur. J. Neurol.* **2020**, *27*, 1017–1022. [CrossRef]
180. Lattante, S.; Conte, A.; Zollino, M.; Luigetti, M.; Del Grande, A.; Marangi, G.; Romano, A.; Marcaccio, A.; Meleo, E.; Bisogni, G.; et al. Contribution of major amyotrophic lateral sclerosis genes to the etiology of sporadic disease. *Neurology* **2012**, *79*, 66–72. [CrossRef]
181. Felbecker, A.; Camu, W.; Valdmanis, P.N.; Sperfeld, A.D.; Waibel, S.; Steinbach, P.; Rouleau, G.A.; Ludolph, A.C.; Andersen, P.M. Four familial ALS pedigrees discordant for two SOD1 mutations: Are all SOD1 mutations pathogenic? *J. Neurol. Neurosurg. Psychiatry* **2010**, *81*, 572–577. [CrossRef]
182. Nogales-Gadea, G.; Garcia-Arumi, E.; Andreu, A.L.; Cervera, C.; Gamez, J. A novel exon 5 mutation (N139H) in the SOD1 gene in a Spanish family associated with incomplete penetrance. *J. Neurol. Sci.* **2004**, *219*, 1–6. [CrossRef]
183. Ferrera, L.; Caponnetto, C.; Marini, V.; Rizzi, D.; Bordo, D.; Penco, S.; Amoroso, A.; Origone, P.; Garrè, C. An Italian dominant FALS Leu144Phe SOD1 mutation: Genotype-phenotype correlation. *Amyotroph. Lateral Scler. Other Mot. Neuron Disord.* **2003**, *4*, 167–170. [CrossRef] [PubMed]
184. Luisa Conforti, F.; Sprovieri, T.; Mazzei, R.; Patitucci, A.; Ungaro, C.; Zoccolella, S.; Magariello, A.; Bella, V.L.; Tessitore, A.; Tedeschi, G.; et al. Further evidence that D90A-SOD1 mutation is recessively inherited in ALS patients in Italy. *Amyotroph. Lateral Scler.* **2009**, *10*, 58–60. [CrossRef] [PubMed]
185. Murakami, T.; Warita, H.; Hayashi, T.; Sato, K.; Manabe, Y.; Mizuno, S.; Yamane, K.; Abe, K. A novel SOD1 gene mutation in familial ALS with low penetrance in females. *J. Neurol. Sci.* **2001**, *189*, 45–47. [CrossRef]
186. Segovia-Silvestre, T.; Andreu, A.L.; Vives-Bauza, C.; Garcia-Arumi, E.; Cervera, C.; Gamez, J. A novel exon 3 mutation (D76V) in the SOD1 gene associated with slowly progressive ALS. *Amyotroph. Lateral Scler. Other Mot. Neuron Disord.* **2002**, *3*, 69–74. [CrossRef]
187. Ticozzi, N.; Tiloca, C.; Mencacci, N.E.; Morelli, C.; Doretti, A.; Rusconi, D.; Colombrita, C.; Sangalli, D.; Verde, F.; Finelli, P.; et al. Oligoclonal bands in the cerebrospinal fluid of amyotrophic lateral sclerosis patients with disease-associated mutations. *J. Neurol.* **2013**, *260*, 85–92. [CrossRef]
188. Mandrioli, J.; Michalke, B.; Solovyev, N.; Grill, P.; Violi, F.; Lunetta, C.; Conte, A.; Sansone, V.A.; Sabatelli, M.; Vinceti, M. Elevated Levels of Selenium Species in Cerebrospinal Fluid of Amyotrophic Lateral Sclerosis Patients with Disease-Associated Gene Mutations. *Neurodegener. Dis.* **2017**, *17*, 171–180. [CrossRef]
189. van der Zee, J.; Gijselinck, I.; Van Mossevelde, S.; Perrone, F.; Dillen, L.; Heeman, B.; Bäumer, V.; Engelborghs, S.; De Bleecker, J.; Baets, J.; et al. TBK1 Mutation Spectrum in an Extended European Patient Cohort with Frontotemporal Dementia and Amyotrophic Lateral Sclerosis. *Hum. Mutat.* **2017**, *38*, 297–309. [CrossRef]
190. Weinreich, M.; Shepheard, S.R.; Verber, N.; Wyles, M.; Heath, P.R.; Highley, J.R.; Kirby, J.; Shaw, P.J. Neuropathological characterization of a novel TANK binding kinase (TBK1) gene loss of function mutation associated with amyotrophic lateral sclerosis. *Neuropathol. Appl. Neurobiol.* **2019**, nan.12578. [CrossRef]

191. Caroppo, P.; Camuzat, A.; De Septenville, A.; Couratier, P.; Lacomblez, L.; Auriacombe, S.; Flabeau, O.; Jornéa, L.; Blanc, F.; Sellal, F.; et al. Semantic and nonfluent aphasic variants, secondarily associated with amyotrophic lateral sclerosis, are predominant frontotemporal lobar degeneration phenotypes in TBK1 carriers. *Alzheimer's Dement. Diagnosis, Assess. Dis. Monit.* **2015**, *1*, 481–486. [CrossRef]
192. Dols-Icardo, O.; García-Redondo, A.; Rojas-García, R.; Borrego-Hernández, D.; Illán-Gala, I.; Muñoz-Blanco, J.L.; Rábano, A.; Cervera-Carles, L.; Juárez-Rufián, A.; Spataro, N.; et al. Analysis of known amyotrophic lateral sclerosis and frontotemporal dementia genes reveals a substantial genetic burden in patients manifesting both diseases not carrying the C9orf72 expansion mutation. *J. Neurol. Neurosurg. Psychiatry* **2018**, *89*, 162–168. [CrossRef]
193. Calvo, A.; Moglia, C.; Canosa, A.; Brunetti, M.; Barberis, M.; Traynor, B.J.; Carrara, G.; Valentini, C.; Restagno, G.; Chiò, A. A de novo nonsense mutation of the FUS gene in an apparently familial amyotrophic lateral sclerosis case. *Neurobiol. Aging* **2014**, *35*, 1513.e7–1513.e11. [CrossRef] [PubMed]
194. Damme, P.V.; Goris, A.; Race, V.; Hersmus, N.; Dubois, B.; Bosch, L.V.D.; Matthijs, G.; Robberecht, W. The occurrence of mutations in FUS in a Belgian cohort of patients with familial ALS. *Eur. J. Neurol.* **2010**, *17*, 754–756. [CrossRef] [PubMed]
195. Zou, Z.-Y.; Liu, M.-S.; Li, X.-G.; Cui, L.-Y. Mutations in FUS are the most frequent genetic cause in juvenile sporadic ALS patients of Chinese origin. *Amyotroph. Lateral Scler. Front. Degener.* **2016**, *17*, 249–252. [CrossRef] [PubMed]
196. Naumann, M.; Peikert, K.; Günther, R.; Kooi, A.J.; Aronica, E.; Hübers, A.; Danel, V.; Corcia, P.; Pan-Montojo, F.; Cirak, S.; et al. Phenotypes and malignancy risk of different FUS mutations in genetic amyotrophic lateral sclerosis. *Ann. Clin. Transl. Neurol.* **2019**, *6*, 2384–2394. [CrossRef]
197. Tümer, Z.; Bertelsen, B.; Gredal, O.; Magyari, M.; Nielsen, K.C.; LuCamp; Grønskov, K.; Brøndum-Nielsen, K. A novel heterozygous nonsense mutation of the OPTN gene segregating in a Danish family with ALS. *Neurobiol. Aging* **2012**, *33*, 208.e1–208.e5.
198. Iida, A.; Hosono, N.; Sano, M.; Kamei, T.; Oshima, S.; Tokuda, T.; Nakajima, M.; Kubo, M.; Nakamura, Y.; Ikegawa, S. Novel deletion mutations of OPTN in amyotrophic lateral sclerosis in Japanese. *Neurobiol. Aging* **2012**, *33*, 1843.e19–1843.e24. [CrossRef]
199. Feng, S.; Che, C.; Feng, S.; Liu, C.; Li, L.; Li, Y.; Huang, H.; Zou, Z. Novel mutation in optineurin causing aggressive ALS+/−frontotemporal dementia. *Ann. Clin. Transl. Neurol.* **2019**, *6*, 2377–2383. [CrossRef]
200. Li, C.; Ji, Y.; Tang, L.; Zhang, N.; He, J.; Ye, S.; Liu, X.; Fan, D. Optineurin mutations in patients with sporadic amyotrophic lateral sclerosis in China. *Amyotroph. Lateral Scler. Front. Degener.* **2015**, *16*, 485–489. [CrossRef]
201. Goldstein, O.; Nayshool, O.; Nefussy, B.; Traynor, B.J.; Renton, A.E.; Gana-Weisz, M.; Drory, V.E.; Orr-Urtreger, A. OPTN 691_692insAG is a founder mutation causing recessive ALS and increased risk in heterozygotes. *Neurology* **2016**, *86*, 446–453. [CrossRef]
202. Weishaupt, J.H.; Waibel, S.; Birve, A.; Volk, A.E.; Mayer, B.; Meyer, T.; Ludolph, A.C.; Andersen, P.M. A novel optineurin truncating mutation and three glaucoma-associated missense variants in patients with familial amyotrophic lateral sclerosis in Germany. *Neurobiol. Aging* **2013**, *34*, 1516.e9–1516.e15. [CrossRef]
203. Del Bo, R.; Tiloca, C.; Pensato, V.; Corrado, L.; Ratti, A.; Ticozzi, N.; Corti, S.; Castellotti, B.; Mazzini, L.; Soraru, G.; et al. Novel optineurin mutations in patients with familial and sporadic amyotrophic lateral sclerosis. *J. Neurol. Neurosurg. Psychiatry* **2011**, *82*, 1239–1243. [CrossRef] [PubMed]
204. Li, J.; He, J.; Tang, L.; Chen, L.; Xu, L.; Ma, Y.; Zhang, N.; Fan, D. TUBA4A may not be a significant genetic factor in Chinese ALS patients. *Amyotroph. Lateral Scler. Front. Degener.* **2016**, *17*, 148–150. [CrossRef] [PubMed]
205. Pensato, V.; Tiloca, C.; Corrado, L.; Bertolin, C.; Sardone, V.; Del Bo, R.; Calini, D.; Mandrioli, J.; Lauria, G.; Mazzini, L.; et al. TUBA4A gene analysis in sporadic amyotrophic lateral sclerosis: Identification of novel mutations. *J. Neurol.* **2015**, *262*, 1376–1378. [CrossRef]
206. Borghero, G.; Pugliatti, M.; Marrosu, F.; Marrosu, M.G.; Murru, M.R.; Floris, G.; Cannas, A.; Parish, L.D.; Occhineri, P.; Cau, T.B.; et al. Genetic architecture of ALS in Sardinia. *Neurobiol. Aging* **2014**, *35*, 2882.e7–2882.e12. [CrossRef] [PubMed]
207. Corcia, P.; Valdmanis, P.; Millecamps, S.; Lionnet, C.; Blasco, H.; Mouzat, K.; Daoud, H.; Belzil, V.; Morales, R.; Pageot, N.; et al. Phenotype and genotype analysis in amyotrophic lateral sclerosis with TARDBP gene mutations. *Neurology* **2012**, *78*, 1519–1526. [CrossRef] [PubMed]

208. Orrù, S.; Manolakos, E.; Orrù, N.; Kokotas, H.; Mascia, V.; Carcassi, C.; Petersen, M.B. High frequency of the TARDBP p.Ala382Thr mutation in Sardinian patients with amyotrophic lateral sclerosis. *Clin. Genet.* **2012**, *81*, 172–178. [CrossRef]
209. Ticozzi, N.; LeClerc, A.L.; van Blitterswijk, M.; Keagle, P.; McKenna-Yasek, D.M.; Sapp, P.C.; Silani, V.; Wills, A.-M.; Brown, R.H.; Landers, J.E. Mutational analysis of TARDBP in neurodegenerative diseases. *Neurobiol. Aging* **2011**, *32*, 2096–2099. [CrossRef]
210. Xu, G.; Hu, W.; Zhan, L.-L.; Wang, C.; Xu, L.-Q.; Lin, M.-T.; Chen, W.-J.; Wang, N.; Zhang, Q.-J. High frequency of the TARDBP p.M337 V mutation among south-eastern Chinese patients with familial amyotrophic lateral sclerosis. *BMC Neurol.* **2018**, *18*, 35. [CrossRef]
211. Caroppo, P.; Camuzat, A.; Guillot-Noel, L.; Thomas-Antérion, C.; Couratier, P.; Wong, T.H.; Teichmann, M.; Golfier, V.; Auriacombe, S.; Belliard, S.; et al. Defining the spectrum of frontotemporal dementias associated with TARDBP mutations. *Neurol. Genet.* **2016**, *2*, e80. [CrossRef]
212. Corrado, L.; Mazzini, L.; Oggioni, G.D.; Luciano, B.; Godi, M.; Brusco, A.; D'Alfonso, S. ATXN-2 CAG repeat expansions are interrupted in ALS patients. *Hum. Genet.* **2011**, *130*, 575–580. [CrossRef]
213. Tavares de Andrade, H.M.; Cintra, V.P.; de Albuquerque, M.; Piccinin, C.C.; Bonadia, L.C.; Duarte Couteiro, R.E.; Sabino de Oliveira, D.; Claudino, R.; Magno Gonçalves, M.V.; Dourado, M.E.T.; et al. Intermediate-length CAG repeat in ATXN2 is associated with increased risk for amyotrophic lateral sclerosis in Brazilian patients. *Neurobiol. Aging* **2018**, *69*, 292.e15–292.e18. [CrossRef] [PubMed]
214. Ross, O.A.; Rutherford, N.J.; Baker, M.; Soto-Ortolaza, A.I.; Carrasquillo, M.M.; DeJesus-Hernandez, M.; Adamson, J.; Li, M.; Volkening, K.; Finger, E.; et al. Ataxin-2 repeat-length variation and neurodegeneration. *Hum. Mol. Genet.* **2011**, *20*, 3207–3212. [CrossRef] [PubMed]
215. Millecamps, S.; Boillée, S.; Le Ber, I.; Seilhean, D.; Teyssou, E.; Giraudeau, M.; Moigneu, C.; Vandenberghe, N.; Danel-Brunaud, V.; Corcia, P.; et al. Phenotype difference between ALS patients with expanded repeats in C9ORF72 and patients with mutations in other ALS-related genes. *J. Med. Genet.* **2012**, *49*, 258–263. [CrossRef] [PubMed]
216. Xi, Z.; Zinman, L.; Grinberg, Y.; Moreno, D.; Sato, C.; Bilbao, J.M.; Ghani, M.; Hernández, I.; Ruiz, A.; Boada, M.; et al. Investigation of C9orf72 in 4 Neurodegenerative Disorders. *Arch. Neurol.* **2012**, *69*, 1583. [CrossRef]
217. Dols-Icardo, O.; Garcia-Redondo, A.; Rojas-Garcia, R.; Sanchez-Valle, R.; Noguera, A.; Gomez-Tortosa, E.; Pastor, P.; Hernandez, I.; Esteban-Perez, J.; Suarez-Calvet, M.; et al. Characterization of the repeat expansion size in C9orf72 in amyotrophic lateral sclerosis and frontotemporal dementia. *Hum. Mol. Genet.* **2014**, *23*, 749–754. [CrossRef]
218. van Blitterswijk, M.; DeJesus-Hernandez, M.; Niemantsverdriet, E.; Murray, M.E.; Heckman, M.G.; Diehl, N.N.; Brown, P.H.; Baker, M.C.; Finch, N.A.; Bauer, P.O.; et al. Association between repeat sizes and clinical and pathological characteristics in carriers of C9ORF72 repeat expansions (Xpansize-72): A cross-sectional cohort study. *Lancet Neurol.* **2013**, *12*, 978–988. [CrossRef]
219. Gijselinck, I.; Van Langenhove, T.; van der Zee, J.; Sleegers, K.; Philtjens, S.; Kleinberger, G.; Janssens, J.; Bettens, K.; Van Cauwenberghe, C.; Pereson, S.; et al. A C9orf72 promoter repeat expansion in a Flanders-Belgian cohort with disorders of the frontotemporal lobar degeneration-amyotrophic lateral sclerosis spectrum: A gene identification study. *Lancet Neurol.* **2012**, *11*, 54–65. [CrossRef]
220. Goldstein, O.; Gana-Weisz, M.; Nefussy, B.; Vainer, B.; Nayshool, O.; Bar-Shira, A.; Traynor, B.J.; Drory, V.E.; Orr-Urtreger, A. High frequency of C9orf72 hexanucleotide repeat expansion in amyotrophic lateral sclerosis patients from two founder populations sharing the same risk haplotype. *Neurobiol. Aging* **2018**, *64*, 160.e1–160.e7. [CrossRef]
221. Fournier, C.; Barbier, M.; Camuzat, A.; Anquetil, V.; Lattante, S.; Clot, F.; Cazeneuve, C.; Rinaldi, D.; Couratier, P.; Deramecourt, V.; et al. Relations between C9orf72 expansion size in blood, age at onset, age at collection and transmission across generations in patients and presymptomatic carriers. *Neurobiol. Aging* **2019**, *74*, 234.e1–234.e8. [CrossRef]
222. Zhang, M.; Tartaglia, M.C.; Moreno, D.; Sato, C.; McKeever, P.; Weichert, A.; Keith, J.; Robertson, J.; Zinman, L.; Rogaeva, E. DNA methylation age-acceleration is associated with disease duration and age at onset in C9orf72 patients. *Acta Neuropathol.* **2017**, *134*, 271–279. [CrossRef]

223. Gendron, T.F.; van Blitterswijk, M.; Bieniek, K.F.; Daughrity, L.M.; Jiang, J.; Rush, B.K.; Pedraza, O.; Lucas, J.A.; Murray, M.E.; Desaro, P.; et al. Cerebellar c9RAN proteins associate with clinical and neuropathological characteristics of C9ORF72 repeat expansion carriers. *Acta Neuropathol.* **2015**, *130*, 559–573. [CrossRef] [PubMed]
224. Kaivorinne, A.-L.; Bode, M.K.; Paavola, L.; Tuominen, H.; Kallio, M.; Renton, A.E.; Traynor, B.J.; Moilanen, V.; Remes, A.M. Clinical Characteristics of C9ORF72-Linked Frontotemporal Lobar Degeneration. *Dement. Geriatr. Cogn. Dis. Extra* **2013**, *3*, 251–262. [CrossRef] [PubMed]
225. van der Burgh, H.K.; Westeneng, H.-J.; Walhout, R.; van Veenhuijzen, K.; Tan, H.H.G.; Meier, J.M.; Bakker, L.A.; Hendrikse, J.; van Es, M.A.; Veldink, J.H.; et al. Multimodal longitudinal study of structural brain involvement in amyotrophic lateral sclerosis. *Neurology* **2020**. [CrossRef] [PubMed]
226. Floris, G.; Borghero, G.; Di Stefano, F.; Melis, R.; Puddu, R.; Fadda, L.; Murru, M.R.; Corongiu, D.; Cuccu, S.; Tranquilli, S.; et al. Phenotypic variability related to C9orf72 mutation in a large Sardinian kindred. *Amyotroph. Lateral Scler. Front. Degener.* **2016**, *17*, 245–248. [CrossRef]
227. Narain, P.; Padhi, A.K.; Dave, U.; Mishra, D.; Bhatia, R.; Vivekanandan, P.; Gomes, J. Identification and characterization of novel and rare susceptible variants in Indian amyotrophic lateral sclerosis patients. *Neurogenetics* **2019**, *20*, 197–208. [CrossRef] [PubMed]
228. Origone, P.; Verdiani, S.; Bandettini Di Poggio, M.; Zuccarino, R.; Vignolo, M.; Caponnetto, C.; Mandich, P. A novel Arg147Trp MATR3 missense mutation in a slowly progressive ALS Italian patient. *Amyotroph. Lateral Scler. Front. Degener.* **2015**, *16*, 530–531. [CrossRef]
229. Leblond, C.S.; Gan-Or, Z.; Spiegelman, D.; Laurent, S.B.; Szuto, A.; Hodgkinson, A.; Dionne-Laporte, A.; Provencher, P.; de Carvalho, M.; Orrù, S.; et al. Replication study of MATR3 in familial and sporadic amyotrophic lateral sclerosis. *Neurobiol. Aging* **2016**, *37*, 209.e17–209.e21. [CrossRef]
230. Lin, K.P.; Tsai, P.C.; Liao, Y.C.; Chen, W.T.; Tsai, C.P.; Soong, B.W.; Lee, Y.C. Mutational analysis of MATR3 in Taiwanese patients with amyotrophic lateral sclerosis. *Neurobiol. Aging* **2015**, *36*, 2005.e1–2005.e4. [CrossRef]
231. Marangi, G.; Lattante, S.; Doronzio, P.N.; Conte, A.; Tasca, G.; Monforte, M.; Patanella, A.K.; Bisogni, G.; Meleo, E.; La Spada, S.; et al. Matrin 3 variants are frequent in Italian ALS patients. *Neurobiol. Aging* **2017**, *49*, 218.e1–218.e7. [CrossRef]
232. Osmanovic, A.; Rangnau, I.; Kosfeld, A.; Abdulla, S.; Janssen, C.; Auber, B.; Raab, P.; Preller, M.; Petri, S.; Weber, R.G. FIG4 variants in central European patients with amyotrophic lateral sclerosis: A whole-exome and targeted sequencing study. *Eur. J. Hum. Genet.* **2017**, *25*, 324–331. [CrossRef]
233. Conforti, F.L.; Sprovieri, T.; Mazzei, R.; Ungaro, C.; La Bella, V.; Tessitore, A.; Patitucci, A.; Magariello, A.; Gabriele, A.L.; Tedeschi, G.; et al. A novel Angiogenin gene mutation in a sporadic patient with amyotrophic lateral sclerosis from southern Italy. *Neuromuscul. Disord.* **2008**, *18*, 68–70. [CrossRef] [PubMed]
234. Paubel, A. Mutations of the ANG Gene in French Patients With Sporadic Amyotrophic Lateral Sclerosis. *Arch. Neurol.* **2008**, *65*, 1333. [CrossRef] [PubMed]
235. Fernández-Santiago, R.; Hoenig, S.; Lichtner, P.; Sperfeld, A.-D.; Sharma, M.; Berg, D.; Weichenrieder, O.; Illig, T.; Eger, K.; Meyer, T.; et al. Identification of novel Angiogenin (ANG) gene missense variants in German patients with amyotrophic lateral sclerosis. *J. Neurol.* **2009**, *256*, 1337–1342. [CrossRef] [PubMed]
236. Nguyen, H.P.; Van Mossevelde, S.; Dillen, L.; De Bleecker, J.L.; Moisse, M.; Van Damme, P.; Van Broeckhoven, C.; van der Zee, J.; Engelborghs, S.; Crols, R.; et al. NEK1 genetic variability in a Belgian cohort of ALS and ALS-FTD patients. *Neurobiol. Aging* **2018**, *61*, 255.e1–255.e7. [CrossRef] [PubMed]
237. He, J.; Liu, X.; Tang, L.; Zhao, C.; He, J.; Fan, D. Whole-exome sequencing identified novel KIF5A mutations in Chinese patients with amyotrophic lateral sclerosis and Charcot-Marie-Tooth type 2. *J. Neurol. Neurosurg. Psychiatry* **2020**, *91*, 326–328. [CrossRef]
238. Tiloca, C.; Ticozzi, N.; Pensato, V.; Corrado, L.; Del Bo, R.; Bertolin, C.; Fenoglio, C.; Gagliardi, S.; Calini, D.; Lauria, G.; et al. Screening of the PFN1 gene in sporadic amyotrophic lateral sclerosis and in frontotemporal dementia. *Neurobiol. Aging* **2013**, *34*, 1517.e9–1517.e10. [CrossRef]
239. Ingre, C.; Landers, J.E.; Rizik, N.; Volk, A.E.; Akimoto, C.; Birve, A.; Hübers, A.; Keagle, P.J.; Piotrowska, K.; Press, R.; et al. A novel phosphorylation site mutation in profilin 1 revealed in a large screen of US, Nordic, and German amyotrophic lateral sclerosis/frontotemporal dementia cohorts. *Neurobiol. Aging* **2013**, *34*, 1708.e1–1708.e6. [CrossRef]

240. Smith, B.N.; Vance, C.; Scotter, E.L.; Troakes, C.; Wong, C.H.; Topp, S.; Maekawa, S.; King, A.; Mitchell, J.C.; Lund, K.; et al. Novel mutations support a role for Profilin 1 in the pathogenesis of ALS. *Neurobiol. Aging* **2015**, *36*, 1602.e17–1602.e27. [CrossRef]
241. Chen, Y.; Zheng, Z.-Z.; Huang, R.; Chen, K.; Song, W.; Zhao, B.; Chen, X.; Yang, Y.; Yuan, L.; Shang, H.-F. PFN1 mutations are rare in Han Chinese populations with amyotrophic lateral sclerosis. *Neurobiol. Aging* **2013**, *34*, 1922.e1–1922.e5. [CrossRef]
242. van Blitterswijk, M.; van Es, M.A.; Koppers, M.; van Rheenen, W.; Medic, J.; Schelhaas, H.J.; van der Kooi, A.J.; de Visser, M.; Veldink, J.H.; van den Berg, L.H. VAPB and C9orf72 mutations in 1 familial amyotrophic lateral sclerosis patient. *Neurobiol. Aging* **2012**, *33*, 2950.e1–2950.e4. [CrossRef]
243. Di, L.; Chen, H.; Da, Y.; Wang, S.; Shen, X.-M. Atypical familial amyotrophic lateral sclerosis with initial symptoms of pain or tremor in a Chinese family harboring VAPB-P56S mutation. *J. Neurol.* **2016**, *263*, 263–268. [CrossRef] [PubMed]
244. Gellera, C.; Tiloca, C.; Del Bo, R.; Corrado, L.; Pensato, V.; Agostini, J.; Cereda, C.; Ratti, A.; Castellotti, B.; Corti, S.; et al. Ubiquilin 2 mutations in Italian patients with amyotrophic lateral sclerosis and frontotemporal dementia. *J. Neurol. Neurosurg. Psychiatry* **2013**, *84*, 183–187. [CrossRef] [PubMed]
245. Dillen, L.; Van Langenhove, T.; Engelborghs, S.; Vandenbulcke, M.; Sarafov, S.; Tournev, I.; Merlin, C.; Cras, P.; Vandenberghe, R.; De Deyn, P.P.; et al. Explorative genetic study of UBQLN2 and PFN1 in an extended Flanders-Belgian cohort of frontotemporal lobar degeneration patients. *Neurobiol. Aging* **2013**, *34*, 1711.e1–1711.e5. [CrossRef] [PubMed]
246. Yang, Y.; Tang, L.; Zhang, N.; Pan, L.; Hadano, S.; Fan, D. Six SQSTM1 mutations in a Chinese amyotrophic lateral sclerosis cohort. *Amyotroph. Lateral Scler. Front. Degener.* **2015**, *16*, 378–384. [CrossRef] [PubMed]
247. Al-Obeidi, E.; Al-Tahan, S.; Surampalli, A.; Goyal, N.; Wang, A.K.; Hermann, A.; Omizo, M.; Smith, C.; Mozaffar, T.; Kimonis, V. Genotype-phenotype study in patients with valosin-containing protein mutations associated with multisystem proteinopathy. *Clin. Genet.* **2018**, *93*, 119–125. [CrossRef]
248. Daoud, H.; Zhou, S.; Noreau, A.; Sabbagh, M.; Belzil, V.; Dionne-Laporte, A.; Tranchant, C.; Dion, P.; Rouleau, G.A. Exome sequencing reveals SPG11 mutations causing juvenile ALS. *Neurobiol. Aging* **2012**, *33*, 839.e5–839.e9. [CrossRef] [PubMed]
249. Cooper-Knock, J.; Moll, T.; Ramesh, T.; Castelli, L.; Beer, A.; Robins, H.; Fox, I.; Niedermoser, I.; Van Damme, P.; Moisse, M.; et al. Mutations in the Glycosyltransferase Domain of GLT8D1 Are Associated with Familial Amyotrophic Lateral Sclerosis. *Cell Rep.* **2019**, *26*, 2298–2306.e5. [CrossRef] [PubMed]
250. Dobson-Stone, C.; Hallupp, M.; Shahheydari, H.; Ragagnin, A.M.G.; Chatterton, Z.; Carew-Jones, F.; Shepherd, C.E.; Stefen, H.; Paric, E.; Fath, T.; et al. CYLD is a causative gene for frontotemporal dementia—Amyotrophic lateral sclerosis. *Brain* **2020**, *143*, 783–799. [CrossRef]
251. Farhan, S.M.K.; Howrigan, D.P.; Abbott, L.E.; Klim, J.R.; Topp, S.D.; Byrnes, A.E.; Churchhouse, C.; Phatnani, H.; Smith, B.N.; Rampersaud, E.; et al. Exome sequencing in amyotrophic lateral sclerosis implicates a novel gene, DNAJC7, encoding a heat-shock protein. *Nat. Neurosci.* **2019**, *22*, 1966–1974. [CrossRef]
252. Cudkowicz, M.E.; McKenna-Yasek, D.; Sapp, P.E.; Chin, W.; Geller, B.; Hayden, D.L.; Schoenfeld, D.A.; Hosler, B.A.; Horvitz, H.R.; Brown, R.H. Epidemiology of mutations in superoxide dismutase in amyotrophic lateal sclerosis. *Ann. Neurol.* **1997**, *41*, 210–221. [CrossRef]
253. Ogasawara, M.; Matsubara, Y.; Narisawa, K.; Aoki, M.; Nakamura, S.; Itoyama, Y.; Abe, K. Mild ALS in Japan associated with novel SOD mutation. *Nat. Genet.* **1993**, *5*, 323–324. [CrossRef] [PubMed]

Journal of *Personalized Medicine*

MDPI

Review

DUX4 Expression in FSHD Muscles: Focus on Its mRNA Regulation

Eva Sidlauskaite [1], Laura Le Gall [1], Virginie Mariot [1] and Julie Dumonceaux [1,2,*]

1 NIHR Biomedical Research Centre, University College London, Great Ormond Street Institute of Child Health and Great Ormond Street Hospital NHS Trust, London WC1N 1EH, UK; e.sidlauskaite@ucl.ac.uk (E.S.); l.gall@ucl.ac.uk (L.L.G.); virginie.mariot@ucl.ac.uk (V.M.)

2 Northern Ireland Center for Stratified/Personalised Medicine, Biomedical Sciences Research Institute, Ulster University, Derry~Londonderry, Northern Ireland BT47 6SB, UK

* Correspondence: j.dumonceaux@ucl.ac.uk; Tel.: +44-(0)-2079052860

Received: 16 June 2020; Accepted: 24 July 2020; Published: 28 July 2020

Abstract: Facioscapulohumeral dystrophy (FSHD) is the most frequent muscular disease in adults. FSHD is characterized by a weakness and atrophy of a specific set of muscles located in the face, the shoulder, and the upper arms. FSHD patients may present different genetic defects, but they all present epigenetic alterations of the D4Z4 array located on the subtelomeric part of chromosome 4, leading to chromatin relaxation and, ultimately, to the aberrant expression of one gene called *DUX4*. Once expressed, DUX4 triggers a cascade of deleterious events, eventually leading to muscle dysfunction and cell death. Here, we review studies on *DUX4* expression in skeletal muscle to determine the genetic/epigenetic factors and regulatory proteins governing *DUX4* expression, with particular attention to the different transcripts and their very low expression in muscle.

Keywords: FSHD; DUX4; transcription; muscle; regulation

1. Introduction

Double homeobox 4 (*DUX4*) is a transcription factor that is normally expressed during embryonic development and in the human testes but suppressed in somatic tissue (for review see [1]). The recent finding of DUX4 in an early cleavage-stage embryo raised the hypothesis that DUX4 might act as a functional transcriptional programmer to activate the cleavage-stage transcriptional platform and might be a key regulator of zygotic genome activation [2–4]. Moreover, the presence of DUX4 in the testis suggests that *DUX4* may be activated in the primary spermatocytes during spermatogenesis [5]. More recently, *DUX4* activation gained a particular interest across cancer research, as *DUX4* expression in tumours results in immune evasion [6].

Despite the awareness of *DUX4* expression in normal germline biology, DUX4 is principally described as a toxic factor involved in facioscapulohumeral dystrophy (FSHD) pathophysiology. Indeed, in FSHD patients, DUX4 is aberrantly expressed in the muscle tissue [5,7]. The role of DUX4 in FSHD pathogenesis is intensively investigated, and several reviews have been published in this topic [8,9] explaining the potential role of DUX4 in cell death and discussing the role of DNA methylation in FSHD1 and 2 patients. The current review focuses on the recent understanding and regulation of *DUX4* mRNA expression at the mRNA level in skeletal muscle and myogenic cells.

2. FSHD

FSHD is the third most common genetic muscular dystrophy with a frequency between 1/8000 to 1/20,000 (www.orpha.net, April 2020). The primary manifestation of FSHD is an asymmetric atrophy of the muscles located in the face, the shoulder, and the upper arm. The pathology often begins during late adolescence; however, the presence of symptoms at an early age is often associated with more severe muscle weakness (reviewed in [10]). The mutation that causes FSHD was identified nearly 30 years ago [11]. FSHD is associated with genetic and epigenetic molecular changes of the D4Z4 microsatellite repeats in the subtelomeric region of chromosome 4 [12,13]. There are two different genetic mechanisms leading to FSHD, and both are associated with the loss of epigenetic marks within the D4Z4 and the aberrant expression of *DUX4* [14]. The first one concerns 95% of FSHD patients (known as FSHD1, OMIM#158900) who show a contraction of a tandemly repeated 3.3 kb microsatellite D4Z4 repeat at the distal end of chromosomal region 4q35. The number of D4Z4 repeats usually varies from 11 to 150, while fewer repeats are observed in less than 3% of the population [15]. In FSHD1 patients, this number is reduced to 10 and below [16]. This reduction of D4Z4 unit number is associated with chromosome relaxation and loss of repression of *DUX4* gene (OMIM#606009), allowing DUX4 transcription in muscle cells [17]. The second one concerns the remaining 5% of FSHD patients (known as FSHD2, OMIM#158901), who do not present a shortened D4Z4 array but carry a mutation in epigenetic modifier genes. The vast majority of FSHD2 cases have been linked to mutations in the *SMCHD1* (structural maintenance of chromosomes flexible hinge domain containing 1) gene [18], encoding a remodelling protein essential for DNA methylation. Few FSHD2 cases present a heterozygous mutation in the *DNMT3B* (DNA methyltransferase 3 beta) gene [19], which is normally responsible for the establishment of the cytosine methylation profile during development. The exact mechanism of how particular mutations cause the FSHD pathology is still under investigation, but the notion of permissive chromosome 4 is now acknowledged for FSHD patients. This "pathological" chromosome 4 is characterized by the following: the presence of specific simple sequence length polymorphism (SSLP) located 3.5 kb proximal to the D4Z4 repeat [20]; the presence of at least one D4Z4 repeat [21]; a chromatin relaxation within the D4Z4 repeat [17]; and the presence of the 4qA haplotype [22,23] containing the polyadenylation signal for DUX4 [14]. Indeed, each D4Z4 contains the open reading frame (ORF) of the *DUX4* retrogene [7,24]. DUX4 protein and mRNA are detected in both FSHD1 and FSHD2 muscle biopsies at very low levels [5] but sufficient to induce a cascade of mis-regulated genes [25] eventually leading to muscle atrophy and muscle fibre death by the disruption of multiple cellular processes (for review see [8]).

3. Regulation of *DUX4* Expression

There is a consensus in the scientific community on *DUX4* expression in FSHD biopsies, but its regulation still needs to be deciphered. Indeed, *DUX4* expression is regulated by several factors including D4Z4 epigenetic modification, chromosome conformation and the presence of myogenic enhancers (Figure 1).

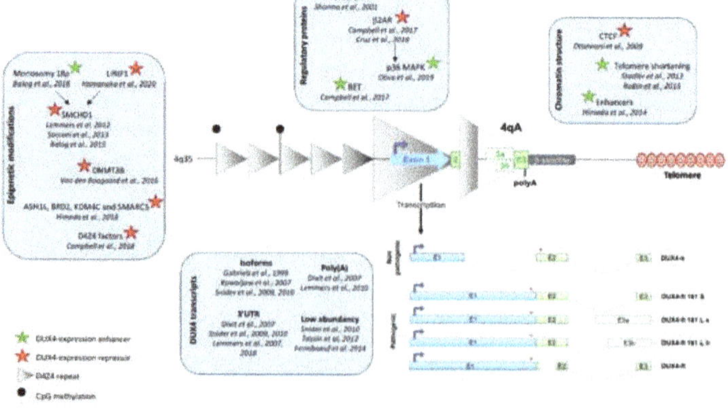

Figure 1. Regulation of *DUX4* expression.

DUX4 expression is regulated by several factors including D4Z4 epigenetic modification, chromatin structure, regulatory proteins, and myogenic enhancers. *DUX4* is composed of 3 exons, exons 1 and 2 are present in each D4Z4 repeat, but exon 3 is located outside of the repeats. Three types of exon 3 have been described: exons 3a and 3b are transcribed from the 4A161L allele (dashed line) and exon 3 from 4A161S allele (plain line). Exon 3 carries the polyadenylation signal. Five *DUX4* isoforms have been characterized. The four leading to the full-length protein (DUX4-fl) are pathogenic, whereas the one leading to a truncated protein (DUX4-s) is non-pathogenic.

3.1. D4Z4 Epigenetic Modification

Because it is well known that epigenetic modifications play a significant role in gene regulation in normal and pathological environments, several studies have evaluated whether or not the epigenetic disruption observed at the 4q35 locus could lead to the expression of *DUX4*. In 2012, Lemmers and colleagues reported that antisense nucleotide-mediated exon skipping of *SMCHD1* in normal human myoblasts led to *DUX4* expression [18]. Combined with the observation that families with FSHD2 present a haploinsufficiency of SMCHD1 and a hypomethylation of the D4Z4 array [18], a link between epigenetic modifications and *DUX4* expression was established. Since then, several articles have reinforced the idea of an epigenetic regulation of DUX4 expression. The consequences of *SMCHD1* expression level on *DUX4* expression were particularly studied, and it was shown that SMCHD1 levels participate in *DUX4* expression in muscle cells. Indeed, depletion of *SMCHD1* in FSHD1 myoblasts increased *DUX4* expression [26] whereas its ectopic overexpression resulted in DUX4 silencing in FSHD1 and FSHD2 myotubes [27]. This is consistent with the fact that *DUX4* expression is increased during muscle differentiation, which correlates with decreased SMCHD1 protein levels at D4Z4 [27]. Moreover, the interaction of SMCHD1 with the chromatin is facilitated by the ligand-dependent nuclear receptor-interacting factor 1 (LRIF1), which binds to the D4Z4 repeat [28]. Interestingly, mutations in *LRIF1* lead to chromatin relaxation and *DUX4* derepression [28], and knockdown of the *LRIF1* long isoform in control myoblasts using siRNA results in the expression of DUX4 [28]. *DUX4* expression in myoblasts was also observed after decreased binding of SMCHD1 to D4Z4 caused by the inhibition of H3K9me3 (repressive mark associated with heterochromatin formation) using drugs [29]. Finally, a recent study has also shown that *DUX4* is expressed in myocytes obtained from patients presenting a 18p hemizygosity with a decreased of *SMCHD1* mRNA [30]. Altogether, these studies suggest a link between *SMCHD1*-mediated epigenetic modifications and DUX4 expression.

Multiple other lines of evidence show a role of epigenetics in DUX4 expression: (i) MyoD-converted fibroblasts isolated from FSHD2 patients carrying a mutation in the *DNA methyltransferase 3B* (*DNMT3B*) gene express DUX4, suggesting a D4Z4 derepression associated with DUX4 expression [19]. (ii) Several epigenetic pathways such as *ASH1L*, *BRD2*, *KDM4C*, and *SMARC5* were found to regulate DUX4 expression in primary FSHD cells after independent knockdown of multiple chromatin regulators [31]. (iii) Human chromosome 4/CHO hybrid cells treated with 5′-aza-2′deoxycytidine (AZA, a cytosine analogue that is incorporated into DNA during DNA replication) and/or trichostatin A (TSA, which inhibits class I and II histone deacetylases) led to *DUX4* expression [32,33]. (iv) Two D4Z4 factors, nucleosome remodelling deacetylase (NuRD) and chromatin assembly gactor 1 (CAF-1) were identified as DUX4 repressors in human skeletal muscle cells using RNA-guided Cas9 nuclease from the microbial clustered regularly interspaced short palindromic repeats (CRISPR/Cas9) engineered chromatin immunoprecipitation (enChIP) locus-specific proteomics to characterize D4Z4-associated proteins [34]. (v) Hemizygous transgenic mice carrying either a 2.5 or 12.5 D4Z4 repeat showed a chromatin relaxation of the D4Z4 repeats in D4Z4-2.5 mice compared to D4Z4-12.5 mice, associated with *DUX4* expression in the D4Z4-2.5 mouse [35].

Altogether, these studies strongly suggest that chromatin relaxation results in inappropriate DUX4 expression in skeletal muscle. However, regulation of *DUX4* expression may be different in other tissues or during development. Indeed, *DUX4* is expressed in early cleavage-stage embryos whereas a high methylation level is found at D4Z4 in pluripotent cells in both FSHD1 and controls [4,36], which goes against a link between D4Z4 hypomethylation and *DUX4* expression.

3.2. Chromatin Conformation

D4Z4 chromatin structure was also associated with *DUX4* expression/repression in muscle. Indeed, the 3D organization of chromatin modulates major biological processes including transcription. In regard of the link between DUX4 expression and chromatin conformation, it was proposed that, as a single repeat, D4Z4 behaves as a CCCTC-binding factor (CTCF) insulator interfering with enhancer–promoter communication [37]. However, both its CTCF binding and insulation properties are suppressed upon multimerization of D4Z4 units, suggesting that FSHD could result from an inappropriate insulation mechanism and a CTCF-gain of function [37]. Because CTCF can mediate transcriptional regulation by creating accessible or inaccessible loops of chromatin at specific sites, the involvement of CTCF in *DUX4* expression was proposed [38]. In this study, the authors found CTCF to be more readily associated with transcriptionally silent arrays, suggesting a role of CTCF in repressing *DUX4* transcription.

D4Z4 was also described as an insulator shielding from telomeric position effect (TPE). Indeed, telomeres can regulate gene expression by trapping adjacent heterochromatin. Using isogenic clones with different telomere lengths, it was demonstrated that telomere shortening led to *DUX4* expression [39]. The likely mechanism is that the epigenetic landscape is altered during telomere shortening resulting in decreased heterochromatin at 4q35 [40,41].

Interestingly, whereas the epigenetic modifications observed in FSHD patients at the D4Z4 array are not restricted to the muscle tissue [42–44], *DUX4* mRNA was found mainly in the skeletal muscle, testis, and thymus [5,45]. Two enhancers upstream of the D4Z4 that upregulate DUX4 expression in skeletal myocytes but not in fibroblasts were described [46]. Importantly, these enhancers participate in *DUX4* expression only when the *DUX4* promoter is hypomethylated. However, the exact role of these enhancers in FSHD onset may be questioned as two FSHD1 patients have been identified with large deletions encompassing this chromosomal region [47]. Moreover, meiotic rearrangements between chromosomes 4 and 10 [14,48] go against a central role of other regions of chromosome 4 in *DUX4* expression.

3.3. Regulatory Proteins of DUX4 Expression

Transcriptional regulation of DUX4 expression may be also controlled by gene regulatory proteins that interact with the DUX4 promoter, and one study identified Poly(ADP-Ribose) Polymerase 1 (*PARP1*) using a DNA pull-down assay coupled with mass spectrometry and chromatin immunoprecipitation [49].

Several inhibitors of *DUX4* have been published, suggesting that the target inhibitors may play a role in *DUX4* expression. It was shown that activation of the Wnt/β-catenin signalling reduced *DUX4* expression whereas knockdown of Wnt/β-catenin signalling pathway components activates DUX4 [50]. The mechanism of *DUX4* regulation by Wnt/β-catenin is likely independent of direct binding of β-catenin at D4Z4. Bromodomain and extra-terminal (BET)- and β2 adrenergic receptor-mediated pathways were also associated with DUX4 expression regulation [51]. Using BET inhibitors (BETi) targeting all proteins of the BET family, *DUX4* and DUX4 target candidates were silenced in primary FSHD muscle cells [51]. The research team suggested that BETi efficiently repressed *DUX4* transcription by lysine deacetylation but not DNA methylation. Similarly, β2 adrenergic receptor agonists activate signalling pathways known to induce chromatin remodelling. *DUX4* and DUX4 target candidates' expression were both repressed following treatment with β2 adrenergic receptor agonists, suggesting the role of BET and β2 adrenergic receptor signalling pathways in DUX4 expression in FSHD patients [51]. Since then, the importance of the β2 adrenergic receptor has been confirmed in additional studies [52], and downstream pathways have been the centre of attention in order to identify therapeutic targets. P38 mitogen-activated protein kinase is activated by the β2 adrenergic receptor signalling pathway [53].

In FSHD muscle cells or in a xenograft model of FSHD, pharmaceutical or siRNA-mediated inhibition of p38 induced a reduction of *DUX4* mRNA levels [54]. This suggests that β2 adrenergic receptor agonist-mediated *DUX4* expression is a consequence of p38 kinase activation. Phosphodiesterases, or PDEs, which are responsible for regulation of available cAMP in the cell, were identified as *DUX4* expression regulators [52] by reducing expression levels of both DUX4 and its target genes *ZSCAN4* and *TRIM43*. β2 adrenergic receptor and PDEs are both implicated in cAMP-mediated signalling that further regulates protein kinase A (PKA) signalling pathways. Both cell-permeable cAMP and catalytic active PKA were sufficient to reduce *DUX4* expression and *ZSCAN4* and *TRIM43* mRNA levels [52] in primary FSHD patients' muscle cells. The authors suggested that β2 adrenergic agonists and PDE inhibitors mediated a c-AMP and PKA-mediated repression of DUX4 gene expression in FSHD muscle cells. However, downstream effectors of cAMP also include PKA-independent pathways, and the results from Campbell et al. suggest a PKA-independent mediated repression of DUX4 [51]. Later, p38α and p38β MAPK inhibitors were identified as suppressors of *DUX4* mRNA transcription in myotubes and in a xenograft model of FSHD [54], suggesting a positive regulation of *DUX4* transcription by both p38α and p38β.

4. *DUX4* mRNA

4.1. DUX4 Transcription

The presence of a large ORF encompassing 2 homeoboxes in each D4Z4 repeat was first described in 1995 [55], but the identification of the DUX4 gene occurred in 1999 by the Belayew group [7]. This group also identified the DUX4 promoter with a variant of TATAA box (TACAA) [7]. The final demonstration that D4Z4 contains a functional DUX4 transcriptional unit leading to the *DUX4* transcription was made few years later after cloning of the D4Z4 region into a promoter-less vector and transfection into myoblasts [56]. 5′ Rapid amplification of cDNA ends (RACE) PCR lead to the identification of the 5′untranslated region (UTR) composed of 97–187 nt [56]. The polyadenylation site was described after 3′ RACE PCR on total RNA extracted from C2C12 mouse myoblasts transfected with a 13.5 kb genomic fragment of a patient with two D4Z4 repeats [57]: It is the ATTAAA hexanucleotide sequence (12852–12858 in GenBank accession no. AF117653).

The *DUX4* mRNA found in the muscle tissue is composed of 3 exons, with the *DUX4* ORF being entirely within exon 1. Importantly, exons 1 and 2 are present in the D4Z4 repeats but not exon 3, which is located in region called pLAM. Notably, the pLAM region is not present on the 4qB haplotype that is classified as non-pathogenic [22,58]. This leads to the hypothesis that DUX4 would only be transcribed for the most telomeric repeat because only this one would give rise to a polyadenylated DUX4. The role of this region in DUX4 expression and stability was highlighted by the report of individuals with a genomic rearrangement between chromosome 4q and 10q. Indeed, the subtelomeric part of these 2 chromosomes is highly homologous and, importantly, chromosome 10 does not carry the ATTAAA poly(A) signal found in chromosome 4, but an ATCAAA sequence that is not known to be a poly(A) signal [14]. Meiotic rearrangements between chromosomes 4 and 10 generated a short hybrid structure on 4qA where the pLAM sequence was conserved but immediately proximal to a 1.5 D4Z4 repeat coming from chromosome 10, resulting in disease presentation. Transfection experiments with genomic D4Z4 constructs derived from permissive or non-permissive chromosomes or in which the poly(A) signals from non-permissive chromosomes are replaced by those from permissive chromosomes established the importance of this poly(A)signal in the stabilization of *DUX4* [14].

Two different *DUX4* mRNAs, resulting from the inclusion or exclusion of an alternatively spliced intron of 136 bp located in the 3′UTR part of mRNA have been described [57]. The two *DUX4* mRNAs have also spliced out a 345 bp intron also located in the 3′UTR region [57]. These two *DUX4* mRNAs were later renamed DUX4-full length (DUX4-fl) [5]. Recently, other *DUX4* mRNAs have been characterized from a common variant of the most prevalent FSHD-permissive haplotype 4A161 (containing an SSLP of 161 nt and the distal 4qA variant [59]). These two variants present a 1.6 kb size difference of the most distal D4Z4 units [60]. Two *DUX4* mRNAs are transcribed from this long allele using 2 alternative 3′ splice sites, leading to either the DUX4-fl 161La or Lb transcripts (Figure 1) (GenBank accession numbers MF693913 and KQ983258.1). The three pathogenic DUX4-fls share the pLAM sequence containing the DUX4 poly(A) and lead to the same DUX4 protein. There is no link between disease severity and transcript variants [60].

4.2. DUX4 Isoforms

DUX4 transcription from the last D4Z4 repeat results in at least 5 different mRNAs, the 4 *DUX4-fls* described above, code for the same protein but differ by an altered splicing of intron 1 in the 3′UTR and by the use 2 alternative 3′ splice sites leading to different types of exon 3. The fifth *DUX4* transcript corresponds to a short version of DUX4 (DUX4-s), in which an alternative donor splice site located in first exon is used [24], leading to a truncated form of DUX4, lacking the C-terminal part of the protein containing the transactivation domain [61] and acting as a dominant negative [25]. *DUX4-fl* isoforms are mainly found in myotubes and muscles biopsies isolated from FSHD patients, whereas *DUX4-s* can be found in both control individuals and FSHD patients [5,62]. DUX4-fl expression increases in myotubes [5,62,63]. An isoform switch may be possible, since it was shown in iPS cells derived from control fibroblasts that *DUX4*-fl is expressed in undifferentiated cells but can switch to *DUX4*-s in embryoid bodies [5]. *DUX4-fl* mRNA is expressed in muscles during development, as both isoforms are found in foetal muscle biopsies and cells derived from foetal muscle [64,65].

Interestingly, *DUX4* mRNA is also found in human testes at a level 100-fold higher compared to FSHD muscle biopsies [5] but does not seem to be toxic. 3′ RACE PCR analysis revealed that both chromosomes 4 and 10 were used for *DUX4* transcription, despite the absence of a permissive poly(A) signal on chromosome 10. Chromosome 10 and some 4qA transcripts use an alternative poly(A) located in exon 7. Surprisingly, *DUX4* transcripts were also found from the 4qB allele, but the poly(A) still need to be identified. Exons 3 and 7 are excluded since they are not present in the 4qB allele. Non-canonical poly(A) signals may be also used in some circumstances, as observed in the presence of antisense oligonucleotides targeting the poly(A) signal [66]. The use of alternative poly(A) signals could also explain the normal embryogenesis observed in individuals carrying non-permissive 4q alleles. Consistent with this hypothesis, studies have also shown that alternative polyadenylation pattern varies among cell types [67] and during embryonic development [68].

5. DUX4 Low Abundancy and Stochastic Expression

DUX4 mRNA is found at a very low level in both biopsies and muscle cells from both FSHD1 and FSHD2 patients. This low abundance could reflect a uniform low level in all nuclei or a high expression in a limited number of nuclei. By pooling a different number of nuclei and after assessment of the presence of *DUX4-fl* by PCR, it has been estimated that about 1 in 1000 FSHD nuclei are positive for *DUX4* mRNA [5]. The question is how could a gene expressed at such low levels be so toxic? The presence of the endogenous DUX4 protein in consecutive myotube nuclei, forming an intensity gradient, suggested a spreading of the protein within the myotubes [69]. This hypothesis was confirmed by co-culture experiments between FSHD myoblasts and murine C2C12 myoblasts. Whereas *DUX4* is transcribed in human nuclei only, the protein was found in both human and murine nuclei showing the spreading of the DUX4 protein [70]. The sporadic and asynchronous burst of expression of DUX4 was confirmed using a *DUX4*-activated reporter [71].

6. Conclusions

During the past decade, our knowledge about FSHD onset considerably improved. Several genetic and epigenetic defects have been clearly identified that cause FSHD, all leading to the aberrant expression of the DUX4 transcription factor. Once expressed, DUX4 triggers a cascade of events that ultimately converge to cell death and impair muscle development and repair (for review see [8]). After years of controversy, DUX4 is now seen as one of most important players in FSHD onset and progression. Some areas remain unelucidated, such as the non-toxic expression of DUX4 during embryogenesis [2,4] or the different splicings observed in the testis [5]: Are they due to a difference between pathogenic and healthy environment or are they tissue-specific?

Multiple studies have deciphered the expression of DUX4 in skeletal muscle and demonstrated that chromatin conformation, DNA methylation and histone modification, myogenic enhancer, and regulatory proteins are involved in the regulation of its expression. Moreover, some other repressor proteins or lncRNA that are associated with the D4Z4 repeat may also play a role [32,72].

Several laboratories are developing therapeutic approaches targeting DUX4 by either blocking *DUX4* mRNA synthesis [31,51,52,54], targeting *DUX4* mRNA using antisense oligonucleotides [66,73–76], or targeting the DUX4 protein or its downstream consequences [77–79]. One phase 2 clinical trial (NCT04003974) aiming at inhibiting or reducing its expression in skeletal muscle is already on-going and may enable a better understanding of the role of DUX4 in the pathophysiology of FSHD.

Author Contributions: The idea of writing this review was done by E.S., L.L.G., V.M. and J.D. All the authors wrote and edited the review. All authors have read and agreed to the published version of the manuscript.

Funding: This research received no external funding.

Acknowledgments: This work was supported by the FSHD society (grant number FSHS-22018-02 for E.S. salary) and L.L.G. is funded by the Association Française contre les Myopathies AFM-Telethon (grant number # #22582). V.M. and J.D. are supported by the National Institute for Health Research Biomedical Research Centre at Great Ormond Street Hospital for Children NHS Foundation Trust and University College London. All research at Great Ormond Street Hospital NHS Foundation Trust and UCL Great Ormond Street Institute of Child Health is made possible by the NIHR Great Ormond Street Hospital Biomedical Research Centre. The views expressed are those of the author(s) and not necessarily those of the NHS, the NIHR, or the Department of Health.

Conflicts of Interest: The authors declare no conflict of interest.

References

1. Greco, A.; Goossens, R.; van Engelen, B.; van der Maarel, S.M. Consequences of epigenetic derepression in facioscapulohumeral muscular dystrophy. *Clin. Genet.* **2020**, *97*, 799–814. [CrossRef]
2. De Iaco, A.; Planet, E.; Coluccio, A.; Verp, S.; Duc, J.; Trono, D. DUX-family transcription factors regulate zygotic genome activation in placental mammals. *Nat. Genet.* **2017**, *49*, 941–945. [CrossRef]
3. Whiddon, J.L.; Langford, A.T.; Wong, C.J.; Zhong, J.W.; Tapscott, S.J. Conservation and innovation in the DUX4-family gene network. *Nat. Genet.* **2017**, *49*, 935–940. [CrossRef]
4. Hendrickson, P.G.; Dorais, J.A.; Grow, E.J.; Whiddon, J.L.; Lim, J.W.; Wike, C.L.; Weaver, B.D.; Pflueger, C.; Emery, B.R.; Wilcox, A.L.; et al. Conserved roles of mouse DUX and human DUX4 in activating cleavage-stage genes and MERVL/HERVL retrotransposons. *Nat. Genet.* **2017**, *49*, 925–934. [CrossRef]
5. Snider, L.; Geng, L.N.; Lemmers, R.J.; Kyba, M.; Ware, C.B.; Nelson, A.M.; Tawil, R.; Filippova, G.N.; van der Maarel, S.M.; Tapscott, S.J.; et al. Facioscapulohumeral dystrophy: Incomplete suppression of a retrotransposed gene. *PLoS Genet.* **2010**, *6*, e1001181. [CrossRef] [PubMed]
6. Chew, G.L.; Campbell, A.E.; De Neef, E.; Sutliff, N.A.; Shadle, S.C.; Tapscott, S.J.; Bradley, R.K. DUX4 Suppresses MHC Class I to Promote Cancer Immune Evasion and Resistance to Checkpoint Blockade. *Dev. Cell* **2019**, *50*, 658–671.e7. [CrossRef] [PubMed]
7. Gabriels, J.; Beckers, M.C.; Ding, H.; De Vriese, A.; Plaisance, S.; van der Maarel, S.M.; Padberg, G.W.; Frants, R.R.; Hewitt, J.E.; Collen, D.; et al. Nucleotide sequence of the partially deleted D4Z4 locus in a patient with FSHD identifies a putative gene within each 3.3 kb element. *Gene* **1999**, *236*, 25–32. [CrossRef]
8. Lim, K.R.Q.; Nguyen, Q.; Yokota, T. DUX4 Signalling in the Pathogenesis of Facioscapulohumeral Muscular Dystrophy. *Int. J. Mol. Sci.* **2020**, *21*, 729. [CrossRef]
9. Salsi, V.; Magdinier, F.; Tupler, R. Does DNA Methylation Matter in FSHD? *Genes* **2020**, *11*, 258. [CrossRef] [PubMed]
10. Tawil, R.; Van Der Maarel, S.r.M. Facioscapulohumeral muscular dystrophy. *Muscle Nerve* **2006**, *34*, 1–15. [CrossRef]
11. Wijmenga, C.; Hewitt, J.E.; Sandkuijl, L.A.; Clark, L.N.; Wright, T.J.; Dauwerse, H.G.; Gruter, A.M.; Hofker, M.H.; Moerer, P.; Williamson, R.; et al. Chromosome 4q DNA rearrangements associated with facioscapulohumeral muscular dystrophy. *Nat. Genet.* **1992**, *2*, 26–30. [CrossRef] [PubMed]

12. van Deutekom, J.C.; Wijmenga, C.; van Tienhoven, E.A.; Gruter, A.M.; Hewitt, J.E.; Padberg, G.W.; van Ommen, G.J.; Hofker, M.H.; Frants, R.R. FSHD associated DNA rearrangements are due to deletions of integral copies of a 3.2 kb tandemly repeated unit. *Hum. Mol. Genet.* **1993**, *2*, 2037–2042. [CrossRef] [PubMed]
13. van der Maarel, S.M.; Frants, R.R. The D4Z4 repeat-mediated pathogenesis of facioscapulohumeral muscular dystrophy. *Am. J. Hum. Genet.* **2005**, *76*, 375–386. [CrossRef] [PubMed]
14. Lemmers, R.J.; van der Vliet, P.J.; Klooster, R.; Sacconi, S.; Camano, P.; Dauwerse, J.G.; Snider, L.; Straasheijm, K.R.; van Ommen, G.J.; Padberg, G.W.; et al. A unifying genetic model for facioscapulohumeral muscular dystrophy. *Science* **2010**, *329*, 1650–1653. [CrossRef]
15. van Overveld, P.G.; Lemmers, R.J.; Deidda, G.; Sandkuijl, L.; Padberg, G.W.; Frants, R.R.; van der Maarel, S.M. Interchromosomal repeat array interactions between chromosomes 4 and 10: A model for subtelomeric plasticity. *Hum. Mol. Genet.* **2000**, *9*, 2879–2884. [CrossRef]
16. van Deutekom, J.C.; Bakker, E.; Lemmers, R.J.; van der Wielen, M.J.; Bik, E.; Hofker, M.H.; Padberg, G.W.; Frants, R.R. Evidence for subtelomeric exchange of 3.3 kb tandemly repeated units between chromosomes 4q35 and 10q26: Implications for genetic counselling and etiology of FSHD1. *Hum. Mol. Genet.* **1996**, *5*, 1997–2003. [CrossRef]
17. van der Maarel, S.M.; Miller, D.G.; Tawil, R.; Filippova, G.N.; Tapscott, S.J. Facioscapulohumeral muscular dystrophy: Consequences of chromatin relaxation. *Curr. Opin. Neurol.* **2012**, *25*, 614–620. [CrossRef]
18. Lemmers, R.J.; Tawil, R.; Petek, L.M.; Balog, J.; Block, G.J.; Santen, G.W.; Amell, A.M.; van der Vliet, P.J.; Almomani, R.; Straasheijm, K.R.; et al. Digenic inheritance of an SMCHD1 mutation and an FSHD-permissive D4Z4 allele causes facioscapulohumeral muscular dystrophy type 2. *Nat. Genet.* **2012**, *44*, 1370–1374. [CrossRef]
19. Van den Boogaard, M.L.; Lemmers, R.; Balog, J.; Wohlgemuth, M.; Auranen, M.; Mitsuhashi, S.; van der Vliet, P.J.; Straasheijm, K.R.; van den Akker, R.F.P.; Kriek, M.; et al. Mutations in DNMT3B Modify Epigenetic Repression of the D4Z4 Repeat and the Penetrance of Facioscapulohumeral Dystrophy. *Am. J. Hum. Genet.* **2016**, *98*, 1020–1029. [CrossRef]
20. Lemmers, R.J.; van der Vliet, P.J.; van der Gaag, K.J.; Zuniga, S.; Frants, R.R.; de Knijff, P.; van der Maarel, S.M. Worldwide population analysis of the 4q and 10q subtelomeres identifies only four discrete interchromosomal sequence transfers in human evolution. *Am. J. Hum. Genet.* **2010**, *86*, 364–377. [CrossRef]
21. Tupler, R.; Berardinelli, A.; Barbierato, L.; Frants, R.; Hewitt, J.E.; Lanzi, G.; Maraschio, P.; Tiepolo, L. Monosomy of distal 4q does not cause facioscapulohumeral muscular dystrophy. *J. Med. Genet.* **1996**, *33*, 366–370. [CrossRef] [PubMed]
22. Thomas, N.S.T.; Wiseman, K.; Spurlock, G.; MacDonald, M.; Ustek, D.; Upadhyaya, M. A large patient study confirming that facioscapulohumeral muscular dystrophy (FSHD) disease expression is almost exclusively associated with an FSHD locus located on a 4qA-defined 4qter subtelomere. *J. Med. Genet.* **2007**, *44*, 215–218. [CrossRef] [PubMed]
23. Lemmers, R.J.; de Kievit, P.; Sandkuijl, L.; Padberg, G.W.; van Ommen, G.J.; Frants, R.R.; van der Maarel, S.M. Facioscapulohumeral muscular dystrophy is uniquely associated with one of the two variants of the 4q subtelomere. *Nat. Genet.* **2002**, *32*, 235–236. [CrossRef] [PubMed]
24. Snider, L.; Asawachaicharn, A.; Tyler, A.E.; Geng, L.N.; Petek, L.M.; Maves, L.; Miller, D.G.; Lemmers, R.J.L.F.; Winokur, S.T.; Tawil, R.; et al. RNA transcripts, miRNA-sized fragments and proteins produced from D4Z4 units: New candidates for the pathophysiology of facioscapulohumeral dystrophy. *Hum. Mol. Genet.* **2009**, *18*, 2414–2430. [CrossRef]
25. Geng, L.N.; Yao, Z.; Snider, L.; Fong, A.P.; Cech, J.N.; Young, J.M.; van der Maarel, S.M.; Ruzzo, W.L.; Gentleman, R.C.; Tawil, R.; et al. DUX4 Activates Germline Genes, Retroelements, and Immune Mediators: Implications for Facioscapulohumeral Dystrophy. *Dev. Cell* **2012**, *22*, 38–51. [CrossRef]
26. Sacconi, S.; Lemmers, R.J.; Balog, J.; van der Vliet, P.J.; Lahaut, P.; van Nieuwenhuizen, M.P.; Straasheijm, K.R.; Debipersad, R.D.; Vos-Versteeg, M.; Salviati, L.; et al. The FSHD2 gene SMCHD1 is a modifier of disease severity in families affected by FSHD1. *Am. J. Hum. Genet.* **2013**, *93*, 744–751. [CrossRef]
27. Balog, J.; Thijssen, P.E.; Shadle, S.; Straasheijm, K.R.; van der Vliet, P.J.; Krom, Y.D.; van den Boogaard, M.L.; de Jong, A.; Lemmers, R.J.L.F.; Tawil, R.; et al. Increased DUX4 expression during muscle differentiation correlates with decreased SMCHD1 protein levels at D4Z4. *Epigenetics* **2015**, *10*, 1133–1142. [CrossRef]

28. Hamanaka, K.; Sikrova, D.; Mitsuhashi, S.; Masuda, H.; Sekiguchi, Y.; Sugiyama, A.; Shibuya, K.; Lemmers, R.; Goossens, R.; Ogawa, M.; et al. Homozygous nonsense variant in LRIF1 associated with facioscapulohumeral muscular dystrophy. *Neurology* **2020**, *94*, e2441–e2447. [CrossRef]
29. Zeng, W.; Chen, Y.Y.; Newkirk, D.A.; Wu, B.; Balog, J.; Kong, X.; Ball, A.R., Jr.; Zanotti, S.; Tawil, R.; Hashimoto, N.; et al. Genetic and Epigenetic Characteristics of FSHD-Associated 4q and 10q D4Z4 that are Distinct from Non-4q/10q D4Z4 Homologs. *Hum. Mutat* **2014**, *35*, 998–1010. [CrossRef]
30. Balog, J.; Goossens, R.; Lemmers, R.; Straasheijm, K.R.; van der Vliet, P.J.; Heuvel, A.V.D.; Cambieri, C.; Capet, N.; Feasson, L.; Manel, V.; et al. Monosomy 18p is a risk factor for facioscapulohumeral dystrophy. *J. Med. Genet.* **2018**, *55*, 469–478. [CrossRef]
31. Himeda, C.L.; Jones, T.I.; Virbasius, C.M.; Zhu, L.J.; Green, M.R.; Jones, P.L. Identification of Epigenetic Regulators of DUX4-fl for Targeted Therapy of Facioscapulohumeral Muscular Dystrophy. *Mol. Ther.* **2018**, *26*, 1797–1807. [CrossRef]
32. Cabianca, D.S.; Casa, V.; Bodega, B.; Xynos, A.; Ginelli, E.; Tanaka, Y.; Gabellini, D. A Long ncRNA Links Copy Number Variation to a Polycomb/Trithorax Epigenetic Switch in FSHD Muscular Dystrophy. *Cell* **2012**, *149*, 819–831. [CrossRef] [PubMed]
33. Huichalaf, C.; Micheloni, S.; Ferri, G.; Caccia, R.; Gabellini, D. DNA methylation analysis of the macrosatellite repeat associated with FSHD muscular dystrophy at single nucleotide level. *PLoS ONE* **2014**, *9*, e115278. [CrossRef] [PubMed]
34. Campbell, A.E.; Shadle, S.C.; Jagannathan, S.; Lim, J.W.; Resnick, R.; Tawil, R.; van der Maarel, S.M.; Tapscott, S.J. NuRD and CAF-1-mediated silencing of the D4Z4 array is modulated by DUX4-induced MBD3L proteins. *Elife* **2018**, *7*. [CrossRef]
35. Krom, Y.D.; Thijssen, P.E.; Young, J.M.; den Hamer, B.; Balog, J.; Yao, Z.; Maves, L.; Snider, L.; Knopp, P.; Zammit, P.S.; et al. Intrinsic Epigenetic Regulation of the D4Z4 Macrosatellite Repeat in a Transgenic Mouse Model for FSHD. *PLoS Genet.* **2013**, *9*, e1003415. [CrossRef]
36. Dion, C.; Roche, S.; Laberthonniere, C.; Broucqsault, N.; Mariot, V.; Xue, S.; Gurzau, A.D.; Nowak, A.; Gordon, C.T.; Gaillard, M.C.; et al. SMCHD1 is involved in de novo methylation of the DUX4-encoding D4Z4 macrosatellite. *Nucleic Acids Res.* **2019**, *47*, 2822–2839. [CrossRef]
37. Ottaviani, A.; Rival-Gervier, S.; Boussouar, A.; Foerster, A.M.; Rondier, D.; Sacconi, S.; Desnuelle, C.; Gilson, E.; Magdinier, F. The D4Z4 macrosatellite repeat acts as a CTCF and A-type lamins-dependent insulator in facio-scapulo-humeral dystrophy. *PLoS Genet.* **2009**, *5*, e1000394. [CrossRef]
38. Haynes, P.; Bomsztyk, K.; Miller, D.G. Sporadic DUX4 expression in FSHD myocytes is associated with incomplete repression by the PRC2 complex and gain of H3K9 acetylation on the contracted D4Z4 allele. *Epigenetics Chromatin* **2018**, *11*, 47. [CrossRef]
39. Stadler, G.; Rahimov, F.; King, O.D.; Chen, J.C.; Robin, J.D.; Wagner, K.R.; Shay, J.W.; Emerson, C.P., Jr.; Wright, W.E. Telomere position effect regulates DUX4 in human facioscapulohumeral muscular dystrophy. *Nat. Struct. Mol. Biol.* **2013**, *20*, 671–678. [CrossRef]
40. Stadler, G.; King, O.D.; Robin, J.D.; Shay, J.W.; Wright, W.E. Facioscapulohumeral muscular dystrophy: Are telomeres the end of the story? *Rare Dis.* **2013**, *1*, e26142. [CrossRef]
41. Robin, J.D.; Ludlow, A.T.; Batten, K.; Gaillard, M.C.; Stadler, G.; Magdinier, F.; Wright, W.; Shay, J.W. SORBS2 transcription is activated by telomere position effect-over long distance upon telomere shortening in muscle cells from patients with facioscapulohumeral dystrophy. *Genome Res.* **2015**, *25*, 1781–1790. [CrossRef]
42. Zeng, W.; de Greef, J.C.; Chen, Y.Y.; Chien, R.; Kong, X.; Gregson, H.C.; Winokur, S.T.; Pyle, A.; Robertson, K.D.; Schmiesing, J.A.; et al. Specific loss of histone H3 lysine 9 trimethylation and HP1gamma/cohesin binding at D4Z4 repeats is associated with facioscapulohumeral dystrophy (FSHD). *PLoS Genet.* **2009**, *5*, e1000559. [CrossRef] [PubMed]
43. Jones, T.I.; Yan, C.; Sapp, P.C.; McKenna-Yasek, D.; Kang, P.B.; Quinn, C.; Salameh, J.S.; King, O.D.; Jones, P.L. Identifying diagnostic DNA methylation profiles for facioscapulohumeral muscular dystrophy in blood and saliva using bisulfite sequencing. *Clin. Epigenetics* **2014**, *6*, 23. [CrossRef]
44. Jones, T.I.; King, O.D.; Himeda, C.L.; Homma, S.; Chen, J.C.; Beermann, M.L.; Yan, C.; Emerson, C.P., Jr.; Miller, J.B.; Wagner, K.R.; et al. Individual epigenetic status of the pathogenic D4Z4 macrosatellite correlates with disease in facioscapulohumeral muscular dystrophy. *Clin. Epigenetics* **2015**, *7*, 37. [CrossRef] [PubMed]
45. Das, S.; Chadwick, B.P. Influence of Repressive Histone and DNA Methylation upon D4Z4 Transcription in Non-Myogenic Cells. *PLoS ONE* **2016**, *11*, e0160022. [CrossRef] [PubMed]

46. Himeda, C.L.; Debarnot, C.; Homma, S.; Beermann, M.L.; Miller, J.B.; Jones, P.L.; Jones, T.I. Myogenic enhancers regulate expression of the facioscapulohumeral muscular dystrophy-associated DUX4 gene. *Mol. Cell Biol.* **2014**, *34*, 1942–1955. [CrossRef] [PubMed]
47. Lemmers, R.J.; Osborn, M.; Haaf, T.; Rogers, M.; Frants, R.R.; Padberg, G.W.; Cooper, D.N.; van der Maarel, S.M.; Upadhyaya, M. D4F104S1 deletion in facioscapulohumeral muscular dystrophy: Phenotype, size, and detection. *Neurology* **2003**, *61*, 178–183. [CrossRef]
48. Nguyen, K.; Broucqsault, N.; Chaix, C.; Roche, S.; Robin, J.D.; Vovan, C.; Gerard, L.; Megarbane, A.; Urtizberea, J.A.; Bellance, R.; et al. Deciphering the complexity of the 4q and 10q subtelomeres by molecular combing in healthy individuals and patients with facioscapulohumeral dystrophy. *J. Med. Genet.* **2019**, *56*, 590–601. [CrossRef]
49. Sharma, V.; Pandey, S.N.; Khawaja, H.; Brown, K.J.; Hathout, Y.; Chen, Y.W. PARP1 Differentially Interacts with Promoter region of DUX4 Gene in FSHD Myoblasts. *J. Genet. Syndr. Gene Ther.* **2016**, *7*. [CrossRef]
50. Block, G.J.; Narayanan, D.; Amell, A.M.; Petek, L.M.; Davidson, K.C.; Bird, T.D.; Tawil, R.; Moon, R.T.; Miller, D.G. Wnt/beta-catenin signaling suppresses DUX4 expression and prevents apoptosis of FSHD muscle cells. *Hum. Mol. Genet.* **2013**, *22*, 390–396. [CrossRef]
51. Campbell, A.E.; Oliva, J.; Yates, M.P.; Zhong, J.W.; Shadle, S.C.; Snider, L.; Singh, N.; Tai, S.; Hiramuki, Y.; Tawil, R.; et al. BET bromodomain inhibitors and agonists of the beta-2 adrenergic receptor identified in screens for compounds that inhibit DUX4 expression in FSHD muscle cells. *Skelet. Muscle* **2017**, *7*, 16. [CrossRef] [PubMed]
52. Cruz, J.M.; Hupper, N.; Wilson, L.S.; Concannon, J.B.; Wang, Y.; Oberhauser, B.; Patora-Komisarska, K.; Zhang, Y.; Glass, D.J.; Trendelenburg, A.U.; et al. Protein kinase A activation inhibits DUX4 gene expression in myotubes from patients with facioscapulohumeral muscular dystrophy. *J. Biol. Chem.* **2018**, *293*, 11837–11849. [CrossRef] [PubMed]
53. Yamauchi, J.; Nagao, M.; Kaziro, Y.; Itoh, H. Activation of p38 mitogen-activated protein kinase by signaling through G protein-coupled receptors. Involvement of Gbetagamma and Galphaq/11 subunits. *J. Biol. Chem.* **1997**, *272*, 27771–27777. [CrossRef] [PubMed]
54. Oliva, J.; Galasinski, S.; Richey, A.; Campbell, A.E.; Meyers, M.J.; Modi, N.; Zhong, J.W.; Tawil, R.; Tapscott, S.J.; Sverdrup, F.M. Clinically Advanced p38 Inhibitors Suppress DUX4 Expression in Cellular and Animal Models of Facioscapulohumeral Muscular Dystrophy. *J. Pharmacol Exp. Ther.* **2019**, *370*, 219–230. [CrossRef]
55. Lee, J.H.; Goto, K.; Matsuda, C.; Arahata, K. Characterization of a tandemly repeated 3.3-kb Kpnl unit in the facioscapulohumeral muscular dystrophy (FSHD) gene region on chromosome 4q35. *Muscle Nerve Suppl.* **1995**, *2*, S6–S13. [CrossRef]
56. Kowaljow, V.; Marcowycz, A.; Ansseau, E.; Conde, C.B.; Sauvage, S.; Matteotti, C.; Arias, C.; Corona, E.D.; Nunez, N.G.; Leo, O.; et al. The DUX4 gene at the FSHD1A locus encodes a pro-apoptotic protein. *Neuromuscul. Disord.* **2007**, *17*, 611–623. [CrossRef]
57. Dixit, M.; Ansseau, E.; Tassin, A.; Winokur, S.; Shi, R.; Qian, H.; Sauvage, S.; Matteotti, C.; van Acker, A.M.; Leo, O.; et al. DUX4, a candidate gene of facioscapulohumeral muscular dystrophy, encodes a transcriptional activator of PITX1. *Proc. Natl. Acad. Sci. USA* **2007**, *104*, 18157–18162. [CrossRef]
58. Lemmers, R.J.; Wohlgemuth, M.; Frants, R.R.; Padberg, G.W.; Morava, E.; van der Maarel, S.M. Contractions of D4Z4 on 4qB subtelomeres do not cause facioscapulohumeral muscular dystrophy. *Am. J. Hum. Genet.* **2004**, *75*, 1124–1130. [CrossRef]
59. Lemmers, R.J.; Wohlgemuth, M.; van der Gaag, K.J.; van der Vliet, P.J.; van Teijlingen, C.M.; de Knijff, P.; Padberg, G.W.; Frants, R.R.; van der Maarel, S.M. Specific sequence variations within the 4q35 region are associated with facioscapulohumeral muscular dystrophy. *Am. J. Hum. Genet.* **2007**, *81*, 884–894. [CrossRef]
60. Lemmers, R.J.; van der Vliet, P.J.; Balog, J.; Goeman, J.J.; Arindrarto, W.; Krom, Y.D.; Straasheijm, K.R.; Debipersad, R.D.; Ozel, G.; Sowden, J.; et al. Deep characterization of a common D4Z4 variant identifies biallelic DUX4 expression as a modifier for disease penetrance in FSHD2. *Eur. J. Hum. Genet.* **2018**, *26*, 94–106. [CrossRef]
61. Mitsuhashi, H.; Ishimaru, S.; Homma, S.; Yu, B.; Honma, Y.; Beermann, M.L.; Miller, J.B. Functional domains of the FSHD-associated DUX4 protein. *Biol. Open* **2018**, *7*. [CrossRef] [PubMed]

62. Jones, T.I.; Chen, J.C.; Rahimov, F.; Homma, S.; Arashiro, P.; Beermann, M.L.; King, O.D.; Miller, J.B.; Kunkel, L.M.; Emerson, C.P., Jr.; et al. Facioscapulohumeral muscular dystrophy family studies of DUX4 expression: Evidence for disease modifiers and a quantitative model of pathogenesis. *Hum. Mol. Genet.* **2012**, *21*, 4419–4430. [CrossRef] [PubMed]
63. Krom, Y.D.; Dumonceaux, J.; Mamchaoui, K.; den Hamer, B.; Mariot, V.; Negroni, E.; Geng, L.N.; Martin, N.; Tawil, R.; Tapscott, S.J.; et al. Generation of isogenic D4Z4 contracted and noncontracted immortal muscle cell clones from a mosaic patient: A cellular model for FSHD. *Am. J. Pathol.* **2012**, *181*, 1387–1401. [CrossRef] [PubMed]
64. Broucqsault, N.; Morere, J.; Gaillard, M.C.; Dumonceaux, J.; Torrents, J.; Salort-Campana, E.; Maues de Paula, A.; Bartoli, M.; Fernandez, C.; Chesnais, A.L.; et al. Dysregulation of 4q35- and muscle-specific genes in fetuses with a short D4Z4 array linked to Facio-Scapulo-Humeral Dystrophy. *Hum. Mol. Genet.* **2013**, *22*, 4206–4214. [CrossRef]
65. Ferreboeuf, M.; Mariot, V.; Bessieres, B.; Vasiljevic, A.; Attie-Bitach, T.; Collardeau, S.; Morere, J.; Roche, S.; Magdinier, F.; Robin-Ducellier, J.; et al. DUX4 and DUX4 downstream target genes are expressed in fetal FSHD muscles. *Hum. Mol. Genet.* **2014**, *23*, 171–181. [CrossRef]
66. Marsollier, A.C.; Ciszewski, L.; Mariot, V.; Popplewell, L.; Voit, T.; Dickson, G.; Dumonceaux, J. Antisense targeting of 3′ end elements involved in DUX4 mRNA processing is an efficient therapeutic strategy for facioscapulohumeral dystrophy: A new gene-silencing approach. *Hum. Mol. Genet.* **2016**, *25*, 10. [CrossRef]
67. Liu, D.; Brockman, J.M.; Dass, B.; Hutchins, L.N.; Singh, P.; McCarrey, J.R.; MacDonald, C.C.; Graber, J.H. Systematic variation in mRNA 3′-processing signals during mouse spermatogenesis. *Nucleic Acids Res.* **2007**, *35*, 234–246. [CrossRef]
68. Ji, Z.; Lee, J.Y.; Pan, Z.; Jiang, B.; Tian, B. Progressive lengthening of 3′ untranslated regions of mRNAs by alternative polyadenylation during mouse embryonic development. *Proc. Natl. Acad. Sci. USA* **2009**, *106*, 7028–7033. [CrossRef]
69. Tassin, A.; Laoudj-Chenivesse, D.; Vanderplanck, C.; Barro, M.; Charron, S.; Ansseau, E.; Chen, Y.W.; Mercier, J.; Coppee, F.; Belayew, A. DUX4 expression in FSHD muscle cells: How could such a rare protein cause a myopathy? *J. Cell Mol. Med.* **2012**, *17*, 76–89. [CrossRef]
70. Ferreboeuf, M.; Mariot, V.; Furling, D.; Butler-Browne, G.; Mouly, V.; Dumonceaux, J. Nuclear protein spreading: Implication for pathophysiology of neuromuscular diseases. *Hum. Mol. Genet.* **2014**, *23*, 4125–4133. [CrossRef]
71. Rickard, A.M.; Petek, L.M.; Miller, D.G. Endogenous DUX4 expression in FSHD myotubes is sufficient to cause cell death and disrupts RNA splicing and cell migration pathways. *Hum. Mol. Genet.* **2015**, *24*, 5901–5914. [CrossRef] [PubMed]
72. Gabellini, D.; Green, M.R.; Tupler, R. Inappropriate gene activation in FSHD: A repressor complex binds a chromosomal repeat deleted in dystrophic muscle. *Cell* **2002**, *110*, 339–348. [CrossRef]
73. Vanderplanck, C.; Ansseau, E.; Charron, S.; Stricwant, N.; Tassin, A.; Laoudj-Chenivesse, D.; Wilton, S.D.; Coppee, F.; Belayew, A. The FSHD Atrophic Myotube Phenotype Is Caused by DUX4 Expression. *PLoS ONE* **2011**, *6*, e26820. [CrossRef] [PubMed]
74. Wallace, L.M.; Liu, J.; Domire, J.S.; Garwick-Coppens, S.E.; Guckes, S.M.; Mendell, J.R.; Flanigan, K.M.; Harper, S.Q. RNA Interference Inhibits DUX4-induced Muscle Toxicity In Vivo: Implications for a Targeted FSHD Therapy. *Mol. Ther.* **2012**, *20*, 1417–1423. [CrossRef] [PubMed]
75. Chen, J.C.; King, O.D.; Zhang, Y.; Clayton, N.P.; Spencer, C.; Wentworth, B.M.; Emerson, C.P., Jr.; Wagner, K.R. Morpholino-mediated Knockdown of DUX4 Toward Facioscapulohumeral Muscular Dystrophy Therapeutics. *Mol. Ther.* **2016**, *24*, 1405–1411. [CrossRef]
76. Marsollier, A.C.; Joubert, R.; Mariot, V.; Dumonceaux, J. Targeting the Polyadenylation Signal of Pre-mRNA: A New Gene Silencing Approach for Facioscapulohumeral Dystrophy. *Int. J. Mol. Sci.* **2018**, *19*, 1347. [CrossRef]
77. Bosnakovski, D.; da Silva, M.T.; Sunny, S.T.; Ener, E.T.; Toso, E.A.; Yuan, C.; Cui, Z.; Walters, M.A.; Jadhav, A.; Kyba, M. A novel P300 inhibitor reverses DUX4-mediated global histone H3 hyperacetylation, target gene expression, and cell death. *Sci. Adv.* **2019**, *5*, eaaw7781. [CrossRef]

78. DeSimone, A.M.; Leszyk, J.; Wagner, K.; Emerson, C.P., Jr. Identification of the hyaluronic acid pathway as a therapeutic target for facioscapulohumeral muscular dystrophy. *Sci. Adv.* **2019**, *5*, eaaw7099. [CrossRef]
79. Klingler, C.; Ashley, J.; Shi, K.; Stiefvater, A.; Kyba, M.; Sinnreich, M.; Aihara, H.; Kinter, J. DNA aptamers against the DUX4 protein reveal novel therapeutic implications for FSHD. *FASEB J.* **2020**, *34*, 4573–4590. [CrossRef]

Journal of Personalized Medicine

Review

Molecular and Cellular Mechanisms Affected in ALS

Laura Le Gall [1,2,†], Ekene Anakor [1,†], Owen Connolly [1,†], Udaya Geetha Vijayakumar [1], William J. Duddy [1] and Stephanie Duguez [1,*]

1 Northern Ireland Center for Stratified/Personalised Medicine, Biomedical Sciences Research Institute, Ulster University, Derry-Londonderry BT47, UK; le_gall-l@ulster.ac.uk (L.L.G.); Anakor-E@ulster.ac.uk (E.A.); Connolly-O4@ulster.ac.uk (O.C.); u.vijayakumar@ulster.ac.uk (U.G.V.); w.duddy@ulster.ac.uk (W.J.D.)

2 NIHR Biomedical Research Centre, University College London, Great Ormond Street Institute of Child Health and Great Ormond Street Hospital NHS Trust, London WC1N 1EH, UK; l.gall@ucl.ac.uk

* Correspondence: s.duguez@ulster.ac.uk

† These authors contributed equally to this work.

Received: 20 July 2020; Accepted: 22 August 2020; Published: 25 August 2020

Abstract: Amyotrophic lateral sclerosis (ALS) is a terminal late-onset condition characterized by the loss of upper and lower motor neurons. Mutations in more than 30 genes are associated to the disease, but these explain only ~20% of cases. The molecular functions of these genes implicate a wide range of cellular processes in ALS pathology, a cohesive understanding of which may provide clues to common molecular mechanisms across both familial (inherited) and sporadic cases and could be key to the development of effective therapeutic approaches. Here, the different pathways that have been investigated in ALS are summarized, discussing in detail: mitochondrial dysfunction, oxidative stress, axonal transport dysregulation, glutamate excitotoxicity, endosomal and vesicular transport impairment, impaired protein homeostasis, and aberrant RNA metabolism. This review considers the mechanistic roles of ALS-associated genes in pathology, viewed through the prism of shared molecular pathways.

Keywords: oxidative stress; mitochondria dysfunction; axonal transport; autophagy; endocytosis; secretion; excitotoxicity; RNA metabolism; MND

1. Introduction

Amyotrophic lateral sclerosis (ALS) is the most frequent motor neuron disease (MND), with an estimated ~223,000 patients being affected globally in 2015 [1]. The pathology affects both upper motor neurons (UMN) in the cortex and lower motor neurons (LMN) in the brainstem and spinal cord [2]. Paralysis and death usually occur between three to four years after symptom onset [3], and there are currently no effective treatments to slow disease progression [4]. Approximately 90% of ALS cases are sporadic, while 10% are familial, defined by the occurrence of ALS in more than one family member [5]. Around 30 different genes are linked with ALS [5,6], explaining ~20% of all ALS cases and associated with different molecular functions and disease phenotypes [7], so that the task of understanding the relationships between affected pathways is complex.

To investigate the different molecular pathways affected in ALS, various in vivo models, including drosophila [8–11], C-elegans [12], zebrafish [13–16], and rodents [17], as well as in vitro cell models such as patient lymphoblastoid cell lines [18] and hybrid [19,20] or primary murine cell lines, [21] have been developed. Most of these models investigate the pathological effects of mutations to ALS genes, including *Fused in Sarcoma* (*FUS*), *Superoxide dismutase* (*SOD1*), *TAR DNA-binding protein 43* (*TDP-43*), and *Chromosome 9 open reading frame 72* (*C9orf72*) [22,23]. Their study has resulted in numerous cellular and molecular mechanisms being proposed to explain motor neuron death. Mechanisms frequently implicated include: reactive oxygen species (ROS)-associated oxidative stress [24–27], mitochondrial dysfunction [24], axonal and vesicular trafficking dysregulation [28,29], glutamate-mediated excitotoxicity [30–33], proteostatic impairments [34–38], and altered RNA metabolism and/or processing [39–42].

Alteration to one or more of these cellular processes may be present, not only in the motor neurons themselves but, also, in neighboring cell populations, such as glial cells, peripheral inflammatory cells, and muscles, as ALS is increasingly considered a multisystemic disease that culminates in motor neuron death [6,24]. For example, astrocytes and microglia have been implicated in the release of proinflammatory mediators that lead to chronic neuroinflammation and motor neuron toxicity [43]. In addition, the selective overexpression of mutant SOD1 in skeletal muscle was shown to cause mitochondrial abnormalities, induce microglial activation in the central nervous system (CNS), and result in severe muscle atrophy in mice [44].

Consensus is yet to be reached regarding the causal mechanisms involved in the onset and propagation of ALS. The aim of this review is to identify and summarize the different molecular mechanisms implicated in various forms of the disease, including sporadic and familial cases. In doing so, it is hoped that new insights may be gained regarding the role of different pathways across different forms of the disease.

2. Oxidative Stress

Oxidative stress results from an imbalance between the production and elimination of reactive oxygen species (ROS) [45], as well as an impaired ability to repair ROS-mediated toxicity [46], and has been of particular interest in ALS pathogenesis ([47] and Figure 1) since the discovery of *SOD1* mutations in familial forms of ALS [48]. Increased levels of oxidized proteins, RNA, DNA, and lipids have been observed in post-mortem tissue from both sporadic and *SOD1* ALS cases [27,49,50], as well as in the cerebrospinal fluid (CSF), serum, and urine of sporadic ALS patients [26].

SOD1 is a major antioxidant enzyme that is ubiquitously expressed and catalyzes radical superoxide anions into molecular oxygen and hydrogen peroxide [51]. Approximately 80 of the 160 *SOD1* mutations reported in ALS are missense mutations that fail to cause a loss of SOD1 activity [52], and many SOD1 mouse models show a progressive, late-onset motor phenotype with concomitant astrogliosis and motor neuron pathology when mutated forms of human SOD1 are overexpressed [17]. Evidence from human samples have shown that there is a 42% reduction in overall SOD1 activity in familial SOD1 patients [53], potentially leading to an imbalance between ROS production and degradation (Figure 1). This imbalance might be exacerbated by the disruption of the NRF2-ARE (Nuclear erythroid 2-Related Factor—antioxidant response element) signaling pathway that is observed in *SOD1* ALS [54], thus affecting the expression of antioxidant proteins [55] (Figure 1). Supporting these hypotheses, oxidative damage such as protein glycoxidation and lipid peroxidation were observed in the motor neurons of the anterior horn from *SOD1* familial ALS (fALS) patients [56] and SOD1^{G93A} mice [57,58].

The generation of ROS could result from the activity of NADPH oxidase in the lipid raft membrane compartment. Interestingly, the *ATXN2* gene encodes the ataxin-2 polyglutamine (PolyQ) protein, and intermediate-length PolyQ expansions (27–33 Qs), which are known to be a significant risk for ALS [59–61], can interact with NADPH oxidase and may lead to an increase in ROS production, DNA damage, and mitochondrial distress [62] (Figure 1).

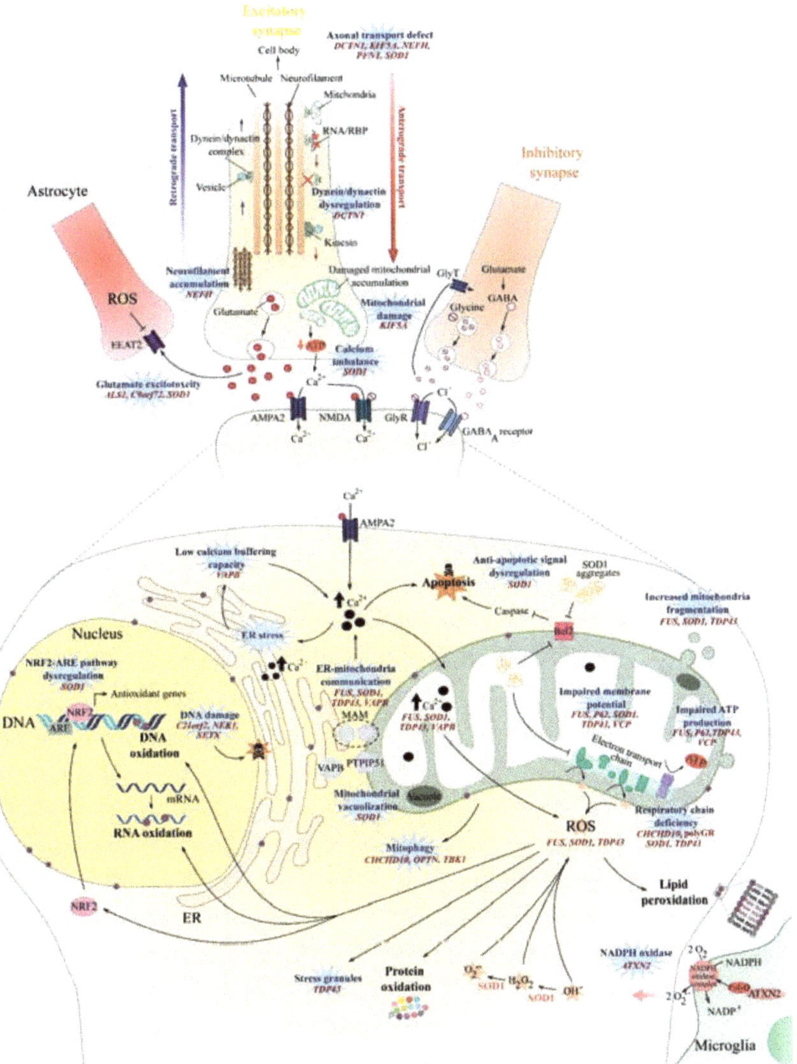

Figure 1. Oxidative stress, mitochondrial dysfunction, axonal transport, and glutamate excitotoxicity in amyotrophic lateral sclerosis (ALS). An increase in oxidative stress can result from defects in detoxifying pathways. Such defects include the loss of SOD1 function, aberrant DNA damage repair machinery, or a decrease in expression of antioxidant genes affecting the NRF2-ARE pathway. Oxidative stress can also be increased by the stimulation of ROS production via increased NADPH oxidase activity or from disrupted mitchondrial respiratory chain activity. Mitochondrial activity can be affected by several ALS mutations, such as those leading to the accumulation of protein aggregates, or to decreased mitochondrial biogenesis and transport, or to increased cytosolic Ca^{2+} (as observed when glutamate receptor activity is stimulated or when the Ca^{2+}-buffering capacity is decreased). Consequently a disruption of the mitochondrial respiratory chain will lead to an increase in ROS production and, thus, to an accumulation of oxidized proteins, lipids, DNA, and RNA. Oxidative damage occurring over time may then stimulate apoptotosis and, thus, cell death. Defective axonal transport affects not only the mitochondria but, also, the transport of other proteins and RNA, with consequences on the axon structure and function being accompanied by neurofilament accumulation.

Defective glutamate uptake by astrocytes, and/or a defect in glutamate receptor clearance or in AMPA or GABA receptors, can lead to increased Ca^{2+} permeability and can impact the post-synaptic hyperexcitability and mitochondrial function. ARE: antioxidant response element, AMPA2: α-amino-3-hydroxy-5-methyl-4-isoxazolepropionic acid receptor 2, ATXN2: ataxin 2, Bcl2: B-cell lymphoma 2, C9orf72: Chromosome 9 open reading frame 72, C21orf2: Chromosome 21 open reading frame 7, CHCHD10: coiled-coil helix coiled-coil helix domain-containing 10, DCTN1: Dynactin 1, EEAT2: Excitatory amino acid transporter, ER: endoplasmic reticulum, FUS: Fused in Sarcoma, GABA: gamma-Aminobutyric acid, GlyR: glycine receptor, GlyT: glycine transporter, KIF5A: Kinesin heavy-chain isoform 5A, MAM: Mitochondria-associated ER membranes, NEFH: heavy-weighted neurofilaments, NEK1: (NIMA)-related kinase 1, NMDA: N-methyl-D-aspartate receptor, NRF2: Nuclear erythroid 2-Related Factor, PFN1: profilin-I, PTPIP51: Protein tyrosine phosphatase-interacting protein 51, SETX: senataxin, SOD: Superoxide dismutase 1, SPG11: Spatacsin, TDP-43: TAR DNA-binding protein 43, VAPB: vesicle-associated membrane protein-associated protein B, VCP: valosin-containing protein, and ROS: reactive oxygen species.

Recurrent oxidative stress and/or mitochondrial dysfunction occurring throughout the life of the cell can lead to DNA damage—damage that can be fixed by activating the DNA damage repair machinery. Several genes known to encode for proteins involved in DNA damage repair [63–65] are also associated with ALS: *NEK1* [66], *C21orf2* [67], and *SETX* [68]. These encode for the proteins never in mitosis-A (NIMA)-related kinase 1 (NEK1), cilia and flagella-associated protein 410, and the DNA/RNA helicase senataxin, respectively. Mutations in these genes may therefore increase the susceptibility to ALS as a result of dysregulated DNA damage repair machinery [67,69,70], leading to an impaired ability of motor neurons to cope with oxidative stress, consequently leading to cell death [64,70] (Figure 1). For example, induced pluripotent stem cell (iPSC) motor neurons derived from $NEK1^{c.2434A>T}$-mutated ALS patients exhibit an increased level of DNA damage, as well as a failure to repair DNA double-strand breaks [70]. Primary motor neurons from $SETX^{R2136H}$ and $SETX^{L389S}$ murine models were unable to cope with induced oxidative stress and showed an increased stress granule formation [71].

Altogether, these studies suggest that oxidative stress might be increased in sporadic and familial ALS patients. Increased oxidative stress may affect mitochondrial function [72], exacerbate endoplasmic reticulum stress [73], and impact protein homeostasis mechanisms [74], ultimately leading to cell damage and neuronal loss.

3. Mitochondrial Dysfunction

Mitochondria are key organelles for ATP generation, calcium buffering, and apoptosis regulation [75], and their dysfunction in the dorsal root ganglion cells of sporadic ALS patients has been described previously [76]. Several mechanisms can trigger mitochondrial dysfunction in ALS (Figure 1).

The maintenance of mitochondrial cristae organization is crucial to ensure respiratory chain function [77] and requires cardiolipin, the ATP synthase dimer, and large protein complexes such as the mitochondrial contact site complex (MICOS) and dynamin-like Opa1/Mgm1 [78,79]. The coiled-coil helix coiled-coil helix domain-containing protein 10 (CHCHD10), known to be associated with ALS [23], is suspected to be either part of [80,81] or interact with MICOS [82]. Consequently, mutations in *CHCHD10* result in the loss of mitochondria cristae [80], mitochondria fragmentation [81], and defective mitochondrial repair [80,83] (Figure 1).

Mitochondrial biogenesis can also be directly affected, as observed in *FUS*-mutated conditions [84,85]. While *FUS* encodes for a DNA/RNA-binding protein [86] predominantly localized to the nucleus, mutated forms of FUS can accumulate in the cytosol and possibly become toxic [87,88] and affect the mitochondrial function. For example, the mutated FUS^{P525L} can interact with mitochondrial chaperone proteins and induce mitochondria fragmentation and elevated ROS production [84,85].

Aberrant swollen mitochondria morphology has also been observed in neuronal, non-neuronal cells, and muscle tissue of other fALS cases, such as *SOD1* [24,89] and *C9orf72* [25,89]-mutated ALS patients but, also, in both SOD1^{G93A} and TDP-43^{A315T} murine models [44,90,91]. The aberrant morphology may result from a cascade of events involving the mutated protein aggregates. For example, insoluble mutant SOD1 can aggregate in mitochondria in the spinal cord of SOD1^{G93A} mice [92], causing the formation of vacuoles in the outer- and inter-mitochondrial membrane [93], affecting mitochondrial respiration, energy production, and ultimately, increasing the level of oxidative stress [94] (Figure 1). ALS patients with the SOD1^{A4V} mutation show significant increases in complex I and III activity of mitochondria in the motor cortex [95,96]. The overactivation of complexes I and III increased the production of mitochondrial ROS [97] and may explain the high level of oxidative stress observed in SOD1 mice and patients.

The G4C2 hexanucleotide repeat expansion mutation (HREM) in the *C9orf72* gene explains 40–50% of familial ALS cases and 5–10% of sporadic cases [98–101]. There are several hypotheses regarding the mechanisms by which this leads to toxicity, and evidence exists for both loss and gain-of-function-mediated toxicity. One hypothesis suggests that the repeat-associated non-AUG (RAN) translation of G4C2 repeats is causal in the expression of toxic dipeptide repeat (DPR) proteins. RAN translation can occur in both sense and antisense reading frames [41], resulting in the production of five different DPRs: glycine-alanine (GA), glycine-arginine (GR), proline-arginine (PR), proline-alanine (PA), and glycine-proline (GP) [38]. Interestingly, the expression of poly-GR results in early abnormalities in the mitochondiral respiratory chain by interacting with ATP5A1, a complex V protein, and leads to its ubiquitination and degradation in *C9orf72* ALS-FTD patients [102]. Mitochondrial dysfunction [103] and an increased oxidative stress [104] are reported in fibroblasts and iPSC-derived astrocytes obtained from *C9orf72* ALS patients (Figure 1).

Nonfunctional and damaged mitochondria can be targeted by NIP3-like protein X (NIX) or PTEN-induced putative kinase protein 1 (PINK1)-E3 ubiquitin ligase parkin, then sequestered into isolation membranes and degraded after fusion with the autophagosome or lysosome [105]. Optineurin (OPTN) and TANK-binding kinase 1 (TBK1) are key actors for mitochondria engulfment [106]. Consequently, ALS mutation in *OPTN* [107] and *TBK1* [23] will affect the mitophagic flux and may lead to an accumulation of nonfunctional mitochondria over time and result in motor neuron death (see [108] for review). Taken together, except for *CHCHD10*, these studies point toward mitochondria dysfunction and damage being a downstream effect of ALS gene mutations that lead to protein aggregations and/or proteostasis dysfunction (see Section 7: Impaired Protein Homeostasis).

In addition, damage to mitochondria and alterations in their functions can disrupt calcium homeostasis, increasing the sensitivity of neurons to glutamate excitotoxicity and the risk of motor neuron damage ([109], Figure 1). Mitochondrial dysfunction can also activate proapoptotic signals [93], such as the caspase-dependent [110] or bcl-2-dependant pathways [93], and might lead to motor neuron degeneration.

4. Axonal Transport

Motor neurons have exceptionally long axons, up to 1 m in length, placing extreme demands on cellular physiological functions that rely on the axonal transport of organelles such as mitochondria or of molecules including proteins, lipids, and RNA to and from the synapse [111]. Axonal transport, as well as the conduction of electrical impulses and the maintenance of the axon structure, are heavily regulated processes linked with control of the neurofilament structure [112,113]. In both sporadic ALS (sALS) and fALS patients, the disorganization of neurofilament networks has been reported [38].

Neurofilaments are neuron-specific intermediate filaments that are stretch-resistant and are major cytoskeleton proteins [114]. They form parallel coiled-coiled heterotetramers composed of light, medium, and heavy-weighted neurofilaments (NF-L, NF-M, and NF-H, respectively) and α-internexin or peripherin [112,114]. Eight heterotetramers form cylindrical structures known as unit-length filaments (ULFs) with the tail domains sticking out [112,114]. A series of ULFs form a filament that matures into neurofilament after a radial compaction of the cylindrical structure [112]. Consequently, variants in the *NEFH* gene affecting the crosslinking properties of the NF-H protein may result in abnormal neurofilament accumulations and in axonal transport defects [115].

Neurofilaments form cross-bridges not only with each other but, also, with actin filaments, actin rings, and microtubules [114], constituing a protein network that might participate to the maintenance of the axon structure [114,116]. Actin polymerization requires the small actin-binding proteins profilin I and II and phosphoinositide islands localized at the membrane [117]. Mutations in the *PFN1* gene encoding profilin-I are associated with ALS [118], and the expression of mutant hPFN1^{G118V} in a murine model resulted in dysregulated actin polymerization [119]. Consequently, the attachement of actin to the microtubules might be affected, probably impacting anterograde and retrograde transport and, thus, leading to an accumulation of fragmented mitochondria and, ultimately, to upper and lower motor neuron death ([119], Section 3 and Figure 1).

Microtubules and motor proteins such as the dynein-dynactin complex [28,120,121] and the kinesins [120,122,123] are involved in the long-distance transport of cellular cargo. Microtubules are composed of dimers of α- and β-tubulin. The alpha tubulin subtype TUBA4A is an ALS-associated protein [124], and ALS-associated mutations of *TUBA4A* lead to microtubule polymerization defects and network destabilization [124].

The dynein-dynactin complex [28,120,121], along with the kinesins [120,122,123], are key drivers of the anterograde and retrograde movements of diverse cargoes along the microtubule cytoskeleton, including organelles, vesicles, neurofilaments, AMPA and GABA receptors, and RNAs. Interestingly, mutations in *dynactin subunit 1 (DCTN1)* affecting the tertiary structure of the dynactin protein and its capacity to bind to microtubules can cause ALS [125]. When the interaction between dynein-dynactin is interrupted by the overexpression of dynamitin, axonal transport is impaired, and mice develop a late-onset motor pathology that recapitulates late-onset progressive ALS [126]. Kinesins form a superfamily of molecular motors that can be divided into three groups [120]. KIF5, a member of kinesin 1-group, is a tetramer with two kinesin heavy chains (KHCs) that contains a motor domain and two kinesin light chains (KLCs) that facilitate connections with cargo. There are three KIF5 isoforms—KIF5A, KIF5B, and KIF5C—all three isoforms being associated with the neuronal function and anterograde transport of proteins and organelles [127]. Mutations in the C-terminal of KIF5A, leading to a loss of function, are associated with ALS [128] and are suspected to disrupt the axonal transport (Figure 1). This hypothesis is supported by the defective axonal transport of mitochondria, the local accumulation of neurofilament, and the reduced axonal growth and survival observed in the primary culture of motor neurons from KIF5A$^{-/-}$ mice [129].

Distal axonal transport is also affected in SOD1^{G93A} mice at an early stage, with an early decrease in kinesin expression in asymptomatic mice, followed by a decrease in dynein expression in older presymptomatic mice [130]. Defective axonal transport may contribute to the accumulation of impaired mitochondria at distal sites ([93], Figure 1), resulting in decreased ATP production and disrupted calcium homeostasisis at the neuromuscular junction, consequently leading to a distal axonopathy in SOD1^{G93A} mice [109,131,132] and *SOD1* patients [24,28]. Kinesin-dynein machineries have been described to be affected in sporadic ALS, where KIF1Bβ and KIF3Aβ, two kinesin-related proteins, were found to be downregulated in motor cortex samples of sporadic patients [133]. However, the expression level of another kinesin-related protein, KIFAP3, is inversely correlated with sporadic ALS patient survival [134].

In conclusion, different mutations associated to ALS can directly alter the architecture and dynamics of the cytoskeleton, affecting the axonal transport machinery. Interestingly, aberrant axonal transport has also been observed in sALS patients and in fALS patients harboring mutations in non-cytoskeletal-related genes. Disrupted transport mechanisms can then affect the mitochondrial metabolism and degeneration (Section 3), as well as protein degradation (Section 7) and RNA transport (Section 8), ultimately leading to motor neuron death.

5. Glutamate Excitotoxicity

Glutamate is the most abundant neurotransmitter in the CNS and is released from presynaptic neurons into the synaptic cleft, resulting in the activation of NMDA and AMPA receptors that mediate calcium and sodium influxes in postsynaptic neurons. Excess glutamate may result in the abnormal activation of glutamate receptors, causing an excessive influx of Ca^{2+} in the postsynaptic neuron (Figure 1), which leads to extreme neuronal firing [135], resulting in excitotoxicity, which is potentially implicated in a number of pathological conditions, including multiple sclerosis [136], Parkinson's diasease [137], and ALS [138,139]. Glutamate excitotoxicity is thought to occur as a result of defective glutamate uptake and transport mechanisms that lead to excessive neuronal Ca^{2+} intake, aberrant Ca^{2+} homeostasis, downstream mitochondrial dysfunction, and increased ROS production [140,141] (Figure 1). Glutamate-gated AMPA receptors are abundant in human and animal motor neurons [142,143] and are made up of four subunits, GluA1 to GluA4 (also GluR1–4) [144]. The overactivation of AMPA receptors has been shown to result in hindlimb paralysis and motor neuron degeneration in wild-type rats, highlighting the susceptibility of motor neurons to Ca^{2+} dysregulation [145]. The Ca^{2+} permeability of AMPA receptors is mediated by the presence of the GluA2 subunit, the absence of which, in addition to impaired transcriptional editing at the Q/R site, confers increased AMPA Ca^{2+} permeability [146]. Interestingly, spinal motor neurons have been reported to display a reduced expression of GluA2 relative to dorsal horn neurons from the same region, providing some explanation for the selective susceptibiltiy of motor neurons in ALS [147], and GluA2 transcriptional editing has been found to be impaired in motor neurons of sporadic ALS patients relative to controls [148]. Furthermore, evidence suggests that GluA2 editing is also impaired in ALS oculomotor neurons, despite their spared function in disease. However, spared functionality has been hypothesized to be the result of increased Ca^{2+}-binding proteins and, in particular, parvalbumin, which is highly abundant in oculomotor neurons and present at low levels in spinal motor neurons [149]. GluA2 transcriptional editing into the Ca^{2+} impermeable subunit is mediated by adenosine deaminase acting on RNA 2 (ADAR2) activity [150,151]. A reduced ADAR2 expression has been reported in sporadic ALS patients and has been shown to result in an increased aggregation of TDP-43 in spinal motor neurons [152]. Taken together, the evidence suggests that a decreased GluA2 expression and impaired transcriptional editing in spinal motor neuron AMPA receptors is a contributing factor to the increased uptake of Ca^{2+} and the downstream susceptibility to excitotoxicity in ALS (Figure 1). Moreover, the finding that AMPA receptor dysfunction can result in the aggregation of misfolded TDP-43 is an important finding for linking ALS pathology with the glutamate excitotoxicity hypothesis.

In addition to dysregulated GluA2 subunit function, research has reported dysfunctional glutamate transport mechanisms in ALS. Under normal physiological conditions, glutamate at the synaptic cleft is cleared by the excitatory amino acid transporter (EAAT2), which functions to maintain low levels of extracellular glutamate and, thus, prevent excessive increases in intracellular Na^{+} and Ca^{2+} levels [89,153]. EAAT2 is found primarely on the synaptic processes of astrocytes, and the loss of EAAT2 has been reported to induce increased extracellular levels of glutamate and cause motor neuron toxicity and muscle paralysis in animal models [154], whilst the pharmalogical stimulation of EAAT2 was found to rescue motor neuron degeneration and delay paralysis in SOD1^{G93A} mice [155,156]. Abnormalaties in EAAT2 have been suggested to occur post-translationally after a post-mortem research highlighted no differences in EAAT2 mRNA expressions between sporadic ALS patients and controls, despite a 95% decrease in protein levels in sALS subjects [157]. Further support of EAAT2 loss and its implications in ALS were reported by a separate group who demonstrated a reduced EAAT2 immunoreactivity in anterior horn cells of sporadic ALS and lower motor neuron disease patients relative to healthy controls [158]. Together, these studies highlight the role of dysfunctional glutamate uptake and transport mechanisms in sporadic cases of ALS (Figure 1).

Excitotoxicity has also been associated with genetic forms of the disease. *SOD1* mutations have been implicated in the glutamate excitotoxicity hypothesis, and research has demonstrated an increased glutamate release [159], as well as motor neuron and inter-neuron hyperexcitability two to three months prior to motor neuron degeneration and phenotype onset in SOD1^{G93A} mice [160]. *SOD1* mutations have also been shown to reduce the expression of astrocytic GluA2 in vitro and in vivo, thereby diminishing their ability to protect against motor neuron excitotoxicity [21]. In patients, the deterioration of neuronal dendrites was observed in sporadic and familial ALS cases, but not in healthy or Alzheimer's disease controls, leading to the suggestion that ALS is a synaptopathy [161], which is perhaps attributable to the excessive levels of glutamate observed in the CSF of patients [138,139]. Indeed, metabolomic analyses suggest that ALS patients show elevated serum levels of glutamate [32], and there is evidence that sporadic and familial ALS cases show heightened levels of cortical excitablity, which can be detected even in the presymptomatic stages in familial *SOD1* mutation carriers [162]. However, other studies have failed to find evidence for elevated glutamate levels in ALS patients [163–165]. *C9orf72* has also been implicated in the glutamate excitotoxicity hypothesis after iPSC motorneurons from ALS patients were found to have impaired autophagosome formation and aberrant accumulations of glutamate receptors [166–168]. This has been supported in vivo with *C9orf72* knockout mice showing GluR1 upregulation in the hippocampus and a greater susceptibility to excitotoxicity compared to controls [166]. In addition, *C9orf72* knockout mice demonstrated a complete loss of SMCR8 [169], a protein that functions in a complex with C9orf72 and WD40 repeat domain 41 (WDR41) to regulate membrane trafficking and autophagy [170]. The concomitant abnormalities in autophagy and aberrant accumulations of GluR1 has led to the hypothesis that *C9orf72* loss-of-function leads to an impaired clearance of excess glutamate receptors, which, in turn, results in a greater glutamate uptake and increased susceptibility to excitotoxicity (Figure 1). *C9orf72* patients have also been reported to demonstrate elevated glutamate levels in their cerebropsinal fluid (CSF), which has been hypothesized to occur as a result of DPR-mediated splicing defects to EAAT2 and subsequent impairments in glutamate clearance [168]. Research has also implicated ALS2 in the glutamate exicitotoxicity hypothesis by virtue of its interaction with Rab5 and the endosomal pathway. ALS2$^{-/-}$ knockout mice have been reported to show significant increases in glutamate receptor degradation, including GluA2 [171], the loss of which is believed to contribute to excitotoxicity and motor neuron degeneration. Similarly, *ALS2*$^{-/-}$ spinal motor neurons were found to be more susceptible to glutamate excitotoxicity as a result of reduced GluA2 at the synapses of neurons, which was attributed to an altered glutamate receptor interacting protein 1 (GRIP1) function, caused by the genetic loss of *ALS2* [171].

Although there is evidence to suggest the presence of glutamate transport and uptake defects in both sporadic and familial cases of ALS, it is unclear how these defects lead to the specific deterioration of motor neurons in disease. Furthermore, it has been 25 years since the approval of the antiglutamatergic drug riluzole for the treatment of ALS, yet there is no understanding as to why it, as well as other antiglutamatergics, including gabapentin, memantine, and ceftriaxone, fail to delay symptom progressions in ALS by more than ~three months [172,173].

6. Endosomal Pathway and Vesicle Secretion

Extracellular vesicles encompass different types such as apoptotic vesicles, microvesicles, and exosomes, all of which can affect the functionality of the recipient cells [174–177]. The last decade has seen several investigations into the relevance of exosomes to ALS, either in propagating the disease or as biomarkers [6,178–181]. In the ALS context, exosomes secreted by astrocytes, neurons, or microglia are suspected to carry neurotoxic elements such as mutated SOD1 or C9orf72-derived DPR and to be responsible for motor neuron death [178,179,182]. Interestingly, sporadic muscle cells present an accumulation of vacuole and multivesicular bodies in their cytosol, suggesting that vesicle trafficking is disrupted in these cells ([183] and personal data) and that extracellular vesicle secretion might have an important role in ALS.

Exosome biogenesis requires the formation of inward buds in the multivesicular body (MVB), followed by their fission and release as vesicles into the MVB lumen. The generation of intraluminal vesicles can be either Endosomal Sorting Complex Required for Transport (ESCRT)-dependent or ESCRT-independent. The ESCRT is composed of four complexes—ESCRT-0, ESCRT-I, ESCRT-II, and ESCRT-III—each complex acting one after the other to form intraluminal vesicles. Interestingly, charged multivesicular protein 2B (CHMP2B) is a component of ESCRT-III involved in the processing of cargo into intraluminal vesicles and is associated with ALS [184]. The dysfunction of ESCRT-III may lead to abnormal and dysmorphic endosomes [185] (Figure 2).

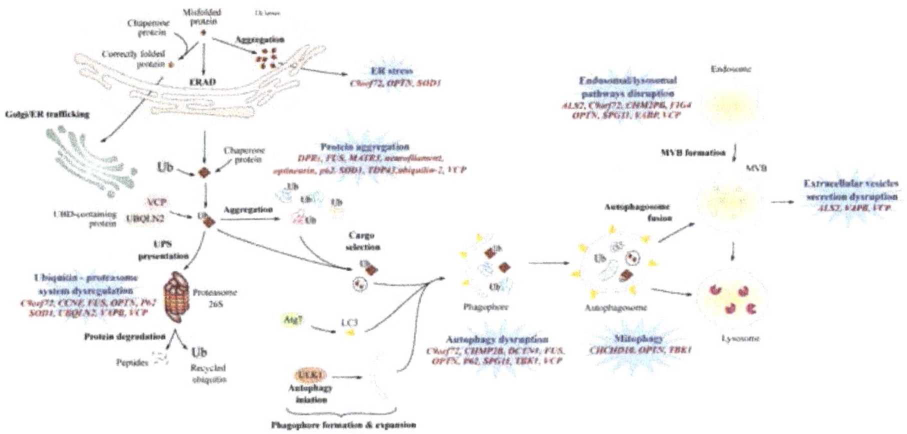

Figure 2. Protein homeostasis dysregulation. Dysregulated protein homeostasis is mediated by multiple pathways encompassing defects in autophagy, the dysregulated ubiquitin-proteasome system (UPS), endo-lysosomal pathway disruptions, or endoplasmic reticulum (ER) stress. The presence of misfolded proteins activates endoplasmic reticulum-associated protein degradation (ERAD), leading to proteasome-mediated degradation to avoid misfolded protein accumulations in the ER lumen and subsequent ER stress. Several ALS-associated gene mutations induce proteasome-mediated toxicity via the sequestration of ubiquilin and chaperone proteins involved in the UPS pathway. The proteolytic activity of the proteasome has also been demonstrated to be targeted by gene mutations in ALS. The autophagic pathway involves the formation and maturation of phagophores that engulf selected transported-cargo and form autophagosomes. Fusion with the lysosome enables the degradation of autophagosome contents. Defects in autophagy initiation and expansion, dysregulated phagophore formation, and/or impaired cargo transport are observed in ALS patients. Mutations in ALS-associated genes also cause defects in mitophagy, a specific form of autophagy. Defects in the endolysosomal have been associated with ALS gene mutations, including defective endolysosomal trafficking and altered lysosomal hoemostasis and degradation. Defects in the autophagy/lysosomal pathway may affect vesicle secretion. Genes implicated in dysregulated protein homeostasis are indicated in red. ER: endoplasmic reticulum, ERAD: endoplasmic reticulum-associated protein degradation, ALS2: Alsin, C9orf72: Chromosome 9 open reading frame 72, CCNF: Cyclin F, CHCHD10: coiled-coil helix coiled-coil helix domain-containing 10, CHMP2B: chromatin-modifying protein 2B, DCTN1: Dynactin 1, FIG4: Phosphoinositide 5-phosphatase, FUS: Fused in Sarcoma, MATR3: matrin 3, MVB: Multivesicular bodies, OPTN: Optineurin, SOD1: Superoxide dismutase 1, SPG11: Spatacsin, TDP-43: TAR DNA-binding protein 43, TBK1: TANK-binding kinase-1, UBQLN2: Ubiquilin-2, VAPB: vesicle-associated membrane protein-associated protein B, and VCP: valosin-containing protein.

Endocytosis and vesicle trafficking from one cellular compartment to another are regulated by small Rab GTPases [186]. For example, Rab5 is associated with the formation of early and late endosomes, while Rabs 11, 35, and 27 have direct roles in exosome biogenesis and secretion. Alsin is an ALS-associated protein [187] and is a guanine nucleotide-exchange factor involved in endosome motility and fusion with the lysosome [171]. Consequently, an absence of alsin expression in hippocampal neurons leads to an accumulation of Rab5-positive vesicles and an enhanced lysosome-mediated degradation, suggesting an enhanced degradation of endosomal vesicles [171], thus probably affecting the production and secretion of exosomal vesicles.

The C9orf72 protein structure presents some similarities with the Differentially Expressed Normal versus Neoplastic (DENN) guanine nucleotide exchange factor and, thus, may activate Rab proteins [101] such as RAB8A and RAB39B [188]; Rab1a [34]; or Rabs 1, 7, and 11 [189], which are associated with autophagy and vesicle-trafficking processes [190–192], as well as exosome biogenesis and secretion [186]. In *C9orf72* knockdown cell lines [189], a *C9orf72* knockout murine model [193], and in ALS patient fibroblasts and iPSC-derived motor neurons [194], transgolgi and endosomal trafficking were reduced, a defective autophagy pathway was observed [34,194], and exosomal secretion was affected [194].

Vesicle-associated membrane protein-associated protein B (VAPB) is involved in vesicle trafficking between the endoplasmic reticulum and the golgi apparatus [195,196]. Interestingly, VAPB has been described to interact with RAB7 and colocalize with CD63 [197], two proteins involved in late-endosome formation and exosome biogenesis [186]. However, the impact of ALS-associated VAPB mutation in exosome biogenesis and secretion still needs to be investigated.

Multivesicular body formation is at a crossroad between the autophagy (Figure 2) and secretion pathways, and an autophagic failure may lead to cell secretion [198]. The *VCP* gene is associated with ALS [199] and encodes for a valosin-containing protein, an ubiquitous AAA+ ATPase that interacts with clathrin to form early endosomes but, also, with the autophagy pathways [200]. In this context, a mutation in a valosin-containing protein (VCP) may affect the endosomal pathway, and one can hypothesize that it has an impact on the formation and secretion of exosomes. Other gene mutations associated to ALS, such as protein polyphosphoinositide 5-phosphatase (FIG4) [201,202] or spastacsin (Spg11) [203,204], are associated with the blockade of lysosomal clearance (see Section 7)—the blockade of which could potentially lead to vesicle secretion [198]. However, futher investigations related to exosome pathways in ALS in vivo are needed.

7. Impaired Protein Homeostasis

Protein aggregates positive for TDP-43 [36,205], neurofilament [41], FUS [87], or SOD1 [206] are observed in the vast majority of ALS patients, with TDP-43 being present in as many as 98% of sporadic and familial cases [207], meaning that the presence of such aggregates is widely regarded as a hallmark feature of ALS pathology. These deposits can occur in the cytoplasm of neurons [208] and skeletal muscle [99,209], and their presence is highly suggestive of an imbalance between protein synthesis and degradation pathways (Figure 2).

7.1. Proteasome and Autophagic Degradation Pathways

In the late 1960s and early 1970s, the presence of protein inclusions in the anterior horn cells of sporadic and familial ALS patients was described [210–212]. Later, these inclusions were found to be ubiquitin positive [35], and SOD1 was the first ALS-associated protein found to be immunoreactive within the inclusions of familial patients [213]. Subsequently, ubiquinated inclusions have often been found to be immunoreactive for the ubiquitin-binding protein p62 [214], and up to 98% of sALS and fALS cases show inclusions that are TDP-43-positive [215], with the exception being *SOD1* [216] and *FUS* [216] patients who do not demonstrate TDP-43 inclusions but do demonstrate SOD1 and FUS immunoreactive inclusions. Other ALS proteins that have been implicated in the formation of cytoplasmic inclusions include optineurin (OPTN) [107], ubiquilin 2 (UBQLN2) [217], dynactin 1 (DNCT1) [218], valosin-containing protein (VCP) [219], and matrin 3 (MATR3) [220]. Studying the structure of the main proteins SOD1, FUS, and TDP-43 helped to unravel potential mechanisms involved in protein misfolding and self-propagation within the cells and in surrounding cells. SOD1 is a stable homodimer, thanks to the intrasubunit disulfide bond and its ability to bind zinc and copper. However, a reducing and metal-poor intracellular environment or mutations [221–227] can abolish these features and destabilize SOD1, leading to the formation of aggregates and amyloid fibril structures [228–231] that can self-propagate in vitro [229,230]. FUS and TDP-43 proteins possess a low complexity domain that presents similarities with yeast prions [209] and can form large aggregates and amyloid fibril structures [209,229,232,233]. Interestingly, mutated forms of FUS and TDP-43 can induce the misfolding of wild type forms of FUS and TDP-43, respectively [229], and have also been shown to induce the misfolding of wild type forms of SOD1 in vitro [234]. Altogether, these studies suggest a potential mechanism for the self-propagation of misfolded proteins in vitro—misfolded proteins that can potentially be transferred from cell to cell via secreted vesicles, thus propagating the misfolding mechanism to neighboring cells (see Section 6, [178,229,235]). The presence of these protein aggregates has been suggested to impair the proteasome and autophagic degradation pathways and could be key mediators in ALS pathogenesis [38,236,237] (Figure 2).

Dysregulation of the Ubiquitin–Proteasome System (UPS) in ALS patients was first suspected following the identification of mutations in genes encoding ubiquilin 2 [238] and VCP [199], two proteins involved in protein clearance via the ubiquitin-proteasome pathway [239]. Mutations in *OPTN* [240] and *SQSTM1/P62* [241] were then identified, and following this, *SOD1* [236], *VABP* [242], *C9orf72* [25,208], and *CCNF* (cyclin F) [243] mutations were all reported to reduce UPS activation. Ubiquitin-positive inclusions were observed in post-mortem neuronal and muscular tissues of fALS and sALS patients [35] and, more specifically, in *C9orf72* patients [244]. Similarly, SOD1 [245], FUS [87], ubiquilin 2 [246], and C9orf72-derived DPR proteins [247] can generate toxic aggregates positive for some proteasome components [248]. These ubiquitin-positive inclusions can also contain and "trap" nonmutated forms of SOD1 [48], TDP-43 [205], optineurin [107], and ubiquilin 2 [249], thus exacerbating the already disrupted cellular homeostasis in ALS.

The degradation of ubiquitinated proteins through the autophagy/lysosomal pathway occurs in four steps: (1) the initiation and extension of the bilayer vacuole into phagophores; (2) the transport of selective cargoes (including ubiquitinated proteins, dysfunctional mitochondria, and protein aggregates); (3) the maturation into autophagosomes; and (4) fusion with low pH lysosomes to form autolysosomes where degradation of the cargoes can proceed [250].

The fusion of the endosome with the lysosome for degradation is an tightly regulated event [251] involving the protein polyphosphoinositide 5-phosphatase, FIG4 [252]. Deleterious mutations in *FIG4* in ALS leads to abnormal lysosomal storage [201,202]. Spastacsin is also involved in lysosomal clearance, and the absence of *Spg11* expression impaired the lysosomal-autophagy pathway and is accompanied by an accumulation of lipid within the lysosomes [203,204]. Nonfunctional TBK1 [253] and p62 [254] inhibit the transport of targeted cargoes toward the autophagosome. Interestingly, impaired autophagosome maturation was also observed in ALS cells mutated for *FUS* [255], *VCP* [89], *CHMP2B* [184], and *OPTN* [256]. The importance of autophagy can be observed in studies that stimulate autophagic activation in the presence of ALS mutations. For instance, the stimulation of autophagy in murine and human iPSC-derived neurons expressing *TARDBP* mutations demonstrated a greater clearance of TDP-43 aggregates relative to nonstimulated cells and resulted in improved motor neuron survival [257].

C9orf72 mutations can interfere with the autophagy pathway at several levels. When *C9orf72* expression is abolished or decreased as suggested by the haploinsufficiency hypothesis [258,259], autophagy is inhibited [34,189,193,194], leading to simultaneous increases in the number of cytoplasmic inclusions immunoreactive for ubiquitin, p62, and TDP-43 [34,101,260]. The impairment of autophagy in *C9orf72* cells may also result in the accumulation of cytotoxic DPR proteins encoded by the G4C2 HREM and, ultimately, lead to neuronal loss [261]. Similarly to cells expressing *TARDBP* mutations [257], the stimulation of autophagy abolished the accumulation of poly-DPR proteins and neuronal toxicity [261].

Altogether, these studies illustrate the importance of autophagy for the efficient clearance of misfolded and aggregated proteins and is indicative of the underlying impairments in proteostatic mechanisms that mediate ALS physiopathology (Figure 2).

7.2. Endoplasmic Reticulum Stress

During the formation of misfolded proteins, the unfolded-protein response (UPR) may be initiated to transport defective proteins to the endoplasmic reticulum (ER), where ER-resident chaperones will properly fold the protein [262]. The accumulation of misfolded proteins in the ER activates the ER stress response pathway, also known as the endoplasmic reticulum-associated protein degradation (ERAD) pathway. The ERAD pathway involves the translocation of misfolded proteins from the ER lumen to the cytosol, where they undergo ubiquitination and degradation through the ubiquitin-proteasome pathway [262]. In ALS patient cells, mutated SOD1 agregates were observed in ER and colocalized with UPR markers, leading to an increase in ER stress [263] by interacting with ER stress response proteins and inhibiting their function in the ERAD response [236]. The presence of poly (GA) aggregates, observed in neuronal post-mortem *C9orf72* ALS patients, can inhibit proteasome activity and induce ER stress, which can be abolished when using ER stress inhibitors such as salubrinal and tauroursodeoxycholic acid (TUDCA) [208]. Concordantly, the cerebropsinal fluid (CSF) of sporadic ALS patients displays an accumulation of ER stress markers [263], and when healthy neurons were exposed to patient CSF, the ER became fragmentated and caspase-dependent apoptosis was activated, suggesting an increase in ER stress [264].

Vesicle-associated membrane protein-associated protein B (VAPB) is localized in the endoplasmic reticulum membrane and has a key role in vesicle trafficking between the endoplasmic reticulum, golgi apparatus, and the nuclear envelope [195–197]. The $VAPB^{P56S}$ mutation associated with ALS leads to a misfolded protein that accumulates in the ER [265] and can cause a defect in nuclear envelope protein transport, leading to an aberrant nuclear envelope structure [266]. Interestingly, the accumulation of VAPB has also been observed in the endoplasmic reticulum of peripheral blood mononuclear cells of sporadic ALS [267].

Optineurin is a TBK1 partner and is involved in mitophagy (Section 3). When the association of optineurin with myosin VI is disrupted, as osberved in fALS cases associated with *OPTN* mutations, optineurin is diffused in the cytosol of neuronal cells and results in ER stress and Golgi apparatus fragmentation, as well as an inhibition of the autophagy pathway ([256], Figures 1 and 2).

Altogether, these studies suggest that protein degradation could be directly and indirectly affected in ALS, causing protein aggregation that leads, in turn, to the disruption of the function of organelles such as nuclei (Section 7.2) and mitochondria (Section 3) or to the blockage of lysosomal activity that can potentially affect cell-cell communication (Section 6).

8. Aberrant RNA Metabolism

FUS [86,268] and TDP-43 [269] are RNA-binding proteins involved in multiples steps of RNA metabolism. In ALS patients, mutations in both genes give rise to the translation of proteins frequently mislocalized to the cytoplasm [87,88,270] and, subsequently, result in downstream complications that affect RNA-processing mechanisms. Dysregulated RNA metabolism is another key feature of ALS pathogenesis and includes transcription defects, alternate splicing changes, miRNA biogenesis, stress granule formation, and RNA nucleocytoplasmic transport (Figure 3).

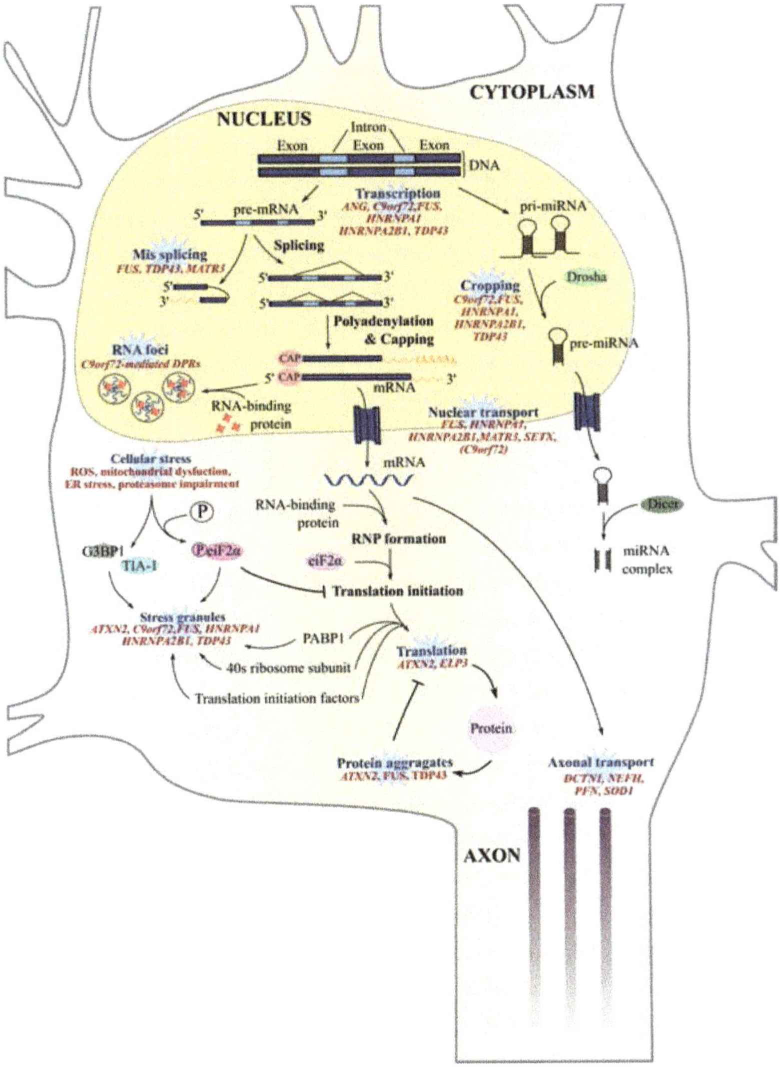

Figure 3. RNA and miRNA biogenesis defects in ALS. Many processes in RNA and miRNA pathways are disrupted in ALS patients, including transcription defects, alternate splicing events, miRNA biogenesis, and nucleus-cytosol transport impairment. RNA metabolism defects are particularly relevant in ALS pathogenesis, since TDP-43 and FUS are both well-known ALS-associated genes involved in RNA processing. Both FUS and TDP-43-mutated proteins mislocalize to the cytoplasm of ALS motor neurons, leading to a probable loss and/or toxic gain-of-function of these proteins. ANG: Angiogenin, ATXN2: Ataxin-2, C9orf72: Chromosome 9 open reading frame 72, DCTN1: Dynactin 1, eIF2α: Eukaryotic translation initiation factor 2A, ELP3: Elongator protein 3, FUS: Fused in Sarcoma, G3BP1: Ras GTPase-activating protein-binding protein 1, hNRNPA1: Heterogeneous nuclear ribonucleoprotein A1, hnRNPA2B1: Heterogeneous nuclear ribonucleoprotein A2B1, MATR3: matrin 3, NEFH: Neurofilament heavy subunit, PABP1: Polyadenylate-binding protein 1, PFN1: Profilin, SETX: Senataxin, SOD1: Superoxide dismutase 1, TDP-43: TAR DNA-binding protein 43, and TIA-1: TIA1 Cytotoxic Granule-Associated RNA-Binding Protein.

8.1. RNA Splicing and Translation

Given the large number of possible protein-protein interactions between *FUS* or *TDP-43* and their partners, it is easy to expect alterations to important RNA-processing mechanisms in ALS patients [89]. *FUS* and *TDP-43* also regulate the expression of multiple proteins involved in neuronal physiology, including components of the synaptic plasticity pathways [39,271,272] and dendritic branching processes [272–274]. In addition, the HREM in *C9orf72* generates repeat RNA and RNA foci, which repress the gene expression of RNA metabolism regulators (such as *hnRNPA3*) [275] or sequester TDP-43 [275,276] and FUS [275] proteins and, thus, indirectly inhibits the transcription of RNA metabolism-associated genes. Similarly to *C9orf72*-mediated RNA-processing defects, *FUS* mutations have been associated with major transcriptional defects [268].

Ataxin-2 is a polyglutamine (polyQ) protein that is involved in mRNA translation, and it interacts with RNA-binding proteins such as TDP-43 and FUS [277]. In ALS spinal cords, ataxin 2 exhibited significant cytoplasmic accumulation and enhanced the toxicity of TDP-43 in Drosophila via RNA binding [59].

TDP-43, *FUS*, *hnRNPA1*, *hnRNPA2B1*, and *MATR3* are associated with ALS and involved in pre-mRNA processing [220,278–281]. Consequently, *TDP-43* knockdown in murine tissues results in the alternate splicing dysregulation of numerous mRNA transcripts [42,282], and the loss-of-function of *FUS* also induces many splicing defects [39], suggesting important alternate splicing events in ALS patients (Figure 3). These downstream complications are not surprising given the ability of the protein FUS to sequester numerous components of the splicing process, such as key splicing factors [283] U1 snRNP and U11/U12 snRNPs [274,284], which are involved in minor intronic splicing. Alternative splicing changes have been identified in neuronal genes involved in cytoskeleton organization, axonal growth, and guidance in *FUS*-mutated ALS patients [285,286], and, interestingly, axonopathy and axon retraction occur in the early stages of ALS [287].

ALS-linked MATR3^{S85C} and MATR3^{P154S} mutations were observed to affect Matrin 3 interactions with the TRanscription and EXport (TREX) protein complex, altering the global nuclear export of mRNA [288]. As a result, mRNA is sequestered within the nucleus, causing export defects of TDP-43 and FUS mRNA [288], which may affect mRNA splicing directly [289] and indirectly [278] (Figure 3). Consequently, as observed in the *MATR3*^{S85C} murine model, dysfunctional MATR3 may lead to astrocyte and microglia activation and result in spinal motor neuron degeneration [290].

The *ELP3* gene encodes for elongator protein 3, a histone acetyltransferase subunit of the RNA polymerase II elongator complex responsible for RNA translation (Figure 3). Mutations in the *ELP3* gene are associated with ALS [291,292] and result in the shortening and abnormal branching of motor neurons, as observed in *ELP3* knockdown in zebrafish embryos [291], and altered tRNA modification, triggering proteome impairment and the subsequent aggregation of susceptible proteins [292].

Angiogenin, encoded by the hypoxia-inducible gene *ANG*, is a member of the pancreatic ribonuclease superfamily [293] and, as well as angiogenesis, is also involved in ribosomal biogenesis [294,295]. Defects in this protein are associated with the impairment of its nuclear localization and diminished ribonucleolytic activity [295] (Figure 3), both of which are essential for normal ANG functioning and motor neuron viability.

Together, these findings suggest that RNA processing is a key pathway affected in ALS, either due to mutations directly affecting proteins involved in RNA processing or as a consequence of protein aggregations.

8.2. RNA Foci

Sense and antisense RNA generated from the bidirectional transcription of G4C2 repeats have been proposed to induce a toxic gain-of-function in ALS *C9orf72* patient cells by forming RNA foci that may sequester RNA-binding proteins, thus disrupting RNA metabolism and processing in cells [101] (Figure 3) widely throughout the central nervous system [296,297]. Both sense and antisense RNA foci are frequently observed in nucleoli, with antisense RNA foci being denser [296]. In addition, the antisense RNA foci correlate with TDP-43 aggregation in the cytosol of *C9orf72* motor neurons [296,297]. An in situ hybridization of post-mortem *C9orf72* ALS tissue revealed that 78.7% of the neurons and 24.9% of the glial cells in the motor brain and spinal cord regions were positive for antisense RNA foci [297]. Interestingly, extra-motor brain regions also show a high percentage of cells positive for antisense RNA foci, with 89.4% of neurons and 46.1% of glia being positive [297].

8.3. Epigenetic Modulation

Epigenetic mechanisms such as microRNA regulation maintain cell type and tissue identity and may be involved in the onset and progression of neurodegenerative diseases, including ALS. The decreased expression of miRNAs, including miRNAs let-7e, miR-148b-5p, miR-577, miR-133b, and miR-140-3p, were observed in post-mortem spinal cords of sporadic ALS patients [298], suggestive of impairment in the genes and pathways associated with miRNA biogenesis, neuroinflammation, and apoptosis.

Interestingly, the class II ribonuclease, Drosha, interacts with TDP-43, FUS, and *C9orf72*-mediated DPRs [299–301], while the Dicer enzyme interacts with TDP-43 protein [301] and FUS can interact with pri-miRNA [302] (Figure 3). Consequently, mutated *TDP-43* may impair the post-transcriptional regulation of miRNAs and lead to an altered expression of miR-132-3p and miR-132-5p (involved in the regulation of neuronal outgrowth [301]), miR-143-3p and miR-143-5p (involved in myoblast cell differentiation [303]), miR-558-3p (involved in neurofilament stability [304]), and miR-574-3p (associated with stroke [305]) [301]. Similarly, the downregulation of FUS in a neuroblastoma cell line had a considerable impact on the biogenesis of miRNAs, with an altered expression of miR-9, miR-125b, and miR-132 implicated in neuronal differentiation, activity, and function [302], while mutated *FUS* affected the expression levels of miR125 and miR192 [302], which are involved in early neural conversion [306] or senescence [307].

Altogether, these findings are consistent with defective miRNA processing in ALS patients, which may affect downstream pathways with an impact on motor neuron survival.

8.4. Stress Granules and Nucleocytoplasmic Transport

In response to stressful conditions, RNA granules, also known as stress granules, are generated and can recruit FUS and TDP-43 [41,308]. Mutations in *FUS* and *TDP-43* can increase the persistence of stress granules in the cytoplasm, resulting in a possible toxic gain-of-function [237] by inhibiting mRNA translation and, thus, contributing to the progression of ALS pathology (Figure 3). The heterogeneous nuclear ribonucleoprotein particle proteins hnRNPA1 and hnRNPA2B1 are RNA-binding proteins and binding partners of TDP-43 and are involved in RNA processing, including miRNA maturation, the nucleocytoplasmic transport of mRNA, and RNA metabolism [300,309]. Mutations in the prion-like domains of hnRNPA2/B1 and hnRNPA1 increase fibril formation and aggregation potential, as well their hyperassembly into stress granules [310,311]. Stress granules are then targeted to the lysosome by the autophagic machinery involving VCP. Indeed, the pharmacological inhibition or RNAi knockdown of VCP is accompanied by reduced stress granule clearance, while Hela cells expressing VCP^{A232E} and VCP^{R155H} mutations showed a constitutive appearance and accumulation of stress granules containing TDP-43 [312]. Concordantly, the ALS-VCP mutation is accompanied by an increase in stress granules [313].

In *C9orf72* patients, stress granules are also involved in the sequestration of proteins required for effective nucleoplasmic transport, such as RAN GAP [40], or importing and exporting proteins [40,41]. The impairment of the nucleoplasmic transport of molecules in *C9orf72* ALS cells is controverisal. Indeed, while some studies observed that newly formed DPRs such as poly-PR can bind to nuclear pore transporters, thereby impairing the subsequent translocation of molecules [314], other studies did not observe any disruption in the nucleocytoplasmic transport with poly-GR or poly-PR [314]. However, with the expression of poly-GA, defects were observed both in import and in export in a SH-SY5Y cell line and in iPSC-derived motor neurons, respectively [101].

9. Concluding Remarks

Over 150 years have passed since ALS was first reported by Charcot, and still, the etiology of the disease remains elusive. Although research is progressing and genetic studies continue to identify novel gene mutations in familial cases of ALS [315], many questions remain surrounding the pathological mechanisms associated with already established mutations, their roles in the disease phenotype, and the as-yet-undiscovered mechanisms that underly sporadic onset. The most investigated mechanisms revolve around neurocentric deficits in dysfunctional mitochondria and oxidative stress, axonal transport, glutamate excitotoxity, protein homeostasis, and RNA processing (Figure 4).

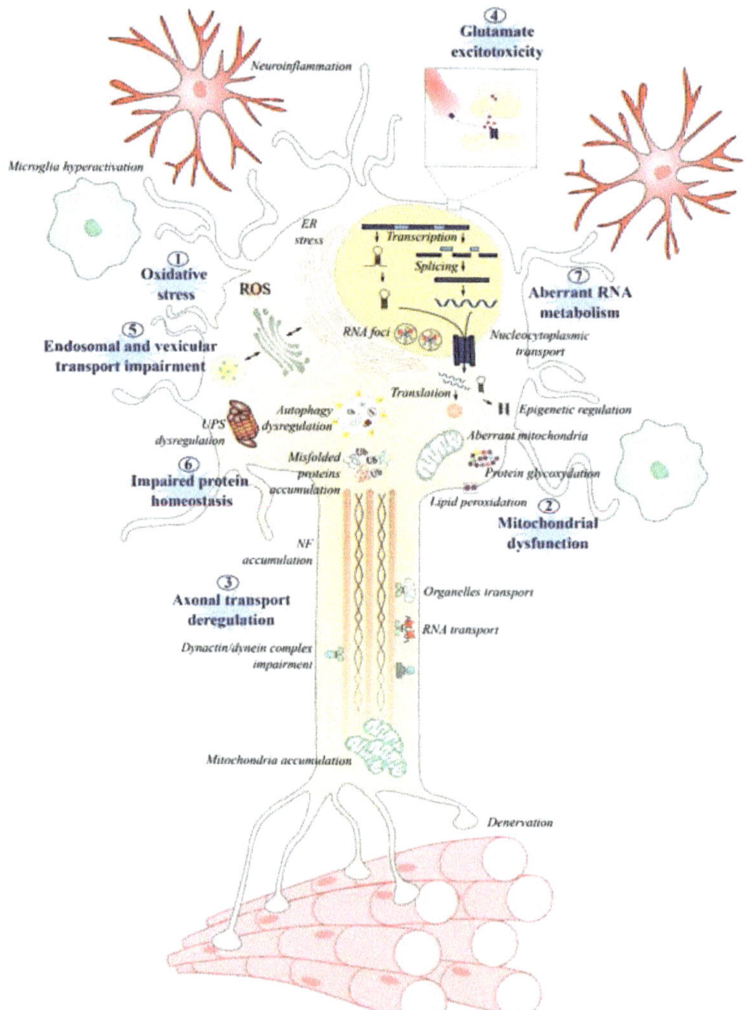

Figure 4. Summary of the different molecular and cellular mechanisms involved in ALS pathogenesis.

Among the most studied and well-established pathways are: oxidative stress, mitochondrial dysfunction, axonal transport, glutamate excitotoxicity, endosomal and vesicle secretions, protein homeostasis, and RNA metabolism. One pathway may lead to another, exacerbating the disruption of cellular homeostasis. The disruption of these pathways can lead to microglia activation, neuroinflamation, astrocytosis, and, ultimately, to motor neuron death and muscle denervation.

By detailing, as has been done in this review, the molecular events of the various pathways that are implicated in ALS, it becomes clear that these pathways can be linked to each other—in some cases, with one leading to another. For example, disrupted axonal transport can lead to an accumulation of nonfuctional mitochondria, while ATP deficiency and increased oxidative stress may damage proteins and DNA, which, in turn, could exacerbate the disruption of cellular homeostasis, leading to motor neuron death (Figure 4). These pathways are disrupted not only in motor neurons [24] but, also, in astrocytes [316,317], microglia [318,319], peripheral blood cells [43,320], and muscle [37,321–323], suggesting multisystemic [6] involvement in motor neuron death. Thus, by considering ALS from the perspective of shared molecular pathways [6,324], a cohesive understanding may yet emerge of the cellular mechanisms driving this pathology. It may be that different molecular pathways correspond to sub-strata of patients, such as among those with known genetic forms of ALS, as suggested in [6]. However, the identification of these strata may prove to be extremely challenging in non-monogenic forms of the disease [324].

Author Contributions: Writing—original draft preparation, L.L.G., E.A., O.C., U.G.V., and S.D.; writing—review and editing, W.J.D. and S.D.; and funding acquisition: S.D. All authors have read and agreed to the published version of the manuscript.

Funding: This work was financed by the European Union Regional Development Fund (ERDF) EU Sustainable Competitiveness Programme for N. Ireland, Northern Ireland Public Health Agency (HSC R&D) & Ulster University (PI: A Bjourson). LLG was a recipient of the ArSLA PhD fellowship, OC was a recipient of the PhD DELL fellowship, EA was a recipient of the Vice-Chancellor's Research Scholarships, and UGV was a recipient of the TargetALS fellowship (PI: S Duguez).

Conflicts of Interest: The authors declare no conflict of interest.

References

1. Arthur, K.C.; Calvo, A.; Price, T.R.; Geiger, J.T.; Chiò, A.; Traynor, B.J. Projected increase in amyotrophic lateral sclerosis from 2015 to 2040. *Nat. Commun.* **2016**, *7*, 12408. [CrossRef] [PubMed]
2. Ragagnin, A.M.G.; Shadfar, S.; Vidal, M.; Jamali, M.S.; Atkin, J.D. Motor Neuron Susceptibility in ALS/FTD. *Front. Neurosci.* **2019**, *13*, 532. [CrossRef] [PubMed]
3. Nicaise, C.; Mitrecic, D.; Pochet, R. Brain and spinal cord affected by amyotrophic lateral sclerosis induce differential growth factors expression in rat mesenchymal and neural stem cells. *Neuropathol. Appl. Neurobiol.* **2011**, *37*, 179–188. [CrossRef] [PubMed]
4. Gibbons, C.; Pagnini, F.; Friede, T.; Young, C.A. Treatment of fatigue in amyotrophic lateral sclerosis/motor neuron disease. *Cochrane Database Syst. Rev.* **2018**. [CrossRef]
5. Turner, M.R.; Al-Chalabi, A.; Chio, A.; Hardiman, O.; Kiernan, M.C.; Rohrer, J.D.; Rowe, J.; Seeley, W.; Talbot, K. Genetic screening in sporadic ALS and FTD. *J. Neurol. Neurosurg. Psychiatry* **2017**, *88*, 1042–1044. [CrossRef]
6. Vijayakumar, U.G.; Milla, V.; Cynthia Stafford, M.Y.; Bjourson, A.J.; Duddy, W.; Duguez, S.M.-R. A Systematic Review of Suggested Molecular Strata, Biomarkers and Their Tissue Sources in ALS. *Front. Neurol.* **2019**, *10*, 400. [CrossRef]
7. Connolly, O.; Le Gall, L.; McCluskey, G.; Donaghy, C.G.; Duddy, W.J.; Duguez, S. A Systematic Review of Genotype–Phenotype Correlation across Cohorts Having Causal Mutations of Different Genes in ALS. *J. Pers. Med.* **2020**, *10*, 58. [CrossRef]
8. Watson, M.R.; Lagow, R.D.; Xu, K.; Zhang, B.; Bonini, N.M. A drosophila model for amyotrophic lateral sclerosis reveals motor neuron damage by human SOD1. *J. Biol. Chem.* **2008**, *283*, 24972–24981. [CrossRef]

9. Estes, P.S.; Boehringer, A.; Zwick, R.; Tang, J.E.; Grigsby, B.; Zarnescu, D.C. Wild-type and A315T mutant TDP-43 exert differential neurotoxicity in a Drosophila model of ALS. *Hum. Mol. Genet.* **2011**, *20*, 2308–2321. [CrossRef]
10. Chen, Y.; Yang, M.; Deng, J.; Chen, X.; Ye, Y.; Zhu, L.; Liu, J.; Ye, H.; Shen, Y.; Li, Y.; et al. Expression of human FUS protein in Drosophila leads to progressive neurodegeneration. *Protein Cell* **2011**, *2*, 477–486. [CrossRef]
11. Xu, Z.; Poidevin, M.; Li, X.; Li, Y.; Shu, L.; Nelson, D.L.; Li, H.; Hales, C.M.; Gearing, M.; Wingo, T.S.; et al. Expanded GGGGCC repeat RNA associated with amyotrophic lateral sclerosis and frontotemporal dementia causes neurodegeneration. *Proc. Natl. Acad. Sci. USA* **2013**, *110*, 7778–7783. [CrossRef] [PubMed]
12. Therrien, M.; Parker, J.A. Worming forward: Amyotrophic lateral sclerosis toxicity mechanisms and genetic interactions in Caenorhabditis elegans. *Front. Genet.* **2014**, *5*, 85. [CrossRef] [PubMed]
13. Sakowski, S.A.; Lunn, J.S.; Busta, A.S.; Oh, S.S.; Zamora-Berridi, G.; Palmer, M.; Rosenberg, A.A.; Philip, S.G.; Dowling, J.J.; Feldman, E.L. Neuromuscular effects of G93A-SOD1 expression in zebrafish. *Mol. Neurodegener.* **2012**, *7*, 44. [CrossRef] [PubMed]
14. Armstrong, G.A.B.; Drapeau, P. Loss and gain of FUS function impair neuromuscular synaptic transmission in a genetic model of ALS. *Hum. Mol. Genet.* **2013**, *22*, 4282–4292. [CrossRef] [PubMed]
15. Ciura, S.; Lattante, S.; Le Ber, I.; Latouche, M.; Tostivint, H.; Brice, A.; Kabashi, E. Loss of function of C9orf72 causes motor deficits in a zebrafish model of amyotrophic lateral sclerosis. *Ann. Neurol.* **2013**, *74*, 180–187. [CrossRef] [PubMed]
16. Schmid, B.; Hruscha, A.; Hogl, S.; Banzhaf-Strathmann, J.; Strecker, K.; van der Zee, J.; Teucke, M.; Eimer, S.; Hegermann, J.; Kittelmann, M.; et al. Loss of ALS-associated TDP-43 in zebrafish causes muscle degeneration, vascular dysfunction, and reduced motor neuron axon outgrowth. *Proc. Natl. Acad. Sci. USA* **2013**, *110*, 4986–4991. [CrossRef]
17. Gurney, M.E.; Pu, H.; Chiu, A.Y.; Dal Canto, M.C.; Polchow, C.Y.; Alexander, D.D.; Caliendo, J.; Hentati, A.; Kwon, Y.W.; Deng, H.X. Motor neuron degeneration in mice that express a human Cu,Zn superoxide dismutase mutation. *Science* **1994**, *264*, 1772–1775. [CrossRef]
18. Pansarasa, O.; Bordoni, M.; Drufuca, L.; Diamanti, L.; Sproviero, D.; Trotti, R.; Bernuzzi, S.; La Salvia, S.; Gagliardi, S.; Ceroni, M.; et al. Lymphoblastoid cell lines as a model to understand amyotrophic lateral sclerosis disease mechanisms. *Dis. Model. Mech.* **2018**, *11*. [CrossRef]
19. Boston-Howes, W.; Williams, E.O.; Bogush, A.; Scolere, M.; Pasinelli, P.; Trotti, D. Nordihydroguaiaretic acid increases glutamate uptake in vitro and in vivo: Therapeutic implications for amyotrophic lateral sclerosis. *Exp. Neurol.* **2008**, *213*, 229–237. [CrossRef]
20. Moreno-Martet, M.; Mestre, L.; Loria, F.; Guaza, C.; Fernandez-Ruiz, J.; de Lago, E. Identification of receptors and enzymes for endocannabinoids in NSC-34 cells: Relevance for in vitro studies with cannabinoids in motor neuron diseases. *Neurosci. Lett.* **2012**, *508*, 67–72. [CrossRef] [PubMed]
21. Van Damme, P.; Bogaert, E.; Dewil, M.; Hersmus, N.; Kiraly, D.; Scheveneels, W.; Bockx, I.; Braeken, D.; Verpoorten, N.; Verhoeven, K.; et al. Astrocytes regulate GluR2 expression in motor neurons and their vulnerability to excitotoxicity. *Proc. Natl. Acad. Sci. USA* **2007**, *104*, 14825–14830. [CrossRef] [PubMed]
22. Volk, A.E.; Weishaupt, J.H.; Andersen, P.M.; Ludolph, A.C.; Kubisch, C. Current knowledge and recent insights into the genetic basis of amyotrophic lateral sclerosis. *Med. Genet.* **2018**, *30*, 252–258. [CrossRef] [PubMed]
23. Chia, R.; Chiò, A.; Traynor, B.J. Novel genes associated with amyotrophic lateral sclerosis: Diagnostic and clinical implications. *Lancet Neurol.* **2018**, *17*, 94–102. [CrossRef]
24. Boillée, S.; Vande Velde, C.; Cleveland, D.W. ALS: A Disease of Motor Neurons and Their Nonneuronal Neighbors. *Neuron* **2006**, *52*, 39–59. [CrossRef]
25. Gupta, R.; Lan, M.; Mojsilovic-Petrovic, J.; Choi, W.H.; Safren, N.; Barmada, S.; Lee, M.J.; Kalb, R. The Proline/Arginine Dipeptide from Hexanucleotide Repeat Expanded C9ORF72 Inhibits the Proteasome. *Eneuro* **2017**, *4*. [CrossRef]
26. Mitsumoto, H.; Santella, R.M.; Liu, X.; Bogdanov, M.; Zipprich, J.; Wu, H.-C.; Mahata, J.; Kilty, M.; Bednarz, K.; Bell, D.; et al. *Oxidative Stress Biomarkers in Sporadic ALS*; Taylor & Francis: Oxfordshire, UK, 2008; Volume 9, pp. 177–183.
27. Chang, Y.; Kong, Q.; Shan, X.; Tian, G.; Ilieva, H.; Cleveland, D.W.; Rothstein, J.D.; Borchelt, D.R.; Wong, P.C.; Lin, C.G. Messenger RNA Oxidation Occurs Early in Disease Pathogenesis and Promotes Motor Neuron Degeneration in ALS. *PLoS ONE* **2008**, *3*, e2849. [CrossRef]

28. De Vos, K.J.; Chapman, A.L.; Tennant, M.E.; Manser, C.; Tudor, E.L.; Lau, K.-F.; Brownlees, J.; Ackerley, S.; Shaw, P.J.; Mcloughlin, D.M.; et al. Familial amyotrophic lateral sclerosis-linked SOD1 mutants perturb fast axonal transport to reduce axonal mitochondria content Europe PMC Funders Group. *Hum. Mol. Genet.* **2007**, *16*, 2720–2728. [CrossRef]
29. Kiaei, M.; Kipiani, K.; Calingasan, N.; Wille, E.; Chen, J.; Heissig, B.; Rafii, S.; Lorenzl, S.; Beal, M. Matrix metalloproteinase-9 regulates TNF-α and FasL expression in neuronal, glial cells and its absence extends life in a transgenic mouse model of amyotrophic lateral sclerosis. *Exp. Neurol.* **2007**, *205*, 74–81. [CrossRef]
30. Laslo, P.; Lipski, J.; Nicholson, L.F.B.; Miles, G.B.; Funk, G.D. GluR2 AMPA Receptor Subunit Expression in Motoneurons at Low and High Risk for Degeneration in Amyotrophic Lateral Sclerosis. *Exp. Neurol.* **2001**, *169*, 461–471. [CrossRef]
31. Foerster, B.R.; Pomper, M.G.; Callaghan, B.C.; Petrou, M.; Edden, R.A.E.; Mohamed, M.A.; Welsh, R.C.; Carlos, R.C.; Barker, P.B.; Feldman, E.L. An imbalance between excitatory and inhibitory neurotransmitters in amyotrophic lateral sclerosis revealed by use of 3-T proton magnetic resonance spectroscopy. *JAMA Neurol.* **2013**, *70*, 1009–1016. [CrossRef]
32. Kumar, A.; Bala, L.; Kalita, J.; Misra, U.K.; Singh, R.L.; Khetrapal, C.L.; Babu, G.N. Metabolomic analysis of serum by (1) H NMR spectroscopy in amyotrophic lateral sclerosis. *Clin. Chim. Acta* **2010**, *411*, 563–567. [CrossRef] [PubMed]
33. Spreux-Varoquaux, O.; Bensimon, G.; Lacomblez, L.; Salachas, F.; Pradat, P.F.; Le Forestier, N.; Marouan, A.; Dib, M.; Meininger, V. Glutamate levels in cerebrospinal fluid in amyotrophic lateral sclerosis: A reappraisal using a new HPLC method with coulometric detection in a large cohort of patients. *J. Neurol. Sci.* **2002**, *193*, 73–78. [CrossRef]
34. Webster, C.P.; Smith, E.F.; Bauer, C.S.; Moller, A.; Hautbergue, G.M.; Ferraiuolo, L.; Myszczynska, M.A.; Higginbottom, A.; Walsh, M.J.; Whitworth, A.J.; et al. The C9orf72 protein interacts with Rab1a and the ULK1 complex to regulate initiation of autophagy. *EMBO J.* **2016**, *35*, 1656–1676. [CrossRef] [PubMed]
35. Leigh, P.N.; Whitwell, H.; Garofalo, O.; Buller, J.; Swash, M.; Martin, J.E.; Gallo, J.M.; Weller, R.O.; Anderton, B.H. Ubiquitin-immunoreactive intraneuronal inclusions in amyotrophic lateral sclerosis. Morphology, distribution, and specificity. *Brain* **1991**, *114 Pt 2*, 775–788. [CrossRef]
36. Sreedharan, J.; Blair, I.P.; Tripathi, V.B.; Hu, X.; Vance, C.; Rogelj, B.; Ackerley, S.; Durnall, J.C.; Williams, K.L.; Buratti, E.; et al. TDP-43 Mutations in Familial and Sporadic Amyotrophic Lateral Sclerosis. *Science* **2008**, *319*, 1668–1672. [CrossRef]
37. Cykowski, M.D.; Dickson, D.W.; Powell, S.Z.; Arumanayagam, A.S.; Rivera, A.L.; Appel, S.H. Dipeptide repeat (DPR) pathology in the skeletal muscle of ALS patients with C9ORF72 repeat expansion. *Acta Neuropathol.* **2019**, *138*, 667–670. [CrossRef]
38. Taylor, J.P.; Brown, R.H.; Cleveland, D.W. Decoding ALS: From genes to mechanism. *Nature* **2016**, *539*, 197–206. [CrossRef]
39. Lagier-Tourenne, C.; Polymenidou, M.; Hutt, K.R.; Vu, A.Q.; Baughn, M.; Huelga, S.C.; Clutario, K.M.; Ling, S.-C.; Liang, T.Y.; Mazur, C.; et al. Divergent roles of ALS-linked proteins FUS/TLS and TDP-43 intersect in processing long pre-mRNAs. *Nat. Neurosci.* **2012**, *15*, 1488–1497. [CrossRef]
40. Haeusler, A.R.; Donnelly, C.J.; Rothstein, J.D. The expanding biology of the C9orf72 nucleotide repeat expansion in neurodegenerative disease. *Nat. Rev. Neurosci.* **2016**, *17*, 383–395. [CrossRef]
41. Butti, Z.; Patten, S.A. RNA Dysregulation in Amyotrophic Lateral Sclerosis. *Front. Genet.* **2019**, *9*, 712. [CrossRef]
42. Polymenidou, M.; Lagier-Tourenne, C.; Hutt, K.R.; Huelga, S.C.; Moran, J.; Liang, T.Y.; Ling, S.-C.; Sun, E.; Wancewicz, E.; Mazur, C.; et al. Long pre-mRNA depletion and RNA missplicing contribute to neuronal vulnerability from loss of TDP-43. *Nat. Neurosci.* **2011**, *14*, 459–468. [CrossRef] [PubMed]
43. Rizzo, F.; Riboldi, G.; Salani, S.; Nizzardo, M.; Simone, C.; Corti, S.; Hedlund, E. Cellular therapy to target neuroinflammation in amyotrophic lateral sclerosis. *Cell. Mol. Life Sci.* **2014**, *71*, 999–1015. [CrossRef] [PubMed]
44. Dobrowolny, G.; Aucello, M.; Rizzuto, E.; Beccafico, S.; Mammucari, C.; Boncompagni, S.; Bonconpagni, S.; Belia, S.; Wannenes, F.; Nicoletti, C.; et al. Skeletal muscle is a primary target of SOD1G93A-mediated toxicity. *Cell Metab.* **2008**, *8*, 425–436. [CrossRef] [PubMed]
45. Betteridge, D.J. What is oxidative stress? *Metabolism* **2000**, *49*, 3–8. [CrossRef]

46. Kandlur, A.; Satyamoorthy, K.; Gangadharan, G. Oxidative Stress in Cognitive and Epigenetic Aging: A Retrospective Glance. *Front. Mol. Neurosci.* **2020**, *13*. [CrossRef] [PubMed]
47. Shaw, P.J.; Ince, P.G.; Falkous, G.; Mantle, D. Oxidative damage to protein in sporadic motor neuron disease spinal cord. *Ann. Neurol.* **1995**, *38*, 691–695. [CrossRef]
48. Rosen, D.R. Mutations in Cu/Zn superoxide dismutase gene are associated with familial amyotrophic lateral sclerosis. *Nature* **1993**, *362*, 59–62. [CrossRef]
49. Zarei, S.; Carr, K.; Reiley, L.; Diaz, K.; Guerra, O.; Altamirano, P.F.; Pagani, W.; Lodin, D.; Orozco, G.; Chinea, A. *A Comprehensive Review of Amyotrophic Lateral Sclerosis*; Wolters Kluwer—Medknow Publications: Mumbai, India, 2015; Volume 6, p. 171.
50. Shibata, N.; Nagai, R.; Uchida, K.; Horiuchi, S.; Yamada, S.; Hirano, A.; Kawaguchi, M.; Yamamoto, T.; Sasaki, S.; Kobayashi, M. Morphological evidence for lipid peroxidation and protein glycoxidation in spinal cords from sporadic amyotrophic lateral sclerosis patients. *Brain Res.* **2001**, *917*, 97–104. [CrossRef]
51. Juarez, J.C.; Manuia, M.; Burnett, M.E.; Betancourt, O.; Boivin, B.; Shaw, D.E.; Tonks, N.K.; Mazar, A.P.; Donate, F. Superoxide dismutase 1 (SOD1) is essential for H2O2-mediated oxidation and inactivation of phosphatases in growth factor signaling. *Proc. Natl. Acad. Sci. USA* **2008**, *105*, 7147–7152. [CrossRef]
52. Crosby, K.; Crown, A.M.; Roberts, B.L.; Brown, H.; Ayers, J.I.; Borchelt, D.R. Loss of charge mutations in solvent exposed Lys residues of superoxide dismutase 1 do not induce inclusion formation in cultured cell models. *PLoS ONE* **2018**, *13*, e0206751. [CrossRef]
53. Saccon, R.A.; Bunton-Stasyshyn, R.K.A.; Fisher, E.M.C.; Fratta, P. Is SOD1 loss of function involved in amyotrophic lateral sclerosis? *Brain* **2013**, *136*, 2342–2358. [CrossRef] [PubMed]
54. Kirby, J.; Halligan, E.; Baptista, M.J.; Allen, S.; Heath, P.R.; Holden, H.; Barber, S.C.; Loynes, C.A.; Wood-Allum, C.A.; Lunec, J.; et al. Mutant SOD1 alters the motor neuronal transcriptome: Implications for familial ALS. *Brain* **2005**, *128*, 1686–1706. [CrossRef] [PubMed]
55. Johnson, J.A.; Johnson, D.A.; Kraft, A.D.; Calkins, M.J.; Jakel, R.J.; Vargas, M.R.; Chen, P.-C. The Nrf2-ARE Pathway. *Ann. N. Y. Acad. Sci.* **2008**, *1147*, 61–69. [CrossRef] [PubMed]
56. Shibata, N.; Hirano, A.; Kato, S.; Nagai, R.; Horiuchi, S.; Komori, T.; Umahara, T.; Asayama, K.; Kobayashi, M. Advanced glycation endproducts are deposited in neuronal hyaline inclusions: A study on familial amyotrophic lateral sclerosis with superoxide dismutase-1 mutation. *Acta Neuropathol.* **1999**, *97*, 240–246. [CrossRef] [PubMed]
57. Andrus, P.K.; Fleck, T.J.; Gurney, M.E.; Hall, E.D. Protein Oxidative Damage in a Transgenic Mouse Model of Familial Amyotrophic Lateral Sclerosis. *J. Neurochem.* **2002**, *71*, 2041–2048. [CrossRef]
58. Hall, E.D.; Andrus, P.K.; Oostveen, J.A.; Fleck, T.J.; Gurney, M.E. Relationship of oxygen radical-induced lipid peroxidative damage to disease onset and progression in a transgenic model of familial ALS. *J. Neurosci. Res.* **1998**, *53*, 66–77. [CrossRef]
59. Elden, A.C.; Kim, H.-J.; Hart, M.P.; Chen-Plotkin, A.S.; Johnson, B.S.; Fang, X.; Armakola, M.; Geser, F.; Greene, R.; Lu, M.M.; et al. Ataxin-2 intermediate-length polyglutamine expansions are associated with increased risk for ALS. *Nature* **2010**, *466*, 1069–1075. [CrossRef]
60. van Blitterswijk, M.; Mullen, B.; Heckman, M.G.; Baker, M.C.; DeJesus-Hernandez, M.; Brown, P.H.; Murray, M.E.; Hsiung, G.Y.R.; Stewart, H.; Karydas, A.M.; et al. Ataxin-2 as potential disease modifier in C9ORF72 expansion carriers. *Neurobiol. Aging* **2014**. [CrossRef]
61. Chiò, A.; Calvo, A.; Moglia, C.; Canosa, A.; Brunetti, M.; Barberis, M.; Restagno, G.; Conte, A.; Bisogni, G.; Marangi, G.; et al. ATXN2 polyQ intermediate repeats are a modifier of ALS survival. *Neurology* **2015**. [CrossRef]
62. Bertoni, A.; Giuliano, P.; Galgani, M.; Rotoli, D.; Ulianich, L.; Adornetto, A.; Santillo, M.R.; Porcellini, A.; Avvedimento, V.E. Early and Late Events Induced by PolyQ-expanded Proteins. *J. Biol. Chem.* **2011**, *286*, 4727–4741. [CrossRef]
63. Polci, R.; Peng, A.; Chen, P.-L.; Riley, D.J.; Chen, Y. NIMA-Related Protein Kinase 1 Is Involved Early in the Ionizing Radiation-Induced DNA Damage Response. *Cancer Res.* **2004**, *64*, 8800–8803. [CrossRef] [PubMed]
64. Fang, X.; Lin, H.; Wang, X.; Zuo, Q.; Qin, J.; Zhang, P. The NEK1 interactor, C21ORF2, is required for efficient DNA damage repair. *Acta Biochim. Biophys. Sin. (Shanghai)* **2015**, *47*, 834–841. [CrossRef] [PubMed]
65. Kannan, A.; Bhatia, K.; Branzei, D.; Gangwani, L. Combined deficiency of Senataxin and DNA-PKcs causes DNA damage accumulation and neurodegeneration in spinal muscular atrophy. *Nucleic Acids Res.* **2018**, *46*, 8326–8346. [CrossRef] [PubMed]

66. Cirulli, E.T.; Lasseigne, B.N.; Petrovski, S.; Sapp, P.C.; Dion, P.A.; Leblond, C.S.; Couthouis, J.; Lu, Y.-F.; Wang, Q.; Krueger, B.J.; et al. Exome sequencing in amyotrophic lateral sclerosis identifies risk genes and pathways. *Science* **2015**, *347*, 1436–1441. [CrossRef] [PubMed]
67. van Rheenen, W.; Shatunov, A.; Dekker, A.M.; McLaughlin, R.L.; Diekstra, F.P.; Pulit, S.L.; van der Spek, R.A.A.; Võsa, U.; de Jong, S.; Robinson, M.R.; et al. Genome-wide association analyses identify new risk variants and the genetic architecture of amyotrophic lateral sclerosis. *Nat. Genet.* **2016**, *48*, 1043–1048. [CrossRef]
68. Chen, Y.-Z.; Bennett, C.L.; Huynh, H.M.; Blair, I.P.; Puls, I.; Irobi, J.; Dierick, I.; Abel, A.; Kennerson, M.L.; Rabin, B.A.; et al. DNA/RNA helicase gene mutations in a form of juvenile amyotrophic lateral sclerosis (ALS4). *Am. J. Hum. Genet.* **2004**, *74*, 1128–1135. [CrossRef]
69. Kenna, K.P.; Van Doormaal, P.T.C.; Dekker, A.M.; Ticozzi, N.; Kenna, B.J.; Diekstra, F.P.; Van Rheenen, W.; Van Eijk, K.R.; Jones, A.R.; Keagle, P.; et al. NEK1 variants confer susceptibility to amyotrophic lateral sclerosis. *Nat. Genet.* **2016**, *48*, 1037–1042. [CrossRef]
70. Higelin, J.; Catanese, A.; Semelink-Sedlacek, L.L.; Oeztuerk, S.; Lutz, A.K.; Bausinger, J.; Barbi, G.; Speit, G.; Andersen, P.M.; Ludolph, A.C.; et al. NEK1 loss-of-function mutation induces DNA damage accumulation in ALS patient-derived motoneurons. *Stem Cell Res.* **2018**, *30*, 150–162. [CrossRef]
71. Bennett, C.L.; Dastidar, S.G.; Ling, S.C.; Malik, B.; Ashe, T.; Wadhwa, M.; Miller, D.B.; Lee, C.; Mitchell, M.B.; van Es, M.A.; et al. Senataxin mutations elicit motor neuron degeneration phenotypes and yield TDP-43 mislocalization in ALS4 mice and human patients. *Acta Neuropathol.* **2018**, *136*, 425–443. [CrossRef]
72. Kowaltowski, A.J.; Vercesi, A.E. Mitochondrial damage induced by conditions of oxidative stress. *Free Radic. Biol. Med.* **1999**, *26*, 463–471. [CrossRef]
73. Malhotra, J.D.; Kaufman, R.J. Endoplasmic Reticulum Stress and Oxidative Stress: A Vicious Cycle or a Double-Edged Sword? *Antioxid. Redox Signal.* **2007**, *9*, 2277–2294. [CrossRef] [PubMed]
74. Reichmann, D.; Voth, W.; Jakob, U. Maintaining a Healthy Proteome during Oxidative Stress. *Mol. Cell* **2018**, *69*, 203–213. [CrossRef] [PubMed]
75. McBride, H.M.; Neuspiel, M.; Wasiak, S. Mitochondria: More Than Just a Powerhouse. *Curr. Biol.* **2006**, *16*, R551–R560. [CrossRef] [PubMed]
76. Sasaki, S.; Horie, Y.; Iwata, M. Mitochondrial alterations in dorsal root ganglion cells in sporadic amyotrophic lateral sclerosis. *Acta Neuropathol.* **2007**, *114*, 633–639. [CrossRef] [PubMed]
77. Cogliati, S.; Frezza, C.; Soriano, M.E.; Varanita, T.; Quintana-Cabrera, R.; Corrado, M.; Cipolat, S.; Costa, V.; Casarin, A.; Gomes, L.C.; et al. Mitochondrial Cristae Shape Determines Respiratory Chain Supercomplexes Assembly and Respiratory Efficiency. *Cell* **2013**, *155*, 160–171. [CrossRef] [PubMed]
78. Zerbes, R.M.; Bohnert, M.; Stroud, D.A.; von der Malsburg, K.; Kram, A.; Oeljeklaus, S.; Warscheid, B.; Becker, T.; Wiedemann, N.; Veenhuis, M.; et al. Role of MINOS in Mitochondrial Membrane Architecture: Cristae Morphology and Outer Membrane Interactions Differentially Depend on Mitofilin Domains. *J. Mol. Biol.* **2012**, *422*, 183–191. [CrossRef]
79. Pánek, T.; Eliáš, M.; Vancová, M.; Lukeš, J.; Hashimi, H. Returning to the Fold for Lessons in Mitochondrial Crista Diversity and Evolution. *Curr. Biol.* **2020**, *30*, R575–R588. [CrossRef]
80. Genin, E.C.; Plutino, M.; Bannwarth, S.; Villa, E.; Cisneros-Barroso, E.; Roy, M.; Ortega-Vila, B.; Fragaki, K.; Lespinasse, F.; Pinero-Martos, E.; et al. CHCHD10 mutations promote loss of mitochondrial cristae junctions with impaired mitochondrial genome maintenance and inhibition of apoptosis. *EMBO Mol. Med.* **2016**, *8*, 58–72. [CrossRef]
81. Genin, E.C.; Bannwarth, S.; Lespinasse, F.; Ortega-Vila, B.; Fragaki, K.; Itoh, K.; Villa, E.; Lacas-Gervais, S.; Jokela, M.; Auranen, M.; et al. Loss of MICOS complex integrity and mitochondrial damage, but not TDP-43 mitochondrial localisation, are likely associated with severity of CHCHD10-related diseases. *Neurobiol. Dis.* **2018**, *119*, 159–171. [CrossRef]
82. Zhou, W.; Ma, D.; Sun, A.X.; Tran, H.-D.; Ma, D.; Singh, B.K.; Zhou, J.; Zhang, J.; Wang, D.; Zhao, Y.; et al. PD-linked CHCHD2 mutations impair CHCHD10 and MICOS complex leading to mitochondria dysfunction. *Hum. Mol. Genet.* **2019**, *28*, 1100–1116. [CrossRef]
83. Purandare, N.; Somayajulu, M.; Hüttemann, M.; Grossman, L.I.; Aras, S. The cellular stress proteins CHCHD10 and MNRR1 (CHCHD2): Partners in mitochondrial and nuclear function and dysfunction. *J. Biol. Chem.* **2018**, *293*, 6517–6529. [CrossRef] [PubMed]
84. Deng, J.; Yang, M.; Chen, Y.; Chen, X.; Liu, J.; Sun, S.; Cheng, H.; Li, Y.; Bigio, E.H.; Mesulam, M.; et al. FUS Interacts with HSP60 to Promote Mitochondrial Damage. *PLoS Genet.* **2015**, *11*. [CrossRef] [PubMed]

85. Deng, J.; Wang, P.; Chen, X.; Cheng, H.; Liu, J.; Fushimi, K.; Zhu, L.; Wu, J.Y. FUS interacts with ATP synthase beta subunit and induces mitochondrial unfolded protein response in cellular and animal models. *Proc. Natl. Acad. Sci. USA* **2018**, *115*, E9678–E9686. [CrossRef] [PubMed]
86. Crozat, A.; Aman, P.; Mandahl, N.; Ron, D. Fusion of CHOP to a novel RNA-binding protein in human myxoid liposarcoma. *Nature* **1993**, *363*, 640–644. [CrossRef]
87. Vance, C.; Rogelj, B.; Hortobagyi, T.; De Vos, K.J.; Nishimura, A.L.; Sreedharan, J.; Hu, X.; Smith, B.; Ruddy, D.; Wright, P.; et al. Mutations in FUS, an RNA Processing Protein, Cause Familial Amyotrophic Lateral Sclerosis Type 6. *Science* **2009**, *323*, 1208–1211. [CrossRef]
88. Kwiatkowski, T.J.; Bosco, D.A.; LeClerc, A.L.; Tamrazian, E.; Vanderburg, C.R.; Russ, C.; Davis, A.; Gilchrist, J.; Kasarskis, E.J.; Munsat, T.; et al. Mutations in the FUS/TLS Gene on Chromosome 16 Cause Familial Amyotrophic Lateral Sclerosis. *Science* **2009**, *323*, 1205–1208. [CrossRef]
89. Hardiman, O.; Al-Chalabi, A.; Chio, A.; Corr, E.M.; Logroscino, G.; Robberecht, W.; Shaw, P.J.; Simmons, Z.; van den Berg, L.H. Amyotrophic lateral sclerosis. *Nat. Rev. Dis. Prim.* **2017**, *3*, 17071. [CrossRef]
90. Magrané, J.; Cortez, C.; Gan, W.-B.; Manfredi, G. Abnormal mitochondrial transport and morphology are common pathological denominators in SOD1 and TDP43 ALS mouse models. *Hum. Mol. Genet.* **2014**, *23*, 1413–1424. [CrossRef]
91. Bernardini, C.; Censi, F.; Lattanzi, W.; Barba, M.; Calcagnini, G.; Giuliani, A.; Tasca, G.; Sabatelli, M.; Ricci, E.; Michetti, F. Mitochondrial network genes in the skeletal muscle of amyotrophic lateral sclerosis patients. *PLoS ONE* **2013**, *8*, e57739. [CrossRef]
92. Higgins, C.M.J.; Jung, C.; Xu, Z. ALS-associated mutant SOD1G93A causes mitochondrial vacuolation by expansion of the intermembrane space and by involvement of SOD1 aggregation and peroxisomes. *BMC Neurosci.* **2003**, *4*, 16. [CrossRef]
93. Pasinelli, P.; Belford, M.E.; Lennon, N.; Bacskai, B.J.; Hyman, B.T.; Trotti, D.; Brown, R.H. Amyotrophic Lateral Sclerosis-Associated SOD1 Mutant Proteins Bind and Aggregate with Bcl-2 in Spinal Cord Mitochondria. *Neuron* **2004**, *43*, 19–30. [CrossRef] [PubMed]
94. Kausar, S.; Wang, F.; Cui, H. The Role of Mitochondria in Reactive Oxygen Species Generation and Its Implications for Neurodegenerative Diseases. *Cells* **2018**, *7*, 274. [CrossRef] [PubMed]
95. Browne, S.E.; Bowling, A.C.; Baik, M.J.; Gurney, M.; Brown, R.H.; Beal, M.F. Metabolic Dysfunction in Familial, but Not Sporadic, Amyotrophic Lateral Sclerosis. *J. Neurochem.* **2002**, *71*, 281–287. [CrossRef] [PubMed]
96. Wiedemann, F.R.; Manfredi, G.; Mawrin, C.; Beal, M.F.; Schon, E.A. Mitochondrial DNA and respiratory chain function in spinal cords of ALS patients. *J. Neurochem.* **2002**, *80*, 616–625. [CrossRef] [PubMed]
97. Dröse, S.; Brandt, U. Molecular Mechanisms of Superoxide Production by the Mitochondrial Respiratory Chain. In *Mitochondrial Oxidative Phosphorylation*; Springer: New York, NY, USA, 2012; pp. 145–169.
98. Umoh, M.E.; Fournier, C.; Li, Y.; Polak, M.; Shaw, L.; Landers, J.E.; Hu, W.; Gearing, M.; Glass, J.D. Comparative analysis of C9orf72 and sporadic disease in an ALS clinic population. *Neurology* **2016**, *87*, 1024–1030. [CrossRef] [PubMed]
99. DeJesus-Hernandez, M.; Mackenzie, I.R.R.; Boeve, B.F.F.; Boxer, A.L.L.; Baker, M.; Rutherford, N.J.J.; Nicholson, A.M.M.; Finch, N.A.A.; Flynn, H.; Adamson, J.; et al. Expanded GGGGCC hexanucleotide repeat in noncoding region of C9ORF72 causes chromosome 9p-linked FTD and ALS. *Neuron* **2011**, *72*, 245–256. [CrossRef]
100. Renton, A.E.; Majounie, E.; Waite, A.; Simón-Sánchez, J.; Rollinson, S.; Gibbs, J.R.; Schymick, J.C.; Laaksovirta, H.; van Swieten, J.C.; Myllykangas, L.; et al. A Hexanucleotide Repeat Expansion in C9ORF72 Is the Cause of Chromosome 9p21-Linked ALS-FTD. *Neuron* **2011**, *72*, 257–268. [CrossRef]
101. Balendra, R.; Isaacs, A.M. C9orf72-mediated ALS and FTD: Multiple pathways to disease. *Nat. Rev. Neurol.* **2018**, *14*, 544–558. [CrossRef]
102. Choi, S.Y.; Lopez-Gonzalez, R.; Krishnan, G.; Phillips, H.L.; Li, A.N.; Seeley, W.W.; Yao, W.-D.; Almeida, S.; Gao, F.-B. C9ORF72-ALS/FTD-associated poly(GR) binds Atp5a1 and compromises mitochondrial function in vivo. *Nat. Neurosci.* **2019**, *22*, 851–862. [CrossRef]
103. Onesto, E.; Colombrita, C.; Gumina, V.; Borghi, M.O.; Dusi, S.; Doretti, A.; Fagiolari, G.; Invernizzi, F.; Moggio, M.; Tiranti, V.; et al. Gene-specific mitochondria dysfunctions in human TARDBP and C9ORF72 fibroblasts. *Acta Neuropathol. Commun.* **2016**, *4*, 47. [CrossRef]

104. Birger, A.; Ben-Dor, I.; Ottolenghi, M.; Turetsky, T.; Gil, Y.; Sweetat, S.; Perez, L.; Belzer, V.; Casden, N.; Steiner, D.; et al. Human iPSC-derived astrocytes from ALS patients with mutated C9ORF72 show increased oxidative stress and neurotoxicity. *EBioMedicine* **2019**, *50*, 274–289. [CrossRef] [PubMed]
105. Youle, R.J.; Narendra, D.P. Mechanisms of mitophagy. *Nat. Rev. Mol. Cell Biol.* **2011**, *12*, 9–14. [CrossRef] [PubMed]
106. Moore, A.S.; Holzbaur, E.L.F. Dynamic recruitment and activation of ALS-associated TBK1 with its target optineurin are required for efficient mitophagy. *Proc. Natl. Acad. Sci. USA* **2016**, *113*, E3349–E3358. [CrossRef] [PubMed]
107. Maruyama, H.; Morino, H.; Ito, H.; Izumi, Y.; Kato, H.; Watanabe, Y.; Kinoshita, Y.; Kamada, M.; Nodera, H.; Suzuki, H.; et al. Mutations of optineurin in amyotrophic lateral sclerosis. *Nature* **2010**, *465*, 223–226. [CrossRef] [PubMed]
108. Fivenson, E.M.; Lautrup, S.; Sun, N.; Scheibye-Knudsen, M.; Stevnsner, T.; Nilsen, H.; Bohr, V.A.; Fang, E.F. Mitophagy in neurodegeneration and aging. *Neurochem. Int.* **2017**, *109*, 202–209. [CrossRef]
109. Damiano, M.; Starkov, A.A.; Petri, S.; Kipiani, K.; Kiaei, M.; Mattiazzi, M.; Flint Beal, M.; Manfredi, G. Neural mitochondrial Ca^{2+} capacity impairment precedes the onset of motor symptoms in G93A Cu/Zn-superoxide dismutase mutant mice. *J. Neurochem.* **2006**, *96*, 1349–1361. [CrossRef]
110. Sathasivam, S.; Grierson, A.J.; Shaw, P.J.; Shaw, P.J. Characterization of the caspase cascade in a cell culture model of SOD1-related familial amyotrophic lateral sclerosis: Expression, activation and therapeutic effects of inhibition. *Neuropathol. Appl. Neurobiol.* **2005**, *31*, 467–685. [CrossRef]
111. Sutton, M.A.; Schuman, E.M. Dendritic Protein Synthesis, Synaptic Plasticity, and Memory. *Cell* **2006**, *127*, 49–58. [CrossRef]
112. Yuan, A.; Rao, M.V.; Veeranna; Nixon, R.A. Neurofilaments and Neurofilament Proteins in Health and Disease. *Cold Spring Harb. Perspect. Biol.* **2017**, *9*, a018309. [CrossRef]
113. Brady, S. Mice overexpressing the human neurofilament heavy gene as a model of ALS. *Neurobiol. Aging* **1995**. [CrossRef]
114. Prokop, A. Cytoskeletal organization of axons in vertebrates and invertebrates. *J. Cell Biol.* **2020**, *219*. [CrossRef] [PubMed]
115. Figlewicz, D.A.; Krizus, A.; Martinoli, M.G.; Meininger, V.; Dib, M.; Rouleau, G.A.; Julien, J.P. Variants of the heavy neurofilament subunit are associated with the development of amyotrophic lateral sclerosis. *Hum. Mol. Genet.* **1994**, *3*, 1757–1761. [CrossRef] [PubMed]
116. Dubey, S.; Bhembre, N.; Bodas, S.; Veer, S.; Ghose, A.; Callan-Jones, A.; Pullarkat, P. The axonal actin-spectrin lattice acts as a tension buffering shock absorber. *Elife* **2020**, *9*. [CrossRef] [PubMed]
117. Witke, W.; Podtelejnikov, A.V.; Di Nardo, A.; Sutherland, J.D.; Gurniak, C.B.; Dotti, C.; Mann, M. In mouse brain profilin I and profilin II associate with regulators of the endocytic pathway and actin assembly. *EMBO J.* **1998**, *17*, 967–976. [CrossRef]
118. Wu, C.-H.; Fallini, C.; Ticozzi, N.; Keagle, P.J.; Sapp, P.C.; Piotrowska, K.; Lowe, P.; Koppers, M.; McKenna-Yasek, D.; Baron, D.M.; et al. Mutations in the profilin 1 gene cause familial amyotrophic lateral sclerosis. *Nature* **2012**, *488*, 499–503. [CrossRef]
119. Fil, D.; DeLoach, A.; Yadav, S.; Alkam, D.; MacNicol, M.; Singh, A.; Compadre, C.M.; Goellner, J.J.; O'Brien, C.A.; Fahmi, T.; et al. Mutant Profilin1 transgenic mice recapitulate cardinal features of motor neuron disease. *Hum. Mol. Genet.* **2017**. [CrossRef]
120. Tempes, A.; Weslawski, J.; Brzozowska, A.; Jaworski, J. Role of dynein-dynactin complex, kinesins, motor adaptors, and their phosphorylation in dendritogenesis. *J. Neurochem.* **2020**. [CrossRef]
121. Uchida, A.; Alami, N.H.; Brown, A. Tight Functional Coupling of Kinesin-1A and Dynein Motors in the Bidirectional Transport of Neurofilaments. *Mol. Biol. Cell* **2009**, *20*, 4997–5006. [CrossRef]
122. Heisler, F.F.; Lee, H.K.; Gromova, K.V.; Pechmann, Y.; Schurek, B.; Ruschkies, L.; Schroeder, M.; Schweizer, M.; Kneussel, M. GRIP1 interlinks N-cadherin and AMPA receptors at vesicles to promote combined cargo transport into dendrites. *Proc. Natl. Acad. Sci. USA* **2014**, *111*, 5030–5035. [CrossRef]
123. Nakajima, K.; Yin, X.; Takei, Y.; Seog, D.-H.; Homma, N.; Hirokawa, N. Molecular motor KIF5A is essential for GABA(A) receptor transport, and KIF5A deletion causes epilepsy. *Neuron* **2012**, *76*, 945–961. [CrossRef]
124. Smith, B.N.; Ticozzi, N.; Fallini, C.; Gkazi, A.S.; Topp, S.; Kenna, K.P.; Scotter, E.L.; Kost, J.; Keagle, P.; Miller, J.W.; et al. Exome-wide Rare Variant Analysis Identifies TUBA4A Mutations Associated with Familial ALS. *Neuron* **2014**, *84*, 324–331. [CrossRef] [PubMed]

125. Puls, I.; Jonnakuty, C.; LaMonte, B.H.; Holzbaur, E.L.F.; Tokito, M.; Mann, E.; Floeter, M.K.; Bidus, K.; Drayna, D.; Oh, S.J.; et al. Mutant dynactin in motor neuron disease. *Nat. Genet.* **2003**, *33*, 455–456. [CrossRef] [PubMed]
126. LaMonte, B.H.; Wallace, K.E.; Holloway, B.A.; Shelly, S.S.; Ascano, J.; Tokito, M.; Van Winkle, T.; Howland, D.S.; Holzbaur, E.L.F. Disruption of dynein/dynactin inhibits axonal transport in motor neurons causing late-onset progressive degeneration. *Neuron* **2002**, *34*, 715–727. [CrossRef]
127. Miki, H.; Setou, M.; Kaneshiro, K.; Hirokawa, N. All kinesin superfamily protein, KIF, genes in mouse and human. *Proc. Natl. Acad. Sci. USA* **2001**, *98*, 7004–7011. [CrossRef] [PubMed]
128. Nicolas, A.; Kenna, K.P.; Renton, A.E.; Ticozzi, N.; Faghri, F.; Chia, R.; Dominov, J.A.; Kenna, B.J.; Nalls, M.A.; Keagle, P.; et al. Genome-wide Analyses Identify KIF5A as a Novel ALS Gene. *Neuron* **2018**, *97*, 1268–1283.e6. [CrossRef] [PubMed]
129. Karle, K.N.; Möckel, D.; Reid, E.; Schöls, L. Axonal transport deficit in a KIF5A(-/-) mouse model. *Neurogenetics* **2012**, *13*, 169–179. [CrossRef]
130. Warita, H.; Itoyama, Y.; Abe, K. Selective impairment of fast anterograde axonal transport in the peripheral nerves of asymptomatic transgenic mice with a G93A mutant SOD1 gene. *Brain Res.* **1999**, *819*, 120–131. [CrossRef]
131. Fischer, L.R.; Culver, D.G.; Tennant, P.; Davis, A.A.; Wang, M.; Castellano-Sanchez, A.; Khan, J.; Polak, M.A.; Glass, J.D. Amyotrophic lateral sclerosis is a distal axonopathy: Evidence in mice and man. *Exp. Neurol.* **2004**, *185*, 232–240. [CrossRef]
132. Tallon, C.; Russell, K.A.; Sakhalkar, S.; Andrapallayal, N.; Farah, M.H. Length-dependent axo-terminal degeneration at the neuromuscular synapses of type II muscle in SOD1 mice. *Neuroscience* **2016**, *312*, 179–189. [CrossRef]
133. Pantelidou, M.; Zographos, S.E.; Lederer, C.W.; Kyriakides, T.; Pfaffl, M.W.; Santama, N. Differential expression of molecular motors in the motor cortex of sporadic ALS. *Neurobiol. Dis.* **2007**, *26*, 577–589. [CrossRef]
134. Landers, J.E.; Melki, J.; Meininger, V.; Glass, J.D.; van den Berg, L.H.; van Es, M.A.; Sapp, P.C.; van Vught, P.W.J.; McKenna-Yasek, D.M.; Blauw, H.M.; et al. Reduced expression of the Kinesin-Associated Protein 3 (KIFAP3) gene increases survival in sporadic amyotrophic lateral sclerosis. *Proc. Natl. Acad. Sci. USA* **2009**, *106*, 9004–9009. [CrossRef] [PubMed]
135. Ferraiuolo, L.; Kirby, J.; Grierson, A.J.; Sendtner, M.; Shaw, P.J. Molecular pathways of motor neuron injury in amyotrophic lateral sclerosis. *Nat. Rev. Neurol.* **2011**, *7*, 616–630. [CrossRef] [PubMed]
136. Pitt, D.; Werner, P.; Raine, C.S. Glutamate excitotoxicity in a model of multiple sclerosis. *Nat. Med.* **2000**, *6*, 67–70. [CrossRef] [PubMed]
137. Vaarmann, A.; Kovac, S.; Holmström, K.M.; Gandhi, S.; Abramov, A.Y. Dopamine protects neurons against glutamate-induced excitotoxicity. *Cell Death Dis.* **2013**, *4*, e455. [CrossRef]
138. Rothstein, J.D.; Tsai, G.; Kuncl, R.W.; Clawson, L.; Cornblath, D.R.; Drachman, D.B.; Pestronk, A.; Stauch, B.L.; Coyle, J.T. Abnormal excitatory amino acid metabolism in amyotrophic lateral sclerosis. *Ann. Neurol.* **1990**, *28*, 18–25. [CrossRef]
139. Shaw, P.J.; Forrest, V.; Ince, P.G.; Richardson, J.P.; Wastell, H.J. CSF and Plasma Amino Acid Levels in Motor Neuron Disease: Elevation of CSF Glutamate in a Subset of Patients. *Neurodegeneration* **1995**, *4*, 209–216. [CrossRef]
140. Wang, Y.; Qin, Z. Molecular and cellular mechanisms of excitotoxic neuronal death. *Apoptosis* **2010**, *15*, 1382–1402. [CrossRef]
141. Netzahualcoyotzi, C.; Tapia, R. Degeneration of spinal motor neurons by chronic AMPA-induced excitotoxicity in vivo and protection by energy substrates. *Acta Neuropathol. Commun.* **2015**, *3*, 27. [CrossRef]
142. Van Den Bosch, L.; Vandenberghe, W.; Klaassen, H.; Van Houtte, E.; Robberecht, W. Ca2+-permeable AMPA receptors and selective vulnerability of motor neurons. *J. Neurol. Sci.* **2000**, *180*, 29–34. [CrossRef]
143. Gregory, J.M.; Livesey, M.R.; McDade, K.; Selvaraj, B.T.; Barton, S.K.; Chandran, S.; Smith, C. Dysregulation of AMPA receptor subunit expression in sporadic ALS post-mortem brain. *J. Pathol.* **2020**, *250*, 67–78. [CrossRef]
144. Traynelis, S.F.; Wollmuth, L.P.; McBain, C.J.; Menniti, F.S.; Vance, K.M.; Ogden, K.K.; Hansen, K.B.; Yuan, H.; Myers, S.J.; Dingledine, R. Glutamate Receptor Ion Channels: Structure, Regulation, and Function. *Pharmacol. Rev.* **2010**, *62*, 405–496. [CrossRef] [PubMed]

145. Corona, J.C.; Tapia, R. AMPA receptor activation, but not the accumulation of endogenous extracellular glutamate, induces paralysis and motor neuron death in rat spinal cord in vivo. *J. Neurochem.* **2004**, *89*, 988–997. [CrossRef] [PubMed]
146. Konen, L.M.; Wright, A.L.; Royle, G.A.; Morris, G.P.; Lau, B.K.; Seow, P.W.; Zinn, R.; Milham, L.T.; Vaughan, C.W.; Vissel, B. A new mouse line with reduced GluA2 Q/R site RNA editing exhibits loss of dendritic spines, hippocampal CA1-neuron loss, learning and memory impairments and NMDA receptor-independent seizure vulnerability. *Mol. Brain* **2020**, *13*, 27. [CrossRef] [PubMed]
147. Heath, P.R.; Tomkins, J.; Ince, P.G.; Shaw, P.J. Quantitative assessment of AMPA receptor mRNA in human spinal motor neurons isolated by laser capture microdissection. *Neuroreport* **2002**, *13*, 1753–1757. [CrossRef] [PubMed]
148. Takuma, H.; Kwak, S.; Yoshizawa, T.; Kanazawa, I. Reduction of GluR2 RNA editing, a molecular change that increases calcium influx through AMPA receptors, selective in the spinal ventral gray of patients with amyotrophic lateral sclerosis. *Ann. Neurol.* **1999**, *46*, 806–815. [CrossRef]
149. Ince, P.; Stout, N.; Shaw, P.; Slade, J.; Hunziker, W.; Heizmann, C.W.; Baimbridge, K.G. Parvalbumin and calbindin D-28k in the human motor system and in motor neuron disease. *Neuropathol. Appl. Neurobiol.* **1993**, *19*, 291–299. [CrossRef]
150. Higuchi, M.; Maas, S.; Single, F.N.; Hartner, J.; Rozov, A.; Burnashev, N.; Feldmeyer, D.; Sprengel, R.; Seeburg, P.H. Point mutation in an AMPA receptor gene rescues lethality in mice deficient in the RNA-editing enzyme ADAR2. *Nature* **2000**, *406*, 78–81. [CrossRef]
151. Hideyama, T.; Kwak, S. When Does ALS Start? ADAR2?GluA2 Hypothesis for the Etiology of Sporadic ALS. *Front. Mol. Neurosci.* **2011**, *4*. [CrossRef]
152. Aizawa, H.; Sawada, J.; Hideyama, T.; Yamashita, T.; Katayama, T.; Hasebe, N.; Kimura, T.; Yahara, O.; Kwak, S. TDP-43 pathology in sporadic ALS occurs in motor neurons lacking the RNA editing enzyme ADAR2. *Acta Neuropathol.* **2010**, *120*, 75–84. [CrossRef]
153. Medina, L.; Figueredo-Cardenas, G.; Rothstein, J.D.; Reiner, A. Differential Abundance of Glutamate Transporter Subtypes in Amyotrophic Lateral Sclerosis (ALS)-Vulnerable versus ALS-Resistant Brain Stem Motor Cell Groups. *Exp. Neurol.* **1996**, *142*, 287–295. [CrossRef]
154. Rothstein, J.D.; Dykes-Hoberg, M.; Pardo, C.A.; Bristol, L.A.; Jin, L.; Kuncl, R.W.; Kanai, Y.; Hediger, M.A.; Wang, Y.; Schielke, J.P.; et al. Knockout of glutamate transporters reveals a major role for astroglial transport in excitotoxicity and clearance of glutamate. *Neuron* **1996**, *16*, 675–686. [CrossRef]
155. Kong, Q.; Chang, L.-C.; Takahashi, K.; Liu, Q.; Schulte, D.A.; Lai, L.; Ibabao, B.; Lin, Y.; Stouffer, N.; Mukhopadhyay, C.D.; et al. Small-molecule activator of glutamate transporter EAAT2 translation provides neuroprotection. *J. Clin. Investig.* **2014**, *124*, 1255–1267. [CrossRef] [PubMed]
156. Rothstein, J.D.; Patel, S.; Regan, M.R.; Haenggeli, C.; Huang, Y.H.; Bergles, D.E.; Jin, L.; Dykes Hoberg, M.; Vidensky, S.; Chung, D.S.; et al. β-Lactam antibiotics offer neuroprotection by increasing glutamate transporter expression. *Nature* **2005**, *433*, 73–77. [CrossRef]
157. Bristol, L.A.; Rothstein, J.D. Glutamate transporter gene expression in amyotrophic lateral sclerosis motor cortex. *Ann. Neurol.* **1996**, *39*, 676–679. [CrossRef] [PubMed]
158. Sasaki, S.; Komori, T.; Iwata, M. Excitatory amino acid transporter 1 and 2 immunoreactivity in the spinal cord in amyotrophic lateral sclerosis. *Acta Neuropathol.* **2000**, *100*, 138–144. [CrossRef] [PubMed]
159. Milanese, M.; Zappettini, S.; Onofri, F.; Musazzi, L.; Tardito, D.; Bonifacino, T.; Messa, M.; Racagni, G.; Usai, C.; Benfenati, F.; et al. Abnormal exocytotic release of glutamate in a mouse model of amyotrophic lateral sclerosis. *J. Neurochem.* **2011**, *116*, 1028–1042. [CrossRef] [PubMed]
160. van Zundert, B.; Peuscher, M.H.; Hynynen, M.; Chen, A.; Neve, R.L.; Brown, R.H.; Constantine-Paton, M.; Bellingham, M.C. Neonatal Neuronal Circuitry Shows Hyperexcitable Disturbance in a Mouse Model of the Adult-Onset Neurodegenerative Disease Amyotrophic Lateral Sclerosis. *J. Neurosci.* **2008**, *28*, 10864–10874. [CrossRef]
161. Rothstein, J.D.; Martin, L.J.; Kuncl, R.W. Decreased Glutamate Transport by the Brain and Spinal Cord in Amyotrophic Lateral Sclerosis. *N. Engl. J. Med.* **1992**, *326*, 1464–1468. [CrossRef]
162. Vucic, S.; Nicholson, G.A.; Kiernan, M.C. Cortical hyperexcitability may precede the onset of familial amyotrophic lateral sclerosis. *Brain* **2008**, *131*, 1540–1550. [CrossRef]
163. Perry, T.L.; Krieger, C.; Hansen, S.; Eisen, A. Amyotrophic lateral sclerosis: Amino acid levels in plasma and cerebrospinal fluid. *Ann. Neurol.* **1990**, *28*, 12–17. [CrossRef]

164. Perry, T.L.; Hansen, S.; Jones, K. Brain glutamate deficiency in amyotrophic lateral sclerosis. *Neurology* **1987**, *37*, 1845–1848. [CrossRef] [PubMed]
165. Wuolikainen, A.; Moritz, T.; Marklund, S.L.; Antti, H.; Andersen, P.M. Disease-related changes in the cerebrospinal fluid metabolome in amyotrophic lateral sclerosis detected by GC/TOFMS. *PLoS ONE* **2011**, *6*, e17947. [CrossRef] [PubMed]
166. Shi, Y.; Hung, S.-T.; Rocha, G.; Lin, S.; Linares, G.R.; Staats, K.A.; Seah, C.; Wang, Y.; Chickering, M.; Lai, J.; et al. Identification and therapeutic rescue of autophagosome and glutamate receptor defects in C9ORF72 and sporadic ALS neurons. *JCI Insight* **2019**, *4*. [CrossRef] [PubMed]
167. Selvaraj, B.T.; Livesey, M.R.; Zhao, C.; Gregory, J.M.; James, O.T.; Cleary, E.M.; Chouhan, A.K.; Gane, A.B.; Perkins, E.M.; Dando, O.; et al. C9ORF72 repeat expansion causes vulnerability of motor neurons to Ca2+-permeable AMPA receptor-mediated excitotoxicity. *Nat. Commun.* **2018**, *9*, 347. [CrossRef] [PubMed]
168. Shi, Y.; Lin, S.; Staats, K.A.; Li, Y.; Chang, W.-H.; Hung, S.-T.; Hendricks, E.; Linares, G.R.; Wang, Y.; Son, E.Y.; et al. Haploinsufficiency leads to neurodegeneration in C9ORF72 ALS/FTD human induced motor neurons. *Nat. Med.* **2018**, *24*, 313–325. [CrossRef]
169. Xiao, S.; McKeever, P.M.; Lau, A.; Robertson, J. Synaptic localization of C9orf72 regulates post-synaptic glutamate receptor 1 levels. *Acta Neuropathol. Commun.* **2019**, *7*, 161. [CrossRef]
170. Tang, D.; Sheng, J.; Xu, L.; Zhan, X.; Liu, J.; Jiang, H.; Shu, X.; Liu, X.; Zhang, T.; Jiang, L.; et al. Cryo-EM structure of C9ORF72–SMCR8–WDR41 reveals the role as a GAP for Rab8a and Rab11a. *Proc. Natl. Acad. Sci. USA* **2020**, *117*, 9876–9883. [CrossRef]
171. Lai, C.; Xie, C.; Shim, H.; Chandran, J.; Howell, B.W.; Cai, H. Regulation of endosomal motility and degradation by amyotrophic lateral sclerosis 2/alsin. *Mol. Brain* **2009**, *2*, 23. [CrossRef]
172. Blasco, H.; Mavel, S.; Corcia, P.; Gordon, P.H. The glutamate hypothesis in ALS: Pathophysiology and drug development. *Curr. Med. Chem.* **2014**, *21*, 3551–3575. [CrossRef]
173. Petrov, D.; Mansfield, C.; Moussy, A.; Hermine, O. ALS clinical trials review: 20 years of failure. Are we any closer to registering a new treatment? *Front. Aging Neurosci.* **2017**, *9*, 68. [CrossRef]
174. Thorley, M.; Malatras, A.; Duddy, W.J.; Le Gall, L.; Mouly, V.; Butler Browne, G.; Duguez, S.M.-R. Changes in communication between muscle stem cells and their environment with aging. *J. Neuromuscul. Dis.* **2015**, *2*. [CrossRef] [PubMed]
175. Paolicelli, R.C.; Bergamini, G.; Rajendran, L. Cell-to-cell Communication by Extracellular Vesicles: Focus on Microglia. *Neuroscience* **2019**, *405*, 148–157. [CrossRef] [PubMed]
176. Théry, C. Exosomes: Secreted vesicles and intercellular communications. *F1000 Biol. Rep.* **2011**, *3*, 15. [CrossRef] [PubMed]
177. Raposo, G.; Stoorvogel, W. Extracellular vesicles: Exosomes, microvesicles, and friends. *J. Cell Biol.* **2013**, *200*, 373–383. [CrossRef]
178. Basso, M.; Pozzi, S.; Tortarolo, M.; Fiordaliso, F.; Bisighini, C.; Pasetto, L.; Spaltro, G.; Lidonnici, D.; Gensano, F.; Battaglia, E.; et al. Mutant copper-zinc superoxide dismutase (SOD1) induces protein secretion pathway alterations and exosome release in astrocytes: Implications for disease spreading and motor neuron pathology in amyotrophic lateral sclerosis. *J. Biol. Chem.* **2013**, *288*, 15699–15711. [CrossRef]
179. Silverman, J.M.; Christy, D.; Shyu, C.C.; Moon, K.-M.; Fernando, S.; Gidden, Z.; Cowan, C.M.; Ban, Y.; Stacey, R.G.; Grad, L.I.; et al. CNS-derived extracellular vesicles from superoxide dismutase 1 (SOD1)G93A ALS mice originate from astrocytes and neurons and carry misfolded SOD1. *J. Biol. Chem.* **2019**, *294*, 3744–3759. [CrossRef]
180. Sproviero, D.; La Salvia, S.; Colombo, F.; Zucca, S.; Pansarasa, O.; Diamanti, L.; Costa, A.; Lova, L.; Giannini, M.; Gagliardi, S.; et al. Leukocyte Derived Microvesicles as Disease Progression Biomarkers in Slow Progressing Amyotrophic Lateral Sclerosis Patients. *Front. Neurosci.* **2019**, *13*, 344. [CrossRef]
181. Chen, Y.; Xia, K.; Chen, L.; Fan, D. Increased Interleukin-6 Levels in the Astrocyte-Derived Exosomes of Sporadic Amyotrophic Lateral Sclerosis Patients. *Front. Neurosci.* **2019**, *13*, 574. [CrossRef]
182. Westergard, T.; Jensen, B.K.; Wen, X.; Cai, J.; Kropf, E.; Iacovitti, L.; Pasinelli, P.; Trotti, D. Cell-to-Cell Transmission of Dipeptide Repeat Proteins Linked to C9orf72 -ALS/FTD. *Cell Rep.* **2016**, *17*, 645–652. [CrossRef]
183. Scaramozza, A.; Marchese, V.; Papa, V.; Salaroli, R.; Soraru, G.; Angelini, C.; Cenacchi, G. Skeletal Muscle Satellite Cells in Amyotrophic Lateral Sclerosis. *Ultrastruct. Pathol.* **2014**, *38*, 1–8. [CrossRef]

184. Parkinson, N.; Ince, P.G.; Smith, M.O.; Highley, R.; Skibinski, G.; Andersen, P.M.; Morrison, K.E.; Pall, H.S.; Hardiman, O.; Collinge, J.; et al. ALS phenotypes with mutations in CHMP2B (charged multivesicular body protein 2B). *Neurology* **2006**, *67*, 1074–1077. [CrossRef] [PubMed]
185. Skibinski, G.; Parkinson, N.J.; Brown, J.M.; Chakrabarti, L.; Lloyd, S.L.; Hummerich, H.; Nielsen, J.E.; Hodges, J.R.; Spillantini, M.G.; Thusgaard, T.; et al. Mutations in the endosomal ESCRTIII-complex subunit CHMP2B in frontotemporal dementia. *Nat. Genet.* **2005**, *37*, 806–808. [CrossRef] [PubMed]
186. Blanc, L.; Vidal, M. New insights into the function of Rab GTPases in the context of exosomal secretion. *Small GTPases* **2018**, *9*, 95–106. [CrossRef]
187. Yang, Y.; Hentati, A.; Deng, H.X.; Dabbagh, O.; Sasaki, T.; Hirano, M.; Hung, W.Y.; Ouahchi, K.; Yan, J.; Azim, A.C.; et al. The gene encoding alsin, a protein with three guanine-nucleotide exchange factor domains, is mutated in a form of recessive amyotrophic lateral sclerosis. *Nat. Genet.* **2001**, *29*, 160–165. [CrossRef]
188. Sellier, C.; Campanari, M.-L.; Julie Corbier, C.; Gaucherot, A.; Kolb-Cheynel, I.; Oulad-Abdelghani, M.; Ruffenach, F.; Page, A.; Ciura, S.; Kabashi, E.; et al. Loss of C9ORF72 impairs autophagy and synergizes with polyQ Ataxin-2 to induce motor neuron dysfunction and cell death. *EMBO J.* **2016**, *35*, 1276–1297. [CrossRef] [PubMed]
189. Farg, M.A.; Sundaramoorthy, V.; Sultana, J.M.; Yang, S.; Atkinson, R.A.K.; Levina, V.; Halloran, M.A.; Gleeson, P.A.; Blair, I.P.; Soo, K.Y.; et al. C9ORF72, implicated in amytrophic lateral sclerosis and frontotemporal dementia, regulates endosomal trafficking. *Hum. Mol. Genet.* **2014**, *23*, 3579–3595. [CrossRef] [PubMed]
190. Sato, T.; Iwano, T.; Kunii, M.; Matsuda, S.; Mizuguchi, R.; Jung, Y.; Hagiwara, H.; Yoshihara, Y.; Yuzaki, M.; Harada, R.; et al. Rab8a and Rab8b are essential for several apical transport pathways but insufficient for ciliogenesis. *J. Cell Sci.* **2014**, *127*, 422–431. [CrossRef] [PubMed]
191. Seto, S.; Sugaya, K.; Tsujimura, K.; Nagata, T.; Horii, T.; Koide, Y. Rab39a interacts with phosphatidylinositol 3-kinase and negatively regulates autophagy induced by lipopolysaccharide stimulation in macrophages. *PLoS ONE* **2013**, *8*, e83324. [CrossRef] [PubMed]
192. Pilli, M.; Arko-Mensah, J.; Ponpuak, M.; Roberts, E.; Master, S.; Mandell, M.A.; Dupont, N.; Ornatowski, W.; Jiang, S.; Bradfute, S.B.; et al. TBK-1 promotes autophagy-mediated antimicrobial defense by controlling autophagosome maturation. *Immunity* **2012**, *37*, 223–234. [CrossRef]
193. O'Rourke, J.G.; Bogdanik, L.; Yáñez, A.; Lall, D.; Wolf, A.J.; Muhammad, A.K.M.G.; Ho, R.; Carmona, S.; Vit, J.P.; Zarrow, J.; et al. C9orf72 is required for proper macrophage and microglial function in mice. *Science* **2016**, *351*, 1324–1329. [CrossRef]
194. Aoki, Y.; Manzano, R.; Lee, Y.; Dafinca, R.; Aoki, M.; Douglas, A.G.L.; Varela, M.A.; Sathyaprakash, C.; Scaber, J.; Barbagallo, P.; et al. C9orf72 and RAB7L1 regulate vesicle trafficking in amyotrophic lateral sclerosis and frontotemporal dementia. *Brain* **2017**, *140*, 887–897. [CrossRef] [PubMed]
195. Hantan, D.; Yamamoto, Y.; Sakisaka, T. VAP-B binds to Rab3GAP1 at the ER: Its implication in nuclear envelope formation through the ER-Golgi intermediate compartment. *Kobe J. Med. Sci.* **2014**, *60*, E48–E56. [PubMed]
196. Lev, S.; Ben Halevy, D.; Peretti, D.; Dahan, N. The VAP protein family: From cellular functions to motor neuron disease. *Trends Cell Biol.* **2008**, *18*, 282–290. [CrossRef] [PubMed]
197. Rocha, N.; Kuijl, C.; van der Kant, R.; Janssen, L.; Houben, D.; Janssen, H.; Zwart, W.; Neefjes, J. Cholesterol sensor ORP1L contacts the ER protein VAP to control Rab7–RILP–p150Glued and late endosome positioning. *J. Cell Biol.* **2009**, *185*, 1209–1225. [CrossRef] [PubMed]
198. Buratta, S.; Tancini, B.; Sagini, K.; Delo, F.; Chiaradia, E.; Urbanelli, L.; Emiliani, C. Lysosomal Exocytosis, Exosome Release and Secretory Autophagy: The Autophagic- and Endo-Lysosomal Systems Go Extracellular. *Int. J. Mol. Sci.* **2020**, *21*, 2576. [CrossRef]
199. Johnson, J.O.; Mandrioli, J.; Benatar, M.; Abramzon, Y.; Van Deerlin, V.M.; Trojanowski, J.Q.; Gibbs, J.R.; Brunetti, M.; Gronka, S.; Wuu, J.; et al. Exome Sequencing Reveals VCP Mutations as a Cause of Familial ALS. *Neuron* **2010**, *68*, 857–864. [CrossRef]
200. Bug, M.; Meyer, H. Expanding into new markets–VCP/p97 in endocytosis and autophagy. *J. Struct. Biol.* **2012**, *179*, 78–82. [CrossRef]
201. Chow, C.Y.; Landers, J.E.; Bergren, S.K.; Sapp, P.C.; Grant, A.E.; Jones, J.M.; Everett, L.; Lenk, G.M.; McKenna-Yasek, D.M.; Weisman, L.S.; et al. Deleterious variants of FIG4, a phosphoinositide phosphatase, in patients with ALS. *Am. J. Hum. Genet.* **2009**, *84*, 85–88. [CrossRef]

202. Martyn, C.; Li, J. Fig4 deficiency: A newly emerged lysosomal storage disorder? *Prog. Neurobiol.* **2013**, *101–102*, 35–45. [CrossRef]
203. Branchu, J.; Boutry, M.; Sourd, L.; Depp, M.; Leone, C.; Corriger, A.; Vallucci, M.; Esteves, T.; Matusiak, R.; Dumont, M.; et al. Loss of spatacsin function alters lysosomal lipid clearance leading to upper and lower motor neuron degeneration. *Neurobiol. Dis.* **2017**, *102*, 21–37. [CrossRef]
204. Pérez-Branguli, F.; Mishra, H.K.; Prots, I.; Havlicek, S.; Kohl, Z.; Saul, D.; Rummel, C.; Dorca-Arevalo, J.; Regensburger, M.; Graef, D.; et al. Dysfunction of spatacsin leads to axonal pathology in SPG11-linked hereditary spastic paraplegia. *Hum. Mol. Genet.* **2014**, *23*, 4859–4874. [CrossRef] [PubMed]
205. Neumann, M.; Sampathu, D.M.; Kwong, L.K.; Truax, A.C.; Micsenyi, M.C.; Chou, T.T.; Bruce, J.; Schuck, T.; Grossman, M.; Clark, C.M.; et al. Ubiquitinated TDP-43 in frontotemporal lobar degeneration and amyotrophic lateral sclerosis. *Science* **2006**, *314*, 130–133. [CrossRef] [PubMed]
206. Gill, C.; Phelan, J.P.; Hatzipetros, T.; Kidd, J.D.; Tassinari, V.R.; Levine, B.; Wang, M.Z.; Moreno, A.; Thompson, K.; Maier, M.; et al. SOD1-positive aggregate accumulation in the CNS predicts slower disease progression and increased longevity in a mutant SOD1 mouse model of ALS. *Sci. Rep.* **2019**. [CrossRef] [PubMed]
207. Feneberg, E.; Gray, E.; Ansorge, O.; Talbot, K.; Turner, M.R. Towards a TDP-43-Based Biomarker for ALS and FTLD. *Mol. Neurobiol.* **2018**, *55*, 7789–7801. [CrossRef] [PubMed]
208. Zhang, Y.-J.; Jansen-West, K.; Xu, Y.-F.; Gendron, T.F.; Bieniek, K.F.; Lin, W.-L.; Sasaguri, H.; Caulfield, T.; Hubbard, J.; Daughrity, L.; et al. Aggregation-prone c9FTD/ALS poly(GA) RAN-translated proteins cause neurotoxicity by inducing ER stress. *Acta Neuropathol.* **2014**, *128*, 505–524. [CrossRef]
209. Picchiarelli, G.; Dupuis, L. Role of RNA Binding Proteins with prion-like domains in muscle and neuromuscular diseases. *Cell Stress* **2020**, *4*, 76–91. [CrossRef]
210. Hirano, A. Familial Amyotrophic Lateral Sclerosis. *Arch. Neurol.* **1967**, *16*, 232. [CrossRef]
211. Sun, C.N.; Araoz, C.; Lucas, G.; Morgan, P.N.; White, H.J. Amyotrophic lateral sclerosis. Inclusion bodies in a case of the classic sporadic form. *Ann. Clin. Lab. Sci.* **1975**, *5*, 38–44.
212. Takahashi, K. Hereditary Amyotrophic Lateral Sclerosis. *Arch. Neurol.* **1972**, *27*, 292. [CrossRef]
213. Shibata, N.; Hirano, A.; Kobayashi, M.; Siddique, T.; Deng, H.-X.; Hung, W.-Y.; Kato, T.; Asayama, K. Intense Superoxide Dismutase-1 Immunoreactivity in Intracytoplasmic Hyaline Inclusions of Familial Amyotrophic Lateral Sclerosis with Posterior Column Involvement. *J. Neuropathol. Exp. Neurol.* **1996**, *55*, 481–490. [CrossRef]
214. Matsumoto, G.; Wada, K.; Okuno, M.; Kurosawa, M.; Nukina, N. Serine 403 Phosphorylation of p62/SQSTM1 Regulates Selective Autophagic Clearance of Ubiquitinated Proteins. *Mol. Cell* **2011**, *44*, 279–289. [CrossRef]
215. Ling, S.-C.; Polymenidou, M.; Cleveland, D.W. Converging Mechanisms in ALS and FTD: Disrupted RNA and Protein Homeostasis. *Neuron* **2013**, *79*, 416–438. [CrossRef]
216. Mackenzie, I.R.A.; Bigio, E.H.; Ince, P.G.; Geser, F.; Neumann, M.; Cairns, N.J.; Kwong, L.K.; Forman, M.S.; Ravits, J.; Stewart, H.; et al. Pathological TDP-43 distinguishes sporadic amyotrophic lateral sclerosis from amyotrophic lateral sclerosis with SOD1 mutations. *Ann. Neurol.* **2007**, *61*, 427–434. [CrossRef] [PubMed]
217. Deng, H.-X.; Chen, W.; Hong, S.-T.; Boycott, K.M.; Gorrie, G.H.; Siddique, N.; Yang, Y.; Fecto, F.; Shi, Y.; Zhai, H.; et al. Mutations in UBQLN2 cause dominant X-linked juvenile and adult-onset ALS and ALS/dementia. *Nature* **2011**, *477*, 211–215. [CrossRef] [PubMed]
218. Levy, J.R.; Sumner, C.J.; Caviston, J.P.; Tokito, M.K.; Ranganathan, S.; Ligon, L.A.; Wallace, K.E.; LaMonte, B.H.; Harmison, G.G.; Puls, I.; et al. A motor neuron disease–associated mutation in p150Glued perturbs dynactin function and induces protein aggregation. *J. Cell Biol.* **2006**, *172*, 733–745. [CrossRef] [PubMed]
219. Ayaki, T.; Ito, H.; Fukushima, H.; Inoue, T.; Kondo, T.; Ikemoto, A.; Asano, T.; Shodai, A.; Fujita, T.; Fukui, S.; et al. Immunoreactivity of valosin-containing protein in sporadic amyotrophic lateral sclerosis and in a case of its novel mutant. *Acta Neuropathol. Commun.* **2014**, *2*, 172. [CrossRef] [PubMed]
220. Johnson, J.O.; Pioro, E.P.; Boehringer, A.; Chia, R.; Feit, H.; Renton, A.E.; Pliner, H.A.; Abramzon, Y.; Marangi, G.; Winborn, B.J.; et al. Mutations in the Matrin 3 gene cause familial amyotrophic lateral sclerosis. *Nat. Neurosci.* **2014**, *17*, 664–666. [CrossRef]
221. Chattopadhyay, M.; Nwadibia, E.; Strong, C.D.; Gralla, E.B.; Valentine, J.S.; Whitelegge, J.P. The Disulfide Bond, but Not Zinc or Dimerization, Controls Initiation and Seeded Growth in Amyotrophic Lateral Sclerosis-linked Cu,Zn Superoxide Dismutase (SOD1) Fibrillation. *J. Biol. Chem.* **2015**, *290*, 30624–30636. [CrossRef]

222. Lang, L.; Kurnik, M.; Danielsson, J.; Oliveberg, M. Fibrillation precursor of superoxide dismutase 1 revealed by gradual tuning of the protein-folding equilibrium. *Proc. Natl. Acad. Sci. USA* **2012**, *109*, 17868–17873. [CrossRef]
223. Chan, P.K.; Chattopadhyay, M.; Sharma, S.; Souda, P.; Gralla, E.B.; Borchelt, D.R.; Whitelegge, J.P.; Valentine, J.S. Structural similarity of wild-type and ALS-mutant superoxide dismutase-1 fibrils using limited proteolysis and atomic force microscopy. *Proc. Natl. Acad. Sci. USA* **2013**, *110*, 10934–10939. [CrossRef]
224. Ivanova, M.I.; Sievers, S.A.; Guenther, E.L.; Johnson, L.M.; Winkler, D.D.; Galaleldeen, A.; Sawaya, M.R.; Hart, P.J.; Eisenberg, D.S. Aggregation-triggering segments of SOD1 fibril formation support a common pathway for familial and sporadic ALS. *Proc. Natl. Acad. Sci. USA* **2014**, *111*, 197–201. [CrossRef] [PubMed]
225. Abdolvahabi, A.; Shi, Y.; Chuprin, A.; Rasouli, S.; Shaw, B.F. Stochastic Formation of Fibrillar and Amorphous Superoxide Dismutase Oligomers Linked to Amyotrophic Lateral Sclerosis. *ACS Chem. Neurosci.* **2016**, *7*, 799–810. [CrossRef] [PubMed]
226. McAlary, L.; Aquilina, J.A.; Yerbury, J.J. Susceptibility of Mutant SOD1 to Form a Destabilized Monomer Predicts Cellular Aggregation and Toxicity but Not In vitro Aggregation Propensity. *Front. Neurosci.* **2016**, *10*. [CrossRef] [PubMed]
227. Wright, G.S.A.; Antonyuk, S.V.; Hasnain, S.S. A faulty interaction between SOD1 and hCCS in neurodegenerative disease. *Sci. Rep.* **2016**, *6*, 27691. [CrossRef] [PubMed]
228. Elam, J.S.; Taylor, A.B.; Strange, R.; Antonyuk, S.; Doucette, P.A.; Rodriguez, J.A.; Hasnain, S.S.; Hayward, L.J.; Valentine, J.S.; Yeates, T.O.; et al. Amyloid-like filaments and water-filled nanotubes formed by SOD1 mutant proteins linked to familial ALS. *Nat. Struct. Biol.* **2003**, *10*, 461–467. [CrossRef]
229. Ayers, J.I.; Cashman, N.R. Prion-like mechanisms in amyotrophic lateral sclerosis. In *Handbook of Clinical Neurology*; Elsevier: Amsterdam, The Netherlands, 2018; pp. 337–354.
230. Baumer, K.M.; Koone, J.C.; Shaw, B.F. Kinetic Variability in Seeded Formation of ALS-Linked SOD1 Fibrils Across Multiple Generations. *ACS Chem. Neurosci.* **2020**, *11*, 304–313. [CrossRef]
231. Healy, E.F.; Roth-Rodriguez, A.; Toledo, S. A model for gain of function in superoxide dismutase. *Biochem. Biophys. Rep.* **2020**, *21*, 100728. [CrossRef]
232. Vogler, T.O.; Wheeler, J.R.; Nguyen, E.D.; Hughes, M.P.; Britson, K.A.; Lester, E.; Rao, B.; Betta, N.D.; Whitney, O.N.; Ewachiw, T.E.; et al. TDP-43 and RNA form amyloid-like myo-granules in regenerating muscle. *Nature* **2018**, *563*, 508–513. [CrossRef]
233. Rhoads, S.N.; Monahan, Z.T.; Yee, D.S.; Shewmaker, F.P. The Role of Post-Translational Modifications on Prion-Like Aggregation and Liquid-Phase Separation of FUS. *Int. J. Mol. Sci.* **2018**, *19*, 886. [CrossRef]
234. Pokrishevsky, E.; Grad, L.I.; Cashman, N.R. TDP-43 or FUS-induced misfolded human wild-type SOD1 can propagate intercellularly in a prion-like fashion. *Sci. Rep.* **2016**, *6*, 22155. [CrossRef]
235. Sproviero, D.; La Salvia, S.; Giannini, M.; Crippa, V.; Gagliardi, S.; Bernuzzi, S.; Diamanti, L.; Ceroni, M.; Pansarasa, O.; Poletti, A.; et al. Pathological Proteins Are Transported by Extracellular Vesicles of Sporadic Amyotrophic Lateral Sclerosis Patients. *Front. Neurosci.* **2018**, *12*, 487. [CrossRef] [PubMed]
236. Nishitoh, H.; Kadowaki, H.; Nagai, A.; Maruyama, T.; Yokota, T.; Fukutomi, H.; Noguchi, T.; Matsuzawa, A.; Takeda, K.; Ichijo, H. ALS-linked mutant SOD1 induces ER stress- and ASK1-dependent motor neuron death by targeting Derlin-1. *Genes Dev.* **2008**, *22*, 1451–1464. [CrossRef] [PubMed]
237. Vance, C.; Scotter, E.L.; Nishimura, A.L.; Troakes, C.; Mitchell, J.C.; Kathe, C.; Urwin, H.; Manser, C.; Miller, C.C.; Hortobágyi, T.; et al. ALS mutant FUS disrupts nuclear localization and sequesters wild-type FUS within cytoplasmic stress granules. *Hum. Mol. Genet.* **2013**, *22*, 2676–2688. [CrossRef] [PubMed]
238. Teyssou, E.; Chartier, L.; Amador, M.-D.-M.; Lam, R.; Lautrette, G.; Nicol, M.; Machat, S.; Da Barroca, S.; Moigneu, C.; Mairey, M.; et al. Novel UBQLN2 mutations linked to amyotrophic lateral sclerosis and atypical hereditary spastic paraplegia phenotype through defective HSP70-mediated proteolysis. *Neurobiol. Aging* **2017**, *58*, 239.e11–239.e20. [CrossRef] [PubMed]
239. Saeki, Y. Ubiquitin recognition by the proteasome. *J. Biochem.* **2017**, *161*, 113–124. [CrossRef]
240. Ryan, T.A.; Tumbarello, D.A. Optineurin: A Coordinator of Membrane-Associated Cargo Trafficking and Autophagy. *Front. Immunol.* **2018**, *9*, 1024. [CrossRef]
241. Katsuragi, Y.; Ichimura, Y.; Komatsu, M. p62/SQSTM1 functions as a signaling hub and an autophagy adaptor. *FEBS J.* **2015**, *282*, 4672–4678. [CrossRef]

242. Chen, H.-J.; Anagnostou, G.; Chai, A.; Withers, J.; Morris, A.; Adhikaree, J.; Pennetta, G.; de Belleroche, J.S. Characterization of the properties of a novel mutation in VAPB in familial amyotrophic lateral sclerosis. *J. Biol. Chem.* **2010**, *285*, 40266–40281. [CrossRef]
243. Tsai, P.C.; Liao, Y.C.; Chen, P.L.; Guo, Y.C.; Chen, Y.H.; Jih, K.Y.; Lin, K.P.; Soong, B.W.; Tsai, C.P.; Lee, Y.C. Investigating CCNF mutations in a Taiwanese cohort with amyotrophic lateral sclerosis. *Neurobiol. Aging* **2018**, *62*, 243–e1. [CrossRef]
244. Türk, M.; Haaker, G.; Winter, L.; Just, W.; Nickel, F.T.; Linker, R.A.; Chevessier, F.; Schröder, R. *C9ORF72*-ALS: P62- and ubiquitin-aggregation pathology in skeletal muscle. *Muscle Nerve* **2014**, *50*, 454–455. [CrossRef]
245. Shibata, N.; Hirano, A.; Kobayashi, M.; Sasaki, S.; Kato, T.; Matsumoto, S.; Shiozawa, Z.; Komori, T.; Ikemoto, A.; Umahara, T. Cu/Zn superoxide dismutase-like immunoreactivity in Lewy body-like inclusions of sporadic amyotrophic lateral sclerosis. *Neurosci. Lett.* **1994**, *179*, 149–152. [CrossRef]
246. Hjerpe, R.; Bett, J.S.; Keuss, M.J.; Solovyova, A.; McWilliams, T.G.; Johnson, C.; Sahu, I.; Varghese, J.; Wood, N.; Wightman, M.; et al. UBQLN2 Mediates Autophagy-Independent Protein Aggregate Clearance by the Proteasome. *Cell* **2016**, *166*, 935–949. [CrossRef] [PubMed]
247. Wen, X.; Tan, W.; Westergard, T.; Krishnamurthy, K.; Markandaiah, S.S.; Shi, Y.; Lin, S.; Shneider, N.A.; Monaghan, J.; Pandey, U.B.; et al. Antisense proline-arginine RAN dipeptides linked to C9ORF72-ALS/FTD form toxic nuclear aggregates that initiate in vitro and in vivo neuronal death. *Neuron* **2014**, *84*, 1213–1225. [CrossRef] [PubMed]
248. Kabashi, E.; Agar, J.N.; Taylor, D.M.; Minotti, S.; Durham, H.D. Focal dysfunction of the proteasome: A pathogenic factor in a mouse model of amyotrophic lateral sclerosis. *J. Neurochem.* **2004**, *89*, 1325–1335. [CrossRef] [PubMed]
249. Williams, K.L.; Warraich, S.T.; Yang, S.; Solski, J.A.; Fernando, R.; Rouleau, G.A.; Nicholson, G.A.; Blair, I.P. UBQLN2/ubiquilin 2 mutation and pathology in familial amyotrophic lateral sclerosis. *Neurobiol. Aging* **2012**, *33*, 2527.e3–2527.e10. [CrossRef] [PubMed]
250. Yu, L.; Chen, Y.; Tooze, S.A. Autophagy pathway: Cellular and molecular mechanisms. *Autophagy* **2018**, *14*, 207–215. [CrossRef]
251. de Araujo, M.E.G.; Liebscher, G.; Hess, M.W.; Huber, L.A. Lysosomal size matters. *Traffic* **2020**, *21*, 60–75. [CrossRef]
252. Wallroth, A.; Haucke, V. Phosphoinositide conversion in endocytosis and the endolysosomal system. *J. Biol. Chem.* **2018**, *293*, 1526–1535. [CrossRef]
253. Oakes, J.A.; Davies, M.C.; Collins, M.O. TBK1: A new player in ALS linking autophagy and neuroinflammation. *Mol. Brain* **2017**, *10*, 5. [CrossRef]
254. Goode, A.; Butler, K.; Long, J.; Cavey, J.; Scott, D.; Shaw, B.; Sollenberger, J.; Gell, C.; Johansen, T.; Oldham, N.J.; et al. Defective recognition of LC3B by mutant SQSTM1/p62 implicates impairment of autophagy as a pathogenic mechanism in ALS-FTLD. *Autophagy* **2016**, *12*, 1094–1104. [CrossRef]
255. Soo, K.Y.; Sultana, J.; King, A.; Atkinson, R.; Warraich, S.T.; Sundaramoorthy, V.; Blair, I.; Farg, M.A.; Atkin, J.D. ALS-associated mutant FUS inhibits macroautophagy which is restored by overexpression of Rab1. *Cell Death Discov.* **2015**, *1*, 15030. [CrossRef] [PubMed]
256. Sundaramoorthy, V.; Walker, A.K.; Tan, V.; Fifita, J.A.; McCann, E.P.; Williams, K.L.; Blair, I.P.; Guillemin, G.J.; Farg, M.A.; Atkin, J.D. Defects in optineurin- and myosin VI-mediated cellular trafficking in amyotrophic lateral sclerosis. *Hum. Mol. Genet.* **2015**. [CrossRef] [PubMed]
257. Barmada, S.J.; Serio, A.; Arjun, A.; Bilican, B.; Daub, A.; Ando, D.M.; Tsvetkov, A.; Pleiss, M.; Li, X.; Peisach, D.; et al. Autophagy induction enhances TDP43 turnover and survival in neuronal ALS models. *Nat. Chem. Biol.* **2014**, *10*, 677–685. [CrossRef] [PubMed]
258. Waite, A.J.; Bäumer, D.; East, S.; Neal, J.; Morris, H.R.; Ansorge, O.; Blake, D.J. Reduced C9orf72 protein levels in frontal cortex of amyotrophic lateral sclerosis and frontotemporal degeneration brain with the C9ORF72 hexanucleotide repeat expansion. *Neurobiol. Aging* **2014**, *35*, 1779.e5–1779.e13. [CrossRef] [PubMed]
259. Mathis, S.; Goizet, C.; Soulages, A.; Vallat, J.-M.; Le Masson, G. Genetics of amyotrophic lateral sclerosis: A review. *J. Neurol. Sci.* **2019**, *399*, 217–226. [CrossRef]
260. Allen, S.P.; Hall, B.; Castelli, L.M.; Francis, L.; Woof, R.; Siskos, A.P.; Kouloura, E.; Gray, E.; Thompson, A.G.; Talbot, K.; et al. Astrocyte adenosine deaminase loss increases motor neuron toxicity in amyotrophic lateral sclerosis. *Brain* **2019**, *142*, 586–605. [CrossRef]

261. Boivin, M.; Pfister, V.; Gaucherot, A.; Ruffenach, F.; Negroni, L.; Sellier, C.; Charlet-Berguerand, N. Reduced autophagy upon C9ORF72 loss synergizes with dipeptide repeat protein toxicity in G4C2 repeat expansion disorders. *EMBO J.* **2020**, *39*, e100574. [CrossRef]
262. Stolz, A.; Wolf, D.H. Endoplasmic reticulum associated protein degradation: A chaperone assisted journey to hell. *Biochim. Biophys. Acta Mol. Cell Res.* **2010**, *1803*, 694–705. [CrossRef]
263. Atkin, J.D.; Farg, M.A.; Walker, A.K.; McLean, C.; Tomas, D.; Horne, M.K. Endoplasmic reticulum stress and induction of the unfolded protein response in human sporadic amyotrophic lateral sclerosis. *Neurobiol. Dis.* **2008**, *30*, 400–407. [CrossRef]
264. Vijayalakshmi, K.; Alladi, P.A.; Ghosh, S.; Prasanna, V.K.; Sagar, B.C.; Nalini, A.; Sathyaprabha, T.N.; Raju, T.R. Evidence of endoplasmic reticular stress in the spinal motor neurons exposed to CSF from sporadic amyotrophic lateral sclerosis patients. *Neurobiol. Dis.* **2011**, *41*, 695–705. [CrossRef]
265. Yamanaka, T.; Nishiyama, R.; Shimogori, T.; Nukina, N. Proteomics-Based Approach Identifies Altered ER Domain Properties by ALS-Linked VAPB Mutation. *Sci. Rep.* **2020**, *10*, 7610. [CrossRef] [PubMed]
266. Tran, D.; Chalhoub, A.; Schooley, A.; Zhang, W.; Ngsee, J.K. A mutation in VAPB that causes amyotrophic lateral sclerosis also causes a nuclear envelope defect. *J. Cell Sci.* **2012**, *125*, 2831–2836. [CrossRef] [PubMed]
267. Cadoni, M.P.L.; Biggio, M.L.; Arru, G.; Secchi, G.; Orrù, N.; Clemente, M.G.; Sechi, G.; Yamoah, A.; Tripathi, P.; Orrù, S.; et al. VAPB ER-Aggregates, A Possible New Biomarker in ALS Pathology. *Cells* **2020**, *9*, 164. [CrossRef] [PubMed]
268. Schwartz, J.C.; Ebmeier, C.C.; Podell, E.R.; Heimiller, J.; Taatjes, D.J.; Cech, T.R. FUS binds the CTD of RNA polymerase II and regulates its phosphorylation at Ser2. *Genes Dev.* **2012**, *26*, 2690–2695. [CrossRef] [PubMed]
269. Buratti, E.; Baralle, F.E. Characterization and functional implications of the RNA binding properties of nuclear factor TDP-43, a novel splicing regulator of CFTR exon 9. *J. Biol. Chem.* **2001**, *276*, 36337–36343. [CrossRef] [PubMed]
270. Colombrita, C.; Onesto, E.; Megiorni, F.; Pizzuti, A.; Baralle, F.E.; Buratti, E.; Silani, V.; Ratti, A. TDP-43 and FUS RNA-binding proteins bind distinct sets of cytoplasmic messenger RNAs and differently regulate their post-transcriptional fate in motoneuron-like cells. *J. Biol. Chem.* **2012**. [CrossRef]
271. Godena, V.K.; Romano, G.; Romano, M.; Appocher, C.; Klima, R.; Buratti, E.; Baralle, F.E.; Feiguin, F. TDP-43 Regulates Drosophila Neuromuscular Junctions Growth by Modulating Futsch/MAP1B Levels and Synaptic Microtubules Organization. *PLoS ONE* **2011**, *6*, e17808. [CrossRef]
272. Herzog, J.J.; Deshpande, M.; Shapiro, L.; Rodal, A.A.; Paradis, S. TDP-43 misexpression causes defects in dendritic growth. *Sci. Rep.* **2017**, *7*, 15656. [CrossRef]
273. Feiguin, F.; Godena, V.K.; Romano, G.; D'Ambrogio, A.; Klima, R.; Baralle, F.E. Depletion of TDP-43 affects *Drosophila motoneurons* terminal synapsis and locomotive behavior. *FEBS Lett.* **2009**, *583*, 1586–1592. [CrossRef]
274. Reber, S.; Stettler, J.; Filosa, G.; Colombo, M.; Jutzi, D.; Lenzken, S.C.; Schweingruber, C.; Bruggmann, R.; Bachi, A.; Barabino, S.M.; et al. Minor intron splicing is regulated by FUS and affected by ALS-associated FUS mutants. *EMBO J.* **2016**, *35*, 1504–1521. [CrossRef]
275. Mori, K.; Lammich, S.; Mackenzie, I.R.A.; Forné, I.; Zilow, S.; Kretzschmar, H.; Edbauer, D.; Janssens, J.; Kleinberger, G.; Cruts, M.; et al. hnRNP A3 binds to GGGGCC repeats and is a constituent of p62-positive/TDP43-negative inclusions in the hippocampus of patients with C9orf72 mutations. *Acta Neuropathol.* **2013**, *125*, 413–423. [CrossRef] [PubMed]
276. Khosravi, B.; Hartmann, H.; May, S.; Mö hl, C.; Ederle, H.; Michaelsen, M.; Schludi, M.H.; Dormann, D.; Edbauer, D. Cytoplasmic poly-GA aggregates impair nuclear import of TDP-43 in C9orf72 ALS/FTLD. *Hum. Mol. Genet.* **2017**, *26*, 790–800. [CrossRef] [PubMed]
277. Chen, Y.P.; Huang, R.; Yang, Y.; Chen, K.; Song, W.; Pan, P.L.; Li, J.P.; Shang, H.F. Ataxin-2 intermediate-length polyglutamine: A possible risk factor for Chinese patients with amyotrophic lateral sclerosis. *Neurobiol. Aging* **2011**. [CrossRef]
278. Jutzi, D.; Akinyi, M.V.; Mechtersheimer, J.; Frilander, M.J.; Ruepp, M.-D. The emerging role of minor intron splicing in neurological disorders. *Cell Stress* **2018**, *2*, 40–54. [CrossRef] [PubMed]
279. Scotti, M.M.; Swanson, M.S. RNA mis-splicing in disease. *Nat. Rev. Genet.* **2016**, *17*, 19–32. [CrossRef]

280. Leblond, C.S.; Gan-Or, Z.; Spiegelman, D.; Laurent, S.B.; Szuto, A.; Hodgkinson, A.; Dionne-Laporte, A.; Provencher, P.; de Carvalho, M.; Orrù, S.; et al. Replication study of MATR3 in familial and sporadic amyotrophic lateral sclerosis. *Neurobiol. Aging* **2016**, *37*, 209.e17–209.e21. [CrossRef]
281. Salton, M.; Elkon, R.; Borodina, T.; Davydov, A.; Yaspo, M.-L.; Halperin, E.; Shiloh, Y. Matrin 3 binds and stabilizes mRNA. *PLoS ONE* **2011**, *6*, e23882. [CrossRef]
282. Yang, C.; Wang, H.; Qiao, T.; Yang, B.; Aliaga, L.; Qiu, L.; Tan, W.; Salameh, J.; McKenna-Yasek, D.M.; Smith, T.; et al. Partial loss of TDP-43 function causes phenotypes of amyotrophic lateral sclerosis. *Proc. Natl. Acad. Sci. USA* **2014**, *111*, E1121–E1129. [CrossRef]
283. Kamelgarn, M.; Chen, J.; Kuang, L.; Arenas, A.; Zhai, J.; Zhu, H.; Gal, J. Proteomic analysis of FUS interacting proteins provides insights into FUS function and its role in ALS. *Biochim. Biophys. Acta Mol. Basis Dis.* **2016**, *1862*, 2004–2014. [CrossRef]
284. Yu, Y.; Chi, B.; Xia, W.; Gangopadhyay, J.; Yamazaki, T.; Winkelbauer-Hurt, M.E.; Yin, S.; Eliasse, Y.; Adams, E.; Shaw, C.E.; et al. U1 snRNP is mislocalized in ALS patient fibroblasts bearing NLS mutations in FUS and is required for motor neuron outgrowth in zebrafish. *Nucleic Acids Res.* **2015**, *43*, 3208–3218. [CrossRef]
285. Rogelj, B.; Easton, L.E.; Bogu, G.K.; Stanton, L.W.; Rot, G.; Curk, T.; Zupan, B.; Sugimoto, Y.; Modic, M.; Haberman, N.; et al. Widespread binding of FUS along nascent RNA regulates alternative splicing in the brain. *Sci. Rep.* **2012**, *2*, 603. [CrossRef] [PubMed]
286. Orozco, D.; Tahirovic, S.; Rentzsch, K.; Schwenk, B.M.; Haass, C.; Edbauer, D. Loss of fused in sarcoma (FUS) promotes pathological Tau splicing. *EMBO Rep.* **2012**, *13*, 759–764. [CrossRef] [PubMed]
287. Moloney, E.B.; de Winter, F.; Verhaagen, J. ALS as a distal axonopathy: Molecular mechanisms affecting neuromuscular junction stability in the presymptomatic stages of the disease. *Front. Neurosci.* **2014**, *8*, 252. [CrossRef] [PubMed]
288. Boehringer, A.; Garcia-Mansfield, K.; Singh, G.; Bakkar, N.; Pirrotte, P.; Bowser, R. ALS Associated Mutations in Matrin 3 Alter Protein-Protein Interactions and Impede mRNA Nuclear Export. *Sci. Rep.* **2017**, *7*, 14529. [CrossRef]
289. Coelho, M.B.; Attig, J.; Bellora, N.; König, J.; Hallegger, M.; Kayikci, M.; Eyras, E.; Ule, J.; Smith, C.W.J. Nuclear matrix protein Matrin3 regulates alternative splicing and forms overlapping regulatory networks with PTB. *EMBO J.* **2015**, *34*, 653–668. [CrossRef]
290. Zhang, X.; Yamashita, S.; Hara, K.; Doki, T.; Tawara, N.; Ikeda, T.; Misumi, Y.; Zhang, Z.; Matsuo, Y.; Nagai, M.; et al. A mutant MATR3 mouse model to explain multisystem proteinopathy. *J. Pathol.* **2019**, *249*, 182–192. [CrossRef]
291. Simpson, C.L.; Lemmens, R.; Miskiewicz, K.; Broom, W.J.; Hansen, V.K.; van Vught, P.W.J.; Landers, J.E.; Sapp, P.; Van Den Bosch, L.; Knight, J.; et al. Variants of the elongator protein 3 (ELP3) gene are associated with motor neuron degeneration. *Hum. Mol. Genet.* **2009**, *18*, 472–481. [CrossRef]
292. Bento-Abreu, A.; Jager, G.; Swinnen, B.; Rué, L.; Hendrickx, S.; Jones, A.; Staats, K.A.; Taes, I.; Eykens, C.; Nonneman, A.; et al. Elongator subunit 3 (ELP3) modifies ALS through tRNA modification. *Hum. Mol. Genet.* **2018**, *27*, 1276–1289. [CrossRef]
293. Greenway, M.J.; Andersen, P.M.; Russ, C.; Ennis, S.; Cashman, S.; Donaghy, C.; Patterson, V.; Swingler, R.; Kieran, D.; Prehn, J.; et al. ANG mutations segregate with familial and 'sporadic' amyotrophic lateral sclerosis. *Nat. Genet.* **2006**, *38*, 411–413. [CrossRef]
294. Wu, D.; Yu, W.; Kishikawa, H.; Folkerth, R.D.; Iafrate, A.J.; Shen, Y.; Xin, W.; Sims, K.; Hu, G. Angiogenin loss-of-function mutations in amyotrophic lateral sclerosis. *Ann. Neurol.* **2007**, *62*, 609–617. [CrossRef]
295. Kirby, J.; Highley, J.R.; Cox, L.; Goodall, E.F.; Hewitt, C.; Hartley, J.A.; Hollinger, H.C.; Fox, M.; Ince, P.G.; Mcdermott, C.J.; et al. Lack of unique neuropathology in amyotrophic lateral sclerosis associated with p.K54E angiogenin (ANG) mutation. *Neuropathol. Appl. Neurobiol.* **2013**. [CrossRef] [PubMed]
296. Aladesuyi Arogundade, O.; Stauffer, J.E.; Saberi, S.; Diaz-Garcia, S.; Malik, S.; Basilim, H.; Rodriguez, M.J.; Ohkubo, T.; Ravits, J. Antisense RNA foci are associated with nucleoli and TDP-43 mislocalization in C9orf72-ALS/FTD: A quantitative study. *Acta Neuropathol.* **2019**, *137*, 527–530. [CrossRef] [PubMed]
297. Mehta, A.R.; Selvaraj, B.T.; Barton, S.K.; McDade, K.; Abrahams, S.; Chandran, S.; Smith, C.; Gregory, J.M. Improved detection of RNA foci in C9orf72 amyotrophic lateral sclerosis post-mortem tissue using BaseScope^{TM} shows a lack of association with cognitive dysfunction. *Brain Commun.* **2020**, *2*. [CrossRef] [PubMed]

298. Figueroa-Romero, C.; Hur, J.; Lunn, J.S.; Paez-Colasante, X.; Bender, D.E.; Yung, R.; Sakowski, S.A.; Feldman, E.L. Expression of microRNAs in human post-mortem amyotrophic lateral sclerosis spinal cords provides insight into disease mechanisms. *Mol. Cell. Neurosci.* **2016**, *71*, 34–45. [CrossRef]
299. Porta, S.; Kwong, L.K.; Trojanowski, J.Q.; Lee, V.M.-Y. Drosha Inclusions Are New Components of Dipeptide-Repeat Protein Aggregates in FTLD-TDP and ALS *C9orf72* Expansion Cases. *J. Neuropathol. Exp. Neurol.* **2015**, *74*, 380–387. [CrossRef]
300. Sobue, G.; Mcalary, L.; Taylor, J.P.; Purice, M.D. Linking hnRNP Function to ALS and FTD Pathology. *Front. Neurosci.* **2018**, *1*, 326. [CrossRef]
301. Kawahara, Y.; Mieda-Sato, A. TDP-43 promotes microRNA biogenesis as a component of the Drosha and Dicer complexes. *Proc. Natl. Acad. Sci. USA* **2012**, *109*, 3347–3352. [CrossRef]
302. Morlando, M.; Dini Modigliani, S.; Torrelli, G.; Rosa, A.; Di Carlo, V.; Caffarelli, E.; Bozzoni, I. FUS stimulates microRNA biogenesis by facilitating co-transcriptional Drosha recruitment. *EMBO J.* **2012**, *31*, 4502–4510. [CrossRef]
303. Waller, R.; Wyles, M.; Heath, P.R.; Kazoka, M.; Wollff, H.; Shaw, P.J.; Kirby, J. Small RNA Sequencing of Sporadic Amyotrophic Lateral Sclerosis Cerebrospinal Fluid Reveals Differentially Expressed miRNAs Related to Neural and Glial Activity. *Front. Neurosci.* **2017**, *11*, 731. [CrossRef]
304. Campos-Melo, D.; Droppelmann, C.A.; He, Z.; Volkening, K.; Strong, M.J. Altered microRNA expression profile in amyotrophic lateral sclerosis: A role in the regulation of NFL mRNA levels. *Mol. Brain* **2013**, *6*, 26. [CrossRef]
305. Salinas, J.; Lin, H.; Aparico, H.J.; Huan, T.; Liu, C.; Rong, J.; Beiser, A.; Himali, J.J.; Freedman, J.E.; Larson, M.G.; et al. Whole blood microRNA expression associated with stroke: Results from the Framingham Heart Study. *PLoS ONE* **2019**, *14*, e0219261. [CrossRef] [PubMed]
306. Boissart, C.; Nissan, X.; Giraud-Triboult, K.; Peschanski, M.; Benchoua, A. miR-125 potentiates early neural specification of human embryonic stem cells. *Development* **2012**, *139*, 1247–1257. [CrossRef] [PubMed]
307. Jauhari, A.; Singh, T.; Pandey, A.; Singh, P.; Singh, N.; Srivastava, A.K.; Pant, A.B.; Parmar, D.; Yadav, S. Differentiation Induces Dramatic Changes in miRNA Profile, Where Loss of Dicer Diverts Differentiating SH-SY5Y Cells Toward Senescence. *Mol. Neurobiol.* **2017**, *54*, 4986–4995. [CrossRef] [PubMed]
308. Ranjith Sama, R.K.; Ward, C.L.; Kaushansky, L.J.; Lemay, N.; Ishigaki, S.; Urano, F.; Bosco, D.A. FUS/TLS Assembles Into Stress Granules and Is a Prosurvival Factor During Hyperosmolar Stress. *J. Cell. Physiol.* **2013**, *228*, 2222–2231. [CrossRef]
309. Guil, S.; Cáceres, J.F. The multifunctional RNA-binding protein hnRNP A1 is required for processing of miR-18a. *Nat. Struct. Mol. Biol.* **2007**, *14*, 591–596. [CrossRef]
310. Zhang, K.; Daigle, J.G.; Cunningham, K.M.; Coyne, A.N.; Ruan, K.; Grima, J.C.; Bowen, K.E.; Wadhwa, H.; Yang, P.; Rigo, F.; et al. Stress Granule Assembly Disrupts Nucleocytoplasmic Transport. *Cell* **2018**, *173*, 958–971.e17. [CrossRef]
311. Kim, H.J.; Kim, N.C.; Wang, Y.-D.; Scarborough, E.A.; Moore, J.; Diaz, Z.; MacLea, K.S.; Freibaum, B.; Li, S.; Molliex, A.; et al. Mutations in prion-like domains in hnRNPA2B1 and hnRNPA1 cause multisystem proteinopathy and ALS. *Nature* **2013**, *495*, 467–473. [CrossRef]
312. Buchan, J.R.; Kolaitis, R.-M.; Taylor, J.P.; Parker, R. Eukaryotic stress granules are cleared by autophagy and Cdc48/VCP function. *Cell* **2013**, *153*, 1461–1474. [CrossRef]
313. Uversky, V.N. The roles of intrinsic disorder-based liquid-liquid phase transitions in the "Dr. Jekyll–Mr. Hyde" behavior of proteins involved in amyotrophic lateral sclerosis and frontotemporal lobar degeneration. *Autophagy* **2017**, *13*, 2115–2162. [CrossRef]
314. Vanneste, J.; Vercruysse, T.; Boeynaems, S.; Sicart, A.; Van Damme, P.; Daelemans, D.; Van Den Bosch, L. C9orf72-generated poly-GR and poly-PR do not directly interfere with nucleocytoplasmic transport. *Sci. Rep.* **2019**, *9*, 15728. [CrossRef]
315. Brenner, D.; Weishaupt, J.H. Update on amyotrophic lateral sclerosis genetics. *Curr. Opin. Neurol.* **2019**, *32*, 735–739. [CrossRef] [PubMed]
316. Marchetto, M.C.N.; Muotri, A.R.; Mu, Y.; Smith, A.M.; Cezar, G.G.; Gage, F.H. Non-Cell-Autonomous Effect of Human SOD1G37R Astrocytes on Motor Neurons Derived from Human Embryonic Stem Cells. *Cell Stem Cell* **2008**, *3*, 649–657. [CrossRef] [PubMed]

317. Haidet-Phillips, A.M.; Hester, M.E.; Miranda, C.J.; Meyer, K.; Braun, L.; Frakes, A.; Song, S.; Likhite, S.; Murtha, M.J.; Foust, K.D.; et al. Astrocytes from familial and sporadic ALS patients are toxic to motor neurons. *Nat. Biotechnol.* **2011**, *29*, 824–828. [CrossRef] [PubMed]
318. Hooten, K.G.; Beers, D.R.; Zhao, W.; Appel, S.H. Protective and Toxic Neuroinflammation in Amyotrophic Lateral Sclerosis. *Neurotherapeutics* **2015**, *12*, 364–375. [CrossRef]
319. Brites, D.; Vaz, A.R. Microglia centered pathogenesis in ALS: Insights in cell interconnectivity. *Front. Cell. Neurosci.* **2014**, *8*, 117. [CrossRef]
320. Mougeot, J.-L.C.; Li, Z.; Price, A.E.; Wright, F.A.; Brooks, B.R. Microarray analysis of peripheral blood lymphocytes from ALS patients and the SAFE detection of the KEGG ALS pathway. *BMC Med. Genom.* **2011**, *4*, 74. [CrossRef]
321. Pradat, P.-F.; Barani, A.; Wanschitz, J.; Dubourg, O.; Lombès, A.; Bigot, A.; Mouly, V.; Bruneteau, G.; Salachas, F.; Lenglet, T.; et al. Abnormalities of satellite cells function in amyotrophic lateral sclerosis. *Amyotroph. Lateral Scler.* **2011**, *12*, 264–271. [CrossRef]
322. Scaricamazza, S.; Salvatori, I.; Giacovazzo, G.; Loeffler, J.P.; Renè, F.; Rosina, M.; Quessada, C.; Proietti, D.; Heil, C.; Rossi, S.; et al. Skeletal-Muscle Metabolic Reprogramming in ALS-SOD1G93A Mice Predates Disease Onset and Is A Promising Therapeutic Target. *iScience* **2020**, *23*, 101087. [CrossRef]
323. Loeffler, J.P.; Picchiarelli, G.; Dupuis, L.; Gonzalez De Aguilar, J.L. The role of skeletal muscle in amyotrophic lateral sclerosis. *Brain Pathol.* **2016**, *26*, 227–236. [CrossRef]
324. Morgan, S.; Duguez, S.; Duddy, W. Personalized Medicine and Molecular Interaction Networks in Amyotrophic Lateral Sclerosis (ALS): Current Knowledge. *J. Pers. Med.* **2018**, *8*, 44. [CrossRef]

Journal of *Personalized Medicine*

MDPI

Review

The Identification of Novel Biomarkers Is Required to Improve Adult SMA Patient Stratification, Diagnosis and Treatment

Piera Smeriglio [1,*], Paul Langard [1], Giorgia Querin [1,2,3] and Maria Grazia Biferi [1,*]

1 Centre of Research in Myology, Institute of Myology, Sorbonne Université, INSERM, 75013 Paris, France; p.langard@institut-myologie.org (P.L.); g.querin@institut-myologie.org (G.Q.)
2 Association Institut de Myologie, Plateforme Essais Cliniques Adultes, 75013 Paris, France
3 APHP, Service de Neuromyologie, Hôpital Pitié-Salpêtrière, 75013 Paris, France
* Correspondence: p.smeriglio@institut-myologie.org (P.S.); mg.biferi@institut-myologie.org (M.G.B.)

Received: 12 June 2020; Accepted: 24 July 2020; Published: 29 July 2020

Abstract: Spinal muscular atrophy (SMA) is currently classified into five different subtypes, from the most severe (type 0) to the mildest (type 4) depending on age at onset, best motor function achieved, and copy number of the *SMN2* gene. The two recent approved treatments for SMA patients revolutionized their life quality and perspectives. However, upon treatment with Nusinersen, the most widely administered therapy up to date, a high degree of variability in therapeutic response was observed in adult SMA patients. These data, together with the lack of natural history information and the wide spectrum of disease phenotypes, suggest that further efforts are needed to develop precision medicine approaches for all SMA patients. Here, we compile the current methods for functional evaluation of adult SMA patients treated with Nusinersen. We also present an overview of the known molecular changes underpinning disease heterogeneity. We finally highlight the need for novel techniques, i.e., -omics approaches, to capture phenotypic differences and to understand the biological signature in order to revise the disease classification and device personalized treatments.

Keywords: spinal muscular atrophy; adult patients; disease heterogeneity; Nusinersen; disease modifiers; functional outcomes; biomarkers; epigenetic changes; -omics approaches

1. Introduction

Spinal muscular atrophy (SMA) is a neurodegenerative disease affecting motoneurons (MN) in the brainstem and spinal cord caused by the homozygous mutation of the Survival of Motor Neuron 1 (*SMN1*) gene [1]. The disease presents with a wide spectrum of clinical severity and patients are classified into five types, depending on the age of onset and motor milestones achieved. Patients with type 0 SMA are the most extreme cases with death early after birth. Severity decreases from type 1 to type 4 SMA, which only presents mild symptoms starting at the adult age and have a very slow evolution over time (reviewed in [2], see Table 1 and Figure 1A). Attempts to explain this phenotypic variability have been made and several disease modifiers are known [3]. The most widely accepted modifier is the copy number of the *SMN2* gene, a centromeric paralog of *SMN1*. Following a cytosine to thymine transition in exon 7, which creates an exon splicing suppressor (ESS), the transcription of *SMN2* leads to 90% of transcripts coding for a truncated SMN protein. The remaining 10% produces a full-length (FL) SMN protein, thus inducing low levels of SMN expression.

SMN2 copy number is directly correlated with the SMN expression level; therefore, a higher copy number is associated with a milder phenotype [4]. *SMN2* copy number is used, together with age of onset and motor abilities achieved, as an additional parameter to stratify SMA patients (Table 1 and Figure 1B). This additional factor is, however, still insufficient to explain the phenotypic variability among siblings carrying the same mutation in *SMN1* and same *SMN2* copy number [5,6].

Table 1. Current classification of spinal muscular atrophy (SMA) patients in 5 types. The subtypes, age of onset (mo.: months; y.: years), level of motor functions, life expectancy, and frequency of *SMN2* copy number (*SMN2*cn) are reported [7]. Background colors are used to differentiate SMA types.

SMA Type	Subtype	Age of Onset	Level of Motor Functions	Life Expectancy	*SMN2*cn (%)
0/1a		pre-natal	Need respiratory assistance	<1 month	
1		0–6 mo.	Cannot sit independently	<2 years	2 (73.4%)
	1b		Absence of head control and ability to roll over		
	1c		Sometimes gain head control or the ability to roll from supine to prone position		
2		<18 mo.	Cannot stand independently	>2 years	3 (81.8%)
	2a		Independent sitting lost		
	2b		Independent sitting conserved		
3		>18 mo.	Able to stand and walk independently	Adulthood	3–4 (50.6%; 45.5%)
	3a	18 mo.–3 y.			
	3b	>3 y.			
4		>20 y.	Weaknesses in lower limbs	Adulthood	

In the last few years, other gene modifiers have been described, such as Plastin 3 (PLS3) [8], Coronin 1C (*CORO1C*) [9], and Neurocalcin Delta (*NCALD*) [10] (Figure 1C). It remains puzzling to note that the expression changes of these genes do not fully explain the phenotypic heterogeneity in the SMA population. Additional modifier conditions to be considered are (i) the environmental factors that may influence the final phenotype of the patients [11] and (ii) the differential vulnerability of MN subtypes that is affected by the level of SMN protein [12,13]. Although all of these factors are known contributors to the differential disease severity, other unknown aspects might need to be considered in the future for a comprehensive description of the disease.

Recent advances in the SMA field have led to the development of therapeutic approaches, aiming to increase the level of SMN protein targeting *SMN2* through an antisense (AS) oligonucleotide [14,15] or a pre-mRNA splicing modifier drug [16], as well as inducing the restoration of the SMN protein via gene therapy [17].

Approved by the Food and Drug Administration in 2016, Nusinersen is an AS oligonucleotide that targets exon 7 of *SMN2* to facilitate its inclusion and trigger a higher production of the full-length SMN transcript. Nusinersen is administered via intrathecal injection, directly into the cerebrospinal fluid (CSF), in a two-phase treatment: (1) a loading phase of four doses 14, 14, and 30 days apart followed by (2) a maintenance phase with a dose every four months. Treatment with Nusinersen showed very promising results in young type 1 patients [14,15] and some positive effects in type 2 and 3 patients (no type 4 patients have been enrolled in clinical trials to date) [18–20]. Unexpectedly, the latter studies which include adult SMA patients (>18 years old) showed that their response to Nusinersen, evaluated as motor function improvement, was highly variable, with 40–50% of responders at best [18–20]. These observations suggest that factors modulating response to treatment, in particular for the older patients, remains to be uncovered. Moreover, due to the lack of natural history data and the higher heterogeneity among the adult SMA patient population, therapeutic monitoring of these patients is particularly challenging [21,22]. Therefore, an exceptional effort is needed to facilitate further characterization of the clinical and molecular profiles of these SMA patient populations.

In this context, the identification of novel biomarkers and precise methods for functional outcomes evaluation are unmet needs. Reaching these goals will provide the means to (i) redefine patients' classification, (ii) characterize the full molecular profile of therapy responders or non-responders, and (iii) determine inclusion criteria in clinical trials and treatment protocols. Therefore, several types of biomarkers have been evaluated in order to capture patients' pre-existing differences and their response to treatment.

As previously reviewed, biomarkers can be classified as biomolecular, mirroring the molecular changes, and functional to picture the phenotypic disease progression [23]. This review will initially focus on the most promising biomarkers for adult SMA (>18 years old). Data on therapeutic monitoring are gathered through the different studies reporting results from treatment with Nusinersen, the most widely used therapy for SMA worldwide up to date. Current available information on the molecular signature of SMA patients will be then presented, including gene modifiers, epigenetic changes, and putative molecular biomarkers. Furthermore, although various molecular biomarkers have been proposed in the last few years as possible candidates, no factor has been proven to date to faithfully follow the disease progression in all SMA types. Therefore, here we will also discuss the potential of unbiased -omics (e.g., proteomics) approaches as valuable tools for the identification of novel SMA-specific biomarkers.

In perspective, an in-depth analysis of both functional outcomes' evaluation and molecular biomarkers, would help determine the differences that define the wide spectrum of clinical features among SMA patients and their disparate response to treatment. Efforts in this direction could lead to the development of personalized medicine approaches in the future.

2. Evaluation of Functional Outcomes in Adult Patients

Functional outcomes are essential to monitor disease progression and to precisely understand where the patient stands in the wide spectrum of the disease phenotypes. For chronic diseases, the longitudinal study of functional outcomes contributes to define the natural history of the disease and can also be used to monitor the therapeutic response. In SMA, several tests have been used for diagnosis, prognosis, or therapeutic monitoring. These tests have frequently turned out to be unadaptable to the adult SMA patient population (>18 years of age) due to its high heterogeneity. This group of patients still needs a precise characterization and patient-specific therapeutic adjustments or development. Therefore, this section of the review will focus on functional outcomes reported to reliably assess the Nusinersen response in SMA adult patients. Data collected in younger patients will be presented as comparison.

Gradual motor function loss is a common feature of all SMA types. While adult patients are generally classified as less severe cases than type 1 children, their motor function also declines progressively [24]. The motor abilities can be monitored by a variety of measurements, spanning from scales assessing the general mobility or muscle function, to the more global walking analysis. Respiratory function is also frequently assessed in adult SMA patients, as respiratory failure is still the most frequent cause of morbidity in SMA patients [25]. In this section, we are documenting functional outcomes to assess (i) general mobility; (ii) motricity of the upper or lower limbs, in conjunction with muscle strength or fatigue; (iii) level of ambulation; and (iv) respiratory function in adult SMA patients.

2.1. General Mobility Tests

The Hammersmith Functional Motor Scale Expanded (HFMSE) is a functional scale used for the assessment of physical abilities. It was adapted from the classic HFMS adding 13 new items that allow to capture a wider range of motor skills [26]. Therefore, the HFMSE has been proven to be a reliable indicator of the wide arrays of movements in adult SMA patients types 2 and 3. In the effort to correlate the HFMSE score with the adult SMA types and subtypes, we collected the most recent available data on SMA types 2 and 3 (Table 2).

Independent studies exploring this aspect showed a consistent range of HFMSE score: between 0 and 20 for type 2 and between 25 and 66 for type 3 patients [27–29]. A larger range was recorded for the type 3 patients followed up by Walter and colleagues [18]. A similar trend, but with lower HFMSE score, for type 3b patients (with symptoms onset after 3 years of age) has been described very recently by Kessler et al. [30]. Despite the variability for subtypes 3 and 3b, HFMSE seems to be a reliable indicator of the differences among SMA types.

Table 2. Summary of representative Hammersmith Functional Motor Scale Expanded (HFMSE) scores by SMA subtype. Graphical representation of the HFMSE scores in SMA patients from five independent studies. Columns are named as follows. **References**: bibliographic citations of the natural history or Nusinersen-related studies. **Cohort**: cohort composition in SMA patients with specified type or subtype when the information is available. **T**: time when the analysis was performed; 0: baseline; +10 mo: 10 months after first Nusinersen dose. **HMSFE**: scores represented as a boxplot; values are indicated as median ± standard deviation (s.d.). Δ HMFSE ($T - T_0$) indicates the increment score between the baseline before treatment (T_0) and the final after treatment timepoint (T). **N**: number of patients included. **Age**: age of the subject in years as reported in the relative study (median (s.d.)). Colors of boxplot and dots are related to the cohort characteristics (SMA type): mustard: SMA type 2; green: type 3; light green: subtype 3b. Background color associates consecutive table entries referring to the same study.

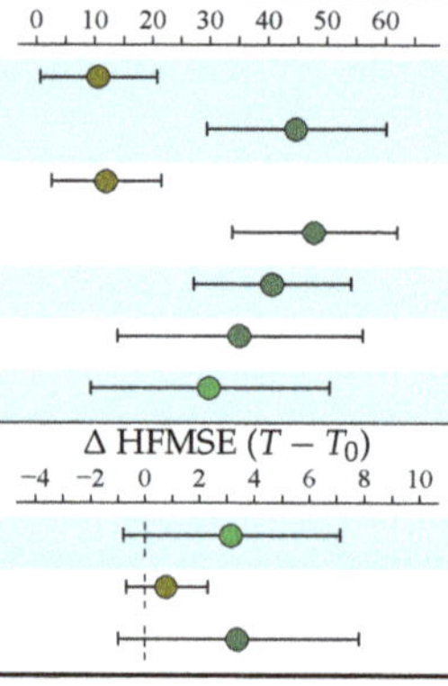

References	Cohort	T	HFMSE	N	Age
Natural history studies					
Kaufmann et al. [27]	type 2	/		41	9.1 (7.4)
	type 3	/		38	13.7 (10.8)
Montes et al. [31]	type 2	/		67	10.9 (8.3)
	type 3	/		59	13.4 (10.7)
Faravelli et al. [29]	type 3	/		12	29 (15–35)
Walter et al. [18]	type 3	/		19	29 (15–35)
Kessler et al. [30]	type 3b	0		7	39 (13)
Nusinersen-related studies			Δ HFMSE ($T - T_0$)		
Kessler et al. [30]	type 3b	+10 mo.		7	39 (13)
Hagenacker et al. [19]	type 2	+10 mo.		30	37 (12)
	type 3	+10 mo.		60	

Furthermore, this scoring system showed that there was a progressive gain during the course of the Nusinersen treatment, with an average gain of 3.12 points at 14 months after the first injection in 57 patients [19]. Results differed between SMA types 2 and 3, with a higher average HFMSE increase for SMA type 3 patients. Interestingly, no correlation was observed between the age at treatment and the evolution of the functional score during the treatment [19,29].

2.2. Revised Upper Limb Module

The revised upper limb module (RULM) (see Table 3), a reconsidered version of the original upper limb module (ULM) [32], has been created to measure upper limb function in a wider range of patients [33] than the non-ambulatory young children and weaker patients [32]. The RULM includes 20 tasks based on a scale of 3 scores: 0 (unable), 1 (able, with modification), and 2 (able, no difficulty) [33]. A large natural history study across three countries analyzed the RULM score for 114 SMA type 2 and 3 patients, ranging from weak non-ambulant to stronger ambulant, over the course of 12 months [34]. This analysis confirms that the RULM scale can detect a wide spectrum of upper motor abilities even in ambulant patients where the ceiling effect was rarely reached (11.4% of the cohort), overcoming the ULM scale limitations. Furthermore, it has been demonstrated that the RULM score is more useful than the HFMSE to picture motor differences in wheelchair-dependent adult SMA type 2 and 3 patients [35], suggesting that it can be used as diagnostic marker. Moreover, in patients treated with Nusinersen, RULM was shown to be more accurate than the HMFSE in detecting the improved motor abilities of type 2 and 3 patients (both young and adults) treated with Nusinersen [18,19] at later time points (10–14 months after the first injection). All these data demonstrate that the RULM scale is a sensitive marker for both diagnostic and therapeutic monitoring purposes and it could be used in combination with a more general scale to reveal the finest motor differences for the large spectrum of adult patients.

Table 3. Summary of representative results for the Revised Upper Limb Module (RULM). Graphical representation of the RULM scores in SMA patients from three independent studies. Columns are named as follows. **References**: bibliographic citations of the natural history or Nusinersen-related studies. **Cohort**: cohort composition in SMA patients with specified type or subtype. **T**: time when the analysis was performed; 0: baseline; +10 mo: 10 months after first Nusinersen dose. **RULM**: scores represented as a boxplot; values are indicated as median ± sd. Δ RULM ($T - T_0$) indicates the increment score between the baseline before treatment (T_0) and the final after treatment timepoint (T). **N**: number of patients included. **Age**: age of the subject in years as reported in the relative study (age range); NA: not available. Colors of boxplot and dots are related to the cohort characteristics (SMA type): mustard: SMA type 2; green: type 3. Background color associates consecutive table entries referring to the same study.

References	Cohort	T	RULM	N	Age
Natural history studies			0 10 20 30 40		
Stolte et al. [36]	type 2	/		9	24 to 48
	type 3	0		19	18 to 61
Walter et al. [18]	type 3	0		19	18 to 59
			Δ RULM ($T - T_0$)		
Nusinersen-related study			−2 0 2 4		
Hagenacker et al. [19]	type 2	+10 mo.		30	*NA*
	type 3	+10 mo.		58	*NA*

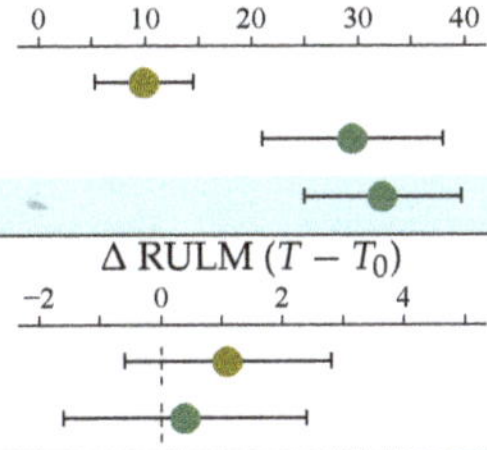

2.3. Quantitative Assessment of Ambulation Capacity

Among the five SMA types, only type 3 and 4 patients can walk unassisted. This is the highest milestone achievable by these patients; however, the probability of remaining ambulant decreases with aging. According to a study of natural history for type 3 SMA patients published by Zerres et al. [37], the probability of preserving ambulation throughout life is tightly linked to the age of disease onset. The authors showed that 83.7% of patients (n = 72) with symptomatic appearance between 18 and 35 months of age—corresponding to SMA type 3a—are able to walk 10 years after disease onset, and this percentage plummeted to 30% at 25 years after disease onset. Accordingly, among patients with a disease onset between 3 and 15 years old (n = 109)—corresponding to SMA type 3b subset—an average 95% were able to walk 10 years after disease onset and 75% at 25 years. These data clearly showed that the loss of the ability to walk declines slowly in patients with a later disease onset, suggesting that this outcome can be monitored to assess disease progression. Several functional scales have been exploited to strictly monitor the ability to walk, such as the 6-min walking test (6MWT).

The 6MWT is a functional test designed for ambulant patients, aiming to measure the distance that a patient can cover during a six-minute timelapse. The test can also document the gait pace by minute and indirectly describes motor fatigue over time. Guidelines for this test were published in 2002 by the American Thoracic Society (ATS) [38]. The 6MWT has been approved as an endpoint outcome in SMA patients motor function assessment [39]. As described in the ATS statement for the 6MWT, individual factors might influence the outcome of the test, including sex, age, height, and weight, but also the personal motivation of the patient. For healthy subjects between 10 and 59 years of age, the 6MWT values range approximately from 600 to 850 meters (m) [40]. In SMA patients, the distance covered in 6 minutes is affected by both age and the type of SMA. Montes et al. [28] showed that the median value is generally higher for SMA subtype 3b averaging at 368 m (n = 28) compared to 253 m for subtype 3a (see Table 4). However, the mean rate of progression in the two SMA subtypes did not differ significantly after 1 years from the baseline obervation. Moreover, one can notice the internal variability into each subtype and the important overlap between subtypes 3a and 3b.

The analysis of the ambulation is a valuable tool to follow up the response to treatment and few studies reported data obtained with the 6MWT. Recently, Hagenacker et al. [19] published the results of a multicenter, observational study on a cohort of 173 adult SMA patients treated with Nusinersen and followed up to 14 months. The 6MWT was used as a secondary endpoint for ambulant patients (n = 46 at baseline; see Table 4). At all time points analyzed (6, 10, and 14 months) patients showed a significant increase in the outcome of the 6MWT with an average gain of 22.1 m (8.7 m–35.6 m; 95% Confidence Interval) after 6 months and 46 m (25.4 m–66.6 m; 95% C.I.) after 14 months. Very interestingly, along the course of treatment, few non-ambulatory patients gained the ability to walk. This large cohort study provides evidence for the efficacy of Nusinersen treatment to improve the ambulatory function in adult SMA patients.

While the 6MWT gave some indications about treatment response in walking adult SMA patients, the variability in the degree of ambulation at both baseline and after treatment should be further explored and correlated to other functional and molecular parameters to better understand the disease.

Table 4. Summary of representative results for the 6-Minute-Walk-Test (6MWT) by ambulatory patients. Graphical representation of the 6MWT in healthy and SMA patients from four independent studies. Columns are named as follows. **References**: bibliographic citations of the natural history or Nusinersen-related studies. **Cohort**: cohort composition in healthy and/or SMA patients with specified type or subtype when the information is available. **T**: time when the analysis was performed. **Distance**: distance reported in meters and represented as a boxplot; values are indicated as median ± standard deviation (s.d.). Δ Distance ($T - T_0$) indicates increment distance covered between the baseline before treatment (T_0) and the final after treatment timepoint (T). **N**: number of patients included. **Age**: age of the subject in years as reported in the relative study (age range or mean (s.d.)); NA: not available. Colors of boxplot and dots are related to the cohort characteristics (SMA type or controls): gray: healthy subject; green: SMA type 3; dark green: SMA subtype 3a; light green: SMA subtype 3b. Background color associates consecutive table entries referring to the same study.

References	Cohort	T	Distance (m)	N	Age
Natural history studies					
Mckay et al. [40]	Healthy	/		400	20 to 59
Montes et al. [28]	type 3a	0		57	10.3 (9.8)
	type 3b	0		28	25.6 (12.5)
			Δ Distance ($T - T_0$)		
Montes et al. [28]	type 3a	+1 y		57	10.3 (9.8)
	type 3b	+1 y		28	25.6 (12.5)
Nusinersen-related studies					
Hagenacker et al. [19]	type 3	+6 mo.		47	NA
		+10 mo.		37	NA
		+14 mo.		25	NA

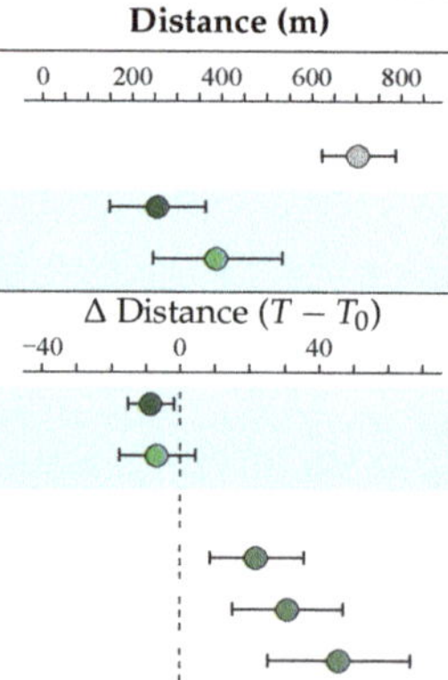

2.4. Lung Function Tests

Respiratory function is impaired in SMA patients as the diaphragm and intercostal muscles are affected by the progression of the disease. The association of this alteration with recurrent scoliosis in patients can result in restrictive lung disease. Difficulty to cough can contribute to a reduced clearance and facilitate the onset of respiratory infections. This decline in the respiratory function is associated with mortality and morbidity in SMA patients of all types [14,25]. For this reason, many efforts have been devoted in the last few years to the search for appropriate parameters to assess this function.

The natural history study of lung function in SMA patients, reported by Wijngaarde et al. [25], analyzed the Forced Expiratory Volume in 1 second (FEV_1) in a cohort of 170 patients SMA types 1c to 4. Using these data, the authors were able to build age-dependent models of the evolution of this measure for 131 patients—subtypes 1c to 3b. The linear models obtained, as expected, showed a progressive annual decrease of the FEV_1 from 100.35% predicted in newborns subtype 3b—almost normal lung function—to 42.12% in 1c newborns. They reported a 1.29% and 1.37% annual rate of decline for subtypes 2a and 2b, respectively, which was associated with an early start of mechanical ventilation (median 12.3 and 16.8 years old, respectively). Subtypes 3a showed a milder 0.73% annual rate of decline, with a median age of 39.9 years old at start of mechanical ventilation. However, this analysis demonstrated that the average annual decline is dependent on age—faster decline in younger patients. Unfortunately, the effect of Nusinersen on the respiratory function in adult patients has been poorly studied so far. An encouraging 5% increase in the average Force Volume Capacity (FVC) of the lungs has been recently reported in Walter et al. [18]. Further studies need to be performed to comprehensively evaluate the impact of Nusinersen treatment on SMA adult population and to define the appropriate outcome measures for lung functional evaluation.

2.5. Additional Tests

Together with the tests discussed above, a series of other functional measures have been widely used to assess the electrophysiological activity of motor neurons, the axon number and reinnervation potential, and the structural changes in spinal cord and muscles. For example, electromyography (EMG), compound muscle action potential (CMAP), and motor neuron number index (MUNIX) have been classically used in patients affected with neuromuscular disorders to assess the functional status of the motor unit [41–44]. The EMG and CMAP methods have been used for SMA diagnosis [43,45,46] and the CMAP has also been demonstrated to be a good prognostic marker [47]. The scarcity of available data on the use of EMG and CMAP for the adult patient population has discouraged the attribution of these tests as ideal outcomes for adult SMA patients at the moment.

Computed from CMAP and electromyography interference pattern, the MUNIX is a quantitative test that aims to estimate the number of functional motor units in a specific muscle [44]. Overall, MUNIX and MUSIX are very interesting tests for the assessment of the number and size of motor units, which have been demonstrated to be directly correlated with motor function in adult type 2 and 3 patients [48]. However, no data is currently available on the reliability of this test to capture adult patients' response to treatment (clinical trial NCT04139343 is currently recruiting patients to monitor MUNE in adults with SMA). Alterations in the cervical spinal cord of adult SMA patients have also been described due to a fine-tuned magnetic resonance imaging (MRI) [49] that can provide structural markers of the disease. Further refinement of the MRI techniques will certainly improve the sensitivity of the method [50] and apply it to a wider number of patients. It remains to be assessed whether it would be a reliable method to follow the response to treatments. Overall, the detailed evaluation of different functional outcomes gives insights regarding the disease progression. These efforts will contribute to a novel description of SMA patients, based on the classical definition of types and a more precise clinical assessment. However, observations made at the macro scale need to be complemented by molecular characterization in order to explain patient's variability and device powerful predictive models of disease progression.

3. Genetic and Epigenetic Etiology of Clinical Heterogeneity

The discordance between genotype and phenotype among patients with same genetic mutation and same *SMN2* copy number can be explained by several factors, including (i) the presence of genetic modifiers and (ii) the epigenetic profile. These parameters need to be considered for both the natural history and to monitor the treatment response of SMA patients.

3.1. SMN2 and Other Genetic Modifiers

Several genes have been identified in the last few years as SMA disease modifiers. These include Survival of Motor Neuron 2 (*SMN2*) [1,51], Plastin 3 (*PLS3*) [8], Coronin 1C (*CORO1C*) [9] and Neurocalcin Delta (*NCALD*) [10], Small EDRK-Rich Factor 1 (*SERF1*) [52], NLR Family Apoptosis

Inhibitory Protein (*NAIP*) [53], General Transcription Factor IIH subunit 2 (i) [54], and Tolloid Like 2 (*TLL2*) [55]. Here, we will focus on the most documented genes.

The *SMN2* gene is the main modifier gene in SMA. Only 10% of the mRNA transcribed from the *SMN2* gene produces the functional FL-SMN protein. The number of copies of the *SMN2* genes is directly correlated to the level of functional FL-SMN and affects the severity of the disease (see Figure 1B) [1,7,47,51,56–75].

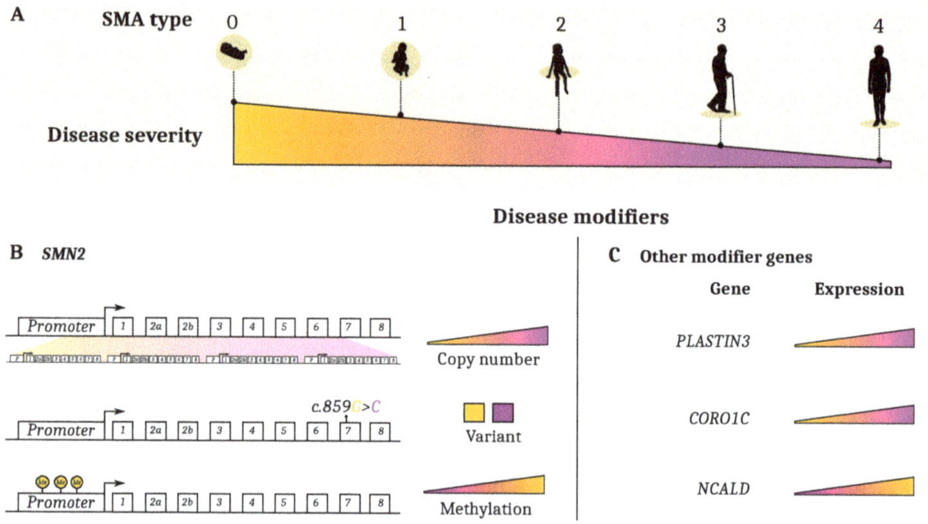

Figure 1. Known molecular disease modifiers. (**A**) Current classification of SMA as a discrete projection of a continuous spectrum of disease severity. This spectrum results from the combination of (**B**) disease modifiers such as *SMN2* copy number, *SMN2* variant, and methylation state; (**C**) level of expression of *CORO1C* and *PLASTIN3* was associated with milder phenotype, while *NCALD* overexpression was associated with more severe phenotype.

However, there are exceptions to this paradigm. Indeed, several type 2 and 3 SMA patients have been described to carry only two copies of *SMN2* instead of the expected three or four copies. This phenotype is often due to a rare single nucleotide variant (*SMN2* c.859G>C) in exon 7 that impacts *SMN2* gene splicing [11,76–78], inducing a higher rate of production of the FL-SMN protein (Figure 1B). Moreover, research from several groups described a variability of clinical phenotypes in siblings carrying the same *SMN1* mutation and identical *SMN2* copy number [5,58,79–81]. These events suggested the presence of modifier factors other than the *SMN2* gene are able to alter the severity of the disease and to be considered to define its full molecular signature.

3.2. Plastin3

F-actin bundling protein plastin 3 (*PLS3*) has been identified as the first gene, besides *SMN2*, to be able to attenuate the disease severity when upregulated (see Figure 1C). *PLS3* encodes a Ca^{2+}-dependent F-actin-binding protein involved in neurotransmitter release and vesicle recycling at the presynaptic site [82]. The seminal work of Oprea et al. [8] demonstrated elevated levels of PLS3 in asymptomatic females of SMA type 2- or 3-affected sibling pairs where the males were always symptomatic. Each analyzed sibling pair carried identical homozygous *SMN1* deletion and the same number of *SMN2* copies. An additional study, described that the *PLS3* higher expression is maintained after generation and differentiation of induced pluripotent stem cells from fibroblasts of asymptomatic but not symptomatic siblings [83]. Unfortunately, no association of PLS3 expression was found in discordant female sibling pairs in an Iranian population [84]. Highly divergent results have also been shown about the role of PLS3 in mouse models of SMA. While two independent studies described an amelioration of the SMA phenotype upon overexpression of PLS3 in mice [82,85], another report showed that a randomly integrated PLS3 allele expressed in the severely affected Δ7-SMA mouse model failed to show rescue of the mice survival or motor function [86]. These discrepancies can be attributed to the fact that PLS3 is directly regulated by the SMN protein [87]; therefore, its overexpression is effective only when used as an adjuvant treatment to the classic SMN induction therapy [9]. It is plausible that the expression of PLS3 is affected by the SMA currently administered therapies and it could be used as a putative molecular biomarker.

3.3. PLS3-Interacting Protein CORO1C and CHP1

As a PLS3 binding partner, CORO1C regulates synaptic vesicles recycling in a calcium-dependent manner [9]. When overexpressed, *CORO1C* induced a significant amount of F-actin, therefore ameliorating endocytosis at the neuromuscular junctions [9,86] (see Figure 1C). Recently identified using a yeast-two-hybrid screen, the calcineurin-like EF-hand protein 1 (CHP1) is a novel PLS3 interacting protein [88] upregulated in SMA mice. Treatment of SMA mice with a low dose of AS oligonucleotide-SMN combined with *CHP1* downregulation improved the survival extension of the animals compared to the single SMN therapy [88]. The authors showed that the negative modulation of *CHP1* induced an activation of calcineurin with consequent restoration of endocytic protein phosphorylation. In the case where *CORO1C* and *CHP1* are differentially expressed in SMA patients, their modulation could also be beneficial to improve therapeutic effects.

3.4. NCALD

NCALD is calcium sensor protein suggested to negatively impact SMA via repression of endocytosis [10]. Reduced expression of NCALD was shown to be protective in a four-generation discordant family with five asymptomatic and two SMA1-affected individuals [10]. Moreover, reducing NCALD levels either in vitro or in vivo significantly ameliorated SMA pathology across SMA species [10]. Based on these encouraging results, Torres-Benito et al. [89] developed an ASO-based therapy to target *Ncald* in mouse spinal cord and used it in combination with low-dose SMN splice switching ASOs to design an efficient combinatorial therapy in SMA mice [89]. Therefore, repression of NCALD is considered protective for SMA.

3.5. NAIP

Another gene located in the same chromosomic region (5q13) of *SMN1* and *-2* is NLR Family Apoptosis Inhibitory Protein (*NAIP*). The functional role of NAIP in the pathogenesis of SMA has not been fully elucidated. However, some reports have demonstrated a correlation between deletion of the *NAIP* gene and severity of SMA [53,90] (see Figure 1). This gene was also found to be frequently deleted in 45% of the Egyptian patients [91]. Further investigation needs to be performed to validate the accuracy of this gene as a biomarker for SMA.

3.6. Epigenetic Modifiers (Methylation)

Epigenetic marks have been demonstrated to modulate the expression of the modifier genes in SMA. In particular, methylation in the promoter of the *SMN2* gene was reported to reduce its transcription and consequently the expression of the SMN protein, independently of the *SMN2* copy number (see Figure 1B). Studies on the methylation level of different CpGs on the *SMN2* genes revealed that hypomethylation was associated with reduced disease severity [92]. In fact, SMA type 3 exhibited a lower degree of methylation in the *SMN2* gene compared to type 1 or 2 patients [93]. After profiling the methylation in SMA patients and healthy controls, Zheleznyakova et al. identified differential degrees of methylation at CpG sites in the following genes; CHM Like Rab Escort Protein (*CHML*), Rho GTPase Activating Protein 22 (*ARHGAP22*), Cytokinesis And Spindle

Organization B (*CYTSB*), Cyclin Dependent Kinase 2 Associated Protein 1 (*CDK2AP1*), and Solute Carrier Family 23 Member 2 (*SLC23A2*) [94]. Moreover, *SLC23A2* was significantly hypomethylated in type 3 and 4 patients compared to type 1 [95]. Thus, DNA methylation may regulate the SMA disease phenotype by modulating gene transcription and could be the molecular mechanism beyond the genotype–phenotype discrepancy often observed in SMA. Whether these changes are induced by environmental factors variations [96] and whether they are dependent or not on the absence of SMN, remain open questions.

3.7. Histone Deacetylases (HDAC) Inhibitors

Many efforts have also been made to target the HDACs and impact the transcription of the *SMN2* gene and its transduction into a functional SMN protein. Inhibition of HDACs impairs the removal of acetyl groups from the histone proteins and promotes gene transcription by mediating a more permissive, open, chromatin. In 2001, Chang et al. first identified an effect of a HDAC inhibitor, sodium butyrate, on SMA. Treatment with this epigenetic drug was able to increase the FL-SMN both in vitro and in vivo [97]. This discovery led to multiple investigations in the attempt to find more potent and stable HDAC inhibitors for SMA treatment. Valproic acid (VPA) [98,99], suberoylanilide hydroxamic acid (SAHA) [100,101], M344 [102], thricostatin A [103], and romidepsin [92] have also been suggested as promising treatments for SMA. The LBH589 molecule showed a higher potential than all other inhibitors, with the ability to induce a strong upregulation of SMN at very low doses [104]. Additionally, a combined therapy using the LBH589 inhibitor together with the Nusinersen drug showed a synergistic effect than the single treatments on SMA cellular models [105]. Despite the number of HDAC inhibitors tested and the encouraging results, only phenylbutyrate and VPA have entered clinical trials for human use. Valproic acid was tested in five Phase I/II clinical trials (NCT00374075, NCT00227266, NCT00481013, NCT00661453, and NCT01033331) [106–111] and showed outcomes variability. All the studies mentioned above were included in the meta-analysis performed by Elshafay et al. [112] suggesting the VPA administration was safe, although some adverse effects were recorded, and it induced a major improvement of motor function. The results of the SMA VALIANT trial (NCT00481013) on adult patients did not show any positive outcome on the motor function at either 6- or 12-month timepoints underlying that this treatment might have limited effects on the adult population [110]. A similar moderate success appeared from the pilot studies based on administration of phenylbutyrate to SMA patients [113–115]. More work needs to be done to establish whether these therapies could be used in combination with other approaches to efficiently ameliorate the SMA condition, even in the adults who seem less responsive to the currently approved therapies. The advantage of using HDAC inhibitors for SMA therapy, as other epigenetic drugs, is their ability to act on the chromatin without permanently affecting the DNA sequence. Moreover, some HDAC inhibitors have been demonstrated to reduce the methylation on the SMN2 promoter [92,101], therefore establishing a link between the two epigenetic modifications could be positively exploited in novel therapies.

4. Molecular Biomarkers

Biomarkers are needed in the field to provide the necessary insights to guide the decision-making in personalized medicine. There are several aspects to be considered in the identification of reliable biomarkers, like reproducibility and accuracy of measurements. We will discuss below the factors that have been identified as putative biomarkers for SMA.

4.1. SMN Protein

The SMN protein is considered the biomarker of choice for SMA, as all the approved therapies are aimed to restore its expression. SMN is ubiquitously expressed [116] and regulates several key processes in neuronal cells including ribonucleoprotein assembly, RNA metabolism [117], actin cytoskeleton dynamics [118], mRNA transport [119], ubiquitin homeostasis [120], bioenergetics pathways [121], synaptic vesicle release [122], and local protein translation [123–125]. Therefore, SMN-reduced expression in SMA has a devastating impact on many aspects of the neuronal homeostasis and survival. In fact, MNs expressing lower amounts of the SMN protein are more vulnerable to cell death [12] and this contributes to disease variability. The goal of the approved therapies for SMA is to re-express SMN in the affected tissues, mainly in MNs, but the scenario has become more complex since multiple studies published in the last few years have reported the importance of re-expressing SMN in peripheral tissues in addition to the CNS [126,127]. However, little is known on the specific tissue requirement for SMN; only recently a study from Ramos et al. described a variable level of SMN expression in different tissues and a general decline with aging [128]. Moreover, no robust correlations between *SMN2* copy number, and therefore SMA types, and SMN protein levels could have been demonstrated probably due to the tissue-specific SMN expression profile [70,128,129]. On the contrary, an analysis of SMN expression in spinal cord resident cells of treated patients revealed a higher amount of SMN protein in motor neurons after Nusinersen therapy [128]. These data suggest that SMN levels could serve as a potential biomarker to follow disease progression, although the reported discrepancies between SMN protein and mRNA expression [128,129] warrant more investigation. Additionally, as mentioned above, the tissue-specific requirement for SMN and its correlation with the patient's response to treatment is largely unexplored and future research should address these problems to obtain a thorough follow-up, in particular for the adult SMA population.

4.2. Neurofilaments

Neurofilaments (NFs) are intermediate filaments of the neuronal axons, and they are abnormally released into the extracellular fluids, namely, CSF and peripheral blood, upon axonal damage in traumatic injuries or neurodegenerative disorders (reviewed in [130,131]). Constituted of three chains, light (NFl), medium (NF-M) and heavy (NF-H), based on their molecular weights, NFs are post-transcriptionally regulated [132]. In particular, the phosphorylation, abundant on the NF-H, ensures protection from degradation [133]. Although NFs seem to be reliable biomarkers for SMA infants, as their level is reduced upon Nusinersen treatment and this is generally correlated with an improved motor function [134], few recent publications have described contradicting results for the use of NFs as biomarkers for adult SMA patients' response to therapy [29,135]. Eleven SMA type 3 patients (all adults—38.5 years mean age) analyzed for their pNF-H content in blood and CSF showed no change after administration of the Nusinersen loading doses [136]. In agreement with this effect, no significant difference was observed in the amount of NfL and pNF-H in another group of SMA patients (types 2 and 3) after the fourth injection of Nusinersen [135]. On the other hand, outcomes of a recently published analysis on a cohort of SMA type 3 patients (including kids, adolescents, and adults) showed a significant reduced amount of NFl and pNF-H after three injections with Nusinersen [29]. These discrepancies could be due to the different duration of follow-up, six months in the latter study against two months post-treatment for the previous ones. Technical limitations might also significantly contribute to the divergent outcomes (e.g., different sensitivity of the ELISA kits used). It is to be noted that at the analyzed time points, despite a decrease in the NFs levels, no locomotor function amelioration was observed in adult patients, suggesting that this aspect should be examined to a greater extent.

4.3. Protein Tau

Like neurofilaments, Tau is a neuron-specific structural protein, generally considered as a cortical neuronal marker. It was observed to be high in the CSF of patients with stroke [137], Alzheimer's disease [138] and Amyothrophic lateral sclerosis (ALS) [139]. In SMA children, the baseline level of Tau in the CSF was significantly higher than in controls and it decreased with Nusinersen treatment [140]. On the contrary, the eleven SMA type 3 patients also analyzed for the NF content, described above by Totzeck et al. [136], did not exhibit alteration in the amount of tau protein. The analyses of tau protein and the discordant outcomes highlight once again the importance of defining bigger cohorts of patients, in particular adults, to study putative molecular biomarkers for SMA.

4.4. Serum Creatinine

Most biomarker studies published to date for SMA are focusing on neuronal-related molecules as previously discussed. If neurofilament and tau levels can mirror neuronal death, it is also known that the skeletal muscle metabolism is altered in neuromuscular disorders, and this could be another aspect to investigate in the search for molecular biomarkers. Indeed, due to either progressive denervation and also to the autonomous reduced level of SMN [141,142], metabolic activity of muscles is impaired as the disease progresses. Few biomarkers of muscle activity are known such as creatinine, a waste product from the phosphorylation of adenosine diphosphate by creatine that mostly occurs in skeletal muscles. Levels of creatinine are relatively lower in patients with SMA than controls [143], as reported in a longitudinal study of a 238 SMA patient cohort. Indeed, in these SMA patients, creatinine levels correlated closely with maximum CMAP and MUNE values, even after correction for confounding factors such age and lean mass (muscle atrophy). These data suggest that serum creatinine could constitute an interesting biomarker of the muscle deterioration and, more generally, of the disease progression. However, the study failed to provide insights regarding the intra-type correlation of creatinine levels and MUNE or CMAP outcomes. A fundamental problem, especially for type 3 patients, is the wide range of phenotype and a very variable disease progression. Additionally, the range of variability from visit to visit, which is what would be needed to guide therapeutic strategies, seems to be very wide. Therefore, additional studies should be performed in clinical trials to explore if the therapies would affect the creatinine levels, and if so, whether these differences would be distinguishable from natural variations in patients.

5. Strategies for the Discoveries of Novel Biomarkers Towards the Development of Personalized Medicine Approaches for SMA

The quest for biomarkers in SMA is still ongoing. Monitoring neurofilaments levels seems to be a very promising biomarker for the prognosis of treated infant patients [134], but does not appear to constitute a robust biomarker for adults patients. Indeed, studies including only adults patients showed comparable values between SMA patients and controls (see results published in [29]), suggesting that the variability observed in SMA samples could be unrelated to the SMA pathogenesis. Consequently, no prognosis biomarker for the outcome of therapies in adult patients is known to date.

Here, we review the strategies that could help tackle this problem, highlighting the approaches used in recent papers and based on the advantages of the unbiased -omics toolbox (Table 5). We then discuss perspectives on the use of these novel techniques for the future development of personalized medicine for SMA.

5.1. Multi-Omics Approaches for the Identification of SMA Biomarkers and Potential Therapeutic Targets

The first comprehensive -omics investigation of unprecedented scale and scope has been BforSMA (Biomarkers for SMA) [144]. The novelty of this study consisted on the use of several unbiased methods (metabolomics, proteomics, and transcriptomics) for the search of biomarkers in a large well-defined SMA patient cohort and age-matched healthy controls. Despite the identification of several putative biomarkers, further validation is needed to confirm these findings. The field is now widely relying on the use of unbiased approaches to highlight potential molecular candidates. Transcriptomic and proteomic approaches are our best hope to grasp the complex mechanism that shapes the landscape of SMA clinical heterogeneity. A genome-wide RNA sequencing analysis on MNs differentiated from patient-derived induced pluripotent stem cells (iPSC) revealed that endoplasmic reticulum stress is upregulated in SMA, representing a novel potential target pathway [145]. A similar transcriptomic analysis on iPSC has been used to identify enriched motifs in differentially expressed genes between SMA and control cells, and pinpoints synaptogamin binding cytoplasmic RNA interacting protein (SYNCRIP) as a key modulator of SMN [146].

Furthermore, an interesting strategy published by Nizzardo et al. [147] took advantage of the known pathophysiological processes in motor neuron disorders (MND) to perform a comprehensive transcriptomic analysis. In particular, the authors started from the observation that specific types of MNs are spared during the progression of the MND, like SMA and ALS. They then compared the transcriptomic profile of affected MNs of the brainstem and spinal cord with the rather unaffected ocular MNs in ALS [147]. This approach led to the identification of several differentially expressed genes (DEGs), including synaptotagmin, which was demonstrated to be a neuroprotective protein in ALS. These results, also confirmed in SMA patient-derived MNs, opened new avenues for the investigation of potential novel biomarkers for SMA. In an independent study, a proteomic approach has been used to screen for candidate proteins regulated by the Nusinersen treatment in a cohort of ten type 2 and 3 patients [30]. Different proteomic clusters were identified containing proteins that are differentially expressed before and after six Nusinersen injections. This analysis allowed the identification of neuronal and non-neuronal proteins that could not only be valuable biomarkers but they could also be putative targets of SMA combined therapies.

Table 5. Overview of multi-omics approaches used to date to characterize SMA and its progression. References: bibliographic citations of the -omics studies. **-Omics:** Specific techniques employed in each study. **Samples' source:** biological material analyzed. **Highlights:** main insights on SMA pathogenesis gained in each study. Background colors are used to differentiate the cited studies.

References	-Omics	Samples' Source	Highlights
Finkel et al. [144]	Metabolomic, transcriptomic, proteomic	Plasma and urine from 108 SMA patients type 1, 2 and 3 (between 2 and 12 years of age)	97 proteins and 59 metabolites in the plasma together with 44 metabolites in the urine correlated with functional score
Rizzo et al. [146]	Transcriptomic	iPSCs-derived motorneurons from SMA patients and healthy controls	NRXN2 protein downregulation was identified as potentially neuroprotective.
Nizzardo et al. [147]	Transcriptomic	Spinal and ocular motoneurons isolated from human central nervous system sections from MND patients	Synaptogamin13 was identified as a putative neuroprotective protein in MND.
Kessler et al. [30]	Proteomic	CSF samples from 10 Nusinersen-treated adults SMA type 2 and 3	No correlation between protein profiling and functional score evolution, over 10 months treatment

5.2. Perspectives for Personalized Medicine in Neuromuscular Disorders

Successful examples of personalized medicine approaches relying on multi-omics tools come from the cancer field. Indeed, a great effort in the last decade was focused in designing targeted therapies based on cancer mutation profiles [148]. The combination of molecular screening and analysis of predictive biomarkers represents a successful emerging strategy for cancer treatment [148]. Similarly, for high-burden bacterial infections, next-generation sequencing (NGS) of the patient's fluids will surely revolutionize the diagnostic process. The pathogen can be easily identified thanks to available bioinformatic data analyzed via artificial intelligence (AI) workflows and, consequently, the appropriate therapy can be quickly determined [149,150]. For SMA and other neuromuscular disorders, these approaches could ideally lead to personalized therapies but major hurdles need to be overcome. First, for these diseases the affected tissues are not readily available for biopsy and, second, the disease spectrum is highly complex, as thoroughly discussed above [151]. Thus, further endeavors are necessary to define personalized medicine approaches for the broad SMA population.

6. Conclusions and Discussion

The complexity of the SMA phenotypes, and the discrepancies between patients' genotype (as per current definition) and phenotype, have fostered the search for functional and molecular biomarkers that could help to better classify patient types. The discovery of several disease modifier genes, starting with *SMN2* [1], and epigenetic factors [92] has revolutionized the traditional classification and account for a wider complexity. This aspect will further change in light of the recently approved therapies that are modifying the course of the disease and increasing patients' survival. For the functional measurements, it is now clear that the available tests to date are inappropriate to grasp the small changes in locomotor abilities or electrophysiological parameters of adult SMA patients. In fact, these changes seem to fall into a different scale compared to the ones form younger patients that have been widely investigated. Therefore, refining the existing scales and finding novel functional measures is a priority [152]. On the other hand, many efforts have been devoted to the identification of reliable molecular biomarkers but all the proposed candidates have revealed some limitations. Therefore, novel screening methods, approaches for an accurate prognosis, and biomarkers for treatment follow-up are being identified. Recent technological advances have allowed the development of many tools that will considerably improve the extent of the analysis. For example, novel sensitive techniques such as NGS can currently identify a broader range of genetic and epigenetic differences that can be explored as possible biomarkers. Therefore, further investigation needs to be performed to unbiasedly identify putative biomarkers for accurate diagnosis, prognosis, and treatment monitoring in SMA. These efforts, combined with refined AI approaches, will represent a milestone for a successful personalized medicine development in SMA.

Author Contributions: Conceptualization, M.G.B and P.S.; writing—original draft preparation, P.S. and P.L.; writing—review and editing, M.G.B. and P.S.; preparation of tables and figure, P.L.; revision, G.Q.; funding acquisition, M.G.B. and P.S. All authors have read and agreed to the published version of the manuscript.

Funding: This research was funded by the Association Française contre les Myopathies (AFM) and the Association Institut de Myologie (AIM), the Sorbonne Université, and the Institut National de la Santé et de la Recherche Médicale (INSERM). P.S. received funding from the Marie Skłodowska-Curie Actions under grant agreement No. 896715.

Conflicts of Interest: The authors declare no conflicts of interest.

References

1. Lefebvre, S.; Bürglen, L.; Reboullet, S.; Clermont, O.; Burlet, P.; Viollet, L.; Benichou, B.; Cruaud, C.; Millasseau, P.; Zeviani, M.; et al. Identification and characterization of a spinal muscular atrophy-determining gene. *Cell* **1995**, *80*, 155–165. [CrossRef]
2. Prior, T.W. Perspectives and diagnostic considerations in spinal muscular atrophy. *Genet. Med.* **2010**, *12*, 145–152. [CrossRef]

3. Maretina, M.A.; Zheleznyakova, G.Y.; Lanko, K.M.; Egorova, A.A.; Baranov, V.S.; Kiselev, A.V. Molecular factors involved in spinal muscular atrophy pathways as possible disease-modifying candidates. *Curr. Genom.* **2018**, *19*, 339–355. [CrossRef] [PubMed]
4. Lunn, M.R.; Wang, C.H. Spinal muscular atrophy. *Lancet* **2008**. doi:10.1016/S0140-6736(08)60921-6. [CrossRef]
5. Cuscó, I.; Barceló, M.; Rojas-García, R.; Illa, I.; Gamez, J.; Cervera, C.; Pou, A.; Izquierdo, G.; Baiget, M.; Tizzano, E. SMN2 copy number predicts acute or chronic spinal muscular atrophy but does not account for intrafamilial variability in siblings. *J. Neurol.* **2006**, *253*, 21–25. [CrossRef]
6. Jones, C.C.; Cook, S.F.; Jarecki, J.; Belter, L.; Reyna, S.P.; Staropoli, J.; Farwell, W.; Hobby, K. Spinal Muscular Atrophy (SMA) Subtype Concordance in Siblings: Findings From the Cure SMA Cohort. *J. Neuromuscul. Dis.* **2020**, *7*, 33–40. [CrossRef]
7. Feldkötter, M.; Schwarzer, V.; Wirth, R.; Wienker, T.F.; Wirth, B. Quantitative analyses of SMN1 and SMN2 based on real-time lightCycler PCR: Fast and highly reliable carrier testing and prediction of severity of spinal muscular atrophy. *Am. J. Hum. Genet.* **2002**, *70*, 358–368. [CrossRef]
8. Oprea, G.E.; Kröber, S.; McWhorter, M.L.; Rossoll, W.; Müller, S.; Krawczak, M.; Bassell, G.J.; Beattie, C.E.; Wirth, B. Plastin 3 is a protective modifier of autosomal recessive spinal muscular atrophy. *Science* **2008**, *320*, 524–527. [CrossRef]
9. Hosseinibarkooie, S.; Peters, M.; Torres-Benito, L.; Rastetter, R.H.; Hupperich, K.; Hoffmann, A.; Mendoza-Ferreira, N.; Kaczmarek, A.; Janzen, E.; Milbradt, J.; et al. The power of human protective modifiers: PLS3 and CORO1C unravel impaired endocytosis in spinal muscular atrophy and rescue SMA phenotype. *Am. J. Hum. Genet.* **2016**, *99*, 647–665. [CrossRef]
10. Riessland, M.; Kaczmarek, A.; Schneider, S.; Swoboda, K.J.; Löhr, H.; Bradler, C.; Grysko, V.; Dimitriadi, M.; Hosseinibarkooie, S.; Torres-Benito, L.; et al. Neurocalcin delta suppression protects against spinal muscular atrophy in humans and across species by restoring impaired endocytosis. *Am. J. Hum. Genet.* **2017**, *100*, 297–315. [CrossRef]
11. Calucho, M.; Bernal, S.; Alías, L.; March, F.; Venceslá, A.; Rodríguez-Álvarez, F.J.; Aller, E.; Fernández, R.M.; Borrego, S.; Millán, J.M.; et al. Correlation between SMA type and SMN2 copy number revisited: An analysis of 625 unrelated Spanish patients and a compilation of 2834 reported cases. *Neuromuscul. Disord.* **2018**, *28*, 208–215. [CrossRef] [PubMed]
12. Rodriguez-Muela, N.; Litterman, N.K.; Norabuena, E.M.; Mull, J.L.; Galazo, M.J.; Sun, C.; Ng, S.Y.; Makhortova, N.R.; White, A.; Lynes, M.M.; et al. Single-cell analysis of SMN reveals its broader role in neuromuscular disease. *Cell Rep.* **2017**, *18*, 1484–1498. [CrossRef] [PubMed]
13. Tu, W.Y.; Simpson, J.E.; Highley, J.R.; Heath, P.R. Spinal muscular atrophy: Factors that modulate motor neurone vulnerability. *Neurobiol. Dis.* **2017**, *102*, 11–20. [CrossRef] [PubMed]
14. Finkel, R.S.; Mercuri, E.; Darras, B.T.; Connolly, A.M.; Kuntz, N.L.; Kirschner, J.; Chiriboga, C.A.; Saito, K.; Servais, L.; Tizzano, E.; et al. Nusinersen versus sham control in infantile-onset spinal muscular atrophy. *N. Engl. J. Med.* **2017**, *377*, 1723–1732. [CrossRef]
15. Chiriboga, C.A.; Swoboda, K.J.; Darras, B.T.; Iannaccone, S.T.; Montes, J.; Darryl, C.; Norris, D.A.; Bennett, C.F.; Bishop, K.M. Results from a phase 1 study of nusinersen (ISIS-SMNRx) in children with spinal muscular atrophy. *Neurology* **2016**, *86*, 890–897. [CrossRef]
16. Ratni, H.; Ebeling, M.; Baird, J.; Bendels, S.; Bylund, J.; Chen, K.S.; Denk, N.; Feng, Z.; Green, L.; Guerard, M.; et al. Discovery of risdiplam, a selective survival of motor neuron-2 (SMN2) gene splicing modifier for the treatment of spinal muscular atrophy (SMA). *J. Med. Chem.* **2018**, *61*, 6501–6517.
17. Mendell, J.R.; Al-Zaidy, S.; Shell, R.; Arnold, W.D.; Rodino-Klapac, L.R.; Prior, T.W.; Lowes, L.; Alfano, L.; Berry, K.; Church, K.; et al. Single-dose gene-replacement therapy for spinal muscular atrophy. *N. Engl. J. Med.* **2017**, *377*, 1713–1722. [CrossRef]
18. Walter, M.C.; Wenninger, S.; Thiele, S.; Stauber, J.; Hiebeler, M.; Greckl, E.; Stahl, K.; Pechmann, A.; Lochmüller, H.; Kirschner, J.; et al. Safety and treatment effects of nusinersen in longstanding adult 5q-SMA type 3—A prospective observational study. *J. Neuromuscul. Dis.* **2019**, *6*, 453–465. doi:10.3233/JND-190416. [CrossRef]

19. Hagenacker, T.; Wurster, C.D.; Günther, R.; Schreiber-Katz, O.; Osmanovic, A.; Petri, S.; Weiler, M.; Ziegler, A.; Kuttler, J.; Koch, J.C.; et al. Nusinersen in adults with 5q spinal muscular atrophy: A non-interventional, multicentre, observational cohort study. *Lancet Neurol.* **2020**, *19*, 317–325. doi:10.1016/S1474-4422(20)30037-5. [CrossRef]
20. Jochmann, E.; Steinbach, R.; Jochmann, T.; Chung, H.Y.; Rödiger, A.; Neumann, R.; Mayer, T.E.; Kirchhof, K.; Loudovici-Krug, D.; Smolenski, U.C.; et al. Experiences from treating seven adult 5q spinal muscular atrophy patients with Nusinersen. *Ther. Adv. Neurol. Disord.* **2020**, *13*, 1756286420907803. [CrossRef] [PubMed]
21. Deymeer, F.; Serdaroglu, P.; Parman, Y.; Poda, M. Natural history of SMA IIIb: Muscle strength decreases in a predictable sequence and magnitude. *Neurology* **2008**, *71*, 644–649. [CrossRef] [PubMed]
22. Tiziano, F.D.; Lomastro, R.; Di Pietro, L.; Pasanisi, M.B.; Fiori, S.; Angelozzi, C.; Abiusi, E.; Angelini, C.; Soraru, G.; Gaiani, A.; et al. Clinical and molecular cross-sectional study of a cohort of adult type III spinal muscular atrophy patients: Clues from a biomarker study. *Eur. J. Hum. Genet.* **2013**, *21*, 630–636. [CrossRef] [PubMed]
23. Kariyawasam, D.S.; D'silva, A.; Lin, C.; Ryan, M.M.; Farrar, M.A. Biomarkers and the development of a personalized medicine approach in spinal muscular atrophy. *Front. Neurol.* **2019**, *10*, 1–12. doi:10.3389/fneur.2019.00898. [CrossRef] [PubMed]
24. Wadman, R.; Wijngaarde, C.; Stam, M.; Bartels, B.; Otto, L.; Lemmink, H.; Schoenmakers, M.; Cuppen, I.; van den Berg, L.; van der Pol, W. Muscle strength and motor function throughout life in a cross-sectional cohort of 180 patients with spinal muscular atrophy types 1c–4. *Eur. J. Neurol.* **2018**, *25*, 512–518. [CrossRef]
25. Wijngaarde, C.A.; Veldhoen, E.S.; van Eijk, R.P.; Stam, M.; Otto, L.A.; Asselman, F.L.; Wösten-van Asperen, R.M.; Hulzebos, E.H.; Verweij-van den Oudenrijn, L.P.; Bartels, B.; et al. Natural history of lung function in spinal muscular atrophy. *Orphanet J. Rare Dis.* **2020**, *15*, 1–11. [CrossRef]
26. O'Hagen, J.M.; Glanzman, A.M.; McDermott, M.P.; Ryan, P.A.; Flickinger, J.; Quigley, J.; Riley, S.; Sanborn, E.; Irvine, C.; Martens, W.B.; et al. An expanded version of the Hammersmith Functional Motor Scale for SMA II and III patients. *Neuromuscul. Disord.* **2007**, *17*, 693–697. [CrossRef]
27. Kaufmann, P.; McDermott, M.P.; Darras, B.T.; Finkel, R.S.; Sproule, D.M.; Kang, P.B.; Oskoui, M.; Constantinescu, A.; Gooch, C.L.; Foley, A.R.; et al. Prospective cohort study of spinal muscular atrophy types 2 and 3. *Neurology* **2012**, *79*, 1889–1897. [CrossRef]
28. Montes, J.; McDermott, M.P.; Mirek, E.; Mazzone, E.S.; Main, M.; Glanzman, A.M.; Duong, T.; Young, S.D.; Salazar, R.; Pasternak, A.; et al. Ambulatory function in spinal muscular atrophy: Age-related patterns of progression. *PLoS ONE* **2018**, *13*, e0199657. [CrossRef]
29. Faravelli, I.; Meneri, M.; Saccomanno, D.; Velardo, D.; Abati, E.; Gagliardi, D.; Parente, V.; Petrozzi, L.; Ronchi, D.; Stocchetti, N.; et al. Nusinersen treatment and cerebrospinal fluid neurofilaments: An explorative study on Spinal Muscular Atrophy type 3 patients. *J. Cell. Mol. Med.* **2020**, 1–6. doi:10.1111/jcmm.14939. [CrossRef]
30. Kessler, T.; Latzer, P.; Schmid, D.; Warnken, U.; Saffari, A.; Ziegler, A.; Kollmer, J.; Möhlenbruch, M.; Ulfert, C.; Herweh, C.; et al. Cerebrospinal fluid proteomic profiling in nusinersen-treated patients with spinal muscular atrophy. *J. Neurochem.* **2020**, e14953. doi:10.1111/jnc.14953. [CrossRef]
31. Montes, J.; Glanzman, A.M.; Mazzone, E.S.; Martens, W.B.; Dunaway, S.; Pasternak, A.; Riley, S.O.; Quigley, J.; Pandya, S.; De Vivo, D.C.; et al. Spinal muscular atrophy functional composite score: A functional measure in spinal muscular atrophy. *Muscle Nerve* **2015**, *52*, 942–947. [CrossRef] [PubMed]
32. Mazzone, E.; Bianco, F.; Martinelli, D.; Glanzman, A.M.; Messina, S.; De Sanctis, R.; Main, M.; Eagle, M.; Florence, J.; Krosschell, K.; et al. Assessing upper limb function in nonambulant SMA patients: Development of a new module. *Neuromuscul. Disord.* **2011**, *21*, 406–412. [CrossRef]
33. Mazzone, E.S.; Mayhew, A.; Montes, J.; Ramsey, D.; Fanelli, L.; Young, S.D.; Salazar, R.; De Sanctis, R.; Pasternak, A.; Glanzman, A.; et al. Revised upper limb module for spinal muscular atrophy: Development of a new module. *Muscle Nerve* **2017**, *55*, 869–874. [CrossRef] [PubMed]
34. Pera, M.C.; Coratti, G.; Mazzone, E.S.; Montes, J.; Scoto, M.; De Sanctis, R.; Main, M.; Mayhew, A.; Muni Lofra, R.; Dunaway Young, S.; et al. Revised upper limb module for spinal muscular atrophy: 12 month changes. *Muscle Nerve* **2019**, *59*, 426–430. [CrossRef] [PubMed]

35. Kizina, K.; Stolte, B.; Totzeck, A.; Bolz, S.; Fleischer, M.; Mönninghoff, C.; Guberina, N.; Oldenburg, D.; Forsting, M.; Kleinschnitz, C.; et al. Clinical implication of dosimetry of computed tomography-and fluoroscopy-guided intrathecal therapy with nusinersen in adult patients with spinal muscular atrophy. *Front. Neurol.* **2019**, *10*, 1166. [CrossRef]
36. Stolte, B.; Totzeck, A.; Kizina, K.; Bolz, S.; Pietruck, L.; Mönninghoff, C.; Guberina, N.; Oldenburg, D.; Forsting, M.; Kleinschnitz, C.; et al. Feasibility and safety of intrathecal treatment with nusinersen in adult patients with spinal muscular atrophy. *Ther. Adv. Neurol. Disord.* **2018**, *11*, 1756286418803246. [CrossRef]
37. Zerres, K.; Rudnik-Schöneborn, S.; Forrest, E.; Lusakowska, A.; Borkowska, J.; Hausmanowa-Petrusewicz, I. A collaborative study on the natural history of childhood and juvenile onset proximal spinal muscular atrophy (type II and III SMA): 569 patients. *J. Neurol. Sci.* **1997**, *146*, 67–72. [CrossRef]
38. ATS statement: Guidelines for the six-minute walk test. *Am. J. Respir. Crit. Care Med.* **2002**, *166*, 111–117. [CrossRef]
39. Dunaway Young, S.; Montes, J.; Kramer, S.S.; Marra, J.; Salazar, R.; Cruz, R.; Chiriboga, C.A.; Garber, C.E.; De Vivo, D.C. Six-minute walk test is reliable and valid in spinal muscular atrophy. *Muscle Nerve* **2016**, *54*, 836–842. [CrossRef]
40. McKay, M.J.; Baldwin, J.N.; Ferreira, P.; Simic, M.; Vanicek, N.; Burns, J.; For the 1000 Norms Project Consortium. Reference values for developing responsive functional outcome measures across the lifespan. *Neurology* **2017**, *88*, 1512–1519. [CrossRef]
41. Mills, K.R. The basics of electromyography. *J. Neurol. Neurosurg. Psychiatry* **2005**, *76*, ii32–ii35. [CrossRef] [PubMed]
42. Kimura, J. Peripheral nerve conduction studies and neuromuscular junction testing. In *Handbook of Clinical Neurophysiology*; Elsevier: Amsterdam, The Netherlands, 2004; Volume 4, pp. 241–270.
43. Arnold, W.D.; Porensky, P.N.; McGovern, V.L.; Iyer, C.C.; Duque, S.; Li, X.; Meyer, K.; Schmelzer, L.; Kaspar, B.K.; Kolb, S.J.; et al. Electrophysiological biomarkers in spinal muscular atrophy: Proof of concept. *Ann. Clin. Transl. Neurol.* **2014**, *1*, 34. [CrossRef] [PubMed]
44. Nandedkar, S.D.; Nandedkar, D.S.; Barkhaus, P.E.; Stalberg, E.V. Motor unit number index (MUNIX). *IEEE Tans. Biomed. Eng.* **2004**, *51*, 2209–2211. [CrossRef] [PubMed]
45. Hausmanowa-Petrusewicz, I. Role of electromyography in the diagnosis of motor neuron disorders. *Neuropatol. Pol.* **1992**, *30*, 187–197.
46. Kolb, S.J.; Coffey, C.S.; Yankey, J.W.; Krosschell, K.; Arnold, W.D.; Rutkove, S.B.; Swoboda, K.J.; Reyna, S.P.; Sakonju, A.; Darras, B.T.; et al. Natural history of infantile-onset spinal muscular atrophy. *Ann. Neurol.* **2017**, *82*, 883–891. [CrossRef]
47. Swoboda, K.J.; Prior, T.W.; Scott, C.B.; McNaught, T.P.; Wride, M.C.; Reyna, S.P.; Bromberg, M.B. Natural history of denervation in SMA: Relation to age, SMN2 copy number, and function. *Ann. Neurol. Offi. J. Am. Neurol. Assoc. Child Neurol. Soc.* **2005**, *57*, 704–712. [CrossRef]
48. Querin, G.; Lenglet, T.; Debs, R.; Stojkovic, T.; Béhin, A.; Salachas, F.; Le Forestier, N.; del Mar Amador, M.; Lacomblez, L.; Meininger, V.; et al. The motor unit number index (MUNIX) profile of patients with adult spinal muscular atrophy. *Clini. Neurophysiol.* **2018**, *129*, 2333–2340. [CrossRef]
49. Querin, G.; El Mendili, M.M.; Lenglet, T.; Behin, A.; Stojkovic, T.; Salachas, F.; Devos, D.; Le Forestier, N.; del Mar Amador, M.; Debs, R.; et al. The spinal and cerebral profile of adult spinal-muscular atrophy: A multimodal imaging study. *NeuroImage Clin.* **2019**, *21*, 101618. [CrossRef]
50. Taso, M.; Le Troter, A.; Sdika, M.; Ranjeva, J.P.; Guye, M.; Bernard, M.; Callot, V. Construction of an in vivo human spinal cord atlas based on high-resolution MR images at cervical and thoracic levels: Preliminary results. *Magn. Reson. Mater. Phys. Biol. Med.* **2014**, *27*, 257–267. [CrossRef]
51. Lefebvre, S.; Burlet, P.; Liu, Q.; Bertrandy, S.; Clermont, O.; Munnich, A.; Dreyfuss, G.; Melki, J. Correlation between severity and SMN protein level in spinal muscular atrophy. *Nat. Genet.* **1997**, *16*, 265–269. [CrossRef]
52. Scharf, J.M.; Endrizzi, M.G.; Wetter, A.; Huang, S.; Thompson, T.G.; Zerres, K.; Dietrich, W.F.; Wirth, B.; Kunkel, L.M. Identification of a candidate modifying gene for spinal muscular atrophy by comparative genomics. *Nat. Genet.* **1998**, *20*, 83–86. [CrossRef] [PubMed]
53. Anderson, K.; Talbot, K. Spinal muscular atrophies reveal motor neuron vulnerability to defects in ribonucleoprotein handling. *Curr. Opin. Neurol.* **2003**, *16*, 595–599. [CrossRef] [PubMed]

54. Bürglen, L.; Seroz, T.; Miniou, P.; Lefebvre, S.; Burlet, P.; Munnich, A.; Pequignot, E.V.; Egly, J.M.; Melki, J. The gene encoding p44, a subunit of the transcription factor TFIIH, is involved in large-scale deletions associated with Werdnig-Hoffmann disease. *Am. J. Hum. Genet.* **1997**, *60*, 72. [PubMed]
55. Jiang, J.; Huang, J.; Gu, J.; Cai, X.; Zhao, H.; Lu, H. Genomic analysis of a spinal muscular atrophy (SMA) discordant family identifies a novel mutation in TLL2, an activator of growth differentiation factor 8 (myostatin): A case report. *BMC Med. Genet.* **2019**, *20*, 204. [CrossRef]
56. Van der Steege, G.; Grootscholten, P.; Van der Vlies, P.; Draaijers, T.; Osinga, J.; Cobben, J.; Scheffer, H.; Buys, C. PCR-based DNA test to confirm clinical diagnosis of autosomal recessive spinal muscular atrophy. *Lancet* **1995**, *345*, 985–986. [CrossRef]
57. Coovert, D.D.; Le, T.T.; McAndrew, P.E.; Strasswimmer, J.; Crawford, T.O.; Mendell, J.R.; Coulson, S.E.; Androphy, E.J.; Prior, T.W.; Burghes, A.H. The survival motor neuron protein in spinal muscular atrophy. *Hum. Mol. Genet.* **1997**, *6*, 1205–1214. [CrossRef]
58. McAndrew, P.; Parsons, D.; Simard, L.; Rochette, C.; Ray, P.; Mendell, J.; Prior, T.; Burghes, A. Identification of proximal spinal muscular atrophy carriers and patients by analysis of SMNT and SMNC gene copy number. *Am. J. Hum. Genet.* **1997**, *60*, 1411–1422. [CrossRef]
59. Taylor, J.E.; Thomas, N.H.; Lewis, C.M.; Abbs, S.J.; Rodrigues, N.R.; Davies, K.E.; Mathew, C.G. Correlation of SMN t and SMN c gene copy number with age of onset and survival in spinal muscular atrophy. *Eur. J. Hum. Genet.* **1998**, *6*, 467–474. [CrossRef]
60. Mailman, M.D.; Heinz, J.W.; Papp, A.C.; Snyder, P.J.; Sedra, M.S.; Wirth, B.; Burghes, A.H.; Prior, T.W. Molecular analysis of spinal muscular atrophy and modification of the phenotype by SMN2. *Genet. Med.* **2002**, *4*, 20–26. [CrossRef]
61. Anhuf, D.; Eggermann, T.; Rudnik-Schöneborn, S.; Zerres, K. Determination of SMN1 and SMN2 copy number using TaqMan™ technology. *Hum. Mutat.* **2003**, *22*, 74–78. [CrossRef]
62. Scarciolla, O.; Stuppia, L.; De Angelis, M.V.; Murru, S.; Palka, C.; Giuliani, R.; Pace, M.; Di Muzio, A.; Torrente, I.; Morella, A.; et al. Spinal muscular atrophy genotyping by gene dosage using multiple ligation-dependent probe amplification. *Neurogenetics* **2006**, *7*, 269–276. [CrossRef] [PubMed]
63. Wirth, B.; Brichta, L.; Schrank, B.; Lochmüller, H.; Blick, S.; Baasner, A.; Heller, R. Mildly affected patients with spinal muscular atrophy are partially protected by an increased SMN2 copy number. *Hum. Genet.* **2006**, *119*, 422–428. [CrossRef] [PubMed]
64. Gómez-Curet, I.; Robinson, K.G.; Funanage, V.L.; Crawford, T.O.; Scavina, M.; Wang, W. Robust quantification of the SMN gene copy number by real-time TaqMan PCR. *Neurogenetics* **2007**, *8*, 271–278. [CrossRef] [PubMed]
65. Huang, C.H.; Chang, Y.Y.; Chen, C.H.; Kuo, Y.S.; Hwu, W.L.; Gerdes, T.; Ko, T.M. Copy number analysis of survival motor neuron genes by multiplex ligation-dependent probe amplification. *Genet. Med.* **2007**, *9*, 241–248. [CrossRef]
66. Tiziano, F.; Bertini, E.; Messina, S.; Angelozzi, C.; Pane, M.; D'Amico, A.; Alfieri, P.; Fiori, S.; Battini, R.; Berardinelli, A.; et al. The Hammersmith functional score correlates with the SMN2 copy number: A multicentric study. *Neuromuscul. Disord.* **2007**, *17*, 400–403. [CrossRef]
67. Elsheikh, B.; Prior, T.; Zhang, X.; Miller, R.; Kolb, S.J.; Moore, D.; Bradley, W.; Barohn, R.; Bryan, W.; Gelinas, D.; et al. An analysis of disease severity based on SMN2 copy number in adults with spinal muscular atrophy. *Muscle Nerve Off. J. Am. Assoc. Electrodiagn. Med.* **2009**, *40*, 652–656. [CrossRef]
68. Alias, L.; Bernal, S.; Barceló, M.J.; Also-Rallo, E.; Martínez-Hernández, R.; Rodriguez-Alvarez, F.J.; Hernández-Chico, C.; Baiget, M.; Tizzano, E.F. Accuracy of marker analysis, quantitative real-time polymerase chain reaction, and multiple ligation-dependent probe amplification to determine SMN2 copy number in patients with spinal muscular atrophy. *Genet. Test. Mol. Biomark.* **2011**, *15*, 587–594. [CrossRef]
69. Amara, A.; Adala, L.; Charfeddine, I.B.; Mamaï, O.; Mili, A.; Lazreg, T.B.; H'mida, D.; Amri, F.; Salem, N.; Boughammura, L.; et al. Correlation of SMN2, NAIP, p44, H4F5 and Occludin genes copy number with spinal muscular atrophy phenotype in Tunisian patients. *Eur. J. Paediatric Neurol.* **2012**, *16*, 167–174. [CrossRef]
70. Crawford, T.O.; Paushkin, S.V.; Kobayashi, D.T.; Forrest, S.J.; Joyce, C.L.; Finkel, R.S.; Kaufmann, P.; Swoboda, K.J.; Tiziano, D.; Lomastro, R.; et al. Evaluation of SMN protein, transcript, and copy number in the biomarkers for spinal muscular atrophy (BforSMA) clinical study. *PLoS ONE* **2012**, *7* e33572. [CrossRef]

71. Dobrowolski, S.F.; Pham, H.T.; Pouch Downes, F.; Prior, T.W.; Naylor, E.W.; Swoboda, K.J. Newborn screening for spinal muscular atrophy by calibrated short-amplicon melt profiling. *Clin. Chem.* **2012**, *58*, 1033–1039. [CrossRef]
72. Kirwin, S.M.; Vinette, K.M.; Gonzalez, I.L.; Abdulwahed, H.A.; Al-Sannaa, N.; Funanage, V.L. A homozygous double mutation in SMN 1: A complicated genetic diagnosis of SMA. *Mol. Genet. Genom. Med.* **2013**, *1*, 113–117. [CrossRef] [PubMed]
73. Brkušanin, M.; Kosać, A.; Jovanović, V.; Pešović, J.; Brajušković, G.; Dimitrijević, N.; Todorović, S.; Romac, S.; Rašić, V.M.; Savić-Pavićević, D. Joint effect of the SMN2 and SERF1A genes on childhood-onset types of spinal muscular atrophy in Serbian patients. *J. Hum. Genet.* **2015**, *60*, 723–728. [CrossRef] [PubMed]
74. Fang, P.; Li, L.; Zeng, J.; Zhou, W.J.; Wu, W.Q.; Zhong, Z.Y.; Yan, T.Z.; Xie, J.S.; Huang, J.; Lin, L.; et al. Molecular characterization and copy number of SMN1, SMN2 and NAIP in Chinese patients with spinal muscular atrophy and unrelated healthy controls. *BMC Musculoskel. Disord.* **2015**, *16*, 11. [CrossRef] [PubMed]
75. Stabley, D.L.; Harris, A.W.; Holbrook, J.; Chubbs, N.J.; Lozo, K.W.; Crawford, T.O.; Swoboda, K.J.; Funanage, V.L.; Wang, W.; Mackenzie, W.; et al. SMN1 and SMN2 copy numbers in cell lines derived from patients with spinal muscular atrophy as measured by array digital PCR. *Mol. Genet. Genom. Med.* **2015**, *3*, 248–257. [CrossRef]
76. Prior, T.W.; Krainer, A.R.; Hua, Y.; Swoboda, K.J.; Snyder, P.C.; Bridgeman, S.J.; Burghes, A.H.; Kissel, J.T. A positive modifier of spinal muscular atrophy in the SMN2 gene. *Am. J. Hum. Genet.* **2009**, *85*, 408–413. [CrossRef]
77. Bernal, S.; Alias, L.; Barcelo, M.; Also-Rallo, E.; Martínez-Hernández, R.; Gámez, J.; Guillén-Navarro, E.; Rosell, J.; Hernando, I.; Rodríguez-Alvarez, F.; et al. The c. 859G> C variant in the SMN2 gene is associated with types II and III SMA and originates from a common ancestor. *J. Med. Genet.* **2010**, *47*, 640–642. [CrossRef]
78. Vezain, M.; Saugier-Veber, P.; Goina, E.; Touraine, R.; Manel, V.; Toutain, A.; Fehrenbach, S.; Frébourg, T.; Pagani, F.; Tosi, M.; et al. A rare SMN2 variant in a previously unrecognized composite splicing regulatory element induces exon 7 inclusion and reduces the clinical severity of spinal muscular atrophy. *Hum. Mutat.* **2010**, *31*, E1110–E1125. [CrossRef]
79. Burghes, A.H.; Ingraham, S.E.; McLean, M.; Thompson, T.G.; McPherson, J.D.; Kote-Jarai, Z.; Carpten, J.D.; DiDonato, C.J.; Ikeda, J.E.; Surh, L.; et al. A multicopy dinucleotide marker that maps close to the spinal muscular atrophy gene. *Genomics* **1994**, *21*, 394–402. [CrossRef]
80. Cobben, J.; Van der Steege, G.; Grootscholten, P.; De Visser, M.; Scheffer, H.; Buys, C. Deletions of the survival motor neuron gene in unaffected siblings of patients with spinal muscular atrophy. *Am. J. Hum. Genet.* **1995**, *57*, 805.
81. Hahnen, E.; Forkert, R.; Marke, C.; Rudnik-Schöneborn, S.; Schönling, J.; Zerres, K.; Creavin, T.; Wirth, B. Molecular analysis of candidate genes on chromosome 5q13 in autosomal recessive spinal muscular atrophy: Evidence of homozygous deletions of the SMN gene in unaffected individuals. *Hum. Mol. Genet.* **1995**, *4*, 1927–1933. [CrossRef]
82. Ackermann, B.; Kröber, S.; Torres-Benito, L.; Borgmann, A.; Peters, M.; Hosseini Barkooie, S.M.; Tejero, R.; Jakubik, M.; Schreml, J.; Milbradt, J.; et al. Plastin 3 ameliorates spinal muscular atrophy via delayed axon pruning and improves neuromuscular junction functionality. *Hum. Mol. Genet.* **2013**, *22*, 1328–1347. [CrossRef]
83. Heesen, L.; Peitz, M.; Torres-Benito, L.; Hölker, I.; Hupperich, K.; Dobrindt, K.; Jungverdorben, J.; Ritzenhofen, S.; Weykopf, B.; Eckert, D.; et al. Plastin 3 is upregulated in iPSC-derived motoneurons from asymptomatic SMN1-deleted individuals. *Cell. Mol. Life Sci.* **2016**, *73*, 2089–2104. [CrossRef] [PubMed]
84. Hasanzad, M.; Azad, M.; Kahrizi, K.; Saffar, B.S.; Nafisi, S.; Keyhanidoust, Z.; Azimian, M.; Refah, A.; Urtizberea, J.A.; Tizzano, E.F.; et al. Carrier frequency of SMA by quantitative analysis of the SMN1 deletion in the Iranian population. *Eur. J. Neurol.* **2010**, *17*, 160. [CrossRef] [PubMed]
85. Kaifer, K.A.; Villalón, E.; Osman, E.Y.; Glascock, J.J.; Arnold, L.L.; Cornelison, D.; Lorson, C.L. Plastin-3 extends survival and reduces severity in mouse models of spinal muscular atrophy. *JCI Insight* **2017**, *2*, e89970. [CrossRef] [PubMed]

86. McGovern, V.L.; Massoni-Laporte, A.; Wang, X.; Le, T.T.; Le, H.T.; Beattie, C.E.; Rich, M.M.; Burghes, A.H. Plastin 3 expression does not modify spinal muscular atrophy severity in the Δ7 SMA mouse. *PLoS ONE* **2015**, *10*, e0132364. [CrossRef]
87. Hao, L.T.; Wolman, M.; Granato, M.; Beattie, C.E. Survival motor neuron affects plastin 3 protein levels leading to motor defects. *J. Neurosci.* **2012**, *32*, 5074–5084. [CrossRef]
88. Janzen, E.; Mendoza-Ferreira, N.; Hosseinibarkooie, S.; Schneider, S.; Hupperich, K.; Tschanz, T.; Grysko, V.; Riessland, M.; Hammerschmidt, M.; Rigo, F.; et al. CHP1 reduction ameliorates spinal muscular atrophy pathology by restoring calcineurin activity and endocytosis. *Brain* **2018**, *141*, 2343–2361. [CrossRef]
89. Torres-Benito, L.; Schneider, S.; Rombo, R.; Ling, K.K.; Grysko, V.; Upadhyay, A.; Kononenko, N.L.; Rigo, F.; Bennett, C.F.; Wirth, B. NCALD Antisense oligonucleotide therapy in addition to nusinersen further ameliorates spinal muscular atrophy in mice. *Am. J. Hum. Genet.* **2019**, *105*, 221–230. [CrossRef]
90. Hassan, H.A.; Zaki, M.S.; Issa, M.Y.; El-Bagoury, N.M.; Essawi, M.L. Genetic pattern of SMN1, SMN2, and NAIP genes in prognosis of SMA patients. *Egypt. J. Med. Hum. Genet.* **2020**, *21*, 1–7. [CrossRef]
91. Essawi, M.; Effat, L.; Shanab, G.; Al-Ettribi, G.; El-Haronui, A.; Karim, A. Molecular analysis of SMN1 and NAIP genes in Egyptian patients with spinal muscular atrophy. *Bratislavske lekarske listy* **2007**, *108*, 133.
92. Hauke, J.; Riessland, M.; Lunke, S.; Eyüpoglu, I.Y.; Blümcke, I.; El-Osta, A.; Wirth, B.; Hahnen, E. Survival motor neuron gene 2 silencing by DNA methylation correlates with spinal muscular atrophy disease severity and can be bypassed by histone deacetylase inhibition. *Hum. Mol. Genet.* **2009**, *18*, 304–317. [CrossRef] [PubMed]
93. Cao, Y.Y.; Qu, Y.J.; He, S.X.; Li, Y.; Bai, J.L.; Jin, Y.W.; Wang, H.; Song, F. Association between SMN2 methylation and disease severity in Chinese children with spinal muscular atrophy. *J. Zhejiang Univ. Sci. B* **2016**, *17*, 76–82. [CrossRef] [PubMed]
94. Zheleznyakova, G.Y.; Voisin, S.; Kiselev, A.V.; Almén, M.S.; Xavier, M.J.; Maretina, M.A.; Tishchenko, L.I.; Fredriksson, R.; Baranov, V.S.; Schiöth, H.B. Genome-wide analysis shows association of epigenetic changes in regulators of Rab and Rho GTPases with spinal muscular atrophy severity. *Eur. J. Hum. Genet.* **2013**, *21*, 988–993. [CrossRef] [PubMed]
95. Zheleznyakova, G.Y.; Nilsson, E.K.; Kiselev, A.V.; Maretina, M.A.; Tishchenko, L.I.; Fredriksson, R.; Baranov, V.S.; Schiöth, H.B. Methylation levels of SLC23A2 and NCOR2 genes correlate with spinal muscular atrophy severity. *PLoS ONE* **2015**, *10* e0121964. [CrossRef]
96. Aguilera, O.; Fernández, A.F.; Muñoz, A.; Fraga, M.F. Epigenetics and environment: A complex relationship. *J. Appl. Physiol.* **2010**, *109*, 243–251. [CrossRef]
97. Chang, J.G.; Hsieh-Li, H.M.; Jong, Y.J.; Wang, N.M.; Tsai, C.H.; Li, H. Treatment of spinal muscular atrophy by sodium butyrate. *Proc. Natl. Acad. Sci. USA* **2001**, *98*, 9808–9813. [CrossRef]
98. Brichta, L.; Hofmann, Y.; Hahnen, E.; Siebzehnrubl, F.; Raschke, H.; Blumcke, I.; Eyupoglu, I.; Wirth, B. Valproic acid increases the SMN2 protein level: A well-known drug as a potential therapy for spinal muscular atrophy. *Hum. Mol. Genet.* **2003**, *12*, 2481–2489. [CrossRef]
99. Sumner, C.J.; Huynh, T.N.; Markowitz, J.A.; Perhac, J.S.; Hill, B.; Coovert, D.D.; Schussler, K.; Chen, X.; Jarecki, J.; Burghes, A.H.; et al. Valproic acid increases SMN levels in spinal muscular atrophy patient cells. *Ann.Neurol. Offi. J. Am. Neurol. Assoc. Child Neurol. Soc.* **2003**, *54*, 647–654. [CrossRef]
100. Hahnen, E.; Eyüpoglu, I.Y.; Brichta, L.; Haastert, K.; Tränkle, C.; Siebzehnrübl, F.A.; Riessland, M.; Hölker, I.; Claus, P.; Romstöck, J.; et al. In vitro and ex vivo evaluation of second-generation histone deacetylase inhibitors for the treatment of spinal muscular atrophy. *J. Neurochem.* **2006**, *98*, 193–202. [CrossRef]
101. Mohseni, J.; Al-Najjar, B.O.; Wahab, H.A.; Zabidi-Hussin, Z.; Sasongko, T.H. Transcript, methylation and molecular docking analyses of the effects of HDAC inhibitors, SAHA and Dacinostat, on SMN2 expression in fibroblasts of SMA patients. *J. Hum. Genet.* **2016**, *61*, 823–830. [CrossRef]
102. Riessland, M.; Brichta, L.; Hahnen, E.; Wirth, B. The benzamide M344, a novel histone deacetylase inhibitor, significantly increases SMN2 RNA/protein levels in spinal muscular atrophy cells. *Hum. Genet.* **2006**, *120*, 101–110. [CrossRef] [PubMed]
103. Avila, A.M.; Burnett, B.G.; Taye, A.A.; Gabanella, F.; Knight, M.A.; Hartenstein, P.; Cizman, Z.; Di Prospero, N.A.; Pellizzoni, L.; Fischbeck, K.H.; et al. Trichostatin A increases SMN expression and survival in a mouse model of spinal muscular atrophy. *J. Clin. Investig.* **2007**, *117*, 659–671. [CrossRef]

104. Garbes, L.; Riessland, M.; Hölker, I.; Heller, R.; Hauke, J.; Tränkle, C.; Coras, R.; Blümcke, I.; Hahnen, E.; Wirth, B. LBH589 induces up to 10-fold SMN protein levels by several independent mechanisms and is effective even in cells from SMA patients non-responsive to valproate. *Hum. Mol. Genet.* **2009**, *18*, 3645–3658. [CrossRef] [PubMed]
105. Pagliarini, V.; Guerra, M.; Di Rosa, V.; Compagnucci, C.; Sette, C. Combined treatment with the histone deacetylase inhibitor LBH589 and a splice-switch antisense oligonucleotide enhances SMN2 splicing and SMN expression in Spinal Muscular Atrophy cells. *J. Neurochem.* **2020**, *153*, 264–275. [CrossRef] [PubMed]
106. Swoboda, K.J.; Scott, C.B.; Reyna, S.P.; Prior, T.W.; LaSalle, B.; Sorenson, S.L.; Wood, J.; Acsadi, G.; Crawford, T.O.; Kissel, J.T.; et al. Phase II open label study of valproic acid in spinal muscular atrophy. *PLoS ONE* **2009**, *4*, e5268. [CrossRef]
107. Swoboda, K.J.; Scott, C.B.; Crawford, T.O.; Simard, L.R.; Reyna, S.P.; Krosschell, K.J.; Acsadi, G.; Elsheik, B.; Schroth, M.K.; D'Anjou, G.; et al. SMA CARNI-VAL trial part I: Double-blind, randomized, placebo-controlled trial of L-carnitine and valproic acid in spinal muscular atrophy. *PLoS ONE* **2010**, *5* e12140. [CrossRef]
108. Krosschell, K.J.; Kissel, J.T.; Townsend, E.L.; Simeone, S.D.; Zhang, R.Z.; Reyna, S.P.; Crawford, T.O.; Schroth, M.K.; Acsadi, G.; Kishnani, P.S.; et al. Clinical trial of L-Carnitine and valproic acid in spinal muscular atrophy type I. *Muscle Nerve* **2018**, *57*, 193–199. [CrossRef]
109. Kissel, J.T.; Scott, C.B.; Reyna, S.P.; Crawford, T.O.; Simard, L.R.; Krosschell, K.J.; Acsadi, G.; Elsheik, B.; Schroth, M.K.; D'Anjou, G.; et al. SMA carni-VAL trial part II: A prospective, single-armed trial of L-carnitine and valproic acid in ambulatory children with spinal muscular atrophy. *PLoS ONE* **2011**, *6*, e21296. [CrossRef]
110. Kissel, J.T.; Elsheikh, B.; King, W.M.; Freimer, M.; Scott, C.B.; Kolb, S.J.; Reyna, S.P.; Crawford, T.O.; Simard, L.R.; Krosschell, K.J.; et al. SMA valiant trial: A prospective, double-blind, placebo-controlled trial of valproic acid in ambulatory adults with spinal muscular atrophy. *Muscle Nerve* **2014**, *49*, 187–192. [CrossRef]
111. Darbar, I.A.; Plaggert, P.G.; Resende, M.B.D.; Zanoteli, E.; Reed, U.C. Evaluation of muscle strength and motor abilities in children with type II and III spinal muscle atrophy treated with valproic acid. *BMC Neurol.* **2011**, *11*, 36. [CrossRef]
112. Elshafay, A.; Hieu, T.H.; Doheim, M.F.; Kassem, M.A.M.; ELdoadoa, M.F.; Holloway, S.K.; Abo-Elghar, H.; Hirayama, K.; Huy, N.T. Efficacy and safety of valproic acid for spinal muscular atrophy: A systematic review and meta-analysis. *CNS Drugs* **2019**, *33*, 239–250. [CrossRef]
113. Brahe, C.; Vitali, T.; Tiziano, F.D.; Angelozzi, C.; Pinto, A.M.; Borgo, F.; Moscato, U.; Bertini, E.; Mercuri, E.; Neri, G. Phenylbutyrate increases SMN gene expression in spinal muscular atrophy patients. *Eur. J. Hum. Genet.* **2005**, *13*, 256–259. [CrossRef]
114. Mercuri, E.; Bertini, E.; Messina, S.; Pelliccioni, M.; D'Amico, A.; Colitto, F.; Mirabella, M.; Tiziano, F.D.; Vitali, T.; Angelozzi, C.; et al. Pilot trial of phenylbutyrate in spinal muscular atrophy. *Neuromuscul. Disord.* **2004**, *14*, 130–135. [CrossRef]
115. Mercuri, E.; Bertini, E.; Messina, S.; Solari, A.; D'amico, A.; Angelozzi, C.; Battini, R.; Berardinelli, A.; Boffi, P.; Bruno, C.; et al. Randomized, double-blind, placebo-controlled trial of phenylbutyrate in spinal muscular atrophy. *Neurology* **2007**, *68*, 51–55. [CrossRef]
116. Liu, Q.; Dreyfuss, G. A novel nuclear structure containing the survival of motor neurons protein. *EMBO J.* **1996**, *15*, 3555–3565. [CrossRef]
117. Li, D.K.; Tisdale, S.; Lotti, F.; Pellizzoni, L. SMN control of RNP assembly: From post-transcriptional gene regulation to motor neuron disease. *Semin. Cell Dev. Biol.* **2014**, *32*, 22–29. [CrossRef]
118. Hensel, N.; Claus, P. The actin cytoskeleton in SMA and ALS: How does it contribute to motoneuron degeneration? *Neuroscientist* **2018**, *24*, 54–72. [CrossRef]
119. Donlin-Asp, P.G.; Bassell, G.J.; Rossoll, W. A role for the survival of motor neuron protein in mRNP assembly and transport. *Curr. Opin. Neurobiol.* **2016**, *39*, 53–61. [CrossRef]
120. Groen, E.J.; Gillingwater, T.H. UBA1: At the crossroads of ubiquitin homeostasis and neurodegeneration. *Trends Mol. Med.* **2015**, *21*, 622–632. [CrossRef]
121. Boyd, P.J.; Tu, W.Y.; Shorrock, H.K.; Groen, E.J.; Carter, R.N.; Powis, R.A.; Thomson, S.R.; Thomson, D.; Graham, L.C.; Motyl, A.A.; et al. Bioenergetic status modulates motor neuron vulnerability and pathogenesis in a zebrafish model of spinal muscular atrophy. *PLoS Genet.* **2017**, *13*, e1006744. [CrossRef]

122. Kong, L.; Wang, X.; Choe, D.W.; Polley, M.; Burnett, B.G.; Bosch-Marcé, M.; Griffin, J.W.; Rich, M.M.; Sumner, C.J. Impaired synaptic vesicle release and immaturity of neuromuscular junctions in spinal muscular atrophy mice. *J. Neurosci.* **2009**, *29*, 842–851. [CrossRef] [PubMed]
123. Akten, B.; Kye, M.J.; Hao, L.T.; Wertz, M.H.; Singh, S.; Nie, D.; Huang, J.; Merianda, T.T.; Twiss, J.L.; Beattie, C.E.; et al. Interaction of survival of motor neuron (SMN) and HuD proteins with mRNA cpg15 rescues motor neuron axonal deficits. *Proc. Natl. Acad. Sci. USA* **2011**, *108*, 10337–10342. [CrossRef] [PubMed]
124. Fallini, C.; Zhang, H.; Su, Y.; Silani, V.; Singer, R.H.; Rossoll, W.; Bassell, G.J. The survival of motor neuron (SMN) protein interacts with the mRNA-binding protein HuD and regulates localization of poly (A) mRNA in primary motor neuron axons. *J. Neurosci.* **2011**, *31*, 3914–3925. [CrossRef] [PubMed]
125. Ottesen, E.W.; Singh, N.N.; Luo, D.; Singh, R.N. High-affinity RNA targets of the Survival Motor Neuron protein reveal diverse preferences for sequence and structural motifs. *Nucleic Acids Res.* **2018**, *46*, 10983–11001. [CrossRef] [PubMed]
126. Hamilton, G.; Gillingwater, T.H. Spinal muscular atrophy: Going beyond the motor neuron. *Trends Mol. Med.* **2013**, *19*, 40–50. [CrossRef]
127. Besse, A.; Astord, S.; Marais, T.; Roda, M.; Giroux, B.; Lejeune, F.X.; Relaix, F.; Smeriglio, P.; Barkats, M.; Biferi, M.G. AAV9-mediated expression of SMN restricted to neurons does not rescue the spinal muscular atrophy phenotype in mice. *Mol. Ther.* **2020**. [CrossRef]
128. Ramos, D.M.; d'Ydewalle, C.; Gabbeta, V.; Dakka, A.; Klein, S.K.; Norris, D.A.; Matson, J.; Taylor, S.J.; Zaworski, P.G.; Prior, T.W.; et al. Age-dependent SMN expression in disease-relevant tissue and implications for SMA treatment. *J. Clin. Investig.* **2019**, *129*, 4817–4831. [CrossRef]
129. Wadman, R.I.; Stam, M.; Jansen, M.D.; van der Weegen, Y.; Wijngaarde, C.A.; Harschnitz, O.; Sodaar, P.; Braun, K.P.; Dooijes, D.; Lemmink, H.H.; et al. A comparative study of SMN protein and mRNA in blood and fibroblasts in patients with spinal muscular atrophy and healthy controls. *PLoS ONE* **2016**, *11*, e0167087. [CrossRef]
130. Khalil, M.; Teunissen, C.E.; Otto, M.; Piehl, F.; Sormani, M.P.; Gattringer, T.; Barro, C.; Kappos, L.; Comabella, M.; Fazekas, F.; et al. Neurofilaments as biomarkers in neurological disorders. *Nat. Rev. Neurol.* **2018**, *14*, 577–589. [CrossRef]
131. Petzold, A. Neurofilament phosphoforms: Surrogate markers for axonal injury, degeneration and loss. *J. Neurol. Sci.* **2005**, *233*, 183–198. [CrossRef]
132. Yuan, A.; Rao, M.; Veeranna Nixon, R. Neurofilaments and neurofilament proteins in health and disease. *Cold Spring Harb. Perspect. Biol.* **2017**, *9*, a018309. [CrossRef] [PubMed]
133. Goldstein, M.E.; Sternberger, N.H.; Sternberger, L.A. Phosphorylation protects neurofilaments against proteolysis. *J. Neuroimmunol.* **1987**, *14*, 149–160. [CrossRef]
134. Darras, B.T.; Crawford, T.O.; Finkel, R.S.; Mercuri, E.; De Vivo, D.C.; Oskoui, M.; Tizzano, E.F.; Ryan, M.M.; Muntoni, F.; Zhao, G.; et al. Neurofilament as a potential biomarker for spinal muscular atrophy. *Ann. Clin. Transl. Neurol.* **2019**, *6*, 932–944. [CrossRef]
135. Wurster, C.D.; Steinacker, P.; Günther, R.; Koch, J.C.; Lingor, P.; Uzelac, Z.; Witzel, S.; Wollinsky, K.; Winter, B.; Osmanovic, A.; et al. Neurofilament light chain in serum of adolescent and adult SMA patients under treatment with nusinersen. *J. Neurol.* **2019**, *267*, 36–44. [CrossRef]
136. Totzeck, A.; Stolte, B.; Kizina, K.; Bolz, S.; Schlag, M.; Thimm, A.; Kleinschnitz, C.; Hagenacker, T. Neurofilament heavy chain and tau protein are not elevated in cerebrospinal fluid of adult patients with spinal muscular atrophy during loading with nusinersen. *Int. J. Mol. Sci.* **2019**, *20*, 5397. doi:10.3390/ijms20215397. [CrossRef]
137. Blennow, K.; Hampel, H.; Weiner, M.; Zetterberg, H. Cerebrospinal fluid and plasma biomarkers in Alzheimer disease. *Nat. Rev. Neurol.* **2010**, *6*, 131. [CrossRef]
138. Olsson, B.; Lautner, R.; Andreasson, U.; Öhrfelt, A.; Portelius, E.; Bjerke, M.; Hölttä, M.; Rosén, C.; Olsson, C.; Strobel, G.; et al. CSF and blood biomarkers for the diagnosis of Alzheimer's disease: A systematic review and meta-analysis. *Lancet Neurol.* **2016**, *15*, 673–684. doi:10.1016/S1474-4422(16)00070-3. [CrossRef]
139. Scarafino, A.; D'Errico, E.; Introna, A.; Fraddosio, A.; Distaso, E.; Tempesta, I.; Morea, A.; Mastronardi, A.; Leante, R.; Ruggieri, M.; et al. Diagnostic and prognostic power of CSF Tau in amyotrophic lateral sclerosis. *J. Neurol.* **2018**, *265*, 2353–2362. [CrossRef] [PubMed]

140. Winter, B.; Guenther, R.; Ludolph, A.C.; Hermann, A.; Otto, M.; Wurster, C.D. Neurofilaments and tau in CSF in an infant with SMA type 1 treated with nusinersen. *J. Neurol. Neurosurg. Psychiatry* **2019**, *90*, 1068–1069. [CrossRef]
141. Kim, J.K.; Jha, N.N.; Feng, Z.; Faleiro, M.R.; Chiriboga, C.A.; Wei-Lapierre, L.; Dirksen, R.T.; Ko, C.P.; Monani, U.R. Muscle-specific SMN reduction reveals motor neuron—Independent disease in spinal muscular atrophy models. *J. Clin. Investig.* **2020**, *130*, 1271–1287. [CrossRef]
142. Fayzullina, S.; Martin, L.J. Skeletal muscle DNA damage precedes spinal motor neuron DNA damage in a mouse model of Spinal Muscular Atrophy (SMA). *PLoS ONE* **2014**, *9*, e93329. [CrossRef]
143. Alves, C.R.; Zhang, R.; Johnstone, A.J.; Garner, R.; Nwe, P.H.; Siranosian, J.J.; Swoboda, K.J. Serum creatinine is a biomarker of progressive denervation in spinal muscular atrophy. *Neurology* **2020**, *94*, e921–e931. [CrossRef] [PubMed]
144. Finkel, R.S.; Crawford, T.O.; Swoboda, K.J.; Kaufmann, P.; Juhasz, P.; Li, X.; Guo, Y.; Li, R.H.; Trachtenberg, F.; Forrest, S.J.; et al. Candidate proteins, metabolites and transcripts in the Biomarkers for Spinal Muscular Atrophy (BforSMA) clinical study. *PLoS ONE* **2012**, *7* e35462. [CrossRef] [PubMed]
145. Ng, S.Y.; Soh, B.S.; Rodriguez-Muela, N.; Hendrickson, D.G.; Price, F.; Rinn, J.L.; Rubin, L.L. Genome-wide RNA-Seq of human motor neurons implicates selective ER stress activation in spinal muscular atrophy. *Cell Stem Cell* **2015**, *17*, 569–584. [CrossRef] [PubMed]
146. Rizzo, F.; Nizzardo, M.; Vashisht, S.; Molteni, E.; Melzi, V.; Taiana, M.; Salani, S.; Santonicola, P.; Di Schiavi, E.; Bucchia, M.; et al. Key role of SMN/SYNCRIP and RNA-Motif 7 in spinal muscular atrophy: RNA-Seq and motif analysis of human motor neurons. *Brain* **2019**, *142*, 276–294. [CrossRef]
147. Nizzardo, M.; Taiana, M.; Rizzo, F.; Benitez, J.A.; Nijssen, J.; Allodi, I.; Melzi, V.; Bresolin, N.; Comi, G.; Hedlund, E.; et al. Synaptotagmin 13 is neuroprotective across motor neuron diseases. *Acta Neuropathol.* **2020**, *139*, 837–853. [CrossRef]
148. Malone, E.R.; Oliva, M.; Sabatini, P.J.; Stockley, T.L.; Siu, L.L. Molecular profiling for precision cancer therapies. *Genome Med.* **2020**, *12*, 8. [CrossRef]
149. Besser, J.; Carleton, H.A.; Gerner-Smidt, P.; Lindsey, R.L.; Trees, E. Next-generation sequencing technologies and their application to the study and control of bacterial infections. *Clin. Microbiol. Infect.* **2018**, *24*, 335–341. [CrossRef]
150. Allicock, O.M.; Guo, C.; Uhlemann, A.C.; Whittier, S.; Chauhan, L.V.; Garcia, J.; Price, A.; Morse, S.S.; Mishra, N.; Briese, T.; et al. BacCapSeq: A platform for diagnosis and characterization of bacterial infections. *MBio* **2018**, *9*, e02007-18. [CrossRef]
151. Gibbs, R.M.; Lipnick, S.; Bateman, J.W.; Chen, L.; Cousins, H.C.; Hubbard, E.G.; Jowett, G.; LaPointe, D.S.; McCredy, M.J.; Odonkor, M.N.; et al. Toward precision medicine for neurological and neuropsychiatric disorders. *Cell Stem Cell* **2018**, *23*, 21–24. [CrossRef]
152. Seferian, A.M.; Moraux, A.; Canal, A.; Decostre, V.; Diebate, O.; Le Moing, A.G.; Gidaro, T.; Deconinck, N.; Van Parys, F.; Vereecke, W.; et al. Upper limb evaluation and one-year follow up of non-ambulant patients with spinal muscular atrophy: An observational multicenter trial. *PLoS ONE* **2015**, *10*, e0121799. [CrossRef] [PubMed]

Journal of
Personalized
Medicine

MDPI

Article

Comparison of Serum Pharmacodynamic Biomarkers in Prednisone-Versus Deflazacort-Treated Duchenne Muscular Dystrophy Boys

Shefa Tawalbeh [1,2], Alison Samsel [2], Heather Gordish-Dressman [3], Yetrib Hathout [2,*], CINRG-DNHS Investigators † and Utkarsh J. Dang [4,*]

1 Department of Biomedical Engineering, Binghamton University, Binghamton, NY 13902, USA; stawalb1@binghamton.edu
2 Department of Pharmaceutical Sciences, School of Pharmacy and Pharmaceutical Sciences, Binghamton University, Binghamton, NY 13902, USA; amsamsel@gmail.com
3 Division of Biostatistics, Children's National Hospital, Washington, DC 20010, USA; HGordish@childrensnational.org
4 Department of Health Outcomes and Administrative Sciences, School of Pharmacy and Pharmaceutical Sciences, Binghamton University, Binghamton, NY 13902, USA
* Correspondence: yhathout@binghamton.edu (Y.H.); udang@binghamton.edu (U.J.D.)
† Names and affiliations of the CINRG-DNHS Investigators are provided in the Acknowledgments.

Received: 7 August 2020; Accepted: 10 October 2020; Published: 12 October 2020

Abstract: Prednisone (Pred) and Deflazacort (Dfz) are commonly used glucocorticoids (GCs) for Duchenne muscular dystrophy (DMD) treatment and management. While GCs are known to delay the loss of ambulation and motor abilities, chronic use can result in onerous side effects, e.g., weight gain, growth stunting, loss of bone density, etc. Here, we use the CINRG Duchenne natural history study to gain insight into comparative safety of Pred versus Dfz treatment through GC-responsive pharmacodynamic (PD) biomarkers. Longitudinal trajectories of SOMAscan® protein data obtained on serum of DMD boys aged 4 to 10 (Pred: $n = 7$; Dfz: $n = 8$) were analyzed after accounting for age and time on treatment. Out of the pre-specified biomarkers, seventeen candidate proteins were differentially altered between the two drugs ($p < 0.05$). These include IGFBP-2 and AGER associated with diabetes complications, and MMP-3 associated with extracellular remodeling. As a follow-up, IGFBP-2, MMP-3, and IGF-I were quantified with an ELISA using a larger sample size of DMD biosamples (Dfz: $n = 17$, Pred: $n = 12$; up to 76 sera samples) over a longer treatment duration. MMP-3 and IGFBP-2 validated the SOMAscan® signal, however, IGF-I did not. This study identified GC-responsive biomarkers, some associated with safety, that highlight differential PD response between Dfz and Pred.

Keywords: Duchenne muscular dystrophy; pharmacodynamic biomarkers; prednisone; deflazacort; glucocorticoids; corticosteroids; safety

---

1. Introduction

Duchenne muscular dystrophy (DMD) is an X-linked recessive disorder affecting the expression of dystrophin protein, an essential protein that maintains muscle fiber integrity and function [1]. The lack of dystrophin expression leads to muscle inflammation at an early stage of the disease, followed by progressive muscle degeneration and wasting [2]. While there are promising gene therapy and exon-skipping treatments, some of which have received conditional approval from the EMA [3] and FDA [4–7], these are mutation-specific and only restore a partial amount of truncated dystrophin protein [8]. Hence, DMD patients will continue to need combination therapy using the current standard of care: glucocorticoids (GCs). GC use helps reduce muscle inflammation and delay the loss of motor abilities [9], delay loss of independent ambulation, improve pulmonary function, and delay the onset of cardiomyopathy [10–12]. However, a GC regimen can have many side effects, such as weight gain, growth stunting, loss of bone density, hirsutism, Cushingoid features, osteoporosis, hypertension, diabetes, behavioral disturbances, and difficulty in sleeping [12–14]. Commonly used steroidal drugs for DMD are prednisone (Pred) and deflazacort (Dfz). Prednisone has been used to treat DMD patients in the USA since the seventies [15]. In February 2017, the FDA approved deflazacort to treat DMD patients aged five years and older [16] although this same drug has been widely used in Europe to treat DMD patients for years. Figure 1 shows structures of prednisone and deflazacort along with their respective active metabolites: prednisolone and 21-desacetyl deflazacort, respectively. Both drugs are given to patients in their prodrug forms, which are then metabolized to their active forms in the liver. The active metabolite prednisolone has a hydroxyl group at carbon 17, while the active 21-desacertyl deflazacort has an oxazoline structure at that same position (red arrow in Figure 1).

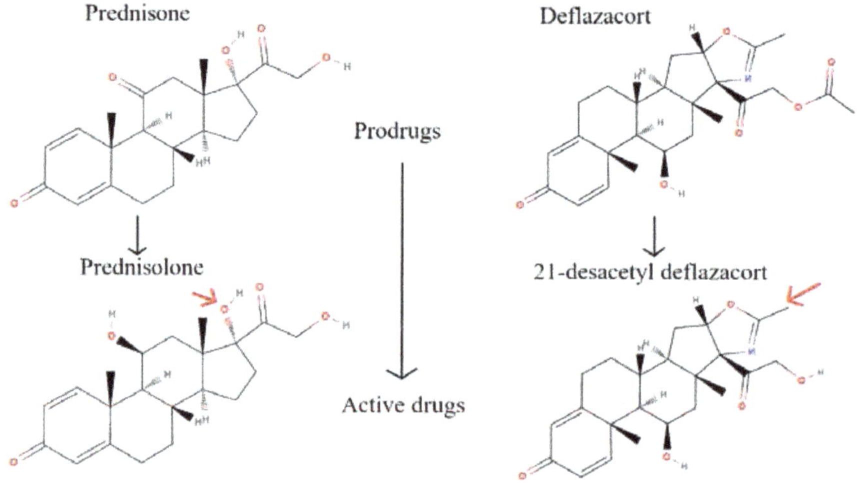

Figure 1. 2D structures of Prednisone and Deflazacort (top panels) and their active drugs: prednisolone and 21-desacetyl deflazacort. The similarities are clear. The red arrows point to the structural differences between the active drugs.

Many efforts have been undertaken to investigate the use of biomarkers in DMD [17–19] and the efficacy and safety of Pred versus Dfz [10–12,20,21]. Both drugs are known to be efficacious in prolonging ambulation [10,11,14,22]. For our purposes here, we provide a small review on safety comparisons from the literature. These studies were compared to clinical outcome measures like changes in height, weight, Cushingoid features, and erythema, among others to evaluate differences in safety. A comparison of high dose Dfz and Pred in 70 systemic lupus patients concluded that Pred was associated with a significant increase in weight gain, Cushingoid severity index, and hirsutism compared to Dfz [23]. In a double-blind, randomized study on 196 DMD subjects, Pred was found to be associated with more weight gain than Dfz [10]. In another randomized study (18 DMD patients), patients treated with Pred showed higher weight gain compared to Dfz-treated patients [13]. As compared to other studies [10,14], a comparison of prednisone/prednisolone versus Dfz treated patients using 340 participants from the Cooperative International Neuromuscular Research Group Duchenne Natural History Study (CINRG-DNHS) showed that participants treated with Dfz had increased frequency of Cushingoid appearance, cataracts, and growth delay. Another study [13] reported that the time of initiating, dosing, and duration of treatments were associated with side effects; longer duration and increased Dfz dosage predicted growth stunting and Dfz was reported to be associated with lighter weight and shorter heights compared to Pred [13]. To summarize, these studies suggest that treatment with Pred is associated with relatively more weight gain, whereas Dfz treatment is associated with relatively more growth stunting. While these studies provide important comparisons between Dfz and Pred, the underlying molecular mechanism leading to these differences, remain unknown. Furthermore, predictive outcome measures, especially for adverse effects are highly desirable.

There are ongoing debates among families, clinicians, and regulatory agencies over which GC drug (Pred or Dfz) and regimen is better for DMD boys. GC responsive biomarkers (for both safety and efficacy) have been defined and confirmed in different diseases and multiple cohorts [17,24,25], using data with pooled corticosteroid drugs not differentiating between Pred and Dfz and different regimens [22]. However, much remains unknown about the comparative effects of Pred vs. Dfz on blood accessible biomarkers and how they can inform clinical decision-making and drug development. Adding pharmacodynamic (PD) biomarkers to the evidence from the aforementioned clinical outcome studies could bring insights into differences between Pred and Dfz at the molecular level. To the best of our knowledge, we are the first to compare serum accessible PD biomarkers in DMD patients (4–16 years old) treated with Dfz or Pred to gain insight into differences at the serum protein level. For this, we use data generated by a high-throughput SOMAScan® technique, followed by confirmation of a specific set of PD biomarkers by ELISAs. The objective here is to define PD biomarkers that differ in their response to Pred and Dfz, and investigate if these biomarkers are known to be associated with reported differences in term of side effects between these two drugs.

2. Materials and Methods

The study protocol was approved by Institutional Review Boards at all participating institutions that provided serum samples. These included, the Office of Research IRB administrations at the University of California Davis, Davis, CA, the University of Pittsburgh, Pittsburgh, PA, Children's National Health System, Washington DC, the Conjoint Health Research Ethics Board, University of Calgary and the Human Subject Research Review Committee at Binghamton University, NY. Informed written consent was obtained from the parents of the participants or their legal guardians for biomarker studies at each site.

2.1. SOMAscan® Dataset

The CINRG-DNHS is a natural history study of 440 DMD boys. The CINRG-DNHS investigators enrolled these subjects at 20 centers in nine countries [11]. In this prospective cohort study, subjects genetically confirmed to have DMD were followed for up to 10 years. Details about the study, including informed consent and others, have previously been published [26]. At entry, some DMD subjects were steroid-naïve and were then treated during follow up visits, some remained steroid-naïve throughout, while others were already on GCs [26]. Characteristics (height, weight, age, time on GC treatment, and type of GC) were recorded on follow-up visits as well as clinical outcome data on time to run/walk, time to stand, and time to climb until loss of ability [26]. Note that six-minute walk distance and the NorthStar Ambulatory Assessment score were measured for a small subset of patients enrolled later during the study [26]. Blood samples for biomarker investigations were collected by the CINRG-DNHS investigators during some visits on a subset of participants. GC-treated patients were on Pred or Dfz with variation in dose and dose schedules (intermittent or daily). In a recent study [17], we generated biomarker data on a subset of these DMD samples ($n = 31$ boys, 4 to 10 years old) and used steroid-naïve and steroid-treated paired samples to define 107 (false discovery rate-adjusted p-values < 0.05) PD biomarkers from 1310 serum proteins (SOMAscan® array).

Based on the above-mentioned study [17], we pre-specified the 107 PD biomarkers of interest and focused on an analysis of 35 longitudinal samples from 8 Pred-treated and 7 Dfz-treated ambulatory patients only (screening dataset A) to identify potential candidate serum proteins differentially altered between Pred and Dfz. These subjects were age-matched (see Table 1 for a summary of their characteristics). For these longitudinal data, only subjects with at least 2 visits were used. The name of the study, type of treatment, duration of treatment, age, and some clinical measurements (height, weight, BMI) are tabulated in Supplementary Table S1.

Table 1. Characteristics of patients/samples for SOMAscan® screening dataset A. All samples were from ambulatory patients.

Treatment	Number of Patients	Average Number of Visits (Min, Max)	Age Range at Sample Collection (Years)	Average Time between Biosample Collection (Days) (Mix, Max)	Regimen
Deflazacort	8	2.5 (2, 4)	4.7–9.4	465 (56, 1268)	Daily on Dfz
Prednisone	7	2.14 (2, 3)	4.3–8.3	582 (157, 1392)	Daily on Pred

2.2. Validation of Key Pharmacodynamic Biomarkers Using ELISA

A larger data set (confirmation dataset B) contained up to 76 longitudinal samples on 29 ambulatory subjects (17 deflazacort-treated and 12 prednisone-treated) from CINRG DNHS. Note that all samples/subjects from screening dataset A were used in confirmation dataset B as well. For both datasets, we focused on ambulatory patients to minimize any possible confounding issues with the stage of disease (after loss of ambulation). Again, these subjects were age-matched. Summary characteristics of the subjects are provided in Table 2.

Table 2. Characteristics of patients/samples for confirmation dataset B. These serum samples were used for ELISA confirmation of MMP3, IGFBP-2 and IGF-1. All the samples were from ambulatory patients.

Treatment	Number of Patients	Average Number of Visits (Min, Max)	Mean Age at Sample Collection in Years (Min-Max)	Average Time between Biosample Collection (Days) (Mix, Max)
Deflazacort	17	2.7 (2, 5)	9 (4.7–15.3)	1392 (370, 3058)
Prednisone	12	2.5 (2, 3)	8.5 (4.2–15.8)	1685 (594, 3391)

A subset of PD serum protein biomarkers, including MMP3, IGFBP-2, and IGF-I, were selected for confirmation analysis using ELISA assays. An ELISA kit from Meso Scale Diagnostics (MSD) was used to measure levels of MMP3. Similarly, ELISA kits from Thermo Fisher Scientific and R&D Systems, Inc. were used to measure levels of IGFBP2 and IGF-I, respectively. These candidates were chosen based on the availability of a validated ELISA assay kit. All three assays were sandwich ELISA, which were highly specific for antigen detection. Dilution factors for serum samples were 1:10, 1:400, and 1:100 for MMP-3, IGFBP-2, and IGF-I, respectively. The three assays were performed following the manufacturer's instructions. Details about serum samples used in ELISA validation are shown in Table 2.

2.3. Data Analysis and Statistical Methods

As depicted in the schematic (Figure 2), we first pre-specified 107 steroidal PD biomarkers previously identified as being significantly altered over time in their levels between GC-naive and paired GC-treated DMD samples from the same subjects [17]. To compare the serum levels and trajectories of these pre-specified 107 PD biomarkers between DMD patients treated with Dfz and Pred, linear mixed effect models were used to analyze longitudinal measurements for 7 boys on Pred and 8 on Dfz (daily regimen). All statistical analyses were performed using R [27]. Linear mixed effect models were run using the lme4 and lmerTest packages [28,29]. Serum protein levels, which had previously been hybridization control and median signal normalized, were log-transformed (see [17] for details). Random intercept linear mixed effect models were used to investigate the association of (mean centered) time on drug, type of GC (Pred or Dfz), and interaction between time on drug and type of GC on log transformed protein RFUs. The samples were age-matched at baseline (average age of Pred-treated subject at baseline = 5.7 (min age = 4.3, max age = 7.3) years; average age of Dfz-treated subject at baseline = 6 (min age = 4.7, max age = 7.4) years) and over time (see Table 1 for age range at sample collection). Some models did not converge (7 out of 107) due to numerical optimization issues (likely due to small sample size). Here, the interaction coefficient represents the difference in estimated slopes of biomarker trajectory over time between Pred-treated and Dfz-treated DMD patients. If the interaction coefficient is significant, this indicates that treatment type (Pred or Dfz) is associated with different trajectories of biomarker over time. The coefficient for the type of GC represents the difference in mean protein levels between Pred-treated and Dfz-treated subjects for a patient with an average treatment duration with similar biomarker trajectories over time (this is considered after the interaction effect). The same model was also fit to log-transformed ELISA data. To study the effect of time on GC on weight, height, and BMI, data analyses were performed using linear mixed effect models on confirmation dataset B. For this, we used a random intercept linear mixed model adjusting for age at baseline, time on GC, and interaction between type of GC and time on GC. Concordance of SOMAscan® and ELISA measurements was investigated on the subset of samples common to screening dataset A and confirmation dataset B.

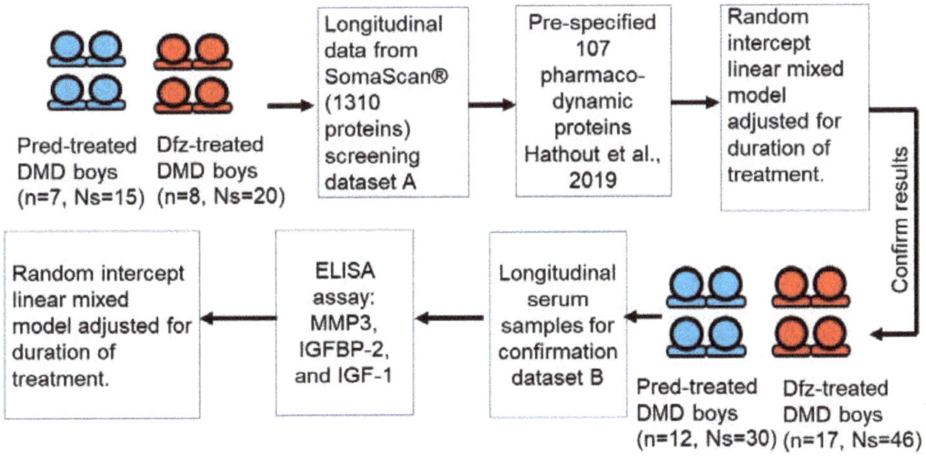

Figure 2. Schematic describing the workflow for the biomarker analyses. Proteins identified as differentially affected by prednisone vs. deflazacort were identified from the SomaScan® based screening dataset. This was followed by confirmation analysis using ELISA assays of three selected biomarkers on a larger set of subjects (confirmation dataset). n: number of subjects, Ns: number of serum samples used in this study.

3. Results

3.1. Longitudinal Trajectory of Serum PD Biomarkers in Prednisone vs. Deflazacort Treated DMD boys

We examined differences in the longitudinal trajectories of pre-specified 107 PD biomarkers in Pred-treated patients (n = 7) versus Dfz-treated DMD patients (n = 8) accounting for duration of treatment (time on GC) and interaction between treatment duration and type of GC (Pred or Dfz). In general, Pred and Dfz seem to engage the PD biomarkers similarly and altered their trajectory in the same direction in screening dataset A. This was not the case for 17 PD biomarkers that were found differentially altered in their average levels and/or longitudinal trajectories (unadjusted p value < 0.05) in Pred-treated vs. Dfz-treated DMD patients. These 17 differentially altered PD biomarkers are listed in Table 3 with fold change between non-GC-treated DMD subjects and healthy controls (from [17]), fold change between Pred and Dfz-treated subjects, p values for the difference in the mean levels and p values for the difference in longitudinal trajectories between Pred and Dfz treated DMD patients, along with potential significance and references to published literature.

Table 3. List of PD biomarkers that were differentially altered between Pred-treated (n = 7) and Dfz-treated (n = 8) groups (SOMAscan® screening dataset A).

Abbreviated Gene Name [1] (Uniprot ID)	Fold Change between Untreated DMD vs. Healthy Controls with p-Value from [17]	Fold-Change between DMD Subjects Treated with Pred and Dfz	p-Value: Difference in Mean Levels [2]	p-Value: Difference in Trajectories [2]	Protein Function—Biological Process	Potential Significance
LILRB1 (Q8NHL6)	0.96 (0.861)	−1.5 (↓) in Pred vs. Dfz	0.005	0.426	Immune response	Side effect (immune suppression)
TNFRSF21 (O75509)	1.14 (0.454)	−1.33 (↓) in Pred vs. Dfz	0.005	0.934	Apoptotic process, adaptive immune response	Side effect (immune suppression)
CHRDL1 (Q9BU40)	1.30 (0.039)	−1.2 (↓) in Pred vs. Dfz	0.007	0.455	Bone Morphogenetic Proteins (BMP) signaling pathway	Efficacy via action on TGF-β signaling [30]
IGF-I (P05019)	0.85 (0.161)	1.14 (↑) in Pred vs. Dfz	0.007	0.096	Promotes growth	Potential efficacy associated with anti-inflammatory propriety [31]
MMP-3 (P08254)	0.77 (0.488)	2 (↑) in Dfz vs. Pred	0.008	0.216	Extracellular matrix degradation	Efficacy/Side effect [17,25,32]
sRAGE/AGER (Q15109)	0.60 (0.005)	−1.82 (↓) in Pred vs. Dfz	0.010	0.640	Inflammatory (causes complications in diabetes)	Side effect may be associated with diabetes risk [33]
ANXA2 (P07355)	1.24 (0.246)	−1.22 (↓) in Pred vs. Dfz	0.011	0.054	Angiogenesis, biomineral tissue development, inflammation	Potential efficacy marker associated with inflammation [34–36]
CD166 (Q13740)	0.81 (0.088)	−1.25 (↓) in Pred vs. Dfz	0.014	0.616	Cell adhesion, adaptive immune response	Side effect (immune suppression)
HJV (Q6ZVN8)	0.83 (0.073)	1.19 (↑) in Dfz vs. Pred	0.017	0.385	BMP signaling pathway, iron ion homeostasis	Unknown
sCD163 (Q86VB7)	0.86 (0.466)	−1.30 (↓) in Pred vs. Dfz	0.025	0.340	Inflammation	Efficacy [37]
Mcl-1 (Q07820)	2.31 (<0.001)	−1.33 (↓) in Pred vs. Dfz	0.029	0.146	Apoptosis, DifferentiationInflammation	Efficacy [38]
PDE3A (Q14432)	1.03 (0.805)	1.22 (↑) in Pred vs. Dfz	0.033	0.956	Cell to cell signaling	Unknown
GPNMB (Q14956)	0.79 (0.111)	−1.32 (↓) in Pred vs. Dfz	0.033	0.944	Cell adhesion, bone mineralization.	Side effect related to bone [39,40]
FCN1 (O00602)	1.02 (0.905)	1.18 (↑) in Pred vs. Dfz	0.038	0.030	Innate immune response	Unknown
MAPK14 (Q16539)	2.10 (<0.001)	−1.18 (↓) in Pred vs. Dfz	0.041	0.982	Potential muscle injury biomarker	Efficacy, muscle injury protein that normalized after GC treatment [17]
NCAM-L1 (P32004)	0.76 (0.072)	−1.28 (↓) in Pred vs. Dfz	0.042	0.691	Cell adhesion and differentiation	Side effect, risk of developing diabetes [41]
IGFBP-2 (P18065)	2.55 (<0.001)	−1.16 (↓) in Dfz vs. Pred	0.329	0.040	Growth regulation	Side effect, growth stunting [42]

[1] LILRB1 = Leukocyte immunoglobulin-like receptor subfamily B member 1; TNFRSF21 = Tumor necrosis factor receptor superfamily member 21; CHRDL1 = Chordin-like protein 1; IGF-I = Insulin-like growth factor I; MMP-3 = Stromelysin-1; sRAGE/AGER = Advanced glycosylation end product-specific receptor, soluble; ANXA2 = Annexin A2; CD166 = CD166 antigen; HJV = Hemojuvelin; sCD163 = Scavenger receptor cysteine-rich type 1 protein M130; Mcl-1 = Induced myeloid leukemia cell differentiation protein; PDE3A = cGMP-inhibited 3′,5′-cyclic phosphodiesterase A; GPNMB = Transmembrane glycoprotein NMB; FCN1 = Ficolin-1; MAPK14 = Mitogen-activated protein kinase 14; NCAM-L1 = Neural cell adhesion molecule L1; IGFBP-2 = Insulin-like growth factor-binding protein 2. [2] The two p-values are for the difference in mean levels and difference in longitudinal trajectories between Pred-treated and Dfz-treated subjects, respectively.

Two major groups of differentially affected steroidal PD biomarkers were thus identified. The first major group consisted of PD biomarkers that were repressed by GC in general, but exhibited a significantly lower mean level in Pred treated group relative to Dfz treated group. These included leukocyte immunoglobulin-like receptor subfamily B member 1 (LILRB1), tumor necrosis factor receptor superfamily member 21 (TNFRSF21), chordin-like protein 1 (CHRDL1), soluble advanced glycosylation end product-specific receptor (sRAGE), annexin A2 (ANXA2), CD166 antigen (CD166), scavenger receptor cysteine-rich type 1 protein M130 (sCD163), induced myeloid leukemia cell differentiation protein (MCL-1), transmembrane glycoprotein NMB (GPNMB), mitogen-activated protein kinase 14 (MAPK14), and neural cell adhesion molecule L1 (NCAM-L1). The second (minor) group consisted of PD biomarkers that increased in their levels following GC treatment and consisted of two subgroups. The first subgroup consisted of those that were significantly elevated in serum samples of Dfz relative to the Pred-treated group such as hemojuvelin (HJV), stromelysin-1 (MMP3) while the second group consists of those biomarkers that were significantly elevated in serum samples of Pred relative to Dfz-treated group such as insulin-like growth factor I (IGF-I), ficolin-1 (FCN1), and cGMP-inhibited 3′,5′-cyclic phosphodiesterase A (PDE3A). One interesting PD biomarker is insulin-like growth factor-binding protein 2 (IGFBP-2); this showed diverging longitudinal trajectories ($p = 0.040$) in the Dfz-treated group, as compared to the Pred-treated group. Similarly, longitudinal trajectories of Ficolin-1 (FCN1; $p = 0.03$) and Annexin A2 (ANXA2; 0.054) also seemingly diverged in addition to their differential levels between the two drugs. Figure 3 shows selected examples of these differentially affected steroidal PD biomarkers. For example, IGFBP-2 sharply decreased over time in Dfz-treated DMD patients, while it remained unchanged or slightly increased over time in Pred treated group ($p = 0.040$). While TNFRSF21 (also known as DR6) was not found to be differentially affected by the two different drugs in terms of longitudinal trajectory slopes but its mean sera levels was significantly lower in the Pred-treated group compared to the Dfz-treated group ($p = 0.005$). Similarly, mean levels of FCN-1 were lower in the Dfz-treated group relative to the Pred-treated group ($p = 0.038$), while MMP3 mean levels were significantly elevated in the Dfz-treated group relative to the Pred-treated group.

3.2. *Effect of Dfz and Pred on Height and Weight of DMD Boys*

We also investigated differences in growth stunting in Dfz treated patients relative to Pred treated patients in confirmation dataset B. DMD patients treated with Dfz had lower height growth rates ($p = 0.006$ for difference in trajectory slopes over time) compared to those treated with Pred (Figure 5). We did not find any significant difference in weight ($p = 0.112$) and BMI ($p = 0.08$) over time between patients treated with Dfz and Pred (Figure 5). Additionally, note that more variation is observed in longitudinal BMI measurements around the estimated model (Figure 5).

3.3. *Data Validation Using ELISA*

For the reported data above, we had a small sample size and we did not adjust for multiple testing, thus a follow-up confirmation analyses on a subset of PD biomarkers was carried out. Unfortunately, from the list of PD biomarkers identified in Table 3 above, a good validated ELISA assay that used low sera volume was available for only three PD biomarker candidates: MMP3, IGFBP2 and IGF-I. There were other ELISA assays for other candidates, but they were either not validated or required a larger volume of sera samples, which we did not have. To confirm the SomaScan® findings obtained for MMP3, IGFBP2, and IGF-1, we used a larger sample size of samples over a longer treatment duration and a wider age range of subjects. The findings from the ELISA runs are summarized in Table 4. Results for MMP-3 validated the SOMAscan® signal (same directionality; p-value for the difference in mean levels = 0.022), IGFBP-2 neared significance (p-value for difference in trajectory slopes = 0.051), while IGF-I did not validate the SOMAscan® signal. Figure 4 shows correlations between SOMAscan® data and ELISA data with the IGF-I measurements having a substantially worse Pearson correlation than MMP-3 and IGFBP-2.

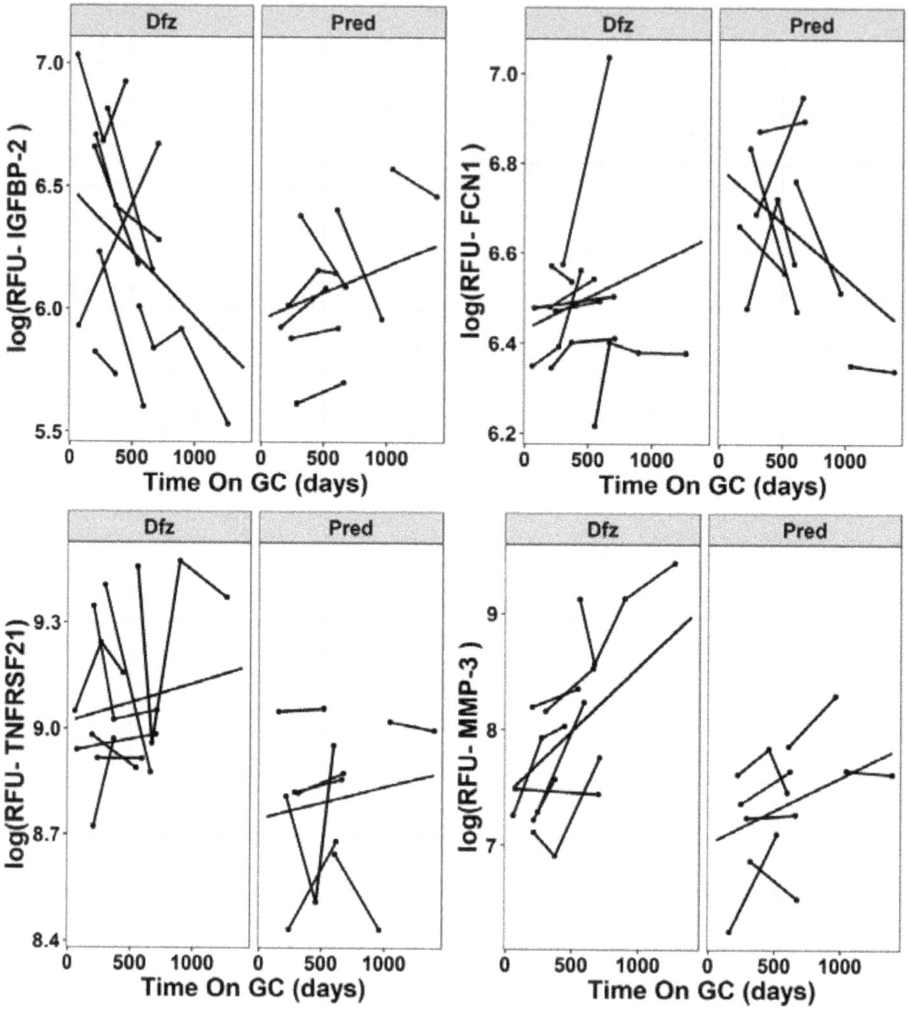

Figure 3. Selected examples of serum protein PD biomarkers showing difference between prednisone (Pred) and deflazacort (Dfz)-treated samples. FCN-1 has relatively higher mean RFU levels in Dfz-treated patients (p = 0.038 for mean levels; p = 0.03 for difference in trajectory slopes over time). TNFRSF21 has higher mean RFU levels in Pred- vs. Dfz-treated DMD patients (p = 0.005 mean levels; p = 0.934 for difference in trajectory slopes over time). The longitudinal trajectories of IGFBP-2 levels are different between the two drugs (p = 0.04). MMP-3 mean RFU level is elevated in DMD boys treated with Dfz, as compared to Pred (p = 0.008).

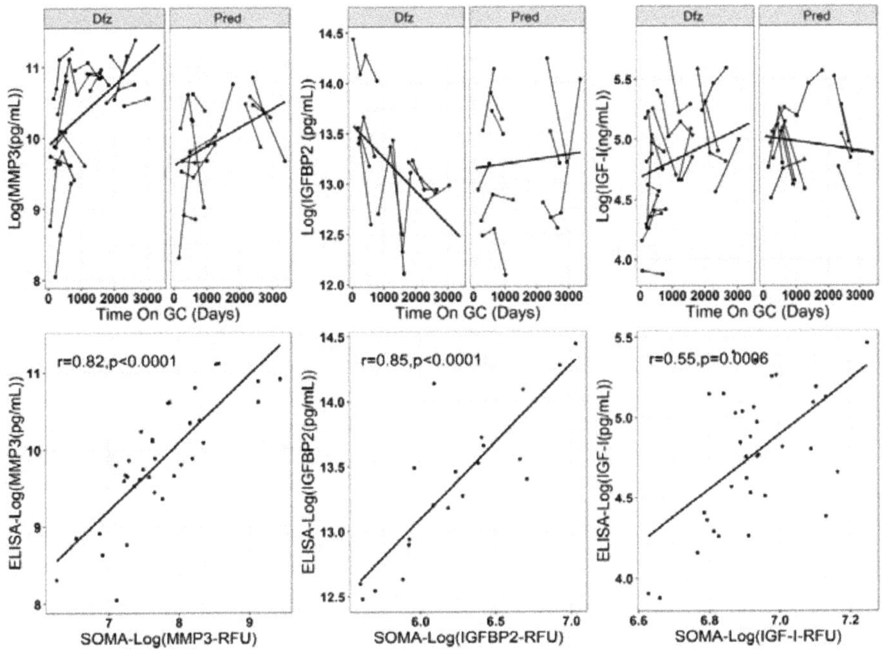

Figure 4. SOMAscan® signal confirmation using ELISA. Upper panel shows longitudinal trajectories of selected DMD serum protein PD biomarkers from ELISA assays. Lower panel shows correlation plots between SOMA and ELISA data for the samples that overlapped between the screening and confirmation data sets.

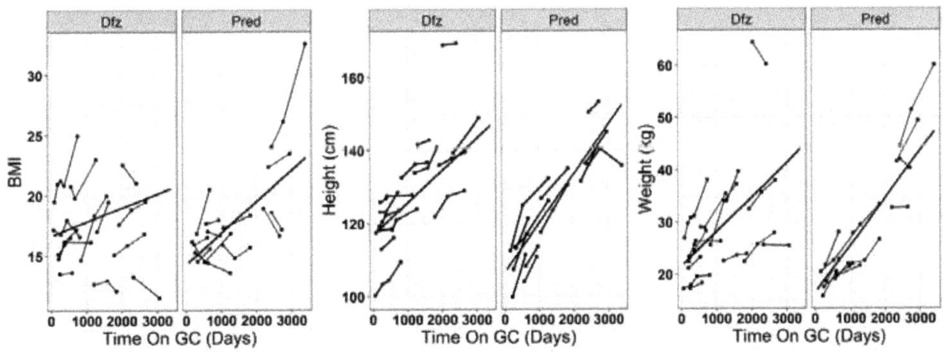

Figure 5. Comparison of longitudinal trajectories between DMD patients treated with deflazacort and prednisone of BMI ($p = 0.08$ for difference in trajectory slopes), height (cm; $p = 0.006$), and weight (kg; $p = 0.112$).

Table 4. Summary for biomarker signal confirmation using ELISA.

Protein Name (Uniprot ID)	SOMAscan® Data			ELISA Data			Function
	p-Value [1]: Difference in Mean Levels	*p*-Value [1]: Difference in Trajectories	Number of Patients/Samples	*p*-Value [1]: Difference in Mean Levels	*p*-Value [1]: Difference in Trajectories	Number of Patients/Samples	
MMP-3 (P08254)	0.008	0.216	8 Dfz, 7 Pred/35 samples	0.022	0.378	17 Dfz, 12 Pred/76 Samples	Extracellular matrix degradation
IGFBP-2 (P18065)	0.328	0.04	8 Dfz, 7 Pred/35 samples	0.744	0.0507	10 Dfz, 10 Pred/49 Samples	Regulates growth
IGF-I (P05019)	0.007	0.096	8 Dfz, 7 Pred/35 samples	0.246	0.137	17 Dfz, 12 Pred/75 Samples	Promotes growth

[1] Two *p*-values are provided for difference in mean levels and difference in longitudinal trajectory slopes between Pred-treated and Dfz-treated subjects for both SOMAscan® and ELISA® analysis, respectively.

4. Discussion

GCs are and continue to be the standard of care for several chronic inflammatory and auto-immune diseases including DMD [11,21,25]. In this study, and due to the interest and ongoing debates between families and clinicians regarding which GC drug is more beneficial and safe for DMD boys, we focused on a subset of Dfz- an Pred-treated DMD patients to define the differential pharmacodynamic response to these two commonly prescribed drugs using blood circulating proteins. We previously defined a set of PD biomarkers that are responsive to GC treatment in DMD [17]. In that previous study, we showed that GC affected the levels of 107 circulating serum proteins. In general, use of GCs decreased the levels of several circulating pro-inflammatory and immune response associated proteins, but caused an increase in certain proteins associated with metabolism and extracellular remodeling [17]. In this current study, we find that, in general, both Dfz and Pred engaged the PD biomarkers in a similar fashion, however, among the 107 PD biomarkers, 17 exhibited a differential longitudinal response to Dfz relative to Pred, in directionality, mean levels, or both.

A close examination of the 17 differentially altered PD biomarkers (from screening dataset A) between Dfz and Pred in term of their levels, longitudinal directionality and their potential physiological function led to the following interpretation. Among these 17 PD biomarkers, 12 were significantly reduced in the Pred-treated group relative to the Dfz-treated group after adjusting for treatment duration (the samples were age matched). These 12 decreased PD biomarkers by Pred can be classified into subgroups with the first subgroup consisting of inflammatory and immune associated proteins such as LILRB1, TNFRSF21, sRAGE, CD166, and sCD163. Previous studies have suggested that Dfz is a stronger immune suppressant than Pred [43]. In contrast, our analysis showed that for the above-mentioned five serum proteins, Pred seemed to be more immunosuppressive than Dfz. However, further studies using a larger sample size and additional cellular biomarkers are needed to claim this finding. Another subgroup that was differentially decreased by Pred relative to Dfz included bone mineralization protein GPNMB and the cell adhesion protein NCAM-L1. GPNMB, also known as osteoactvin in rats, has been shown to be implicated in bone mineralization and bone regeneration [39,40] and the larger decrease in its mean level in the Pred treated group relative to the Dfz treated group could suggest that Pred treated patients might have a higher risk of losing bone density than Dfz treated patients. This agrees with earlier studies showing that decreases in bone density was markedly spared by Dfz, as compared to Pred in both adult and children cohorts [44]. However, further studies correlating low levels of circulating GPNMB to loss of bone density in Pred treated group relative to Dfz treated group are needed to support this hypothesis. The differential decrease in the circulating levels of NCAM-L1 by Pred could be indicative of another potential side effect of Pred over Dfz. Indeed, a recent study linked low levels of circulating NCAM-L1 to risk of developing type 2 diabetes [41], which is in agreement with an earlier study showing that Pred was more diabetogenic than Dfz in a pediatric population [45]. The differential decrease of sRAGE by Pred over Dfz could also be linked to the diabetogenic effect of Pred since sRAGE has been suggested to act as a decoy that dampens the advanced glycation end product signaling (e.g., RAGE signaling) and thus increases the risk of developing diabetes [33].

However, the differential decrease in the mean levels of MCL-1, MAPK14, and CHRDL1 by Pred relative to Dfz could be considered to be evidence of possible efficacy in DMD. Indeed, MCL-1 and MAPK14 were previously reported by us [17] to be elevated in blood of untreated DMD boys relative to age matched healthy controls then decreased by GC treatment toward the levels in healthy controls. CHRDL1, however, is known to bind to BMP4 and antagonizes its function [30]. BMP4 is a member of the member of the transforming growth factor-β (TGF-β), a well-known pathway reported to be involved in DMD pathogenesis [46]. Thus, a differential decrease of CRDL1 by Pred relative to Dfz could suggest a beneficial effect but further experiments using a CRDL1 knockout animal model are needed to verify this hypothesis.

Another relevant PD biomarker that was differentially altered over time between the two drugs is IGFBP-2. It sharply decreased in its longitudinal trajectory in Dfz-treated group compared to the Pred-treated group. This observation was further validated by ELISA assay ($p = 0.0507$; near significance). Note that, in our ELISA confirmations, we had the smallest number of additional samples available for IGFBP-2. IGFBP-2 binds to IGF-1 and regulates growth. While IGFBP-2 was decreased by Dfz treatment, IGF-1 was relatively increased by Dfz over Pred. Unfortunately, we were unable to validate the IGF-1 SomaScan® signal using ELISA; nevertheless, an increase in IGF-1 by GC treatment might be associated with antiinflammation efficacy [31]. However, the selective longitudinal decrease of IGFBP2 by Dfz could be associated with significant growth stunting by Dfz over the Pred reported by others [20,47,48] and confirmed by us in this study. Furthermore, a gene knockout study, conducted on mouse model comparing Igfbp2-/- mice colony with Igfbp2+/+ mice colony determined the role of IGFBP2 in bone turnover and showed that Igfbp2-/- males had shorter femurs and were heavier than controls but were not insulin resistant [42]. The decrease in IGFBP2 could have dual side effects on both growth stunting and risk of bone fracture. Indeed, a recent study compared a cohort of DMD boys treated with Dfz to a cohort treated with hydrocortisone and concluded that DMD boys receiving daily dose of Dfz had a higher incidence of bone fractures with greater risk of growth stunting [47].

The second group of steroidal PD biomarkers that were differentially affected by the two drugs are those that increased following treatment with GC. FCN1, IGF-I and PDE3A were relatively increased by Pred over Dfz, while MMP3 and HJV were relatively increased by Dfz over Pred. While FCN1 mean levels were higher in the Pred treated group relative to Dfz treated group, it decreased over time with Pred treatment while remaining relatively unchanged over time following Dfz treatment. This is intriguing and requires further validation using orthogonal methods such as ELISA. Unfortunately, the available ELISA assay required the use of a larger sample volume, which was a limitation of our study.

A comparison of the relative effects of Pred and dexamethasone (Dex), another glucocorticoid, on short-term growth and bone turnover confirmed that both drugs affected short-term bone turnover and growth [20]. However, Dex may be more potent in suppressing linear growth, simulating weight gain and bone turnover compared to prednisone. Dex was more potent at depressing IGF-I levels than prednisolone. This is consistent with our SOMAscan® data, IGF-1 was significantly more depressed or decreased in Dfz treated group relative to Pred treated group. However, we were not able to confirm this finding using ELISA. The discrepancy between the SOMAscan® data and ELISA data could be associated with an epitope effect. The two techniques recognize different epitopes on the IGF-I. Furthermore, IGF-I might co-exist as free form and bound form to circulating insulin growth factor binding proteins and these might further interfere with both these affinity-based assays that relies on epitope binding. Future analyses using denatured condition and targeted mass spectrometry analysis are needed to examine the true levels of circulating IGF-I.

MMP3, also known as stromelysin-1, was dramatically increased in the blood circulation following GC treatment, as previously shown by us [17] and others in both DMD patients [24] and inflammatory bowel disease patients [25] and anti-neutrophil cytoplasmic antibody-associated vasculitis, and juvenile dermatomyositis [25]. In this study, we further show that this increase is associated more with Dfz use than Pred use. This was further confirmed by an ELISA assay on a confirmation dataset using a larger sample size with longer treatment duration. The mechanism by which GC induces levels of circulating MMP3 remains unclear. Further studies using cell culture expressing MMP3 are needed to explain the difference of action of Dfz and Pred on the expression of MMP3. Owing to the function of MMP3, i.e., degradation of extracellular matrix, this differential increase might result in more extracellular matrix remodeling in Dfz-treated DMD patients relative to Pred-treated patients. Whether this extracellular matrix remodeling is adverse or beneficial remain to be carefully examined.

5. Conclusions

Here, we investigated less invasive and objective PD biomarkers that might prove useful to monitor disease progression and response to GC therapies in DMD. We identified differences in serum levels of PD biomarkers between Pred- and Dfz-treated subjects, some of which may be associated with safety. Such blood accessible biomarkers in DMD could play an important role in clinical trials and decision making [49]. In our comparisons, we adjusted for duration of treatment, the comparison groups were age matched, and for a subset of proteins, we conducted ELISA validation testing of SOMAscan® signal. IGFB2 had a decreasing longitudinal trend associated with Dfz, at odds with the observation for Pred. This may be associated with differential growth stunting seen between the two drugs. The dramatic increase of MMP3 by Dfz relative to Pred remains to be interpreted. Further studies are needed to test the physiological significance of these Dfz and Pred differentially affected PD biomarkers. The study's limitations include the small sample size, no adjustments for multiple testing, and that the data come from a natural history study with lots of observed variability. We have tried to overcome these limitations by performing lab validation using ELISA and using a larger sample size and longer treatment duration. For biological validation, an external cohort is needed to validate our findings. Note also that we did not investigate direct associations of biomarkers with efficacy/safety outcomes; larger sample sizes are required for such studies and are the objective of ongoing research. While preliminary, this study identified serum proteins that were altered between the Dfz and Pred groups using SOMAscan® array data and we successfully validated two biomarkers using ELISA that may be associated with adverse effects. However, further studies using larger sample sizes collected from well controlled cohorts enrolled in ongoing and future clinical trials comparing the safety and efficacy of Dfz versus Pred are needed to validate and test these differentially altered pharmacodynamic biomarkers identified herein.

Supplementary Materials: The following are available online at http://www.mdpi.com/2075-4426/10/4/164/s1, Supplemental Table S1: A subset of DMD patients from the Cooperative International Neuromuscular Research Group Duchenne Natural History Study (CINRG-DNHS).

Author Contributions: Conceptualization, Y.H. and U.J.D.; Methodology, Y.H., U.J.D. and S.T.; Validation, S.T. and A.S.; Formal Analysis, S.T.; Data Curation, Y.H., U.J.D., H.G.-D. and S.T.; Writing—Original Draft Preparation, S.T.; Writing—Review & Editing, Y.H. and U.J.D.; Funding Acquisition, Y.H. The CINRG DNHS study was run by the CINRG DNHS investigators who also collected the bio-samples. All authors have read and agreed to the published version of the manuscript.

Funding: This work was supported by DoD grant W81XWH-16-1-0557, NIH grant P50HD090254 and the Muscular Dystrophy Association USA (MDA353094).

Acknowledgments: The authors would like to thank the patients and their families for their participation in the study. The authors would also like to thank study team members for our participating CINRG sites: University of California Davis: C. McDonald, E. Henricson, M. Cregan, L. Johnson, J. Han, N. Joyce, A. Nicorici, D. Reddy; Sundaram Medical Foundation and Apollo Children's Hospital, Alberta Children's Hospital, Calgary: J. Mah, A. Chiu, T. Haig, M. Harris, M. Kornelsen, N. Rincon, K. Sanchez, L. Walker; Queen Silvia Children's Hospital: M. Tulinius, A. Alhander, A. Ekstrom, A. Gustafsson, A. Kroksmark, U. Sterky, L. Wahlgren; Children's National Health System: R. Leshner, N. Brody, B. Drogo, M. Leach, C. Tesi-Rocha, M. Birkmeier, B. Tadese, A. Toles and M. Thangarajh; Royal Children's Hospital: A. Kornberg, K. Carroll, K. DeValle, R. Kennedy, V. Rodriguez, D. Villano; Hadassah Hebrew University Hospital: Y. Nevo, R. Adani, A. Bar Leve, L. Chen-Joseph, M. Daana, V. Panteleyev-Yitshak, E. Simchovitz, D. Yaffe; Instituto de Neurosciencias Fundacion Favaloro: L. Andreone, F. Bonaudo, J. Corderi, L. Levi, L. Mesa, P. Marco; Children's Hospital of Pittsburgh of UPMC and the University of Pittsburgh: P. Clemens, H. Abdel-Hamid, R. Bendixen, C. Bise, A. Craig, K. Karnavas, C. Matthews, G. Niizawa, A. Smith, J. Weimer; Washington University: J. Anger, T. Christenson, J. Florence, R. Gadeken, P. Golumbak, B.

Malkus, A. Pestronk, R. Renna, J. Schierbecker, C. Seiner, C. Wulf; Children's Hospital of Richmond at VCU: J. Teasley, S. Blair, B. Grillo, E. Monasterio; University of Tennessee: T. Bertorini, M. Barrett-Adair, C. Benzel, K. Carter, J. Clift, B. Gatlin, R. Henegar, J. Holloway, M. Igarashi, F. Kiphut, A. Parker, A. Phillips, R. Young; Children's Hospital of Westmead: K. North, K. Cornett, N. Gabriel, M. Harman, C. Miller, K. Rose, S. Wicks; University of Alberta: H. Kolski, L. Chen, C. Kennedy; Centro Clinico Nemo: M. Beneggi, L. Capone, A. Molteni, V. Morettini; Texas Children's Hospital: T. Lotze, A. Gupta, A. Knight, B. Lott, R. McNeil, G. Orozco, R. Schlosser; University of Minnesota: G. Chambers, J. Day, J. Dalton, A. Erickson, M. Margolis, J. Marsh, C. Naughton; Mayo Clinic: K. Coleman-Wood, A. Hoffman, W. Korn-Petersen, N. Kuntz; University of Puerto Rico: B. Deliz, S. Espada, P. Fuste, C. Luciano, J. Torres; and the CINRG Coordinating Center: Lauren Morgenroth, M. Ahmed, A. Arrieta, N. Bartley, T. Brown-Caines, C. Carty, T. Duong, J. Feng, F. Hu, L. Hunegs, Z. Sund, W. Tang, and A. Zimmerman. The CINRG-DNHS was funded by the U.S. Department of Education/NIDRR (#H133B031118, #H133B090001); U.S. Department of Defense (#W81XWH-12-1-0417); National Institutes of Health/NIAMS (#R01AR061875); Parent Project Muscular Dystrophy.

Conflicts of Interest: The authors declare no conflict of interest.

References

1. Hoffman, E.P.; Brown, R.H.; Kunkel, L.M. Dystrophin: The protein product of the Duchenne muscular dystrophy locus. *Cell* **1987**, *51*, 919–928. [CrossRef]
2. Wallace, G.Q.; McNally, E.M. Mechanisms of muscle degeneration, regeneration, and repair in the muscular dystrophies. *Annu. Rev. Physiol.* **2009**, *71*, 37–57. [CrossRef] [PubMed]
3. Ryan, N.J. Ataluren: First global approval. *Drugs* **2014**, *74*, 1709–1714. [CrossRef] [PubMed]
4. Mendell, J.R.; Rodino-Klapac, L.R.; Sahenk, Z.; Roush, K.; Bird, L.; Lowes, L.P.; Alfano, L.; Gomez, A.M.; Lewis, S.; Kota, J.; et al. Eteplirsen for the treatment of Duchenne muscular dystrophy. *Ann. Neurol.* **2013**, *74*, 637–647. [CrossRef]
5. Syed, Y.Y. Eteplirsen: First global approval. *Drugs* **2016**, *76*, 1699–1704. [CrossRef]
6. Heo, Y.A. Golodirsen: First approval. *Drugs* **2020**, *80*, 329–333. [CrossRef]
7. Frank, D.E.; Schnell, F.J.; Akana, C.; El-Husayni, S.H.; Desjardins, C.A.; Morgan, J.; Charleston, J.S.; Sardone, V.; Domingos, J.; Dickson, G.; et al. Increased dystrophin production with golodirsen in patients with Duchenne muscular dystrophy. *Neurology* **2020**, *94*, e2270–e2282. [CrossRef]
8. Hoffman, E.P.; Bronson, A.; Levin, A.A.; Takeda, S.; Yokota, T.; Baudy, A.R.; Connor, E.M. Restoring dystrophin expression in Duchenne muscular dystrophy muscle: Progress in exon skipping and stop codon read through. *Am. J. Pathol.* **2011**, *179*, 12–22. [CrossRef]
9. Bushby, K.; Finkel, R.; Birnkrant, D.J.; Case, L.E.; Clemens, P.R.; Cripe, L.; Kaul, A.; Kinnett, K.; McDonald, C.; Pandya, S.; et al. Diagnosis and management of Duchenne muscular dystrophy, part 1: Diagnosis, and pharmacological and psychosocial management. *Lancet Neurol.* **2010**, *9*, 77–93. [CrossRef]
10. Griggs, R.C.; Miller, J.P.; Greenberg, C.R.; Fehlings, D.L.; Pestronk, A.; Mendell, J.R.; Moxley, R.T.; King, W.; Kissel, J.T.; Cwik, V.; et al. Efficacy and safety of deflazacort vs prednisone and placebo for Duchenne muscular dystrophy. *Neurology* **2016**, *87*, 2123–2131. [CrossRef]
11. McDonald, C.M.; Henricson, E.K.; Abresch, R.T.; Duong, T.; Joyce, N.C.; Hu, F.; Clemens, P.R.; Hoffman, E.P.; Cnaan, A.; Gordish-Dressman, H.; et al. Long-term effects of glucocorticoids on function, quality of life, and survival in patients with Duchenne muscular dystrophy: A prospective cohort study. *Lancet* **2018**, *391*, 451–461. [CrossRef]
12. Bello, L.; Gordish-Dressman, H.; Morgenroth, L.P.; Henricson, E.K.; Duong, T.; Hoffman, E.P.; Cnaan, A.; McDonald, C.M. Prednisone/prednisolone and deflazacort regimens in the CINRG Duchenne Natural History Study. *Neurology* **2015**, *85*, 1048–1055. [CrossRef] [PubMed]
13. Lamb, M.M.; West, N.A.; Ouyang, L.; Yang, M.; Weitzenkamp, D.; James, K.; Ciafaloni, E.; Pandya, S.; DiGuiseppi, C.; Cunniff, C.; et al. Corticosteroid treatment and growth patterns in ambulatory males with Duchenne muscular dystrophy. *J. Pediatr.* **2016**, *173*, 207–213.e3. [CrossRef] [PubMed]
14. Bonifati, M.D.; Ruzza, G.; Bonometto, P.; Berardinelli, A.; Gorni, K.; Orcesi, S.; Lanzi, G.; Angelini, C. A multicenter, double-blind, randomized trial of deflazacort versus prednisone in Duchenne muscular dystrophy. *Muscle Nerve* **2000**, *23*, 1344–1347. [CrossRef]
15. Cohen, L.; Morgan, J.; Bozyk, M. Variable effects of corticosteroid treatment of serum enzyme activities in Duchenne's muscular dystrophy. *Res. Commun. Chem. Pathol. Pharmacol.* **1977**, *17*, 529–538.

16. FDA News Release (2017) FDA Approves Drug to Treat Duchenne Muscular Dystrophy. Available online: https://www.fda.gov/news-events/press-announcements/fda-approves-drug-treat-duchenne-muscular-dystrophy (accessed on 1 September 2020).
17. Hathout, Y.; Liang, C.; Ogundele, M.; Xu, G.; Tawalbeh, S.M.; Dang, U.J.; Hoffman, E.P.; Gordish-Dressman, H.; Conklin, L.S.; van den Anker, J.N.; et al. Disease-specific and glucocorticoid-responsive serum biomarkers for Duchenne muscular dystrophy. *Sci. Rep.* **2019**, *9*, 12167. [CrossRef]
18. Dang, U.J.; Ziemba, M.; Clemens, P.R.; Hathout, Y.; Conklin, L.S.; Hoffman, E.P. Serum biomarkers associated with baseline clinical severity in young steroid-naïve Duchenne muscular dystrophy boys. *Hum. Mol. Genet.* **2020**, *29*, 2481–2495. [CrossRef]
19. Spitali, P.; Hettne, K.; Tsonaka, R.; Charrout, M.; van den Bergen, J.; Koeks, Z.; Kan, H.E.; Hooijmans, M.T.; Roos, A.; Straub, V.; et al. Tracking disease progression non-invasively in Duchenne and Becker muscular dystrophies. *J. Cachexia. Sarcopenia Muscle* **2018**, *9*, 715–726. [CrossRef]
20. Ahmed, S.F.; Tucker, P.; Mushtaq, T.; Wallace, A.M.; Williams, D.M.; Hughes, I.A. Short-term effects on linear growth and bone turnover in children randomized to receive prednisolone or dexamethasone. *Clin. Endocrinol.* **2002**, *57*, 185–191. [CrossRef]
21. Matthews, E.; Brassington, R.; Kuntzer, T.; Jichi, F.; Manzur, A.Y. Corticosteroids for the treatment of Duchenne muscular dystrophy. *Cochrane Database Syst. Rev.* **2016**. [CrossRef]
22. McDonald, C.M.; Sajeev, G.; Yao, Z.; McDonnell, E.; Elfring, G.; Souza, M.; Peltz, S.W.; Darras, B.T.; Shieh, P.B.; Cox, D.A.; et al. Deflazacort vs prednisone treatment for Duchenne muscular dystrophy: A meta-analysis of disease progression rates in recent multicenter clinical trials. *Muscle Nerve.* **2020**, *61*, 26–35. [CrossRef] [PubMed]
23. Ganapati, A.; Ravindran, R.; David, T.; Yadav, B.; Jeyaseelan, V.; Jeyaseelan, L.; Danda, D. Head to head comparison of adverse effects and efficacy between high dose deflazacort and high dose prednisolone in systemic lupus erythematosus: A prospective cohort study. *Lupus* **2018**, *27*, 890–898. [CrossRef] [PubMed]
24. Hathout, Y.; Conklin, L.S.; Seol, H.; Gordish-Dressman, H.; Brown, K.J.; Morgenroth, L.P.; Nagaraju, K.; Heier, C.R.; Damsker, J.M.; van den Anker, J.N.; et al. Serum pharmacodynamic biomarkers for chronic corticosteroid treatment of children. *Sci. Rep.* **2016**, *6*, 31727. [CrossRef] [PubMed]
25. Conklin, L.S.; Merkel, P.A.; Pachman, L.M.; Parikh, H.; Tawalbeh, S.; Damsker, J.M.; Cuthbertson, D.D.; Morgan, G.A.; Monach, P.A.; Hathout, Y.; et al. Serum biomarkers of glucocorticoid response and safety in anti-neutrophil cytoplasmic antibody-associated vasculitis and juvenile dermatomyositis. *Steroids* **2018**, *140*, 159–166. [CrossRef]
26. McDonald, C.M.; Henricson, E.K.; Abresch, R.T.; Han, J.J.; Escolar, D.M.; Florence, J.M.; Duong, T.; Arrieta, A.; Clemens, P.R.; Hoffman, E.P.; et al. The cooperative international neuromuscular research group duchenne natural history study-a longitudinal investigation in the era of glucocorticoid therapy: Design of protocol and the methods used. *Muscle Nerve* **2013**, *48*, 32–54. [CrossRef] [PubMed]
27. R Core Team. *R: A Language and Environment for Statistical Computing*; R Foundation for Statistical Computing: Vienna, Austria, 2019.
28. Bates, D.; Mächler, M.; Bolker, B.; Walker, S. Fitting linear mixed-effects models using lme4. *J. Stat. Softw.* **2015**, *67*, 1–48. [CrossRef]
29. Kuznetsova, A.; Brockhoff, P.B.; Christensen, R.H.B. lmerTest package: Tests in linear mixed effects models. *J. Stat. Softw.* **2017**, *82*, 1–26. [CrossRef]
30. Lin, J.; Patel, S.R.; Cheng, X.; Cho, E.A.; Levitan, I.; Ullenbruch, M.; Phan, S.H.; Park, J.M.; Dressler, G.R. Kielin/chordin-like protein, a novel enhancer of BMP signaling, attenuates renal fibrotic disease. *Nat. Med.* **2005**, *11*, 387–393. [CrossRef]
31. Zatorski, H.; Marynowski, M.; Fichna, J. Is insulin-like growth factor 1 (IGF-1) system an attractive target inflammatory bowel diseases? Benefits and limitation of potential therapy. *Pharmacol. Rep.* **2016**, *68*, 809–815. [CrossRef]
32. Serrano, A.L.; Muñoz-Cánoves, P. Fibrosis development in early-onset muscular dystrophies: Mechanisms and translational implications. *Semin. Cell Dev. Biol.* **2017**, *64*, 181–190. [CrossRef]
33. Selvin, E.; Halushka, M.K.; Rawlings, A.M.; Hoogeveen, R.C.; Ballantyne, C.M.; Coresh, J.; Astor, B.C. SRAGE and risk of diabetes, cardiovascular disease, and death. *Diabetes* **2013**, *62*, 2116–2121. [CrossRef] [PubMed]

34. Tanida, S.; Mizoshita, T.; Ozeki, K.; Katano, T.; Kataoka, H.; Kamiya, T.; Joh, T. Advances in refractory ulcerative colitis treatment: A new therapeutic target, Annexin A2. *World J. Gastroenterol.* **2015**, *21*, 8776–8786. [CrossRef] [PubMed]
35. Liu, N.; Jiang, Y.; Chung, J.Y.; Li, Y.; Yu, Z.; Kim, J.W.; Lok, J.M.; Whalen, M.J.; Wang, X. Annexin A2 deficiency exacerbates neuroinflammation and long-term neurological deficits after traumatic brain injury in mice. *Int. J. Mol. Sci.* **2019**, *20*, 6125. [CrossRef] [PubMed]
36. Defour, A.; Medikayala, S.; Van der Meulen, J.H.; Hogarth, M.W.; Holdreith, N.; Malatras, A.; Duddy, W.; Boehler, J.; Nagaraju, K.; Jaiswal, J.K. Annexin A2 links poor myofiber repair with inflammation and adipogenic replacement of the injured muscle. *Hum. Mol. Genet.* **2017**, *26*, 1979–1991. [CrossRef]
37. Grønbæk, H.; Kreutzfeldt, M.; Kazankov, K.; Jessen, N.; Sandahl, T.; Hamilton-Dutoit, S.; Vilstrup, H.; Møller, H.J. Single-centre experience of the macrophage activation marker soluble (s)CD163–associations with disease activity and treatment response in patients with autoimmune hepatitis. *Aliment. Pharmacol. Ther.* **2016**, *44*, 1062–1070.
38. Espinasse, M.-A.; Pépin, A.; Virault-Rocroy, P.; Szely, N.; Chollet-Martin, S.; Pallardy, M.; Biola-Vidamment, A. Glucocorticoid-induced leucine zipper is expressed in human neutrophils and promotes apoptosis through Mcl-1 down-regulation. *J. Innate Immun.* **2016**, *8*, 81–96. [CrossRef]
39. Hu, X.; Zhang, P.; Xu, Z.; Chen, H.; Xie, X. GPNMB enhances bone regeneration by promoting angiogenesis and osteogenesis: Potential role for tissue engineering bone. *J. Cell. Biochem.* **2013**, *114*, 2729–2737. [CrossRef]
40. Huang, Y.; Bai, B.; Yao, Y. Prospects of osteoactivin in tissue regeneration. *Expert Opin. Ther. Targets* **2016**, *20*, 1357–1364. [CrossRef]
41. Belongie, K.J.; Ferrannini, E.; Johnson, K.; Andrade-Gordon, P.; Hansen, M.K.; Petrie, J.R. Identification of novel biomarkers to monitor β-cell function and enable early detection of type 2 diabetes risk. *PLoS ONE* **2017**, *12*, e0182932. [CrossRef]
42. DeMambro, V.E.; Clemmons, D.R.; Horton, L.G.; Bouxsein, M.L.; Wood, T.L.; Beamer, W.G.; Canalis, E.; Rosen, C.J. Gender-specific changes in bone turnover and skeletal architecture in Igfbp-2-null mice. *Endocrinology* **2008**, *149*, 2051–2061. [CrossRef]
43. Hahn, B.H.; MacDermott, R.P.; Jacobs, S.B.; Pletscher, L.S.; Beale, M.G. Immunosuppressive effects of low doses of glucocorticoids: Effects on autologous and allogeneic mixed leukocyte reactions. *J. Immunol.* **1980**, *124*, 2812–2817. [PubMed]
44. Lippuner, K.; Casez, J.-P.; Horber, F.F.; Jaeger, P. Effects of deflazacort versus prednisone on bone mass, body composition, and lipid profile: A randomized, double blind study in kidney transplant patients. *J. Clin. Endocrinol. Metab.* **1998**, *83*, 3795–3802. [CrossRef] [PubMed]
45. Pagano, G.; Lombardi, A.; Ferraris, G.M.; Imbimbo, B.; Cavallo Perin, P. Acute effect of prednisone and deflazacort on glucose tolerance in prediabetic subjects. *Eur. J. Clin. Pharmacol.* **1982**, *22*, 469–471. [CrossRef] [PubMed]
46. Kemaladewi, D.U.; Pasteuning, S.; Van Der Meulen, J.W.; Van Heiningen, S.H.; Van Ommen, G.J.; Ten Dijke, P.; Aartsma-Rus, A.; 'T Hoen, P.A.C.; Hoogaars, W.M. Targeting TGF-β signaling by antisense oligonucleotide-mediated knockdown of TGF-β type I receptor. *Mol. Ther.-Nucleic Acids* **2014**, *3*, e156. [CrossRef] [PubMed]
47. Joseph, S.; Wang, C.; Bushby, K.; Guglieri, M.; Horrocks, I.; Straub, V.; Ahmed, S.F.; Wong, S.C. Fractures and linear growth in a nationwide cohort of boys with Duchenne muscular dystrophy with and without glucocorticoid treatment: Results from the UK Northstar database. *JAMA Neurol.* **2019**, *76*, 701–709. [CrossRef] [PubMed]
48. Marden, J.R.; Freimark, J.; Yao, Z.; Signorovitch, J.; Tian, C.; Wong, B.L. Real-world outcomes of long-term prednisone and deflazacort use in patients with Duchenne muscular dystrophy: Experience at a single, large care center. *J. Comp. Eff. Res.* **2020**, *9*, 177–189. [CrossRef]
49. Conklin, L.S.; Damsker, J.M.; Hoffman, E.P.; Jusko, W.J.; Mavroudis, P.D.; Schwartz, B.D.; Mengle-Gaw, L.J.; Smith, E.C.; Mah, J.K.; Guglieri, M.; et al. Phase IIa trial in Duchenne muscular dystrophy shows vamorolone is a first-in-class dissociative steroidal anti-inflammatory drug. *Pharmacol. Res.* **2018**, *136*, 140–150. [CrossRef]

Journal of Personalized Medicine

MDPI

Article

Multi-Omics Identifies Circulating miRNA and Protein Biomarkers for Facioscapulohumeral Dystrophy

Christopher R. Heier [1,*], Aiping Zhang [2], Nhu Y Nguyen [2], Christopher B. Tully [2], Aswini Panigrahi [2], Heather Gordish-Dressman [1,2], Sachchida Nand Pandey [2], Michela Guglieri [3], Monique M. Ryan [4], Paula R. Clemens [5], Mathula Thangarajh [6], Richard Webster [7], Edward C. Smith [8], Anne M. Connolly [9], Craig M. McDonald [10], Peter Karachunski [11], Mar Tulinius [12], Amy Harper [13], Jean K. Mah [14], Alyson A. Fiorillo [1,2], Yi-Wen Chen [2,*] and Cooperative International Neuromuscular Research Group (CINRG) Investigators †

1. Department of Genomics and Precision Medicine, George Washington University School of Medicine and Health Sciences, Washington, DC 20037, USA; HGordish@childrensnational.org (H.G.-D.); afiorillo@childrensnational.org (A.A.F.)
2. Center for Genetic Medicine Research, Children's National Hospital, Washington, DC 20010, USA; AZhang@childrensnational.org (A.Z.); nnguyen@childrensnational.org (N.Y.N.); ctully2@childrensnational.org (C.B.T.); APANIGRAHI@childrensnational.org (A.P.); spandey@childrensnational.org (S.N.P.)
3. Newcastle Upon Tyne Hospitals, Newcastle NE1 3BZ, UK; michela.guglieri@newcastle.ac.uk
4. The Royal Children's Hospital, Melbourne University, Parkville, Victoria 3052, Australia; monique.ryan@rch.org.au
5. Department of Neurology, University of Pittsburgh School of Medicine, Pittsburgh, PA 15261, USA; pclemens@pitt.edu
6. Department of Neurology, Virginia Commonwealth University School of Medicine, Richmond, VA 23298, USA; mathula.thangarajh@vcuhealth.org
7. Children's Hospital at Westmead, Sydney 2145, Australia; richard.webster@health.nsw.gov.au
8. Department of Pediatrics, Duke University Medical Center, Durham, NC 27705, USA; edward.smith@duke.edu
9. Nationwide Children's Hospital, The Ohio State University, Columbus, OH 43205, USA; anne.connolly@nationwidechildrens.org
10. Department of Physical Medicine and Rehabilitation, University of California at Davis Medical Center, Sacramento, CA 95817, USA; cmmcdonald@ucdavis.edu
11. Department of Neurology, University of Minnesota, Minneapolis, MN 55455, USA; karac001@umn.edu
12. Department of Pediatrics, Gothenburg University, Queen Silvia Children's Hospital, 41685 Göteborg, Sweden; mar.tulinius@vgregion.se
13. Department of Neurology, Virginia Commonwealth University, Richmond, VA 23298, USA; amy.harper@vcuhealth.org
14. Deparment of Pediatrics and Clinical Neurosciences, Cumming School of Medicine, University of Calgary, T2N T3B, Calgary, AB 6A81N4, Canada; Jean.Mah@albertahealthservices.ca

* Correspondence: cheier@childrensnational.org (C.R.H.); ychen@childrensnational.org (Y.-W.C.)

† University of California Davis.

Received: 25 September 2020; Accepted: 6 November 2020; Published: 19 November 2020

Abstract: The development of therapeutics for muscle diseases such as facioscapulohumeral dystrophy (FSHD) is impeded by a lack of objective, minimally invasive biomarkers. Here we identify circulating miRNAs and proteins that are dysregulated in early-onset FSHD patients to develop blood-based molecular biomarkers. Plasma samples from clinically characterized individuals with early-onset FSHD provide a discovery group and are compared to healthy control volunteers. Low-density quantitative polymerase chain reaction (PCR)-based arrays identify 19 candidate miRNAs, while mass spectrometry proteomic analysis identifies 13 candidate proteins. Bioinformatic analysis of chromatin immunoprecipitation (ChIP)-seq data shows that the FSHD-dysregulated DUX4 transcription factor binds to regulatory regions of several candidate miRNAs. This panel of miRNAs also shows ChIP signatures consistent with regulation by additional transcription factors which are up-regulated in FSHD (FOS, EGR1, MYC, and YY1). Validation studies in a separate group of patients with FSHD show consistent up-regulation of miR-100, miR-103, miR-146b, miR-29b, miR-34a, miR-454, miR-505, and miR-576. An increase in the expression of S100A8 protein, an inflammatory regulatory factor and subunit of calprotectin, is validated by Enzyme-Linked Immunosorbent Assay (ELISA). Bioinformatic analyses of proteomics and miRNA data further support a model of calprotectin and toll-like receptor 4 (TLR4) pathway dysregulation in FSHD. Moving forward, this panel of miRNAs, along with S100A8 and calprotectin, merit further investigation as monitoring and pharmacodynamic biomarkers for FSHD.

Keywords: FSHD; biomarkers; miRNA; proteomics; calprotectin; dystrophy; muscle

---

1. Introduction

Facioscapulohumeral muscular dystrophy (FSHD) is an autosomal dominant muscle disorder with no current therapy, a variable prognosis, and complex genetic and molecular mechanisms. FSHD is caused by aberrant expression of *double homeobox 4* (*DUX4*) due to epigenetic changes of the *D4Z4* repeat region at chromosome 4q35 [1–3]. Roughly 95% of patients have Type 1 FSHD (FSHD1) due to contraction of the D4Z4 array; a small portion (~5%) of patients have Type 2 FSHD (FSHD2) caused by mutations in *the structural maintenance of chromosomes flexible hinge domain containing 1 (SMCHD1)* gene, the *DNA methyltransferase 3B (DNMT3B)* gene, or the *ligand-dependent nuclear receptor-interacting factor 1 (LRIF1)* gene [4–6]. The aberrant expression of DUX4 protein causes mis-regulation of genes involved in germline function, oxidative stress responses, myogenesis, post-transcriptional regulation, and additional cellular functions [7–13]. These downstream molecular changes are believed to cause FSHD, although the exact mechanisms are not clear.

Although the onset of FSHD is generally around adolescent years, a small portion (~4%) of patients present with an early-onset or infantile form of FSHD [14]. Previous studies have shown that the disease severity of FSHD1 is negatively correlated with the size of *D4Z4* repeats [15,16]. Individuals with early-onset FSHD1 tend to have smaller *D4Z4* repeats and more severe disease phenotypes, including more profound muscle weakness, younger age at loss of independent ambulation, and extramuscular manifestations such as retinal vasculopathy or hearing loss [14,15,17,18].

In clinical practice, particularly with pediatric-onset FSHD, there is a low use of serial histological assessments because they require painful biopsies of muscle tissue that typically reveal patchy or uneven pathology. Given this, many patients no longer undergo muscle biopsy once a genetic diagnosis is made. Functional motor scales provide a non-invasive alternative to study neuromuscular disease progression; however, they can show great variability, can be age- or disease stage-limited, and they can be subject to placebo or coaching effects in clinical trials [19,20]. Circulating molecular biomarkers provide a promising alternative to these clinical assessments because they are objective measurements that can be assayed repeatedly over time using minimally invasive methods. Blood-based miRNAs or proteins that measure the progression of disease or a patient response to therapy over time are known as a monitoring biomarker [21]. In clinical trials, monitoring biomarkers may also be used as pharmacodynamic biomarkers to identify patients who are early responders to therapy, to demonstrate exposure-response relationships, or to improve statistical power and modeling. As patient populations are sensitive and limited for this relatively rare pediatric disease, less invasive monitoring or pharmacodynamic biomarkers are important for early-onset FSHD, as frequent serial biopsies are especially problematic in this population.

Recently, circulating miRNAs have emerged as exciting potential diagnostic, prognostic, and drug-responsive biomarkers. This is a class of small non-coding ribonucleic acid (RNA) molecules

(~22 nucleotides in length) that can help to regulate gene expression [22], and which are highly stable in biofluids such as blood and urine [23,24]. In rare diseases with highly variable symptoms, such as multiple acyl-coenzyme A dehydrogenase deficiency (MADD), the serum-based detection of muscle-specific miRNAs termed myomiRs can signal the presence of underlying muscle-specific pathologies [25]. In Duchenne and Becker muscular dystrophies, myomiRs are up-regulated in serum from both patient populations, while detection of miR-206 up-regulation can be used to differentially diagnose severe Duchenne versus Becker patients [26–28]. In addition to myomiRs, inflammatory miRNAs such as miR-146a, miR-146b, miR-221 and miR-155 have been found to be dysregulated in multiple forms of muscular dystrophies [29–31]. These two classes of miRNA show potential as pharmacodynamic biomarkers, with myomiRs proposed for muscle-stabilizing treatments such as gene therapy [32,33], and inflammatory microRNAs proposed for current steroids [34,35] as well as newly emerging dissociative anti-inflammatory drugs such as vamorolone [36–38] or edasalonexent [39,40]. In parallel to development of miRNA monitoring biomarkers, new advances in whole exome sequencing are enabling clinicians to diagnose novel mutations in over 60 genes known to be responsible for muscular dystrophies such as FSHD and limb-girdle muscular dystrophy (LGMD) [41–44]. Together, these advances will help to improve the diagnosis, monitoring, and treatment of a diverse number of diseases affecting muscle.

The development of circulating biomarkers for FSHD has the potential to improve clinical management and to facilitate the development of new treatments. In this study, we test plasma samples from a cohort of individuals with early-onset FSHD1 using both miRNA and proteomic profiling approaches. Our goal is to identify molecules that can be used to monitor FSHD disease activity and that may ultimately facilitate future therapeutic trials. Initial analysis of a discovery group identifies a panel of miRNAs and proteins as biomarker candidates. Bioinformatic analyses of ChIP-seq data provide a rationale for the changes in candidate biomarkers, as their behavior is consistent with changes in transcription factor pathways that are disrupted in FSHD1. Subsequent characterization in separate, non-overlapping groups of FSHD1 patients provides validation of nine biomarkers whose expression can be conveniently assayed by qRT-PCR or Enzyme-Linked Immunosorbent Assay (ELISA), and are increased in early-onset FSHD.

2. Materials and Methods

2.1. Ethics Statement

We obtained institutional ethics and research review boards approval for these clinical studies from the Institutional Review Board of Children's National Hospital and at all participating Cooperative International Neuromuscular Research Group (CINRG) sites, in accordance with all requirements, as previously described in Mah et al. 2018 [45]. Written informed consent was obtained from all the participants before the study procedures. Where applicable, informed consent and/or assent was obtained from all patients or legal guardians before enrollment.

2.2. Patients and Sample Collection

Plasma samples were collected and biobanked from a previous early-onset FSHD study conducted by CINRG as described by Mah et al. [45]. For the discovery experiments, FSHD1 patients aged 10 to 51 years old were included (n = 16 for miRNA discovery, n = 25 for proteomics discovery), along with healthy control volunteers (n = 8 for miRNA discovery, n = 17 for proteomics discovery) aged 16 to 54 years old. All patients had Type 1 FSHD caused by epigenetic changes due to *D4Z4* contraction which results in up-regulation of *DUX4*.

2.3. miRNA Profiling

RNA was isolated and quantified from the discovery cohort of patients as described previously [34]. Briefly, RNA was isolated from 150 μL of plasma using Trizol liquid sample (LS) reagent (ThermoFisher,

Waltham, MA, USA), then converted to cDNA using the High Capacity Reverse Transcription Kit with multiplexed reverse transcription (RT) primers (ThermoFisher). Synthesized cDNA was then pre-amplified using PreAmp MasterMix with multiplexed TaqMan (TM) primers corresponding to the RT primers used in initial cDNA reaction. Quantitative analysis of miRNA was performed via TaqMan Low-Density Array Cards (TaqMan™ Array Human MicroRNA A Cards v2.0; ThermoFisher). The ThermoFisher Cloud software suite with the Relative quantification (Rq) application was used to perform statistical analysis and determine expression of miRNA in either mild or severe FSHD1 patient groups versus healthy controls. A value > 1 indicates an increase and a value < 1 indicates a decrease in miRNA expression in FSHD1 versus healthy controls, with p-values ≤ 0.05 considered significant. To reduce false-positive discovery in this setting, we used an evidence-based approach where candidate miRNAs that significantly increased in the discovery groups were cross-referenced to a separate set of non-overlapping CINRG patients used as a validation group.

2.4. Bioinformatics of miRNA Regulation via DUX4 and FSHD-Associated Factors

Surrounding DNA regulatory regions of candidate miRNA genes were queried in ChIP-seq datasets for binding by transcription factors known to be impacted by FSHD. These analyses were performed using the University of California Santa Cruz (UCSC) Genome Browser with alignment to the GECh37/hg19 genome build. For primary effects, due to the underlying mutation that causes FSHD, DUX4 binding was queried. For this, we uploaded a user-supplied DUX4 ChIP-seq track published by Geng et al. [9] to determine which candidate miRNAs displayed physical binding of DUX4 at potential regulatory regions within 100 kb of the gene for each miRNA.

To investigate secondary factors whose dysregulation is associated with FSHD-causing mutations, we investigated DNA binding by transcription factors shown to be significantly up-regulated in cultured human muscle cells using microarray data by Geng et al. [9]. For this, we used ChIP-seq data from the Encyclopedia of DNA Elements (ENCODE) [46,47]. From a master list of DUX4-regulated genes published in [9], we identified a list of 34 transcription factors with ChIP-seq data from ENCODE available within the UCSC Txn Factor ChIP Track and 47 transcription factors from the Txn Factor ChIP E3 Track [48–50]. After an initial survey of these full transcription factor lists for the 19 candidate miRNAs, we narrowed down to a shorter focus list of 9 transcription factors whose binding was most frequently associated with the candidate miRNAs. DNA binding by transcription factors was queried in datasets produced using ChIP-seq from all 9 available cell line tracks, including GM12878 (lymphoblasts), H1-hESC (embryonic stem cells), HeLa-S3 (cervical cancer cells), HepG2 (liver cancer cells), HSMM (skeletal muscle myoblasts), HUVEC (umbilical vein endothelial cells), K562 (immortalized myelogenous leukemia cells), NHEK (epidermal keratinocytes), and normal human lung fibroblasts (NHLF).

In addition to binding by DUX4 and the transcription factors described above, ChIP-seq data for histone modifications were queried to gain insight into potential promoter or enhancer regulatory functions for the identified transcription factor binding sites. For this, histone H3K4 tri-methylation (found near promoters), H3K4 mono-methylation (found near regulatory elements), and H3K27 acetylation (found near active regulatory elements) were included. These histone modifications were queried in ChIP-seq datasets using all 9 available cell line tracks.

Pathway analysis was performed using Ingenuity Pathway Analysis software version 52912811. Candidate miRNAs from these studies were uploaded along with transcription factors whose dysregulation is associated with FSHD. Defined network connections were identified using the Pathway Builder application. Molecules confirmed to have established relationships were used to visualize a novel network built from these FSHD expression data.

2.5. Expression of Individual miRNAs in a Validation Sample Set

Circulating miRNAs that were significantly up-regulated in individuals affected by FSHD1 were examined in a separate set of non-overlapping CINRG patients used as a validation group.

For this group, FSHD patients had a confirmed diagnosis of FSHD1 ($n = 12$; 9 females, 3 males) and were compared to healthy volunteer control samples ($n = 7$; 4 females, 3 males). RNA was isolated from 150 µL of plasma using Trizol LS liquid extraction. Total RNA was converted to cDNA using a High Capacity Reverse Transcription Kit with multiplexed RT primers, pre-amplified using PreAmp MasterMix with multiplexed TM primers, and quantified with individual TaqMan assays on an ABI QuantStudio 7 real time PCR machine (Applied Biosystems; Foster City, CA, USA). Assay IDs used are: miR-32-002109, miR-103-000439, miR-505-002089, miR-146b-001097, miR-29b-000413, miR-34a-000426, miR-141-000463, miR-98-000577, miR-576-3p-002351, miR-9-000583, and miR-142-3p-000464. Expression levels of all miRNAs were normalized to the geometric mean of multiple control genes (miR-150 and miR-342-3p) determined previously to be stable circulating miRNA controls [35,51]. Expression was analyzed in FSHD1 versus healthy control patients via *t*-test analysis, including assessment of directionality. A *p*-value of ≤ 0.05 was considered significant. Data are presented as mean ± SEM unless otherwise noted.

2.6. Proteomics Profiling

Plasma samples were first processed using Pierce™ Top 12 Abundant Protein Depletion Spin Columns (Thermo Scientific) before mass spectrometry analyses using the Q Exactive HF mass spectrometer. Briefly, the 12 most abundant proteins from 5 µL of plasma sample were affinity depleted by incubating with Top 12 protein depletion resin. Following this, the unbound fraction was collected according to the manufacturer's protocol. Proteins were precipitated with pre-cooled acetone (1:5 vol) for 30 min at −20 °C and centrifuged at 4 °C for 15 min at max speed in a micro-centrifuge. The liquid was decanted and the pellet was air dried briefly and resuspended with 8 M Urea, followed by reduction and alkylation with 5 mM DDT and 15 mM idodoacetamide for 30 min at room temperature. Samples were diluted with 100 mM ammonia bicarbonate to final urea concentration of less than 2 M. Afterwards, the samples were digested with 1 µg of trypsin (Promega) at 37 °C overnight. Trypsin was inactivated by 0.1% TFA and samples were desalted by capturing the peptides onto C18 100 µL bed tips (Pierce®C18 tips, Thermo Scientific) following the manufacture's protocol. The bound peptides were eluted with 60% acetonitrile, 0.1% TFA, then dried using a SpeedVac, and resuspended in 20 µL buffer containing 2% acetonitrile with 0.1% acetic acid.

The peptide mixtures from each fraction were sequentially analyzed by liquid chromatography tandem mass spectrometry (LC-MS/MS) using Thermo Ultimate 3000 RSLCnano-Q Exactive mass spectrometry platform nano-LC system (Easy nLC1000) connected to Q Exactive HF mass spectrometer (Thermo Scientific). This platform is configured with nano-electrospray ion source (Easy-Spray, Thermo Scientific), Acclaim PepMap 100 C18 nanoViper trap column (3 µm particle size, 75 µm ID × 20 mm length), EASY-Spray C18 analytical column (2 µm particle size, 75 µm ID × 500 mm length). The data from each sample was collected in triplicate at 2 µL per injection, following which the peptides were eluted at a flow rate of 300 nL/min using linear gradients of 7–25% Acetonitrile (in aqueous phase and 0.1% Formic Acid) for 80 min, followed by to 45% for 25 min, and static flow at 90% for 15 min. The mass spectrometry data was collected in data-dependent manner switching between one full scan MS mode (*m/z* 380–1600, resolution 70,000, AGC 3e6) and 10 MS/MS mode (resolution 17,500); where MS/MS analysis of the top 10 target ions were performed once and dynamically excluded from the list for 30 s.

The MS raw data sets were searched against UniProt human database that included common contaminants using MaxQuant software (version 1.5.5.1) [52]. We used default parameters for the searches; first search peptide tolerance 20 ppm, main search peptide tolerance 4.5 ppm, maximum two missed cleavage; and the peptide and resulting protein assignments were allowed at 0.01 FDR (thus 99% confidence level). Protein levels were quantified in 25 FSHD1 patients and 17 healthy controls and reported for each protein as the number of unique peptides detected and the intensity measured. Proteins with altered abundance with greater than 2-fold were selected for further inquiry.

Several pre-processing steps were performed on the raw data values before statistical analysis. Each sample had either 2 or 3 replicates which were averaged to yield a single quantification for each subject for each protein. When a value of zero occurs, it can indicate either a true zero or an assay that did not detect that protein. To accurately reflect protein levels, we incorporated zeroes into our analysis in the following way. If one replicate yielded a zero value, that zero was left as is and treated as a true zero. If two replicates yielded a zero, all values for that protein/sample were set to missing as we cannot distinguish true zeroes from artificial ones. We then applied a normalization factor to the average values to account for differences in the amount assayed per sample. We summed the protein counts for all proteins for each sample and used the maximum value to normalize all other samples. This allowed us to ensure that the amount of proteins assayed were proportional for all samples.

All values were log-transformed for analysis. We assessed the relationship between protein levels and disease severity in the FSHD1 patients using a linear regression model where protein level was the dependent variable, severity was the independent variable, and age and gender were covariates. Regression models were performed only for proteins found in 5 or more samples. Model estimates were reported for each protein and included the coefficient and p-value for all terms in the model (severity, age and gender) along with an indication of the direction of each effect. This same method was used to assess the relationship between protein level and the number of D4Z4 repeats. We assessed the difference in protein expression between FSHD1 patients and healthy controls using a linear regression model where protein level was the dependent variable, a categorical indicator of disease was the independent variable, and age and gender were covariates. Again, regression models were performed only for proteins found in 5 or more samples. Model estimates were reported for each protein and included the coefficient and p-value for all terms in the model (disease status, age and gender), an indication of the direction of each effect, and age and gender adjusted means for each disease group. As this part of the analysis was discovery in nature, we did not adjust resulting p-values for multiple testing. Our intention was to find those proteins showing some evidence of an effect and to move those proteins forward for an additional evidence-based validation experiment. The significance level for all analyses was set at 0.05.

2.7. Enzyme-Linked Immunosorbent Assay (ELISA)

Five proteins were chosen for further validation in a separate set of patients via protein-specific ELISA assays. Human specific protein ELISA kits for human insulin-like growth factor-1 (IGF1) (R&D Systems, Minneapolis, MN, USA), profilin 1 (PFN1) (LSBio, Seattle, WA, USA), S100 Calcium-Binding Protein A8 (S100-A8) (Biotechne, Minneapolis, MN, USA), Proteoglycan 4 (PRG4) (AVIVA Systems Biology, San Diego, CA, USA), Human Tropomyosin alpha-4 chain (TPM4) (MyBioSource, San Diego, CA, USA) were performed to determine protein level in FSHD1 and unaffected controls. Plasma (20 μL) from FSHD1 patients (n = 19) and healthy volunteer (n = 13) controls (age and gender matched) were tested in duplicate following the manufacturer's recommended protocols. ELISA values were assessed for normality and a log-transformation applied where appropriate. We assessed the relationship between protein level and severity using, as described above, a linear regression model where protein level was the dependent variable, severity was the independent variable, and age and gender were covariates. We assessed the difference in protein expression between FSHD1 and healthy controls using a linear regression model where protein level was the dependent variable, a categorical indicator of disease was the independent variable, and age and gender were covariates. All analyses were performed at the 0.05 significance level.

3. Results

3.1. Discovery of Novel Candidate miRNA Biomarkers Associated with FSHD

Sixteen FSHD1 patients with pediatric-onset, matched for sex and age, were selected into two groups of a discovery sample set for circulating biomarker studies: one mild FSHD1 group (n = 8), and one severe FSHD1 group (n = 8), as determined by an FSHD disease severity score. These two groups were each compared to a group of healthy control volunteers (n = 8). Demographics are displayed in Table 1. Patients with severe FSHD1 showed a significantly higher FSHD severity score (12.25 ± 2.76; $p \leq 0.00001$) than patients with mild FSHD1 (4.88 ± 1.46), with any value of nine or higher being classified as severe FSHD.

Table 1. Clinical characteristics of study group patients.

	Healthy Control	Mild FSHD	Severe FSHD
N	8	8	8
Age in years (mean ± SD)	28.29 ± 15.82	24.84 ± 10.46	27.58 ± 15.11
Males:Females	4:4	4:4	4:4
FSHD severity score	N/A	4.88 ± 1.46	12.25 ± 2.76 **

** $p \leq 0.00001$, t-test of mild FSHD versus severe FSHD severity score.

Ten miRNAs showed a significant change in expression level in mild FSHD1 plasma versus healthy controls, and 12 miRNAs showed a significant change in expression level in severe FSHD samples versus controls (Table 2). Of these, three miRNAs showed a significant increase in both mild and severe FSHD1 in comparison to healthy controls: miR-32, miR-505, and miR-29b. Each of these three miRNAs showed an approximately two-fold higher change in expression in severe FSHD1 patients than in mild FSHD1 patients versus healthy controls. Of the 19 unique miRNAs identified, several have been previously found to play a role in muscle disease pathways. miR-29b, which is associated with TGFβ-signaling and fibrosis, was up-regulated in both mild and severe FSHD1 patients. Both miR-146b and miR-142-3p, which are known to be up-regulated in inflammatory disease states, were up-regulated in mild FSHD1 patients and have previously been shown to be up-regulated in dystrophinopathy (Becker and Duchenne muscular dystrophy) patients and/or animal models [30,36]. miR-486 has previously been defined as a muscle-enriched microRNA or "myomiR" [53], and was found here to be down-regulated in mild FSHD1 patients ($p < 0.005$).

Table 2. Discovery of 19 circulating miRNAs with altered expression in mild or severe FSHD.

Mild FSHD Versus Healthy Controls				
miRNA	**↑ or ↓**	***p*-Value**	**Rq ***	**Known Roles in Muscle/Disease Pathways**
138	↓	0.004	0.05	Heart development; hypoxia and S100A1 [54–56]
486	↓	0.009	0.26	myomiR; steroid-response in IBD blood [35,53]
9	↑	0.017	9.58	Inhibits satellite cells; COPD weakness [57,58]
32	↑	0.020	8.45	Cardiac fibrosis; VSMC calcification [59,60]
146b	↑	0.034	2.18	Up-regulated in DMD and BMD [30,36]
92a	↓	0.039	0.31	Inhibits myogenic differentiation via Sp1 [61]
576	↑	0.043	3.64	Up-regulated in smooth muscle tumors [62]
142-3p	↑	0.044	2.69	Elevated in models of DMD and myositis [31,36]
505	↑	0.046	9.69	Cardiac development and regeneration [63]
29b	↑	0.050	17.48	Muscle atrophy, therapeutic target [64,65]

Table 2. *Cont.*

		Severe FSHD versus Healthy Controls		
32	↑	*0.001*	*17.09*	*Cardiac fibrosis; VSMC calcification* [59,60]
505	↑	*0.007*	*19.51*	*Cardiac development and regeneration* [63]
502-3p	↓	0.009	0.36	Myogenic differentiation; ACAD marker [66,67]
103	↑	0.013	4.29	Myogenic differentiation [67]
98	↑	0.014	21.65	Muscle differentiation [68]
141	↑	0.016	7.52	Biomarker for prostate and bladder cancer [69]
29b	↑	*0.018*	*28.78*	*Muscle atrophy, therapeutic target* [64,65]
34a	↑	0.024	8.12	Up in FSHD and myotonic dystrophy [70,71]
140-3p	↓	0.028	0.54	Plasma biomarker of myotonic dystrophy [72,73]
100	↑	0.029	3.58	Up-regulated in LMNA dystrophy biopsies [74]
329	↑	0.030	4.63	Counteracts muscle hypertrophy [75]
454	↑	0.046	2.02	Plasma biomarker of myotonic dystrophy [72,73]
		Severe FSHD versus Mild FSHD		
502-3p	↓	0.041	0.45	Myogenic differentiation; ACAD marker [66,67]
95	↑	0.042	2.21	Up in DMD patient and dog model serum [76]
886-3p	↑	0.048	3.27	Up in plasma of myotonic dystrophy patients [73]

Italics = dysregulated in both mild and severe FSHD; ACAD = acute coronary artery disease, BMD = Becker muscular dystrophy, COPD = chronic obstructive pulmonary disease, DMD = Duchenne muscular dystrophy, IBD = inflammatory bowel disease, LMNA = Lamin A/C, TGFβ = Transforming Growth Factor β, VSMC = vascular smooth muscle cell. * $p < 0.005$.

3.2. Bioinformatic Analysis of miRNA Regulation and Pathways

To examine their regulation by transcription factors which are dysregulated by the FSHD disease process, we next performed bioinformatic analyses of ChIP-seq data for DNA binding by transcription factors in proximity to each candidate miRNA's genomic locus. To gain insight into direct consequences of *DUX4*-up-regulating mutations that cause FSHD, we analyzed ChIP-seq data for DUX4. To do this, we analyzed DUX4 binding via a user-supplied DUX4 ChIP-seq track published by Geng et al. [9]. Genes for 16 of the candidate miRNAs had at least one binding site within distances capable of providing gene enhancer functions (Figure 1). Examination of the miR-100 home gene (*MIR100HG*) locus was particularly interesting. In total, we found 18 DUX4 binding sites in the area surrounding *MIR100HG*, and many of these clearly overlapped with histone modifications associated with active promoters (H3K4 tri-methylation) and regulatory elements (H3K27Ac). These data are consistent with regulation of miR-100 expression by DUX4 (Figure 1b).

To gain insight into additional pathways that may drive expression of candidate miRNAs and contribute to FSHD molecular pathophysiology, we performed bioinformatic analyses of ChIP-seq data for transcription factors that are dysregulated as a result of DUX4 mutations. For this, we obtained a list of transcription factors which are expressed at significantly different levels in human skeletal muscle cells as a result of DUX4 overexpression [9]. We then queried publicly available ChIP-seq datasets to identify which of these transcription factors had ChIP-seq datasets available through the ENCODE public research consortium [46,47]. Of the transcription factors in this dataset, 34 had ChIP-seq datasets available in the Factorbook repository and 47 had ChIP-seq datasets available in the ENCODE 3 repository [48–50]. Genomic binding by each of these transcription factors was surveyed for each of these transcription factors for all candidate miRNAs (Table S1). Transcription factors that increased in response to overexpression of toxic, full-length DUX4 but did not increase in response to a non-toxic, truncated isoform of DUX4 were considered to be of particular interest (Figure 2a). Of these factors, four showed a particularly high number of binding sites within regulatory distance of the candidate miRNAs: early growth response protein 1 (EGR1), FOS, MYC, and yin yang 1 (YY1). As an example of these findings, miR-576 was up-regulated in FSHD patients, has five DUX4 binding sites neighboring its home gene (SEC24B), and has a high number of binding sites for the secondary transcription factors described here (Figure 2b). EGR1, FOS, MYC and YY1 all showed a large number of binding sites around miR-576, and these frequently overlapped with histone modifications which mark active promoter and enhancer regions, consistent with these four transcription factors driving gene expression signatures in FSHD.

a

miRNA	Locus	Home Gene	DUX4 sites
32	9q31	TMEM245	6
138-1	3p21	N/A	3
103a-1	5q34	PANK3	1
505	Xq27	ATP11C	0
146b	10q24	N/A	0
29b-1	7q32	LOC646329	2
486	8p11	ANK1, NKX6-3	4
34a	1p36	miR-34a HG	1
141	12p13	miR-200C HG	5
98	Xp11	HUWE1	0
576	4q25	SEC24B	5
9-1	1q22	C1orf61	1
142	17q22	N/A	3
92a-1	13q31	miR-17 HG	11
502	Xp11	CLCN5	0
140	16q22	WWP2	3
100	11q24	miR-100 HG	18
329-1	14q32	N/A	7
454	17q22	SKA2	3

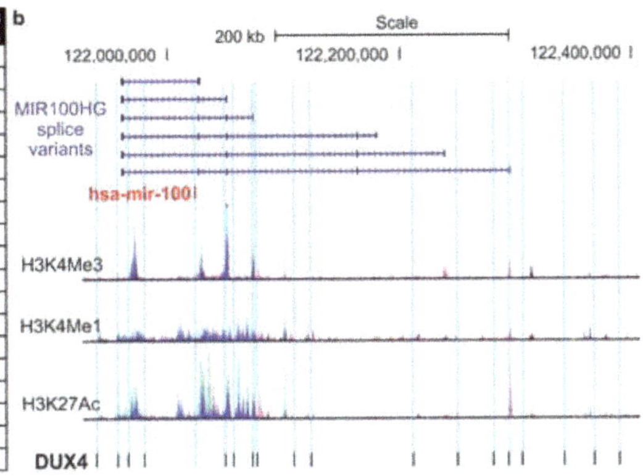

Figure 1. DUX4 binding sites at loci surrounding miRNAs dysregulated in FSHD patients. The 19 miRNAs dysregulated in FSHD1 patient plasma samples were queried for potential regulation by the DUX4 transcription factor, which aberrantly expressed in FSHD, using a DUX4 ChIP-seq dataset [9]. (**a**) Overview of all DUX4 binding sites within regions capable of acting as regulatory elements (100 kb) of the 19 miRNAs and their home genes. (**b**) Schematic of DUX4 binding sites within the miR-100 locus and its surrounding home gene (*MIR100HG*) variants. Note, miR-100 is transcribed from right to left on this image. Corresponding epigenetic modification maps display the location of histone modifications associated with active promoters (H3K4me3) and poised/active enhancers (H3K4me1 and H3K27Ac, respectively).

a

Txn Factor	# binding sites
EGR1	104
FOS	147
MYC	127
YY1	164
EZH2	46
RBBP5	66
TAF7	39
KDM5B	53
ATF3	93

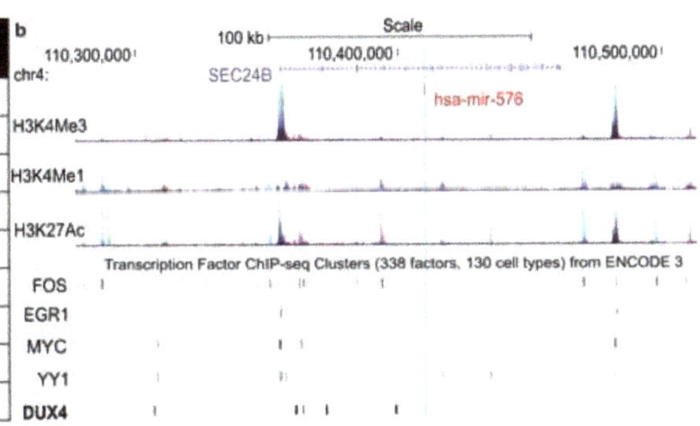

Figure 2. Candidate miRNA loci are consistent with regulation via transcription factors dysregulated in FSHD. (**a**) Table listing a subset of transcription factors which are each increased in human skeletal muscle cells in response to DUX4 overexpression [9], along with the number (#) of binding sites they show within potential regulatory distance (100 kb) of the 19 candidate miRNAs. (**b**) The miR-576 locus shows binding consistent with regulation by FOS, EGR1, MYC, YY1, and DUX4. Corresponding epigenetic modification maps display the location of histone modifications associated with active promoters (H3K4me3) and poised/active enhancers (H3K4me1 and H3K27Ac) in the vicinity of the miR-576 locus and its surrounding home gene, *SEC24 homolog B* (*SEC24B*). (DUX4 binding sites identified using ChIP-seq data uploaded from Geng et al. [9]; binding sites for additional transcription factors identified using UCSC Genome Browser and respective ChIP-seq datasets accessed via the ENCODE3 regulation track [46–50]).

Additionally, we used Ingenuity Pathway Analysis software to perform a bioinformatic analysis on the candidate miRNAs identified in this study, together with transcription factors previously published to be dysregulated in FSHD [9], to see if there are defined signaling pathways or interactions shared by these factors. Interestingly, this analysis showed that there are previously established connections between many of the miRNAs and transcription factors examined, with 15 of the miRNAs and 18 of the transcription factors found to make up a network with previously defined interactions (Figure 3). For example, increased levels of miR-34a are known to decease cAMP response element-binding protein (CREB) to drive neuronal dysfunction in HIV-induced neurocognitive disorders, and to increase AMP-dependent transcription factor 3 (ATF3) levels in colon cancer [77,78]. MYC binds to ATF3 as well as to lysine-specific demethylase 5B (KDM5B) and YY1, all four of which are elevated in FSHD [9,79–81]; in addition, MYC is known to activate transcription of both enhancer of zeste homolog 2 (EZH2) and miR-9 [82,83], both of which are also increased in FSHD. Together, these bioinformatics data show our candidate miRNA markers are consistent with a change in transcriptional programming that results from FSHD-causing DUX4 overexpression mutations.

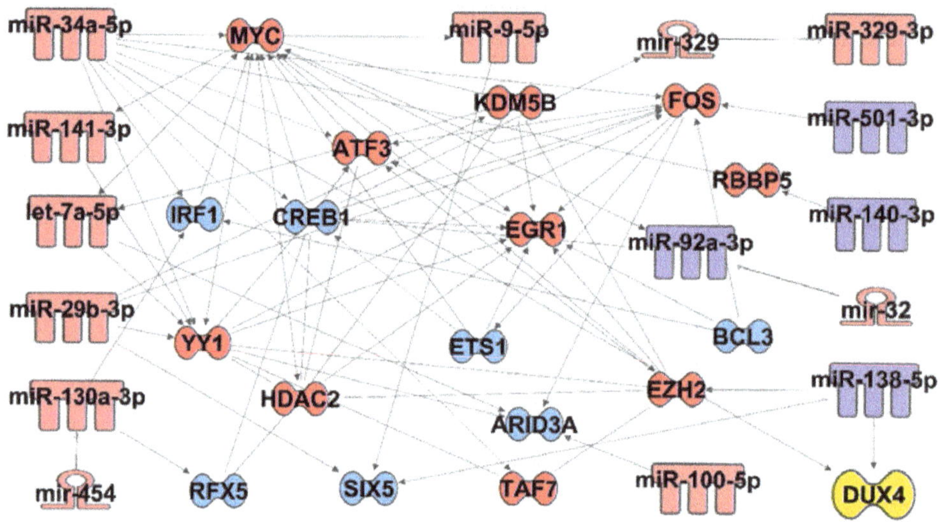

Figure 3. Pathway analysis of miRNAs and transcription factors dysregulated by FSHD mutations. Ingenuity Pathway Analysis software was used to identify established connections between candidate miRNAs from this study with transcription factors known to be dysregulated by FSHD-causing overexpression of DUX4 [9]. Red-shaded miRNAs and transcription factors were observed to increase, while those shaded blue were observed to decrease. Solid arrows denote direct relationships, while dashed arrows denote indirect relationships.

3.3. *Confirmation of miRNA Increases in FSHD1 Patients*

Next, we assayed expression of candidate miRNA biomarkers in samples from a separate and non-overlapping group of patients. Upon clinical examination, all patients in this validation group were determined to have FSHD1. We selected 14 miRNAs that significantly increased in the discovery experiments for follow-up study in the validation group. We found three of these miRNAs (miR-9, miR-32 and miR-329) were not expressed at consistently high enough levels for detection within plasma from the validation set of FSHD1 patients, leaving 11 miRNAs for validation. Here, these 11 individual candidate miRNAs were quantified in FSHD1 (n = 12; 9 females, 3 males) versus healthy volunteer control samples (n = 7; 4 females, 3 males).

Upon quantification, we found 8 of these 11 candidate miRNAs also showed a clear increase in samples from the FSHD validation group in comparison to healthy controls (Figure 4). miR-100, miR-103, miR-29b, miR-34a, miR-454, miR-505 and miR-576 were all expressed at significantly higher levels ($p \leq 0.05$) in FSHD1 serum. miR-100, miR-29b, miR-34a, miR-505, and miR-576 were the most highly up-regulated in FSHD1, showing up-regulation from approximately 4- to 20-fold higher than healthy controls. miR-146b was also expressed at an approximately 2-fold higher level in this set of FSHD1 patients; however it did not reach significance ($p = 0.06$). Of the remaining three miRNA candidates, miR-98 showed no apparent change, while miR-141 and miR-142-3p showed an approximately 50% increase that did not reach significance. As most candidate miRNAs showed consistent behavior in this separate validation set of FSHD samples, this panel of miRNAs merits further investigation as biomarkers moving forward.

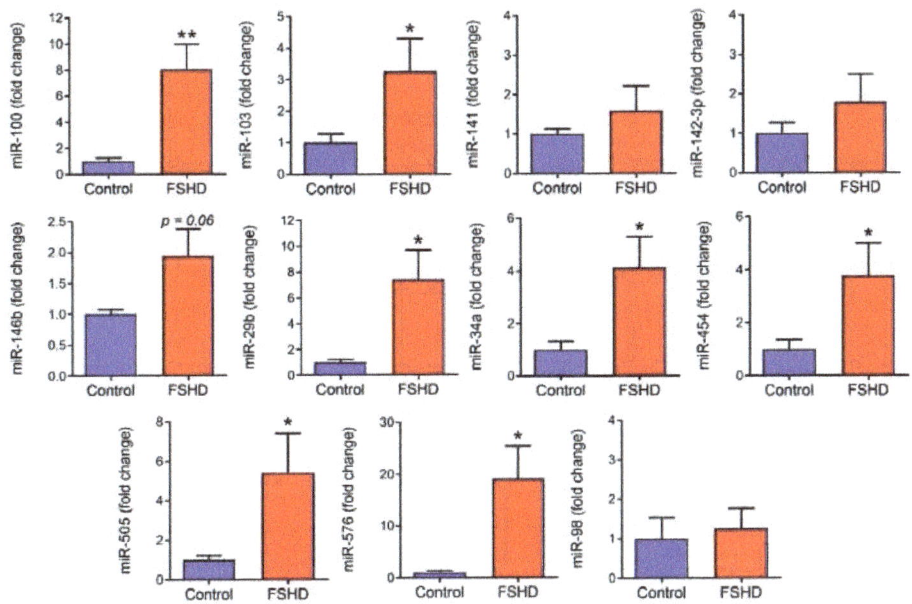

Figure 4. Expression of candidate miRNAs in a validation group of FSHD patients. Candidate miRNAs that increased in the FSHD discovery experiment were assayed via individual qRT-PCR assay in a separate validation group of FSHD1 patient plasma samples. Expression levels of each miRNA are expressed as fold change versus healthy control volunteers. (values are mean ± SEM, * $p \leq 0.05$, ** $p \leq 0.01$, one-tailed t-test comparing FSHD1 to control in direction of Discovery experiment; one outlier removed from miR-34a and miR-576 after significant Grubb's outlier test; $n = 7$ healthy control volunteers, 12 FSHD1).

3.4. Proteomics Profiling

To identify protein candidate biomarkers, we performed LC-MS/MS-based proteomic profiling of samples from a discovery group of FSHD patients (Table 3). For this, plasma from FSHD1 patients ($n = 25$) was compared to healthy volunteer controls ($n = 17$), with a roughly even mix of males and females, and an average age of early- to mid-twenties for each group. All FSHD patients were confirmed to have FSHD1 resulting from D4Z4 contraction mutations that alter epigenetic regulation of DUX4.

Table 3. Clinical characteristics of patients in proteomics discovery group.

	Healthy Control	FSHD
N	17	25
Age in years (mean± SD)	23.45 ± 13.18	25.68 ± 14.71
Males:Females	9:8	13:12
FSHD severity score	N/A	8.54 ± 4.10

Based on signal intensity, we identified 32 proteins that were significantly different between FSHD1 and healthy control samples (Table S2). To further filter the protein list, we used unique peptide count data to identify proteins that had significantly different counts between FSHD1 and control samples. This narrowed the candidates down to 13 proteins (Table 4). Within these protein markers, fibulin-1 (FBLN1) and insulin-like growth factor 1 (IGF1) showed potential effects of sex and age, while keratin 16 (KRT16) displayed a potential age effect and profilin-1 (PFN1) showed a potential sex effect (Table S3). Among the 13 total protein biomarker candidates, 11 proteins were higher in FSHD1 samples versus healthy controls, while two proteins were lower in the FSHD1 samples versus healthy controls.

Table 4. Thirteen circulating proteins identified as dysregulated in FSHD plasma via LC-MS/MS.

Gene Name	UniProt ID	↑ or ↓	*p*-Value	Known Roles in Muscle/Disease
F13A1	P00488	↑	0.031	Hypertension, angiotensin II, coagulation
IGF1	P05019	↑	0.043	hypertrophy, development, satellite cells, regeneration
S100A8	P05109	↑	0.009	TLR4; pro-inflammation, up in rheumatic diseases [84–87]
PFN1	P07737	↑	0.010	actin cytoskeleton organization
FBLN1	p23142	↑	0.011	positive regulation of fibroblast proliferation
CFL1	P23528	↑	0.031	actin filament organization and depolymerization
TMSB4X	P62328	↑	0.017	actin filament organization
TPM4	P67936	↑	0.015	actin organization, muscle contraction
EFEMP1	Q12805	↑	0.001	plasma biomarker for mesothelioma; retinal dystrophy [88]
KRT16	P08779	↑	0.009	elevated with S100A8 in skin disorders, psoriasis [85,89–91]
SPP2	Q13103	↑	0.017	pro-inflammatory, NF-κB; blood pressure; bone health [92]
PROC	P04070	↓	0.048	anti-inflammatory, down in chronic inflammation [93,94]
PRG4	Q92954	↓	0.024	TLR4; anti-inflammatory, down in arthritis [95,96]

CFL1 = Cofilin 1, EFEMP1 = EGF-containing fibulin-like extracellular matrix protein 1, F13A1 = Coagulation factor XIII A chain, FBLN1 = fibulin-1, IBD = inflammatory bowel disease, IGFI = insulin-like growth factor 1, KRT16 = Keratin 16, PFN1 = Profilin-1, PRG4 = Proteoglycan 4 or lubricin, PROC = Protein C, S100A8 = S100 calcium-binding protein A8, SPP2 = Secreted phosphoprotein 24, TLR4 = Toll-like receptor 4, TMSB4X = Thymosin beta-4, TPM4 = Tropomyosin alpha-4 chain.

We selected five candidate protein markers for subsequent quantification via protein-specific ELISA analysis of a non-overlapping validation group of FSHD1 samples. These included insulin-like growth factor 1 (IGF1), proteoglycan 4 (PRG4), profilin 1 (PFN1), tropomyosin 4 (TPM4), and S100 calcium-binding protein A8 (S100A8). Of these candidate proteins, S100A8 showed a significant increase in FSHD1 plasma of approximately 4.5-fold over healthy controls in the validation group (Figure 5a), consistent with its behavior in the discovery experiment. To determine if elevated S100A8 signaling was consistent with the overall proteomic and miRNA profiling results, we performed bioinformatic pathway analyses focused on the S100A8 pathway along with the full list of candidate protein (Figure 5b) and miRNA (Figure 5c) markers. Nine proteins and 13 miRNAs were shown to have previously established connections to the toll-like receptor 4 (TLR4) signaling pathway, which is activated by S100A8 and drives increased inflammatory (NF-κB and AP-1) gene expression. As miRNAs can reflect a direct readout of transcription factor activity, we also surveyed ChIP-seq data to analyze DNA regions encoding miRNAs elevated in FSHD for binding by the NF-κB and AP-1 transcription factors activated by S100A8 (Figure 5d). All miRNAs except for one (miR-329) showed binding by NF-κB and/or AP-1 subunits at DNA regions capable of acting as regulatory promoter or enhancer elements. As S100A8 is a well-established biomarker of inflammatory disease processes (reviewed in [86]) and these can be up-regulated in the muscular dystrophies, this protein merits further investigation as a biomarker for FSHD moving forward.

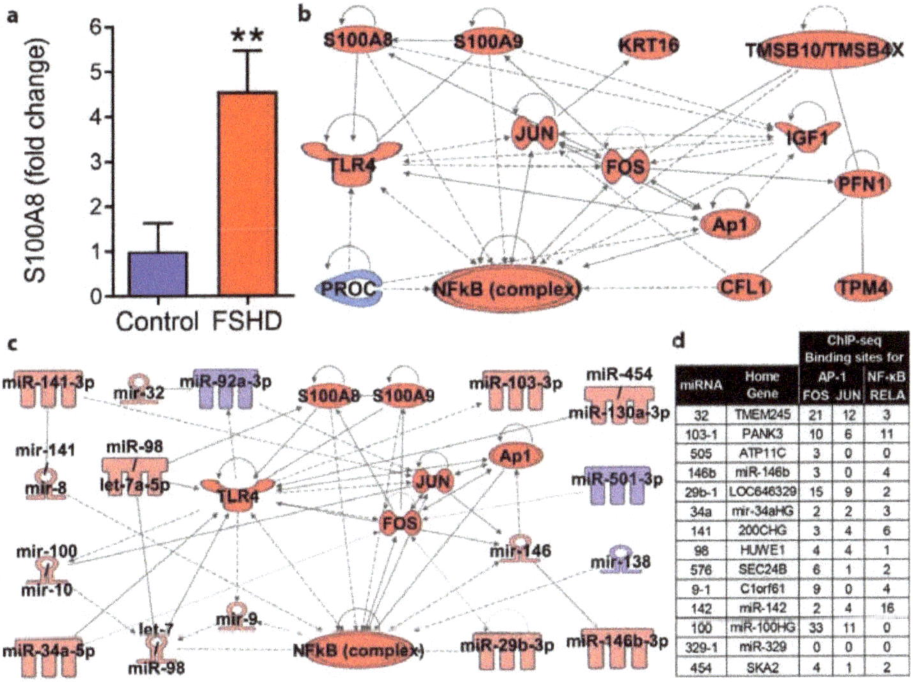

miRNA	Home Gene	ChIP-seq Binding sites for AP-1 FOS	AP-1 JUN	NF-κB RELA
32	TMEM245	21	12	3
103-1	PANK3	10	6	11
505	ATP11C	3	0	0
146b	miR-146b	3	0	4
29b-1	LOC646329	15	9	2
34a	mir-34aHG	2	2	3
141	200CHG	3	4	6
98	HUWE1	4	4	1
576	SEC24B	6	1	2
9-1	C1orf61	9	0	4
142	miR-142	2	4	16
100	miR-100HG	33	11	0
329-1	miR-329	0	0	0
454	SKA2	4	1	2

Figure 5. Validation and pathway analysis of elevated S100A8 protein in FSHD. (**a**) ELISA of S100A8 protein in plasma from a separate validation set of FSHD1 patients. (**b**) Bioinformatic pathway analysis was used to identify known connections between candidate protein markers with S100A8 pathway proteins involved in TLR4 signaling. (**c**) Bioinformatic pathway analysis was used to identify established connections between candidate miRNAs with S100A8 pathway proteins involved in TLR4 signaling. (**d**) Bioinformatic analysis of ChIP-seq defined binding sites for the key S100A8 pathway transcription factors AP-1 (FOS and JUN) and NF-κB (RELA), at potential regulatory regions of the candidate miRNAs that were found to increase in FSHD plasma. Binding sites represent the combined number of potential promoter (within 2 kb of promoter) and enhancer (within 10 kb) regulatory regions with ChIP-seq-confirmed transcription factor binding for each miRNA home gene. (** $p \leq 0.01$; $n = 13$ healthy control volunteers, 19 FSHD1; panels (**b**,**c**) produced using Ingenuity Pathway Analysis software, red = increased, blue = decreased; data for panel (**d**) produced using the Factorbook ChIP-seq data repository from ENCODE and the UCSC genome browser).

4. Discussion

There is currently no effective treatment available for FSHD. However, research advances in FSHD are now beginning to yield promising and novel therapeutic strategies that will require well-designed clinical trials to evaluate effectiveness. Potential therapeutic strategies including antisense oligonucleotides (AON) and small molecules have been reported or are being actively pursued [12,97–100]. Changes in biomarkers following a treatment can be a powerful tool for evaluating the efficacy and safety of the treatment. Previous studies seeking to identify circulating miRNA biomarkers in muscular dystrophy have focused exclusively on assaying myomiRs, which are a defined group of miRNAs with muscle-specific or muscle-enhanced expression [101,102]. Previously, a study by Statland et al. identified 7 potential protein biomarkers in 22 FSHD serum samples, using a commercial multiplex assay [103]. A multi-site study using aptamer-based SomaScan proteomics to assay two FSHD populations identified a total of 115 proteins that were dysregulated, four of which behaved consistently between the two independent cohorts (creatine kinase MM, creatine kinase MB, carbonic anhydrase III, and troponin I type 2) [104]. In this study, we used -omics approaches to identify additional circulating miRNA and protein biomarker candidates using samples collected from individuals with early-onset FSHD.

There is an intriguing potential for developing miRNAs as biomarkers in diseases affecting muscle, as they are stable in biofluids, objective, minimally invasive, and well-conserved between human patients and preclinical animal models [23,24]. Recently, the utility of serum miRNAs to detect muscle involvement in complex diseases with highly variable symptoms has been demonstrated, as in patients with MADD [25]. Muscle-specific miRNAs are also elevated in Duchenne and Becker muscular dystrophy, along with a set of inflammatory miRNAs reflecting the chronic inflammatory pathology of these diseases [29,30,105]. Here we identify eight circulating miRNAs that are associated with FSHD in patient plasma samples. The prevalence of DNA binding by DUX4 and FSHD-associated transcription factors, within regions capable of regulating the candidate miRNAs, provides a molecular rationale for their up-regulation in FSHD. Several of the markers have also been previously shown to play a role in muscle diseases and associated pathological pathways. These candidate biomarkers hold potential as monitoring biomarkers in early-onset FSHD.

Several candidate miRNAs we identified have previously been proposed as circulating biomarkers and have shown similar behavior in other diseases. Plasma miR-454 has been identified as a biomarker of myotonic dystrophy [72,73]. Serum miR-146b is a pharmacodynamic biomarker in inflammatory bowel disease (IBD) [34,35]. Intriguingly, miR-146b is also known to down-regulate dystrophin in multiple muscle diseases, is increased in dystrophinopathies and in myositis, and is also drug-responsive in the *mdx* mouse model of DMD [30,31]. Urinary miR-141 provides a promising diagnostic biomarker for the identification of both prostate and bladder cancers [69]; it will be interesting to determine if this or other candidate miRNAs are also dysregulated in urine from dystrophic patients, as this sampling method could provide a completely non-invasive biomarker.

Increases in circulating S100A8, a subunit of calprotectin, are consistent with an inflammatory signature playing a role in FSHD. The inflammatory calprotectin protein consists of a heterodimer (S100A8/S100A9) which binds to toll-like receptor 4 (TLR4) to activate pro-inflammatory gene expression pathways through the NF-κB and AP-1 transcription factors. Consistent with such an inflammatory gene signature in FSHD, bioinformatic analyses here show five of the candidate miRNAs have established connections with TLR4 signaling, are increased in FSHD patients, and have gene promoters that are bound by AP-1 and/or NF-κB. Outside of FSHD, calprotectin is already a well-established biomarker across rheumatic diseases. Fecal calprotectin is a widely used diagnostic, monitoring and pharmacodynamic biomarker for IBD, and recent studies indicate serum calprotectin levels are also well-correlated with IBD disease state [87,106]. Serum calprotectin is used as a monitoring and pharmacodynamic biomarker for rheumatoid arthritis, and intriguingly S100A8/S100A9 may have further utility in arthritis as a molecular imaging marker of inflammatory activity [84,107,108]. Of particular relevance to the present study, calprotectin in both muscle and serum is a biomarker for disease activity in juvenile dermatomyositis [109]. Moving forward, it will be interesting to see if S100A8 or calprotectin can show further utility as completely non-invasive or local biomarker for FSHD and other muscle diseases such as myositis.

Several of the molecular markers we identified here as elevated in FSHD may provide a new therapeutic target. In various states of muscle atrophy miR-29b is also up-regulated, while preventing its expression shows efficacy in mouse models of muscle atrophy [64,65]. In myositis and Becker muscular dystrophy, the inflammatory marker miR-146b is known to down-regulate dystrophin expression, whereas the reduction of miR-146b via anti-inflammatory drugs or via miRNA-targeting oligos is proposed as a method to increase dystrophin levels to help improve muscle health [30,31]. In various rheumatological disease states, the inhibition of S100A8 or calprotectin via small molecule inhibitors or antibodies is a very attractive therapeutic strategy; early studies of such inhibitors are already showing therapeutic efficacy in both human trials and/or in mouse models, including in studies for arthritis, asthma, IBD, and multiple sclerosis (reviewed in [86]). Similarly, decreases in PROC seen here in FSHD are also seen in several rheumatological disorders, where treatment with PROC activators are already being pursued as a therapeutic option (reviewed in [94]).

Bioinformatic analyses of the -omics results support muscle and inflammatory gene expression pathways as being dysregulated in FSHD. As expected, several muscle pathology-associated miRNAs are dysregulated in FSHD patients: miR-486 is a defined myomiR, miR-29b up-regulation promotes muscle atrophy, miR-146b is dysregulated in dystrophinopathies and myositis, miR-329 counteracts muscle hypertrophy, and three others are known to be dysregulated in myotonic dystrophy, lamin A (LMNA) dystrophy, and/or FSHD (miR-34a, miR-140-3p, miR-100, and miR-454). Consistent with these findings, several of the proteins that were dysregulated are known to function in muscle contraction, actin filament organization and/or muscle regeneration (TOM4, PFN1, CFL1, TMSB4X, and IGF1).

S100A8 and its associated inflammatory signaling pathway (TLR4, NF-κB and AP-1) appear to be a substantial hub for dysregulated expression of the candidate markers we identified. Nine of the candidate miRNAs have previously established connections to this TLR4-centered pathway. ChIP-seq analysis of the miRNAs up-regulated in FSHD shows all but one have promoters bound by NF-κB or AP-1, which are activated by S100A8-induced TLR4. In the proteomics data, several of the proteins that increased are pro-inflammatory (S100A8, KRT16 and SPP2) while in contrast the two proteins that decreased have anti-inflammatory (PROC and PRG4) roles. Consistent with our FSHD findings, KRT16 and S100A8 are also up-regulated together in inflammatory skin disorders; additionally, the pattern of increased S100A8 with decreased PROC is seen here in FSHD as well as in IBD and several other chronic inflammatory disorders [93,94]. Pathway analysis further establishes a link between the protein markers, as nine out of 14 have established connections to the S100A8 and TLR4 signaling pathway. Together these data confirm that circulating FSHD biomarkers reflect muscle pathogenesis, and suggest inflammatory S100A8/TLR4 signaling plays a role in pediatric-onset FSHD as well.

5. Conclusions

FSHD is chronic genetic muscle disease with a variable prognosis. There is no cure, and no pharmaceuticals for FSHD have shown efficacy in altering the disease course. Development of objective biomarkers will facilitate the clinical and preclinical development of novel therapies, as well as our ability to monitor disease activity. We identified eight circulating miRNAs (miR-100, miR-103, miR-146b, miR-29b, miR-34a, miR-454, miR-505, and miR-576) which may be developed as biomarkers for FSHD. Additionally, we identified the S100A8 subunit of calprotectin as a primary protein marker of interest for FSHD, consistent with its utility in numerous rheumatic diseases. These molecular markers warrant further investigation in additional cohorts, preclinical drug testing, and prospective clinical trials.

Supplementary Materials: The following are available online at http://www.mdpi.com/2075-4426/10/4/236/s1, Table S1: ChIP-seq of FSHD-disrupted transcription factors at candidate miRNA loci, Table S2: Total proteomic changes in FSHD plasma as detected via LC-MS/MS, Table S3: Proteins affected by sex or age.

Author Contributions: Conceptualization, C.R.H., A.A.F., Y.-W.C.; methodology, C.R.H., A.Z., N.Y.N., C.B.T., A.P., H.G.-D., A.A.F., M.G., M.M.R., P.R.C., M.T. (Mathula Thangarajh), R.W., E.C.S., A.M.C., C.M.M., P.K., M.T. (Mar Tulinius), A.H., J.K.M., A.A.F., Y.-W.C.; data analysis, C.R.H., A.Z., N.Y.N., H.G.-D., A.A.F., Y.-W.C.; experimental procedure carried out by, C.R.H., A.Z., N.Y.N., C.B.T., S.N.P., A.P.; writing-original draft preparation, C.R.H., A.Z., N.Y.N., C.B.T., A.P., H.G.-D., A.A.F., Y.-W.C.; writing-review and editing, C.R.H., A.Z., N.Y.N., C.B.T., H.G.-D., S.N.P., A.P., M.G., M.M.R., P.R.C., M.T. (Mathula Thangarajh), R.W., E.C.S., A.M.C., C.M.M., P.K., M.T. (Mar Tulinius), A.H., J.K.M., A.A.F., Y.-W.C.; project administration, C.R.H., M.G., M.M.R., P.R.C., M.T. (Mathula Thangarajh), R.W., E.C.S., A.M.C., C.M.M., P.K., M.T. (Mar Tulinius), A.H., J.K.M., A.A.F., Y.-W.C. All authors have read and agreed to the published version of the manuscript.

Funding: This study is supported by funding from the Friends of FSH Research, the FSHD Society, Muscular Dystrophy Canada, and the FSHD Global Research Foundation. C.R.H. and this work is supported by the National Institutes of Health (NIH) grants R01HL153054, P50HD090254, and R00HL130035, as well as the Foundation to Eradicate Duchenne and the A. James & Alice B. Clark Foundation. Y.W.C. is partially supported by NIH/NINDS 1R03NS116444, Muscular Dystrophy Association and NIH/NCATS UL1TR001876.

Acknowledgments: The authors would like to thank participants and families for their support, and the FSH Society for assistance with patient travel and care. The authors would like to thank the Cooperative International Neuromuscular Research Group (CINRG) for patient recruitment and care, as well as the CINRG Early-onset FSHD investigators who provided patient samples for this study. The authors would also like to thank the ENCODE Consortium, the ENCODE ChIP-seq production laboratories, and the ENCODE Data Coordination Center for generating and processing ChIP-seq datasets used here.

Conflicts of Interest: The authors declare no conflict of interest.

References

1. Dixit, M.; Ansseau, E.; Tassin, A.; Winokur, S.; Shi, R.; Qian, H.; Sauvage, S.; Matteotti, C.; van Acker, A.M.; Leo, O.; et al. DUX4, a candidate gene of facioscapulohumeral muscular dystrophy, encodes a transcriptional activator of PITX1. *Proc. Natl. Acad. Sci. USA* **2007**, *104*, 18157–18162. [CrossRef]
2. Lemmers, R.J.; van der Vliet, P.J.; Klooster, R.; Sacconi, S.; Camano, P.; Dauwerse, J.G.; Snider, L.; Straasheijm, K.R.; van Ommen, G.J.; Padberg, G.W.; et al. A unifying genetic model for facioscapulohumeral muscular dystrophy. *Science* **2010**, *329*, 1650–1653. [CrossRef] [PubMed]
3. van Overveld, P.G.; Lemmers, R.J.; Sandkuijl, L.A.; Enthoven, L.; Winokur, S.T.; Bakels, F.; Padberg, G.W.; van Ommen, G.J.; Frants, R.R.; van der Maarel, S.M. Hypomethylation of D4Z4 in 4q-linked and non-4q-linked facioscapulohumeral muscular dystrophy. *Nat. Genet.* **2003**, *35*, 315–317. [CrossRef] [PubMed]
4. Lemmers, R.J.; Tawil, R.; Petek, L.M.; Balog, J.; Block, G.J.; Santen, G.W.; Amell, A.M.; van der Vliet, P.J.; Almomani, R.; Straasheijm, K.R.; et al. Digenic inheritance of an SMCHD1 mutation and an FSHD-permissive D4Z4 allele causes facioscapulohumeral muscular dystrophy type 2. *Nat. Genet.* **2012**, *44*, 1370–1374. [CrossRef] [PubMed]
5. van den Boogaard, M.L.; Lemmers, R.; Balog, J.; Wohlgemuth, M.; Auranen, M.; Mitsuhashi, S.; van der Vliet, P.J.; Straasheijm, K.R.; van den Akker, R.F.P.; Kriek, M.; et al. Mutations in DNMT3B Modify Epigenetic Repression of the D4Z4 Repeat and the Penetrance of Facioscapulohumeral Dystrophy. *Am. J. Hum. Genet.* **2016**, *98*, 1020–1029. [CrossRef] [PubMed]
6. Hamanaka, K.; Sikrova, D.; Mitsuhashi, S.; Masuda, H.; Sekiguchi, Y.; Sugiyama, A.; Shibuya, K.; Lemmers, R.; Goossens, R.; Ogawa, M.; et al. Homozygous nonsense variant in LRIF1 associated with facioscapulohumeral muscular dystrophy. *Neurology* **2020**, *94*, e2441–e2447. [CrossRef] [PubMed]
7. Bosnakovski, D.; Lamb, S.; Simsek, T.; Xu, Z.; Belayew, A.; Perlingeiro, R.; Kyba, M. DUX4c, an FSHD candidate gene, interferes with myogenic regulators and abolishes myoblast differentiation. *Exp. Neurol.* **2008**, *214*, 87–96. [CrossRef]
8. Feng, Q.; Snider, L.; Jagannathan, S.; Tawil, R.; van der Maarel, S.M.; Tapscott, S.J.; Bradley, R.K. A feedback loop between nonsense-mediated decay and the retrogene DUX4 in facioscapulohumeral muscular dystrophy. *Elife* **2015**, *4*, e04996. [CrossRef]
9. Geng, L.N.; Yao, Z.; Snider, L.; Fong, A.P.; Cech, J.N.; Young, J.M.; van der Maarel, S.M.; Ruzzo, W.L.; Gentleman, R.C.; Tawil, R.; et al. DUX4 activates germline genes, retroelements, and immune mediators: Implications for facioscapulohumeral dystrophy. *Dev. Cell* **2012**, *22*, 38–51. [CrossRef]

10. Sharma, V.; Harafuji, N.; Belayew, A.; Chen, Y.W. DUX4 differentially regulates transcriptomes of human rhabdomyosarcoma and mouse C2C12 cells. *PLoS ONE* **2013**, *8*, e64691. [CrossRef]
11. Tassin, A.; Laoudj-Chenivesse, D.; Vanderplanck, C.; Barro, M.; Charron, S.; Ansseau, E.; Chen, Y.W.; Mercier, J.; Coppee, F.; Belayew, A. DUX4 expression in FSHD muscle cells: How could such a rare protein cause a myopathy? *J. Cell Mol. Med.* **2013**, *17*, 76–89. [CrossRef] [PubMed]
12. Vanderplanck, C.; Ansseau, E.; Charron, S.; Stricwant, N.; Tassin, A.; Laoudj-Chenivesse, D.; Wilton, S.D.; Coppee, F.; Belayew, A. The FSHD atrophic myotube phenotype is caused by DUX4 expression. *PLoS ONE* **2011**, *6*, e26820. [CrossRef] [PubMed]
13. Tassin, A.; Leroy, B.; Laoudj-Chenivesse, D.; Wauters, A.; Vanderplanck, C.; Le Bihan, M.C.; Coppee, F.; Wattiez, R.; Belayew, A. FSHD myotubes with different phenotypes exhibit distinct proteomes. *PLoS ONE* **2012**, *7*, e51865. [CrossRef] [PubMed]
14. Brouwer, O.F.; Padberg, G.W.; Wijmenga, C.; Frants, R.R. Facioscapulohumeral muscular dystrophy in early childhood. *Arch. Neurol.* **1994**, *51*, 387–394. [CrossRef]
15. Lunt, P.W.; Jardine, P.E.; Koch, M.C.; Maynard, J.; Osborn, M.; Williams, M.; Harper, P.S.; Upadhyaya, M. Correlation between fragment size at D4F104S1 and age at onset or at wheelchair use, with a possible generational effect, accounts for much phenotypic variation in 4q35-facioscapulohumeral muscular dystrophy (FSHD). *Hum. Mol. Genet.* **1995**, *4*, 951–958. [CrossRef] [PubMed]
16. Tawil, R.; Forrester, J.; Griggs, R.C.; Mendell, J.; Kissel, J.; McDermott, M.; King, W.; Weiffenbach, B.; Figlewicz, D. Evidence for anticipation and association of deletion size with severity in facioscapulohumeral muscular dystrophy. The FSH-DY Group. *Ann. Neurol.* **1996**, *39*, 744–748. [CrossRef]
17. Klinge, L.; Eagle, M.; Haggerty, I.D.; Roberts, C.E.; Straub, V.; Bushby, K.M. Severe phenotype in infantile facioscapulohumeral muscular dystrophy. *Neuromuscul. Disord.* **2006**, *16*, 553–558. [CrossRef]
18. Ricci, E.; Galluzzi, G.; Deidda, G.; Cacurri, S.; Colantoni, L.; Merico, B.; Piazzo, N.; Servidei, S.; Vigneti, E.; Pasceri, V.; et al. Progress in the molecular diagnosis of facioscapulohumeral muscular dystrophy and correlation between the number of KpnI repeats at the 4q35 locus and clinical phenotype. *Ann. Neurol.* **1999**, *45*, 751–757. [CrossRef]
19. Hoffman, E.P.; Connor, E.M. Orphan drug development in muscular dystrophy: Update on two large clinical trials of dystrophin rescue therapies. *Discov. Med.* **2013**, *16*, 233–239.
20. Mercuri, E.; Messina, S.; Pane, M.; Bertini, E. Current methodological issues in the study of children with inherited neuromuscular disorders. *Dev. Med. Child. Neurol* **2008**, *50*, 417–421. [CrossRef]
21. Califf, R.M. Biomarker definitions and their applications. *Exp. Biol. Med. (Maywood)* **2018**, *243*, 213–221. [CrossRef] [PubMed]
22. Valencia-Sanchez, M.A.; Liu, J.; Hannon, G.J.; Parker, R. Control of translation and mRNA degradation by miRNAs and siRNAs. *Genes Dev.* **2006**, *20*, 515–524. [CrossRef] [PubMed]
23. Mitchell, P.S.; Parkin, R.K.; Kroh, E.M.; Fritz, B.R.; Wyman, S.K.; Pogosova-Agadjanyan, E.L.; Peterson, A.; Noteboom, J.; O'Briant, K.C.; Allen, A.; et al. Circulating microRNAs as stable blood-based markers for cancer detection. *Proc. Natl. Acad. Sci. USA* **2008**, *105*, 10513–10518. [CrossRef] [PubMed]
24. Mall, C.; Rocke, D.M.; Durbin-Johnson, B.; Weiss, R.H. Stability of miRNA in human urine supports its biomarker potential. *Biomark. Med.* **2013**, *7*, 623–631. [CrossRef] [PubMed]
25. Missaglia, S.; Pegoraro, V.; Marozzo, R.; Tavian, D.; Angelini, C. Correlation between ETFDH mutations and dysregulation of serum myomiRs in MADD patients. *Eur. J. Transl. Myol.* **2020**, *30*, 8880. [CrossRef]
26. Zaharieva, I.T.; Calissano, M.; Scoto, M.; Preston, M.; Cirak, S.; Feng, L.; Collins, J.; Kole, R.; Guglieri, M.; Straub, V.; et al. Dystromirs as serum biomarkers for monitoring the disease severity in Duchenne muscular Dystrophy. *PLoS ONE* **2013**, *8*, e80263. [CrossRef]
27. Hu, J.; Kong, M.; Ye, Y.; Hong, S.; Cheng, L.; Jiang, L. Serum miR-206 and other muscle-specific microRNAs as non-invasive biomarkers for Duchenne muscular dystrophy. *J. Neurochem.* **2014**, *129*, 877–883. [CrossRef]
28. Trifunov, S.; Natera-de Benito, D.; Exposito Escudero, J.M.; Ortez, C.; Medina, J.; Cuadras, D.; Badosa, C.; Carrera, L.; Nascimento, A.; Jimenez-Mallebrera, C. Longitudinal Study of Three microRNAs in Duchenne Muscular Dystrophy and Becker Muscular Dystrophy. *Front. Neurol.* **2020**, *11*, 304. [CrossRef]
29. Eisenberg, I.; Eran, A.; Nishino, I.; Moggio, M.; Lamperti, C.; Amato, A.A.; Lidov, H.G.; Kang, P.B.; North, K.N.; Mitrani-Rosenbaum, S.; et al. Distinctive patterns of microRNA expression in primary muscular disorders. *Proc. Natl. Acad. Sci. USA* **2007**, *104*, 17016–17021. [CrossRef]

30. Fiorillo, A.A.; Heier, C.R.; Novak, J.S.; Tully, C.B.; Brown, K.J.; Uaesoontrachoon, K.; Vila, M.C.; Ngheim, P.P.; Bello, L.; Kornegay, J.N.; et al. TNF-alpha-Induced microRNAs Control Dystrophin Expression in Becker Muscular Dystrophy. *Cell Rep.* **2015**, *12*, 1678–1690. [CrossRef]
31. Kinder, T.B.; Heier, C.R.; Tully, C.B.; Van der Muelen, J.H.; Hoffman, E.P.; Nagaraju, K.; Fiorillo, A.A. Muscle Weakness in Myositis: MicroRNA-Mediated Dystrophin Reduction in a Myositis Mouse Model and Human Muscle Biopsies. *Arthritis Rheumatol.* **2020**, *72*, 1170–1183. [CrossRef] [PubMed]
32. Brusa, R.; Magri, F.; Bresolin, N.; Comi, G.P.; Corti, S. Noncoding RNAs in Duchenne and Becker muscular dystrophies: Role in pathogenesis and future prognostic and therapeutic perspectives. *Cell Mol. Life Sci.* **2020**, *77*, 4299–4313. [CrossRef] [PubMed]
33. Coenen-Stass, A.M.L.; Wood, M.J.A.; Roberts, T.C. Biomarker Potential of Extracellular miRNAs in Duchenne Muscular Dystrophy. *Trends Mol. Med.* **2017**, *23*, 989–1001. [CrossRef] [PubMed]
34. Batra, S.K.; Heier, C.R.; Diaz-Calderon, L.; Tully, C.B.; Fiorillo, A.A.; van den Anker, J.; Conklin, L.S. Serum miRNAs Are Pharmacodynamic Biomarkers Associated With Therapeutic Response in Pediatric Inflammatory Bowel Disease. *Inflamm. Bowel. Dis.* **2020**, *26*, 1597–1606. [CrossRef]
35. Heier, C.R.; Fiorillo, A.A.; Chaisson, E.; Gordish-Dressman, H.; Hathout, Y.; Damsker, J.M.; Hoffman, E.P.; Conklin, L.S. Identification of Pathway-Specific Serum Biomarkers of Response to Glucocorticoid and Infliximab Treatment in Children with Inflammatory Bowel Disease. *Clin. Transl. Gastroenterol.* **2016**, *7*, e192. [CrossRef]
36. Fiorillo, A.A.; Tully, C.B.; Damsker, J.M.; Nagaraju, K.; Hoffman, E.P.; Heier, C.R. Muscle miRNAome shows suppression of chronic inflammatory miRNAs with both prednisone and vamorolone. *Physiol. Genom.* **2018**, *50*, 735–745. [CrossRef]
37. Heier, C.R.; Damsker, J.M.; Yu, Q.; Dillingham, B.C.; Huynh, T.; Van der Meulen, J.H.; Sali, A.; Miller, B.K.; Phadke, A.; Scheffer, L.; et al. VBP15, a novel anti-inflammatory and membrane-stabilizer, improves muscular dystrophy without side effects. *EMBO Mol. Med.* **2013**, *5*, 1569–1585. [CrossRef]
38. Heier, C.R.; Yu, Q.; Fiorillo, A.A.; Tully, C.B.; Tucker, A.; Mazala, D.A.; Uaesoontrachoon, K.; Srinivassane, S.; Damsker, J.M.; Hoffman, E.P.; et al. Vamorolone targets dual nuclear receptors to treat inflammation and dystrophic cardiomyopathy. *Life Sci. Alliance* **2019**, *2*, e201800186. [CrossRef]
39. Hammers, D.W.; Sleeper, M.M.; Forbes, S.C.; Coker, C.C.; Jirousek, M.R.; Zimmer, M.; Walter, G.A.; Sweeney, H.L. Disease-modifying effects of orally bioavailable NF-kappaB inhibitors in dystrophin-deficient muscle. *JCI Insight* **2016**, *1*, e90341. [CrossRef]
40. Finanger, E.; Vandenborne, K.; Finkel, R.S.; Lee Sweeney, H.; Tennekoon, G.; Yum, S.; Mancini, M.; Bista, P.; Nichols, A.; Liu, H.; et al. Phase 1 Study of Edasalonexent (CAT-1004), an Oral NF-kappaB Inhibitor, in Pediatric Patients with Duchenne Muscular Dystrophy. *J. Neuromuscul. Dis.* **2019**, *6*, 43–54. [CrossRef]
41. Fichna, J.P.; Macias, A.; Piechota, M.; Korostynski, M.; Potulska-Chromik, A.; Redowicz, M.J.; Zekanowski, C. Whole-exome sequencing identifies novel pathogenic mutations and putative phenotype-influencing variants in Polish limb-girdle muscular dystrophy patients. *Hum. Genom.* **2018**, *12*, 34. [CrossRef] [PubMed]
42. Mitsuhashi, S.; Boyden, S.E.; Estrella, E.A.; Jones, T.I.; Rahimov, F.; Yu, T.W.; Darras, B.T.; Amato, A.A.; Folkerth, R.D.; Jones, P.L.; et al. Exome sequencing identifies a novel SMCHD1 mutation in facioscapulohumeral muscular dystrophy 2. *Neuromuscul. Disord.* **2013**, *23*, 975–980. [CrossRef] [PubMed]
43. Leidenroth, A.; Sorte, H.S.; Gilfillan, G.; Ehrlich, M.; Lyle, R.; Hewitt, J.E. Diagnosis by sequencing: Correction of misdiagnosis from FSHD2 to LGMD2A by whole-exome analysis. *Eur. J. Hum. Genet.* **2012**, *20*, 999–1003. [CrossRef] [PubMed]
44. Ghaoui, R.; Cooper, S.T.; Lek, M.; Jones, K.; Corbett, A.; Reddel, S.W.; Needham, M.; Liang, C.; Waddell, L.B.; Nicholson, G.; et al. Use of Whole-Exome Sequencing for Diagnosis of Limb-Girdle Muscular Dystrophy: Outcomes and Lessons Learned. *JAMA Neurol.* **2015**, *72*, 1424–1432. [CrossRef]
45. Mah, J.K.; Feng, J.; Jacobs, M.B.; Duong, T.; Carroll, K.; de Valle, K.; Carty, C.L.; Morgenroth, L.P.; Guglieri, M.; Ryan, M.M.; et al. A multinational study on motor function in early-onset FSHD. *Neurology* **2018**, *90*, e1333–e1338. [CrossRef]
46. Kent, W.J.; Sugnet, C.W.; Furey, T.S.; Roskin, K.M.; Pringle, T.H.; Zahler, A.M.; Haussler, D. The human genome browser at UCSC. *Genome Res.* **2002**, *12*, 996–1006. [CrossRef]
47. Mathelier, A.; Fornes, O.; Arenillas, D.J.; Chen, C.Y.; Denay, G.; Lee, J.; Shi, W.; Shyr, C.; Tan, G.; Worsley-Hunt, R.; et al. JASPAR 2016: A major expansion and update of the open-access database of transcription factor binding profiles. *Nucleic Acids Res.* **2016**, *44*, D110–D115. [CrossRef]

48. Wang, J.; Zhuang, J.; Iyer, S.; Lin, X.Y.; Greven, M.C.; Kim, B.H.; Moore, J.; Pierce, B.G.; Dong, X.; Virgil, D.; et al. Factorbook.org: A Wiki-based database for transcription factor-binding data generated by the ENCODE consortium. *Nucleic Acids Res.* **2013**, *41*, D171–D176. [CrossRef]
49. Davis, C.A.; Hitz, B.C.; Sloan, C.A.; Chan, E.T.; Davidson, J.M.; Gabdank, I.; Hilton, J.A.; Jain, K.; Baymuradov, U.K.; Narayanan, A.K.; et al. The Encyclopedia of DNA elements (ENCODE): Data portal update. *Nucleic Acids Res.* **2018**, *46*, D794–D801. [CrossRef]
50. Consortium, E.P. An integrated encyclopedia of DNA elements in the human genome. *Nature* **2012**, *489*, 57–74. [CrossRef]
51. Zahm, A.M.; Thayu, M.; Hand, N.J.; Horner, A.; Leonard, M.B.; Friedman, J.R. Circulating microRNA is a biomarker of pediatric Crohn disease. *J. Pediatr. Gastroenterol. Nutr.* **2011**, *53*, 26–33. [CrossRef] [PubMed]
52. Cox, J.; Mann, M. MaxQuant enables high peptide identification rates, individualized p.p.b.-range mass accuracies and proteome-wide protein quantification. *Nat. Biotechnol.* **2008**, *26*, 1367–1372. [CrossRef] [PubMed]
53. Small, E.M.; O'Rourke, J.R.; Moresi, V.; Sutherland, L.B.; McAnally, J.; Gerard, R.D.; Richardson, J.A.; Olson, E.N. Regulation of PI3-kinase/Akt signaling by muscle-enriched microRNA-486. *Proc. Natl. Acad. Sci. USA* **2010**, *107*, 4218–4223. [CrossRef] [PubMed]
54. Yan, Y.; Shi, R.; Yu, X.; Sun, C.; Zang, W.; Tian, H. Identification of atrial fibrillation-associated microRNAs in left and right atria of rheumatic mitral valve disease patients. *Genes Genet. Syst.* **2019**, *94*, 23–34. [CrossRef] [PubMed]
55. Sen, A.; Ren, S.; Lerchenmuller, C.; Sun, J.; Weiss, N.; Most, P.; Peppel, K. MicroRNA-138 regulates hypoxia-induced endothelial cell dysfunction by targeting S100A1. *PLoS ONE* **2013**, *8*, e78684. [CrossRef]
56. Yu, J.; Lu, Y.; Li, Y.; Xiao, L.; Xing, Y.; Li, Y.; Wu, L. Role of S100A1 in hypoxia-induced inflammatory response in cardiomyocytes via TLR4/ROS/NF-kappaB pathway. *J. Pharm. Pharmacol.* **2015**, *67*, 1240–1250. [CrossRef]
57. Duan, Y.; Zhou, M.; Xiao, J.; Wu, C.; Zhou, L.; Zhou, F.; Du, C.; Song, Y. Prediction of key genes and miRNAs responsible for loss of muscle force in patients during an acute exacerbation of chronic obstructive pulmonary disease. *Int. J. Mol. Med.* **2016**, *38*, 1450–1462. [CrossRef]
58. Yin, H.; He, H.; Shen, X.; Zhao, J.; Cao, X.; Han, S.; Cui, C.; Chen, Y.; Wei, Y.; Xia, L.; et al. miR-9-5p Inhibits Skeletal Muscle Satellite Cell Proliferation and Differentiation by Targeting IGF2BP3 through the IGF2-PI3K/Akt Signaling Pathway. *Int. J. Mol. Sci.* **2020**, *21*, 1655. [CrossRef]
59. Shen, J.; Xing, W.; Liu, R.; Zhang, Y.; Xie, C.; Gong, F. MiR-32-5p influences high glucose-induced cardiac fibroblast proliferation and phenotypic alteration by inhibiting DUSP1. *BMC Mol. Biol.* **2019**, *20*, 21. [CrossRef]
60. Liu, J.; Xiao, X.; Shen, Y.; Chen, L.; Xu, C.; Zhao, H.; Wu, Y.; Zhang, Q.; Zhong, J.; Tang, Z.; et al. MicroRNA-32 promotes calcification in vascular smooth muscle cells: Implications as a novel marker for coronary artery calcification. *PLoS ONE* **2017**, *12*, e0174138. [CrossRef]
61. Lee, S.Y.; Yang, J.; Park, J.H.; Shin, H.K.; Kim, W.J.; Kim, S.Y.; Lee, E.J.; Hwang, I.; Lee, C.S.; Lee, J.; et al. The MicroRNA-92a/Sp1/MyoD Axis Regulates Hypoxic Stimulation of Myogenic Lineage Differentiation in Mouse Embryonic Stem Cells. *Mol. Ther.* **2020**, *28*, 142–156. [CrossRef] [PubMed]
62. Lazzarini, R.; Caffarini, M.; Delli Carpini, G.; Ciavattini, A.; Di Primio, R.; Orciani, M. From 2646 to 15: Differentially regulated microRNAs between progenitors from normal myometrium and leiomyoma. *Am. J. Obstet. Gynecol.* **2020**, *222*, 596.e1–596.e9. [CrossRef] [PubMed]
63. Liu, H.L.; Zhu, J.G.; Liu, Y.Q.; Fan, Z.G.; Zhu, C.; Qian, L.M. Identification of the microRNA expression profile in the regenerative neonatal mouse heart by deep sequencing. *Cell Biochem. Biophys.* **2014**, *70*, 635–642. [CrossRef] [PubMed]
64. Li, J.; Chan, M.C.; Yu, Y.; Bei, Y.; Chen, P.; Zhou, Q.; Cheng, L.; Chen, L.; Ziegler, O.; Rowe, G.C.; et al. miR-29b contributes to multiple types of muscle atrophy. *Nat. Commun.* **2017**, *8*, 15201. [CrossRef]
65. Li, J.; Wang, L.; Hua, X.; Tang, H.; Chen, R.; Yang, T.; Das, S.; Xiao, J. CRISPR/Cas9-Mediated miR-29b Editing as a Treatment of Different Types of Muscle Atrophy in Mice. *Mol. Ther.* **2020**, *28*, 1359–1372. [CrossRef]
66. Wang, J.; Pei, Y.; Zhong, Y.; Jiang, S.; Shao, J.; Gong, J. Altered serum microRNAs as novel diagnostic biomarkers for atypical coronary artery disease. *PLoS ONE* **2014**, *9*, e107012. [CrossRef]

67. Dmitriev, P.; Barat, A.; Polesskaya, A.; O'Connell, M.J.; Robert, T.; Dessen, P.; Walsh, T.A.; Lazar, V.; Turki, A.; Carnac, G.; et al. Simultaneous miRNA and mRNA transcriptome profiling of human myoblasts reveals a novel set of myogenic differentiation-associated miRNAs and their target genes. *BMC Genom.* **2013**, *14*, 265. [CrossRef]
68. Kropp, J.; Degerny, C.; Morozova, N.; Pontis, J.; Harel-Bellan, A.; Polesskaya, A. miR-98 delays skeletal muscle differentiation by down-regulating E2F5. *Biochem. J.* **2015**, *466*, 85–93. [CrossRef]
69. Ghorbanmehr, N.; Gharbi, S.; Korsching, E.; Tavallaei, M.; Einollahi, B.; Mowla, S.J. miR-21-5p, miR-141-3p, and miR-205-5p levels in urine-promising biomarkers for the identification of prostate and bladder cancer. *Prostate* **2019**, *79*, 88–95. [CrossRef]
70. Greco, S.; Perfetti, A.; Fasanaro, P.; Cardani, R.; Capogrossi, M.C.; Meola, G.; Martelli, F. Deregulated microRNAs in myotonic dystrophy type 2. *PLoS ONE* **2012**, *7*, e39732. [CrossRef]
71. Portilho, D.M.; Alves, M.R.; Kratassiouk, G.; Roche, S.; Magdinier, F.; de Santana, E.C.; Polesskaya, A.; Harel-Bellan, A.; Mouly, V.; Savino, W.; et al. miRNA expression in control and FSHD fetal human muscle biopsies. *PLoS ONE* **2015**, *10*, e0116853. [CrossRef] [PubMed]
72. Perfetti, A.; Greco, S.; Cardani, R.; Fossati, B.; Cuomo, G.; Valaperta, R.; Ambrogi, F.; Cortese, A.; Botta, A.; Mignarri, A.; et al. Validation of plasma microRNAs as biomarkers for myotonic dystrophy type 1. *Sci. Rep.* **2016**, *6*, 38174. [CrossRef] [PubMed]
73. Perfetti, A.; Greco, S.; Bugiardini, E.; Cardani, R.; Gaia, P.; Gaetano, C.; Meola, G.; Martelli, F. Plasma microRNAs as biomarkers for myotonic dystrophy type 1. *Neuromuscul. Disord.* **2014**, *24*, 509–515. [CrossRef] [PubMed]
74. Sylvius, N.; Bonne, G.; Straatman, K.; Reddy, T.; Gant, T.W.; Shackleton, S. MicroRNA expression profiling in patients with lamin A/C-associated muscular dystrophy. *FASEB J.* **2011**, *25*, 3966–3978. [CrossRef] [PubMed]
75. Gao, Y.Q.; Chen, X.; Wang, P.; Lu, L.; Zhao, W.; Chen, C.; Chen, C.P.; Tao, T.; Sun, J.; Zheng, Y.Y.; et al. Regulation of DLK1 by the maternally expressed miR-379/miR-544 cluster may underlie callipyge polar overdominance inheritance. *Proc. Natl. Acad. Sci. USA* **2015**, *112*, 13627–13632. [CrossRef] [PubMed]
76. Jeanson-Leh, L.; Lameth, J.; Krimi, S.; Buisset, J.; Amor, F.; Le Guiner, C.; Barthelemy, I.; Servais, L.; Blot, S.; Voit, T.; et al. Serum profiling identifies novel muscle miRNA and cardiomyopathy-related miRNA biomarkers in Golden Retriever muscular dystrophy dogs and Duchenne muscular dystrophy patients. *Am. J. Pathol.* **2014**, *184*, 2885–2898. [CrossRef]
77. Mukerjee, R.; Chang, J.R.; Del Valle, L.; Bagashev, A.; Gayed, M.M.; Lyde, R.B.; Hawkins, B.J.; Brailoiu, E.; Cohen, E.; Power, C.; et al. Deregulation of microRNAs by HIV-1 Vpr protein leads to the development of neurocognitive disorders. *J. Biol. Chem.* **2011**, *286*, 34976–34985. [CrossRef]
78. Tazawa, H.; Tsuchiya, N.; Izumiya, M.; Nakagama, H. Tumor-suppressive miR-34a induces senescence-like growth arrest through modulation of the E2F pathway in human colon cancer cells. *Proc. Natl. Acad. Sci. USA* **2007**, *104*, 15472–15477. [CrossRef]
79. Kalkat, M.; Resetca, D.; Lourenco, C.; Chan, P.K.; Wei, Y.; Shiah, Y.J.; Vitkin, N.; Tong, Y.; Sunnerhagen, M.; Done, S.J.; et al. MYC Protein Interactome Profiling Reveals Functionally Distinct Regions that Cooperate to Drive Tumorigenesis. *Mol. Cell* **2018**, *72*, 836–848. [CrossRef]
80. Wong, P.P.; Miranda, F.; Chan, K.V.; Berlato, C.; Hurst, H.C.; Scibetta, A.G. Histone demethylase KDM5B collaborates with TFAP2C and Myc to repress the cell cycle inhibitor p21(cip) (CDKN1A). *Mol. Cell Biol.* **2012**, *32*, 1633–1644. [CrossRef]
81. Zhao, J.H.; Inoue, T.; Shoji, W.; Nemoto, Y.; Obinata, M. Direct association of YY-1 with c-Myc and the E-box binding protein in regulation of glycophorin gene expression. *Oncogene* **1998**, *17*, 1009–1017. [CrossRef] [PubMed]
82. Koh, C.M.; Iwata, T.; Zheng, Q.; Bethel, C.; Yegnasubramanian, S.; De Marzo, A.M. Myc enforces overexpression of EZH2 in early prostatic neoplasia via transcriptional and post-transcriptional mechanisms. *Oncotarget* **2011**, *2*, 669–683. [CrossRef] [PubMed]
83. Ma, L.; Young, J.; Prabhala, H.; Pan, E.; Mestdagh, P.; Muth, D.; Teruya-Feldstein, J.; Reinhardt, F.; Onder, T.T.; Valastyan, S.; et al. miR-9, a MYC/MYCN-activated microRNA, regulates E-cadherin and cancer metastasis. *Nat. Cell Biol.* **2010**, *12*, 247–256. [CrossRef] [PubMed]
84. Jarlborg, M.; Courvoisier, D.S.; Lamacchia, C.; Martinez Prat, L.; Mahler, M.; Bentow, C.; Finckh, A.; Gabay, C.; Nissen, M.J. Physicians of the Swiss Clinical Quality Management, r. Serum calprotectin: A promising biomarker in rheumatoid arthritis and axial spondyloarthritis. *Arthritis Res. Ther.* **2020**, *22*, 105. [CrossRef]

85. Metz, M.; Torene, R.; Kaiser, S.; Beste, M.T.; Staubach, P.; Bauer, A.; Brehler, R.; Gericke, J.; Letzkus, M.; Hartmann, N.; et al. Omalizumab normalizes the gene expression signature of lesional skin in patients with chronic spontaneous urticaria: A randomized, double-blind, placebo-controlled study. *Allergy* **2019**, *74*, 141–151. [CrossRef]
86. Wang, S.; Song, R.; Wang, Z.; Jing, Z.; Wang, S.; Ma, J. S100A8/A9 in Inflammation. *Front. Immunol.* **2018**, *9*, 1298. [CrossRef]
87. Kalla, R.; Kennedy, N.A.; Ventham, N.T.; Boyapati, R.K.; Adams, A.T.; Nimmo, E.R.; Visconti, M.R.; Drummond, H.; Ho, G.T.; Pattenden, R.J.; et al. Serum Calprotectin: A Novel Diagnostic and Prognostic Marker in Inflammatory Bowel Diseases. *Am. J. Gastroenterol.* **2016**, *111*, 1796–1805. [CrossRef]
88. Pass, H.I.; Levin, S.M.; Harbut, M.R.; Melamed, J.; Chiriboga, L.; Donington, J.; Huflejt, M.; Carbone, M.; Chia, D.; Goodglick, L.; et al. Fibulin-3 as a blood and effusion biomarker for pleural mesothelioma. *N. Engl. J. Med.* **2012**, *367*, 1417–1427. [CrossRef]
89. Zhang, X.; Yin, M.; Zhang, L.J. Keratin 6, 16 and 17-Critical Barrier Alarmin Molecules in Skin Wounds and Psoriasis. *Cells* **2019**, *8*, 807. [CrossRef]
90. Rojahn, T.B.; Vorstandlechner, V.; Krausgruber, T.; Bauer, W.M.; Alkon, N.; Bangert, C.; Thaler, F.M.; Sadeghyar, F.; Fortelny, N.; Gernedl, V.; et al. Single-cell transcriptomics combined with interstitial fluid proteomics defines cell type-specific immune regulation in atopic dermatitis. *J. Allergy Clin. Immunol.* **2020**, *146*, 1056–1069. [CrossRef]
91. Zouboulis, C.C.; Nogueira da Costa, A.; Makrantonaki, E.; Hou, X.X.; Almansouri, D.; Dudley, J.T.; Edwards, H.; Readhead, B.; Balthasar, O.; Jemec, G.B.E.; et al. Alterations in innate immunity and epithelial cell differentiation are the molecular pillars of hidradenitis suppurativa. *J. Eur. Acad. Dermatol. Venereol.* **2020**, *34*, 846–861. [CrossRef] [PubMed]
92. Mechtcheriakova, D.; Wlachos, A.; Sobanov, J.; Kopp, T.; Reuschel, R.; Bornancin, F.; Cai, R.; Zemann, B.; Urtz, N.; Stingl, G.; et al. Sphingosine 1-phosphate phosphatase 2 is induced during inflammatory responses. *Cell Signal.* **2007**, *19*, 748–760. [CrossRef] [PubMed]
93. Vetrano, S.; Ploplis, V.A.; Sala, E.; Sandoval-Cooper, M.; Donahue, D.L.; Correale, C.; Arena, V.; Spinelli, A.; Repici, A.; Malesci, A.; et al. Unexpected role of anticoagulant protein C in controlling epithelial barrier integrity and intestinal inflammation. *Proc. Natl. Acad. Sci. USA* **2011**, *108*, 19830–19835. [CrossRef] [PubMed]
94. Danese, S.; Vetrano, S.; Zhang, L.; Poplis, V.A.; Castellino, F.J. The protein C pathway in tissue inflammation and injury: Pathogenic role and therapeutic implications. *Blood* **2010**, *115*, 1121–1130. [CrossRef]
95. Alquraini, A.; Garguilo, S.; D'Souza, G.; Zhang, L.X.; Schmidt, T.A.; Jay, G.D.; Elsaid, K.A. The interaction of lubricin/proteoglycan 4 (PRG4) with toll-like receptors 2 and 4: An anti-inflammatory role of PRG4 in synovial fluid. *Arthritis Res. Ther.* **2015**, *17*, 353. [CrossRef]
96. Kosinska, M.K.; Ludwig, T.E.; Liebisch, G.; Zhang, R.; Siebert, H.C.; Wilhelm, J.; Kaesser, U.; Dettmeyer, R.B.; Klein, H.; Ishaque, B.; et al. Articular Joint Lubricants during Osteoarthritis and Rheumatoid Arthritis Display Altered Levels and Molecular Species. *PLoS ONE* **2015**, *10*, e0125192. [CrossRef]
97. Block, G.J.; Narayanan, D.; Amell, A.M.; Petek, L.M.; Davidson, K.C.; Bird, T.D.; Tawil, R.; Moon, R.T.; Miller, D.G. Wnt/beta-catenin signaling suppresses DUX4 expression and prevents apoptosis of FSHD muscle cells. *Hum. Mol. Genet.* **2013**, *22*, 4661–4672. [CrossRef]
98. Pandey, S.N.; Cabotage, J.; Shi, R.; Dixit, M.; Sutherland, M.; Liu, J.; Muger, S.; Harper, S.Q.; Nagaraju, K.; Chen, Y.W. Conditional over-expression of PITX1 causes skeletal muscle dystrophy in mice. *Biol. Open* **2012**, *1*, 629–639. [CrossRef]
99. Wallace, L.M.; Garwick-Coppens, S.E.; Tupler, R.; Harper, S.Q. RNA interference improves myopathic phenotypes in mice over-expressing FSHD region gene 1 (FRG1). *Mol. Ther.* **2011**, *19*, 2048–2054. [CrossRef]
100. Lim, K.R.Q.; Maruyama, R.; Echigoya, Y.; Nguyen, Q.; Zhang, A.; Khawaja, H.; Sen Chandra, S.; Jones, T.; Jones, P.; Chen, Y.W.; et al. Inhibition of DUX4 expression with antisense LNA gapmers as a therapy for facioscapulohumeral muscular dystrophy. *Proc. Natl. Acad. Sci. USA* **2020**, *117*, 16509–16515. [CrossRef]
101. Cacchiarelli, D.; Legnini, I.; Martone, J.; Cazzella, V.; D'Amico, A.; Bertini, E.; Bozzoni, I. miRNAs as serum biomarkers for Duchenne muscular dystrophy. *EMBO Mol. Med.* **2011**, *3*, 258–265. [CrossRef] [PubMed]

102. Matsuzaka, Y.; Kishi, S.; Aoki, Y.; Komaki, H.; Oya, Y.; Takeda, S.; Hashido, K. Three novel serum biomarkers, miR-1, miR-133a, and miR-206 for Limb-girdle muscular dystrophy, Facioscapulohumeral muscular dystrophy, and Becker muscular dystrophy. *Environ. Health Prev. Med.* **2014**, *19*, 452–458. [CrossRef] [PubMed]
103. Statland, J.; Donlin-Smith, C.M.; Tapscott, S.J.; van der Maarel, S.; Tawil, R. Multiplex Screen of Serum Biomarkers in Facioscapulohumeral Muscular Dystrophy. *J. Neuromuscul. Dis.* **2014**, *1*, 181–190. [CrossRef] [PubMed]
104. Petek, L.M.; Rickard, A.M.; Budech, C.; Poliachik, S.L.; Shaw, D.; Ferguson, M.R.; Tawil, R.; Friedman, S.D.; Miller, D.G. A cross sectional study of two independent cohorts identifies serum biomarkers for facioscapulohumeral muscular dystrophy (FSHD). *Neuromuscul. Disord.* **2016**, *26*, 405–413. [CrossRef] [PubMed]
105. Marozzo, R.; Pegoraro, V.; Angelini, C. MiRNAs, Myostatin, and Muscle MRI Imaging as Biomarkers of Clinical Features in Becker Muscular Dystrophy. *Diagnostics* **2020**, *10*, 713. [CrossRef] [PubMed]
106. Konikoff, M.R.; Denson, L.A. Role of fecal calprotectin as a biomarker of intestinal inflammation in inflammatory bowel disease. *Inflamm Bowel. Dis.* **2006**, *12*, 524–534. [CrossRef]
107. Foell, D.; Wulffraat, N.; Wedderburn, L.R.; Wittkowski, H.; Frosch, M.; Gerss, J.; Stanevicha, V.; Mihaylova, D.; Ferriani, V.; Tsakalidou, F.K.; et al. Methotrexate withdrawal at 6 vs 12 months in juvenile idiopathic arthritis in remission: A randomized clinical trial. *JAMA* **2010**, *303*, 1266–1273. [CrossRef]
108. Vogl, T.; Eisenblatter, M.; Voller, T.; Zenker, S.; Hermann, S.; van Lent, P.; Faust, A.; Geyer, C.; Petersen, B.; Roebrock, K.; et al. Alarmin S100A8/S100A9 as a biomarker for molecular imaging of local inflammatory activity. *Nat. Commun.* **2014**, *5*, 4593. [CrossRef]
109. Nistala, K.; Varsani, H.; Wittkowski, H.; Vogl, T.; Krol, P.; Shah, V.; Mamchaoui, K.; Brogan, P.A.; Roth, J.; Wedderburn, L.R. Myeloid related protein induces muscle derived inflammatory mediators in juvenile dermatomyositis. *Arthritis Res. Ther.* **2013**, *15*, R131. [CrossRef]

Publisher's Note: MDPI stays neutral with regard to jurisdictional claims in published maps and institutional affiliations.

Journal of
*Personalized
Medicine*

MDPI

Article

Clinical and Laboratory Associations with Methotrexate Metabolism Gene Polymorphisms in Rheumatoid Arthritis

Leon G. D'Cruz [1,2], Kevin G. McEleney [1], Kyle B. C. Tan [1], Priyank Shukla [1], Philip V. Gardiner [3], Patricia Connolly [4], Caroline Conway [5], Diego Cobice [5] and David S. Gibson [1,*]

1 Northern Ireland Centre for Stratified Medicine (NICSM), Biomedical Sciences Research Institute, Ulster University, C-TRIC Building, Londonderry BT47 6SB, UK; darthcruz@gmail.com (L.G.D.); kmceleney29@gmail.com (K.G.M.); Tan-BC@ulster.ac.uk (K.B.C.T.); p.shukla@ulster.ac.uk (P.S.)
2 Respiratory Medicine Department and Clinical Trials Unit, Queen Alexandra Hospital, Portsmouth PO6 3LY, UK
3 Rheumatology Department, Western Health and Social Care Trust, Londonderry BT47 6SB, UK; Philip.Gardiner@westerntrust.hscni.net
4 Cardiac Assessment Unit, Western Health and Social Care Trust, Omagh BT79 0NR, UK; Patricia.Connolly@westerntrust.hscni.net
5 Mass Spectrometry Centre, Biomedical Sciences Research Institute (BMSRI), School of Biomedical Sciences, Ulster University, Cromore Road, Coleraine BT52 1SA, UK; c.conway@ulster.ac.uk (C.C.); d.cobice@ulster.ac.uk (D.C.)
* Correspondence: d.gibson@ulster.ac.uk

Received: 20 August 2020; Accepted: 24 September 2020; Published: 26 September 2020

Abstract: Rheumatoid arthritis (RA) is a chronic systemic autoimmune disease that causes loss of joint function and significantly reduces quality of life. Plasma metabolite concentrations of disease-modifying anti-rheumatic drugs (DMARDs) can influence treatment efficacy and toxicity. This study explored the relationship between DMARD-metabolising gene variants and plasma metabolite levels in RA patients. DMARD metabolite concentrations were determined by tandem mass-spectrometry in plasma samples from 100 RA patients with actively flaring disease collected at two intervals. Taqman probes were used to discriminate single-nucleotide polymorphism (SNP) genotypes in cohort genomic DNA: rs246240 (*ABCC1*), rs1476413 (*MTHFR*), rs2231142 (*ABCG2*), rs3740065 (*ABCC2*), rs4149081 (*SLCO1B1*), rs4846051 (*MTHFR*), rs10280623 (*ABCB1*), rs16853826 (*ATIC*), rs17421511 (*MTHFR*) and rs717620 (*ABCC2*). Mean plasma concentrations of methotrexate (MTX) and MTX-7-OH metabolites were higher ($p < 0.05$) at baseline in rs4149081 GA genotype patients. Patients with rs1476413 SNP TT or CT alleles have significantly higher ($p < 0.001$) plasma poly-glutamate metabolites at both study time points and correspondingly elevated disease activity scores. Patients with the rs17421511 SNP AA allele reported significantly lower pain scores ($p < 0.05$) at both study intervals. Genotyping strategies could help prioritise treatments to RA patients most likely to gain clinical benefit whilst minimizing toxicity.

Keywords: rheumatoid arthritis; SNP; DMARD; methotrexate; pharmacogenomics

1. Introduction

Rheumatoid arthritis (RA) is the most common chronic autoimmune inflammatory arthritis, affecting approximately 0.3–1% of the world's population [1,2]. The disease primarily affects the articular joints, causing swelling, stiffness, joint destruction [3], loss of function in joints [4], disability and a significantly lower quality of life. To prevent irreversible joint damage resulting in substantial disability, it is important to introduce disease-modifying anti-rheumatic drugs (DMARDs) early after onset and failure of non-steroidal anti-inflammatory treatment.

Conventional synthetic disease-modifying anti-rheumatic drugs (csDMARDs) such as methotrexate (MTX), hydroxychloroquine (HCQ), cyclosporin, sulfasalazine (SSZ) and leflunomide are commonly used mainstays of the disease; however, it is widely known that a significant proportion of patients with RA often show poor or inadequate therapeutic response to csDMARDs [5]. The anti-folate MTX is the cheapest drug in treatment of RA and is often the first-line treatment [6]; however, only 55% of patients remain on this drug for more than 2 years due to a build-up of non-response or the accumulation of various adverse side effects [6,7]. MTX is subject to significant metabolic activity in the body; the polyglutamated derivatives of MTX are selectively retained in cells, therefore lengthening the activity of the drug which complicates treatment management, since patients would continue taking their daily drug dosage oblivious to the fact that their circulating drug levels are still high, potentially contributing to undesirable cytotoxic effects [8,9]. MTX is converted in hepatic parenchymal cells resulting in the 2- through 4-glutamate residues derivatives or the drug is catabolised to the 7-hydroxy-methotrexate (MTX-7-OH) form [10]. More than 10% of a dose of methotrexate is oxidised to MTX-7-OH, irrespective of the route of administration [11]. The MTX-7-OH metabolite is extensively (91 to 93%) bound to plasma proteins, in contrast to the parent drug (only 35 to 50% bound) and contributes to inactivity of the drug or poor response to treatment [11].

When non-response has been confirmed, NICE clinical guidelines recommend switching to the more costly biological disease-modifying anti-rheumatic drugs (bDMARDs) [5,12,13]. Various studies indicate that treat-to-target strategies which aim to reduce disease activity shortly after diagnosis result in better long term outcomes and can minimise permanent joint damage, thus there is a genuine need for earlier identification of patients who do not respond well to csDMARDs treatments [6].

It is estimated that 15–30% of variation in drug responses are attributable to genetic or single-nucleotide polymorphisms or SNPs [14]. Not all SNPs are functional; some are in non-coding areas (introns) and there is a variety of ways that a SNP can affect or inhibit downstream transcription factor, gene or protein function [15]. The promise of pharmacogenomics is that identification of SNPs and associated risk alleles could identify patients who may be susceptible to accumulating cytotoxic levels of a drug during therapy (as in the polyglutamate derivatives of MTX) or when certain metabolite levels accumulate rendering a drug as inactive (as in MTX-7-OH).

In this study, we sought to determine the metabolite levels in RA patients taking DMARDs. We then carried out genotyping of 10 SNPs known to influence the metabolic pathways of DMARDs in arthritis. Our aim was to determine if genetic variations or polymorphisms associate with metabolite levels. This could help design studies to improve clinical management, which risk stratify patients at greater predisposition of forming ineffective or potentially harmful metabolite levels, by adequately planning ahead the appropriate drug and dosage.

2. Materials and Methods

2.1. Participant Recruitment

The research team at Ulster University collaborated with rheumatologists from the Western Health and Social Care Trust (WHSCT) to design, conduct and recruit patients to the study. Informed consent to participate was obtained from all RA patients enrolled to the study. One hundred patients identified using following inclusion/exclusion criteria were recruited into the prospective observational cohort study: Remote Arthritis Disease Activity MonitoR (RADAR); ClinicalTrials.gov Identifier: NCT02809547. Inclusion criteria: aged between 18–90 years, diagnosed with RA (according to American College of Rheumatology criteria [5,16]), diagnosed with RA for a minimum of 1 year and maximum 10 year duration, active disease flares on a regular basis, and receiving a disease-modifying anti-rheumatic drug (DMARD). Exclusion criteria: any other inflammatory conditions, any infections or trauma during study period, and have restricted hand function (determined by clinical team). Office for Research Ethics Committees Northern Ireland (ORECNI) (16/NI/0039), Ulster University Research Ethics Committee (UREC) (REC/16/0019) and WHSCT (WT/14/27) approvals were obtained for the study. Informed consent was obtained from all RA patients enrolled to the study.

2.2. Whole Blood and Dried Blood Spot Sample Collection

Venepuncture whole blood samples as part of the normal routine care pathway were forwarded to the hospital laboratories for multiple tests including CRP, ESR, Bilirubin, Liver enzymes and full blood count. An additional 5-mL EDTA tube of blood was collected from each of the 100 participants for DNA genotype and drug metabolite analyses within the RADAR study. Samples were collected at both study baseline and at a 6-week follow-up appointment at an outpatient rheumatology clinic for all participants. Additionally, a sub cohort of 30 of the above participants were supplied with a kit containing sufficient dried blood spot (DBS) cards [17], finger lancets and pre-paid and addressed postal envelopes with desiccant and biohazard sealable pouch to send a weekly samples (approximately 3–5 droplets of blood ~20 µL each) from home to the NICSM laboratory for drug metabolite analysis. Finger lancet blood droplets were deposited onto dried blood spot (DBS) Protein Saver 903TM cards (Whatman, GE Healthcare Life Sciences, Buckinghamshire, UK), pre-treated with a protein stabiliser coating.

2.3. Nucleic Acid Isolation from Peripheral Blood

Total DNA was isolated from peripheral blood samples using TRIzol reagent (TRIzol LS Reagent, Thermo Fisher Scientific, Basingstoke, UK. cat. No 10296-010) according to manufacturer's directions. Total DNA concentration was estimated by spectrophotometry (NanoVue Plus—GE Health Care, Buckinghamshire, UK).

2.4. Mass-Spectrometry Analysis

Determination of methotrexate (MTX) metabolites was performed using a liquid extraction surface analysis (LESA) coupled with nanoESI-triple quadruple mass spectrometer (QQQ) using Triversa nanomate (Advion, New York, NY, USA) and API 4000 QQQ Mass Spectrometer (AB Sciex, Cheshire, UK). Control MTX metabolites and internal standards were from Schircks Laboratories (Jona, Switzerland).

Quantitation for MTX and MTX metabolites was performed by the matrix-matched standards approach using an intensity ratio (ISTD/MTXs) calibration (10–2000 nM). Signal for each metabolite was the average of $n = 2$ (duplicate injection). A total of 5 nM was selected as LLOD (S/N ~ 3) for MTX, 7-OH MTX and MTX-PG2 and 8 nM was selected for MTX-PG3 to PG5; 10 nM was selected as LLOQ for MTX and all metabolites (S/N ~ 10). Intra- and inter-day precision was assessed at both 50 and 500 nM and coefficient of variation (CV) for MTX metabolites ranged from 2.0–7.2%. Linear regression coefficient (R2) of the back-calculated concentration against the nominal concentration for MTX and its metabolites was above 0.995.

Determination of sulfasalazine metabolites and teriflunomide, analysis was performed by liquid chromatography tandem mass spectrometry (LC-MS/MS) using a HP 1200 LC (Agilent, Palo Alto, CA, USA) and a Quattro micro mass spectrometer (Micromass, Manchester, UK). Control sulfasalazine metabolites, teriflunomide and internal standard were obtained from Sigma Aldrich, Gillingham, UK. Quantitation of sulfasalazine, its metabolites and teriflunomide was performed by the matrix-matched standards approach using an intensity ratio (ISTD/Analyte) calibration (5–500 µg/L). Signal for each metabolite was the average of $n = 2$ (duplicate injection). 5 µg/L was selected as LLOD (S/N > 5) for metabolites and 10 µg/L was selected as LLOQ for all metabolites (S/N > 10).

Intra- and inter-day precision was assessed at both 20 and 100 µg/L and coefficient of variation (CV) for sulfasalazine metabolites ranged between 1.4–5.8%. Linear regression coefficient (R2) of the back-calculated concentration against the nominal concentration for sulfasalazine, its metabolites and teriflunomide was above 0.992.

Separation of targeted analytes was carried out by reverse phase chromatography using a C18 column in gradient mode. Quantitation of all analytes were performed in positive ion mode multiple reaction monitoring (MRM) using matrix-matched standards and stable isotope ratios. All Mass Spectrometry methods were validated according to ICH Guidance for selectivity/specificity, limit of detection/quantitation (LLOD/LLOQ), linearity and precision [18]

2.5. Endpoint-Genotyping Using Taqman Assay

Endpoint genotyping analysis was carried out using the LightCycler 480 real-time PCR system (ROCHE). The assay is based on the competition during annealing between probes detecting the wild type and the mutant allele. The 5′-exonuclease activity of DNA polymerase cleaves the doubly labelled Taqman probe hybridised to the SNP-containing sequence, once cleaved, the 5′-fluorophore is separated from a 3′-quencher. Two allele-specific probes carrying different fluorophores (VIC®, emission: 554 nm and FAMTM, emission: 518 nm) permits SNP determination in a single well without any post-PCR processing. Genotype is determined from the ratio of intensities of the two fluorescent probes at the end of amplification (endpoint instead of the entire cycle in conventional PCR).

2.6. Taqman Probes Used for Single-Nucleotide Polymorphism Genotyping

The concentration and integrity of the genomic DNA were assessed by microvolume spectrophotometer (NanoDrop 2000; Thermo Fisher Scientific). DNA samples were genotyped by the following TaqMan SNP genotyping assays [*MTHFR*-rs1476413, Assay Identification (ID) C_8861304_10; RS1-rs 2231142, Assay Identification (ID) C_354526997_10; *ABCC2*-rs3740065, C_22271640_10; *SLCO1B1*-rs4149081, Assay Identification (ID) C_1901759_20; *MTHFR*-rs4846051, Assay Identification (ID) C_25763411_10; *ABCB1*-rs10280623, Assay Identification (ID) C_30537012_10; *ATIC*-rs16853826, C_33295728_10; *MTHFR*-rs17421511, Assay Identification (ID) C_32800189_20; *ABCC2*-rs717620, C_2814642_10; *ABCC1*-rs246240, Assay Identification (ID) C_1003698_10; Life Technologies Ltd.).

2.7. Validation of Polymorphisms by Pyrosequencing

Validation of SNP genotyping results from the Taqman assays was performed on a subset of samples using pyrosequencing. Due to the high number of SNPs to cover, a method using a universal biotinylated primer was employed [19]. Briefly, this method involves the use of standard target specific primer pairs with a universal M13 sequence at the 5′ end of one of the primers. A third, biotinylated M13-targeting primer is included in the PCR amplification reaction, leading to incorporation of biotin into the PCR product without the need for individual biotin labelling of each individual primer pair and thus lowering the cost of pyrosequencing considerably. A list of primers used for pyrosequencing are shown in Table S3.

PyroMark Assay Design Software 2.0 (Qiagen) was used for primer design in the SNP calling assay design format. PCR amplification was carried out using the Pyromark PCR kit (Qiagen) in 25 μL total volumes with 10–20 ng DNA and final concentrations of 0.2 mM for each primer. Standard PCR cycling conditions were used as per manufacturer's instructions and were consistent for all samples and targets. PCR products were checked by agarose gel electrophoresis and those with a positive single band of the expected size were taken forward into pyrosequencing on the Pyromark Q48 (Qiagen) using standard manufacturers protocols and the instrument run setting for SNP calling.

2.8. Statistics and Sample Size Calculations

Statistical analysis was carried out with SPSS ver.25 (IBM Corp, NY, USA), SciPy module (ver. 1.3) for Python (version 3.7.2) and R (version 3.60) with $p < 0.05$ considered as statistically significant, all within 95% confidence intervals. Descriptive statistics were used to characterise the variability in mean MTX, MTX-7-OH and MTX2PG–5PG and MTXtotal concentrations between different genotype groups of patients. Normality of data was determined using the Shapiro-Wilks test in SPSS (ver. 25) prior to employing the Kruskal–Wallis non-parametric (distribution free) one-way ANOVA, with

Dunn–Bonferroni post hoc test to assess differences between genotype group means using GraphPad Prism version 8.0.0 for Windows (GraphPad Software, San Diego, CA, USA, www.graphpad.com).

The Hardy-Weinberg equilibrium was assessed for SNPs with significant clinical associations in the methotrexate treated cohort. A Chi-square test with Benjamini-Hochberg adjusted p values was used to assess if there were significant differences between the genotype frequencies expected from dbGAP European population and those observed. Power was calculated for the same SNPs using the GENPWR package [20] within R (v 4.0.2), using the linear regression model (with alpha at 0.05) since the goal was to calculate power in a continuous outcome (metabolite levels) between genotypes.

3. Results

3.1. Study Population

A total of 100 participants, $n = 68$ female and $n = 32$ male, with active rheumatoid arthritis were enrolled to this study (Table 1). The mean age of study participants was 59.5 years with a mean disease duration of 6 years and mean baseline disease activity score (DAS28ESR) of 3.6.

Table 1. RADAR Study Cohort Demographics. Clinical and laboratory feature summary across $n = 100$ participants. yr: years; Anti-CCP: anti-cyclic citrullinated peptide; ESR: erythrocyte sedimentation rate; DAS28: disease activity score across 28 joints; RBC, red blood cell (count); Hb: haemoglobin; WBC: white blood cells; ALT: alanine aminotransferase; AST: aspartate aminotransferase; ALP: alkaline phosphatase; SD: Standard deviation.

	Mean	SD
Gender, male/female	32/68	-
Age, yr	59.5	12.6
Disease duration, yr	6.0	3.9
C-reactive protein, mg/L	7.9	11
Rheumatoid factor, n positive	63	-
Anti-CCP, n positive	53	-
ESR (mm/h)	15	15
DAS-28 ESR	3.6	1.5
RBC (cells/mL)	4.5	1.2
Hb (g/100 mL)	134	13
WBC (cells/mL)	6.8	2.3
Neutrophils (cells/mL)	4.4	1.9
Lymphocytes (cells/mL)	1.7	1.4
Platelets (cells/mL)	254	68
ALT (u/L)	23	11
AST (u/L)	24	9.4
ALP (u/L)	80	22
Bilirubin mg/L	8.2	5.6

A subgroup of $n = 66$ participants ($n = 46$ female) were treated with weekly methotrexate at baseline was identified for subsequent genotype association analyses. Baseline and 6-week follow-up drug dose information is summarized in Table S2A,B for this main subgroup. A smaller, partially overlapping, subgroup of $n = 27$ participants ($n = 20$ female) being treated with daily sulfasalazine at baseline was also identified for subsequent analyses (Table S2B).

3.2. Single-Nucleotide Polymorphisms Analysed in this Study

Ten SNPs were analysed in this study, previous studies have linked these SNPs to various clinical consequences observed in RA patients being treated with DMARDs, documented in the PharmGKB database [21,22]. SNP genotypes were determined by endpoint PCR assay and allele specific probes (Figure 1). The SNPs characteristics and frequencies in the study cohort are summarised in Table 2.

Table 2. Characteristics of the single nucleotide polymorphisms analysed in the RADAR study cohort. Ref: reference; Alt: alternative; dbGAP: database of Genotypes and Phenotypes; Genotype frequency for $n = 100$ participants, −/− or +/+: homozygote; +/−: heterozygote.

SNP	Variant Locale	Gene	Clinical Consequences	Alleles (Ref > Alt)	Minor Allele Frequency (European; dbGAP Sample Size)	Genotype Frequency		
						−/−	+/−	+/+
rs10280623	Intron	ABCB1	Associated with MTX toxicity	T > C	C = 0.1933 (n = 5,216)	64	21	14
rs246240	Intron	ABCC1	Low response to MTX	A > G	G = 0.153169 (n = 131,156)	6	25	69
rs717620	Non-coding Transcript	ABCC2	Increased time needed to reach therapeutic level	C > T	T = 0.198186 (n = 137,144)	11	24	65
rs3740065	Intron	ABCC2	Increased risk of MTX toxicity	A > G	G = 0.108811 (n = 128,976)	1	10	87
rs2231142	Intron Missense	ABCG2	SNP causes higher toxicity with combination treatments	G > T	T = 0.10454 (n = 88,504)	8	21	71
rs16853826	Intron	ATIC	ATIC rs16853826 variant associated with toxicity	G > A	A = 0.12231 (n = 10,972)	72	20	8
rs4846051	Codon Synonymous	MTHFR	Increased risk of MTX toxicity	A > G	G = 0.023185 (n = 117,404)	0	0	100
rs17421511	Intron	MTHFR	Positive response to MTX treatment (GG)	G > A	A = 0.1674 (n = 5990)	36	46	16
rs1476413	Intron	MTHFR	Positive response to MTX treatment (CC)	C > T	T = 0.270570 (n = 131,256)	15	24	61
rs4149081	Intron	SLC081	Increased risk of MTX toxicity	G > A	A = 0.168487 (n = 136,218)	51	22	27

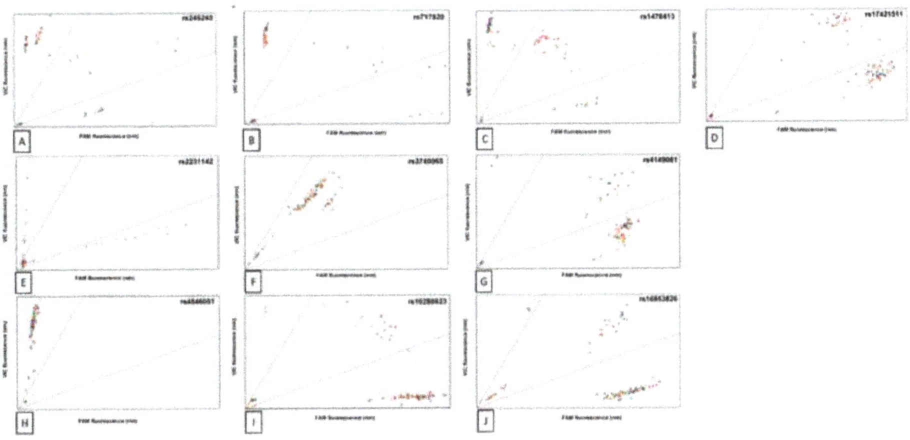

Figure 1. Endpoint polymorphism genotype assay. Genotype were determined for the listed polymorphisms in the RADAR study cohort (n = 100) from the ratio of fluorescent intensities (nm) of the two-allele specific Taqman probes (VIC and FAM) at the end of PCR amplification. Clusters in upper and lower quadrants represent groups of individuals with a homozygous genotype for either allele; the middle quadrant represents individuals with heterozygous genotype. (**A**) rs246240, (**B**) rs717620, (**C**) rs1476413, (**D**) rs17421511, (**E**) rs2231142. (**F**) rs3740065, (**G**) rs4149081, (**H**) rs4846051, (**I**) rs10280623, (**J**) rs16853826. Genotypes were also confirmed by pyrosequencing in selected individuals. Genotype frequency is summarized in Table 1.

3.3. Methotrexate and Sulfasalazine Metabolite Polymorphism Associations

The data strongly suggest plasma concentrations of methotrexate and sulfasalzine metabolites are associated with the allelic genotype for 2 particular polymorphisms, rs4149081 and rs1476413 (see Figure 2). Table 3 indicates that within the n = 66 methotrexate treated subgroup, n = 17 participants with the minor homozygote genotype AA in rs4149081 have a significantly lower mean plasma MTX-7-OH concentration compared to the GA genotype group (p = 0.002) at baseline. Although a similar trend is observed at the 6-week follow-up, this was not statistically significant. The GA genotype group mean concentrations of MTX and MTX-7-OH are significantly higher than those observed in the GG genotype group (p = 0.01 and p = 0.038, respectively; Figure 2A,B). The baseline mean blood bilirubin concentration was the only feature observed at significantly higher levels in the rs4149081 AA genotype, relative to the GG genotype group (p = 0.020; Figure 2C).

A total of n = 8 participants with the rs1476413 homozygous major allele genotype CC have significantly lower (p = 0.012) group mean plasma MTX-7-OH concentration, compared to the CT genotype group (Table 3) at the 6-week follow-up sessions. No significant difference was observed at baseline. The mean plasma concentration of tetraglutamate MTX metabolites are also significantly lower in the rs1476413 CC genotype group at both baseline (p = 0.02) and 6-week follow-up appointments (p = 0.008; Figure 2F,G and Table 3).

A total of n = 6 participants with the minor allele genotype AA in the SNP rs17421511 (Supplementary Figure S1A,B) show significantly higher mean plasma concentration (p = 0.013) of sulfapyridine at the 6-week follow-up appointment period only, relative to GG and GA genotype groups.

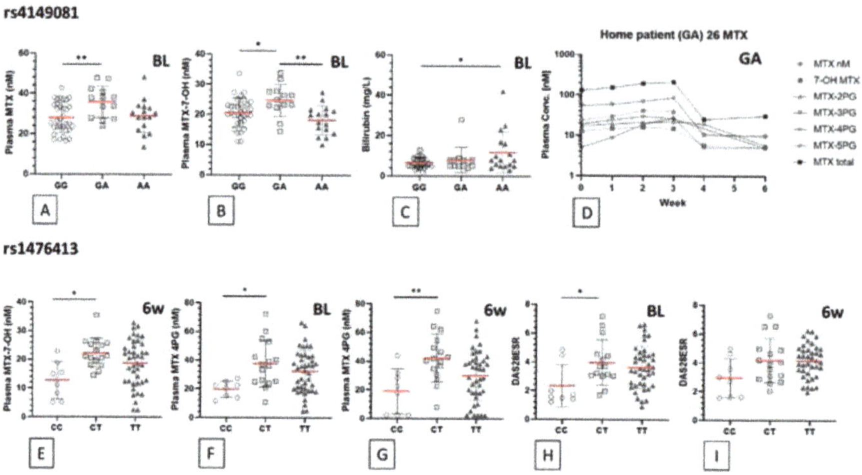

Figure 2. rs4149081 and rs1476413 genotype associations. Statistically significant associations between two polymorphisms, plasma drug metabolite concentration, other laboratory and clinical outcome measures are shown for individual genotypes in methotrexate treated participants (n = 66). Each symbol represents an individual participant of the genotype indicated on the x axes. Data grouped by rs4149081 genotypes: (**A**) baseline plasma concentration of unmetabolised methotrexate, (**B**) baseline plasma concentration of 7-hydroxy-methotrexate, (**C**) baseline bilirubin blood concentration, (**D**) weekly plasma concentrations (log scale) of listed methotrexate metabolites (PGs: polyglutamate subtypes) of a rs4149081 GA genotype participant. Data grouped by rs1476413 genotypes: (**E**) 6-week follow-up plasma concentration of 7-hydroxy-methotrexate, (**F**) baseline and (**G**) 6-week follow-up plasma concentrations of long-chain methotrexate 4-glutamate, (**H**) baseline and (**I**) 6-week follow-up disease activity (DAS28ESR) scores. Statistically significant differences between genotype group means are indicated by horizontal bars and an asterisk used to summarise p values adjusted by Bonferroni's multiple comparison test: (*) $p < 0.05$; (**) $p < 0.005$ (descriptive statistics data shown in Table 3). Red horizontal bar represents genotype group mean; error bars represent standard deviation. MTX: methotrexate; BL: baseline; 6w: 6-week follow-up.

Table 3. Methotrexate Treated Cohort SNP associations. Features including plasma drug metabolite, blood cell counts and clinical outcome measures which had statistically significant associations for the four polymorphisms rs4149081, rs1476413, rs2231142 and rs17421511 in $n = 66$ methotrexate treated participants. Mean data for each genotype group with number of individuals and females per group (n = females/total) indicated. Statistically significant differences between genotype group means were initially assessed by ANOVA and then an adjusted by Bonferroni's multiple comparison test performed for specified genotype group mean comparisons; asterisk used to summarise p values: (*) $p < 0.05$; (**) $p < 0.005$. Features with significant differences between genotype means are graphed in Figures 2 and 3. Unlisted features had no statistically significant association with any SNP (see Table S1). BL: baseline; 6wk: six-week follow-up; SD: standard deviation; ns: not significant.

SNP	Feature (Time pt.)	Units	Mean	±SD	Mean	±SD	Mean	±SD	ANOVA *p* Value	Genotype Comparison	Bonferroni *p* Value	Summary
rs4149081			GG (n = 25/36)		GA (n = 11/13)		AA (n = 10/17)					
	MTX (BL)	nM	28.0	7.5	35.8	7.8	28.9	8.2	0.011	GG vs. GA	**0.010**	**
	MTX (6wk)	nM	26.0	10.6	34.1	11.5	25.9	10.2	0.058	GG vs. GA	0.071	ns
	MTX-7-OH (BL)	nM	20.6	4.8	24.8	5.3	18.2	4.8	0.003	GG vs. GA	**0.038**	*
										GA vs. AA	**0.002**	**
	MTX-7-OH (6wk)	nM	19.1	6.8	23.3	7.2	17.3	7.6	0.079	GG vs. GA	0.239	ns
										GA vs. AA	0.084	ns
	Bilirubin (BL)	mg/L	6.7	2.623	8.1	6.211	11.9	10.15	0.024	GG vs. AA	**0.020**	*
	Bilirubin (6wk)	mg/L	6.6	2.392	7.8	3.76	10.2	8.64	0.059	GG vs. AA	0.054	ns
rs1476413			CC (n = 6/8)		CT (n = 13/16)		TT (n = 26/42)					
	MTX-7-OH (BL)	nM	19.8	5.6	21.2	5.7	20.1	5.9	0.791	CC vs. CT	>0.999	ns
	MTX-7-OH (6wk)	nM	12.7	6.5	22.3	5.2	18.7	8.2	0.015	CC vs. CT	**0.012**	*
	MTX 4PG (BL)	nM	19.8	5.6	37.7	16.2	32.4	15.1	0.024	CC vs. CT	**0.020**	*
	MTX 4PG (6wk)	nM	19.0	15.6	42.4	16.8	30.2	17.5	0.007	CC vs. CT	**0.008**	**
	DAS28ESR (BL)		2.3	1.5	4.0	1.6	3.6	1.5	0.047	CC vs. CT	**0.049**	*
	DAS28ESR (6wk)		3.0	1.4	4.2	1.6	4.2	1.2	0.051	CC vs. CT	0.090	ns
rs2231142			GG (n = 3/6)		GT (n = 12/15)		TT (n = 30/45)					
	RBC (BL)	mill./uL	4.4	0.4	4.1	0.3	4.4	0.4	0.050	GT vs. TT	**0.044**	*
	RBC (6wk)	mill./uL	4.4	0.3	4.1	0.3	4.4	0.4	0.042	GT vs. TT	**0.044**	*
	ALP (BL)	u/L	78.3	26.5	66.6	20.3	84.2	20.2	0.023	GT vs. TT	**0.019**	*
	ALP (6wk)	u/L	84.3	31.4	72.7	15.2	85.1	23.0	0.184	GT vs. TT	0.208	ns
	Lymphocytes (BL)	(cells/ml)	2.1	1.2	1.4	0.3	1.6	0.5	0.048	GG vs. GT	**0.043**	*
	Lymphocytes (6wk)	(cells/ml)	2.2	1.2	1.4	0.4	1.6	0.6	0.023	GG vs. GT	**0.019**	*
rs17421511			GG (n = 18/26)		GA (n = 24/31)		AA (n = 4/9)					
	PgPain (BL)	%	44.8	30.4	53.8	31.5	22.1	26.6	0.037	GA vs. AA	**0.033**	*
	PgPain (6wk)	%	45.4	27.9	55.9	32.8	22.2	23.4	0.016	GA vs. AA	**0.013**	*
	DAS28ESR (BL)		3.7	1.6	3.9	1.5	1.8	0.6	0.002	GA vs. AA	**0.002**	**
										GG vs. AA	**0.005**	**
	DAS28ESR (6wk)		4.1	1.1	4.2	1.4	3.1	1.5	0.088	GA vs. AA	0.097	ns
										GG vs. AA	0.159	ns
	ALT (BL)	u/L	20.6	10.1	21.8	11.1	32.0	15.9	0.047	GG vs. AA	**0.048**	*
	ALT (6wk)	u/L	21.7	10.7	20.5	9.3	27.1	12.4	0.252	GG vs. AA	0.534	ns

3.4. Clinical and Laboratory Feature Polymorphism Associations

While the average red blood cell counts (RBC) remain within reference ranges in both men and women in this study (Table 1), there is a modest but statistically significant decrease in mean RBC levels in the n = 15 participants with the rs2231142 heterozygous genotype GT compared to the TT genotype group at both study time points (p = 0.044; Table 3, Figure 3A,B). Mean alkaline phosphatase concentration is also significantly lower at baseline in the rs2231142 GT genotype group, relative to the TT genotype participants (p = 0.019; Figure 3C). In the smaller group of n = 6 rs2231142 GG genotype participants, mean lymphocyte counts are significantly higher than the GT genotype group, again at both time points (p = 0.043, p = 0.019; Figure 3D,E).

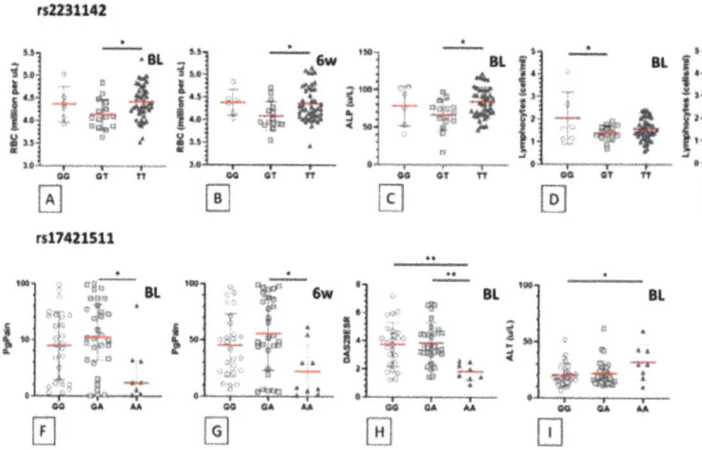

Figure 3. rs2231142 and rs17421511 genotype associations. Statistically significant associations between two polymorphisms and other laboratory and clinical outcome measures are shown for individual genotypes in methotrexate treated participants (n = 66). Each symbol represents an individual participant of the genotype indicated on the x axes. Data grouped by rs2231142 genotypes: (**A**) baseline and (**B**) 6-week follow-up red blood cell (RBC) count, (**C**) baseline blood alkaline phosphatase (ALP) concentration, (**D**) baseline and (**E**) 6-week follow-up lymphocyte count. Data grouped by rs17421511 genotypes: (**F**) baseline and (**G**) 6-week follow-up patient assessed pain (PgPain) levels, (**H**) baseline disease activity (DAS28ESR) scores, (**I**) baseline blood alanine aminotransferase (ALT) concentration. Statistically significant differences between genotype group means are indicated by horizontal bars and an asterisk used to summarise p values adjusted by Bonferroni's multiple comparison test: (*) $p < 0.05$; (**) $p < 0.005$ (descriptive statistics data shown in Table 3). Red horizontal bar represents genotype group mean; error bars represent standard deviation. BL: baseline; 6w: 6 weeks follow-up.

Mean patient-reported general pain (PgPain) scores are significantly lower in n = 9 participants carrying the rs17421511 AA genotype at baseline (p = 0.033) and at the 6-week follow-up appointment (p = 0.013) compared to those with the GA genotype (Figure 3F,G). This trend is also reflected in significantly lower mean baseline DAS28ESR scores in participants carrying the rs17421511 AA genotype, relative to the GA and GG genotype groups (p = 0.002 and p = 0.005 respectively; Table 3, Figure 3H), though scores even-out at the 6-week follow-up period among all three genotypes. The baseline mean alanine aminotransferase (ALT) was recorded at significantly higher blood concentrations in the rs17421511 AA genotype group, relative to the GG genotype participants (p = 0.048; Figure 3I).

The Benjamini–Hochberg-adjusted Chi-square test p-values showed no statistically significant differences between the genotype frequencies observed and those expected from dbGAP European population: rs4149081, p_{adj} = 0.515; rs 1476413, p_{adj} = 0.945; rs17421511, p_{adj} = 0.711; rs2231142,

p_{adj} = 0.9454. The power calculated for each of these SNPs at $p < 0.05$ was: rs4149081, 0.96; rs 1476413, 0.95; rs17421511, 0.97; rs2231142, 0.94.

4. Discussion

This study investigates the influence of ten well-characterised SNPs in RA and we have tried to correlate this with the appearance and accumulation of metabolites measured in the plasma of patients taking DMARDs such as methotrexate and or sulfasalazine. The in vivo pharmacotherapy of DMARDS and potential response biomarkers in RA have been previously described [6,23–25], however there studies of potential associations between circulating csDMARD levels and specific genetic variants remain limited in RA patients.

Typically, methotrexate treatment may cause elevations in serum AST and ALT, long term therapy has also been linked to development of fatty liver disease, fibrosis, cirrhosis, nephrotoxicity, and renal failure [26]. However, under active consultant-led clinical management, these effects are largely minimised. The mean values of all clinical biomarkers, liver enzyme and blood component cell-counts (Table 1) are within recommended normal reference ranges when viewed across all of the study participants. However, mean ALT was significantly higher at baseline in the methotrexate treated subgroup of participants with the rs17421511 AA genotype, albeit the potential effect of multiple drug combinations was not investigated in this subgroup.

Although the average RBC count when taken from all participants appear to be within normal ranges (Table 1), participants with the GT allele in the rs2231142 SNP have significantly decreased erythrocyte counts in their circulation compared to those with the homozygous alleles methotrexate affects folic acid metabolism, thus patients taking MTX may show variations in their mean corpuscular volume (MCV) of red blood cells (RBC), therefore resulting in megaloblastic anaemia.

RBCs retain MTX as the polyglutamate derivatives throughout their lifespan [27,28]. While normal RBC levels are between 4.7 to 6.1 million cells per microlitre (mill.c/μL) for men and between 4.2 to 5.4 mill.c/μL in women, the slight decrease shown in heterozygous rs2231142 SNP patients is statistically significant. However, the lower mean haemoglobin levels observed in the rs2231142 GT genotype is not statistically significant and there is no correlation with disease activity score as may have been anticipated in anaemia of chronic disease.

Apart from the impact of sex-linked genes in RA, the diversity in our genomes are partially accountable for the heterogeneity in the clinical presentation of synovitis among patients [29]. The genetic influence in RA is particularly strong, the heritability in RA is estimated to be around 60% [29] and with the high diversity of clinical presentations observed in RA, the goal in treatment would be to stratify patients according to their genetic profile and clinical outcome, eventually formulating a genetic-risk based personalised treatment management protocol.

MTX is an anti-folate drug, with anti-proliferative and anti-inflammatory effects, by inhibition of folate and adenosine pathways and inhibition of purines and pyrimidines synthesis [30–32]. Approximately 80–90% of methotrexate is primarily excreted by the kidneys [33]. MTX is converted in hepatic parenchymal cells of some patients resulting in the 2- through 4-glutamate residues derivatives or the drug is catabolised to the MTX-7-OH form.

Though not observed consistently on both study time points, participants with the AA allele in rs4149081 and CC allele in rs1476413 can have significantly lower mean plasma levels of MTX-7-OH in their plasma circulation. Since some genotype groups are modest in size, the potential for differences in the mean MTX dose between genotype groups was analysed, though no significant difference was observed (Table S2). Furthermore, only a weak correlation exists between MTX dose and circulating MTX-7-OH ($r2 = 0.08213$). Increasing levels of MTX-7-OH is known to inhibit the clinical responsiveness of RA patients to the MTX drug and therefore, reduced levels of this metabolite could signify a better clinical response to MTX [34]. Thus, with genetic profiling of expanded csDMARD naïve RA cohorts, it would be interesting to further investigate clinical responsiveness to MTX in these genotypes.

The 2- through 4-polyglutamate MTX metabolite-derivatives are selectively retained in cells and participants with the TT or CT alleles in the rs1476413 SNP tend to show significantly higher mean plasma levels of the tetraglutamate metabolite. A significantly higher mean DAS28 is observed only at study baseline in the rs1476413 CT genotype group relative to CC genotype, though due to modest numbers in the latter group this would require independent verification. It is likely that folic acid supplementation in the study cohort to mitigate toxicity of MTX has reduced the frequency of observable side effects.

Sulfasalazine is metabolized by intestinal bacteria, resulting in the release of sulfapyridine (SPY) and 5-aminosalicylate or 5-ASA (SPY and 5ASA are linked by an azo bond) [35]. Sulfapyridine is almost completely absorbed by the colon, metabolized by the liver, and renally excreted [1]. Commonly reported side-effects of sulfapyridine are minor gastrointestinal (GI) and central nervous system (CNS) abnormalities, and uncommon serious haematological and hepatic side-effects [36,37]. Although study participants with the AA and GA alleles of the rs17421511 SNP indicate higher mean plasma levels of sulfapyridine compared to those with the GG allele (Figure S1), no significant adverse phenotypic effects were observed in these subgroups.

The modest number of patients with the AA genotype of the rs17421511 SNP in our study report significantly lower levels of pain and disease activity, relative to the remaining methotrexate treated cohort. In future research with expanded patient cohorts, it would be pertinent to see if this phenomenon is observed in other patient groups carrying this particular genotype. As a general observation, a limitation of the current study is the low number of participants in particular genotype groups and the smoking status was not recorded, which may impact upon methotrexate metabolism. Furthermore, the findings for the methotrexate treated cohort are only generalizable to the European population, as no significant differences in Hardy–Weinberg equilibrium were found by Chi-square test between the dbGAP frequencies and those observed in this study.

While it is challenging to find a clear-cut relationship between genotype and circulating drug levels which translates through to a clear prediction of phenotypic consequence, useful leads are presented in the current study. The rs1476413 and rs17421511 *MTHFR* variants and the rs2231142 *ABCG2* variant display significant changes which are consistent at both study time points.

With further carefully powered studies of variability in both csDMARD response and predisposition to side effects, there is considerable potential to personalise effective treatments whilst avoiding any toxicity.

Supplementary Materials: The following are available online at http://www.mdpi.com/2075-4426/10/4/149/s1. Table S1: Methotrexate treated cohort SNP association with laboratory and clinical features; Table S2A: Methotrexate treated cohort prescription data; Table S2B: Methotrexate and sulfasalazine treated subgroup co-prescription data; Table S3: Pyrosequencing SNP primers; Figure S1: Sulfasalzine treated cohort SNP associations.

Author Contributions: Conceptualization, D.S.G.; methodology, L.G.D., K.G.M., C.C., D.C., K.B.C.T.; patient recruitment and data collection, P.V.G., P.C.; formal analysis, L.G.D.; data curation, K.G.M.; writing—original draft preparation, D.S.G., L.G.D.; writing—review and editing, P.S., P.V.G., D.C., C.C.; visualization, D.S.G.; supervision, D.S.G.; project administration, D.S.G.; funding acquisition, D.S.G. All authors have read and agreed to the published version of the manuscript.

Funding: This research was funded by the Research and Development Office, Health and Social Care NI (RADAR study), The Northern Ireland Rheumatism Trust, and Invest Northern Ireland (PoC 631/731) in support of this work.

Conflicts of Interest: The authors declare no conflict of interest. The funders had no role in the design of the study; in the collection, analyses, or interpretation of data; in the writing of the manuscript, or in the decision to publish the results.

References

1. Firestein, G.S.; Budd, R.C.; Gabriel, S.E.; McInnes, I.B.; O'Dell, J.R.; Kelley, W.N. *Kelley's Textbook of Rheumatology*, 11th ed.; Koretzky, G., Ed.; Elsevier/Saunders: Philadelphia, PA, USA, 2020.
2. World Health Organisation. Chronic Diseases and Health Promotion: Chronic Rheumatic Conditions. Available online: https://www.who.int/chp/topics/rheumatic/en/ (accessed on 17 August 2020).

3. Rea, I.M.; Gibson, D.S.; McGilligan, V.; McNerlan, S.E.; Alexander, H.D.; Ross, O.A. Age and Age-Related Diseases: Role of Inflammation Triggers and Cytokines. *Front. Immunol.* **2018**, *9*, 586. [CrossRef] [PubMed]
4. Noack, M.; Miossec, P. Selected cytokine pathways in rheumatoid arthritis. *Semin. Immunopathol.* **2017**, *39*, 365–383. [CrossRef] [PubMed]
5. Aletaha, D.; Smolen, J.S. Diagnosis and Management of Rheumatoid Arthritis: A Review. *JAMA* **2018**, *320*, 1360–1372. [CrossRef] [PubMed]
6. Gibson, D.S.; Bustard, M.J.; McGeough, C.M.; Murray, H.A.; Crockard, M.A.; McDowell, A.; Blayney, J.K.; Gardiner, P.V.; Bjourson, A.J. Current and future trends in biomarker discovery and development of companion diagnostics for arthritis. *Expert Rev. Mol. Diagn.* **2015**, *15*, 219–234. [CrossRef]
7. Barrera, P.; van der Maas, A.; van Ede, A.E.; Kiemeney, B.A.; Laan, R.F.; van de Putte, L.B.; van Riel, P.L.C.M. Drug survival, efficacy and toxicity of monotherapy with a fully human anti-tumour necrosis factor-alpha antibody compared with methotrexate in long-standing rheumatoid arthritis. *Rheumatology* **2002**, *41*, 430–439. [CrossRef]
8. Eektimmerman, F.; Swen, J.J.; Madhar, M.B.; Allaart, C.F.; Guchelaar, H.J. Predictive genetic biomarkers for the efficacy of methotrexate in rheumatoid arthritis: A systematic review. *Pharm. J.* **2019**, *20*, 159–168. [CrossRef]
9. Eleff, M.; Franks, P.E.; Wampler, G.L.; Collins, J.M.; Goldman, I.D. Analysis of "early" thymidine/inosine protection as an adjunct to methotrexate therapy. *Cancer Treat Rep.* **1985**, *69*, 867–874.
10. Inoue, K.; Yuasa, H. Molecular basis for pharmacokinetics and pharmacodynamics of methotrexate in rheumatoid arthritis therapy. *Drug Metab. Pharmacokinet.* **2013**, *29*, 12–19. [CrossRef]
11. Bannwarth, B.; Labat, L.; Moride, Y.; Schaeverbeke, T. Methotrexate in rheumatoid arthritis. An update. *Drugs* **1994**, *47*, 25–50. [CrossRef]
12. Cairns, A.P.; Patton, J.; Gardiner, P.V.; Liggett, N.; McKane, R.; Rooney, M.; Whitehead, E.; Taggart, A.J. The use of biological agents for severe inflammatory arthritis in Northern Ireland. *Rheumatology* **2002**, *41*, 92.
13. Lai, J.H.; Ling, X.C.; Ho, L.J. Useful message in choosing optimal biological agents for patients with autoimmune arthritis. *Biochem. Pharmacol.* **2019**, *165*, 99–111. [CrossRef] [PubMed]
14. Eichelbaum, M.; Ingelman-Sundberg, M.; Evans, W.E. Pharmacogenomics and individualized drug therapy. *Annu. Rev. Med.* **2006**, *57*, 119–137. [CrossRef] [PubMed]
15. Pang, G.S.; Wang, J.; Wang, Z.; Lee, C.G. Predicting potentially functional SNPs in drug-response genes. *Pharmacogenomics* **2009**, *10*, 639–653. [CrossRef] [PubMed]
16. Aletaha, D.; Neogi, T.; Silman, A.J.; Funovits, J.; Felson, D.T.; Bingham, C.O., 3rd; Birnbaum, N.S.; Burmester, G.R.; Bykerk, V.P.; Cohen, M.D.; et al. Rheumatoid arthritis classification criteria: An American College of Rheumatology/European League Against Rheumatism collaborative initiative. *Arthritis Rheum.* **2010**, *62*, 2569–2581. [CrossRef]
17. Edelbroek, P.M.; van der Heijden, J.; Stolk, L.M. Dried blood spot methods in therapeutic drug monitoring: Methods, assays, and pitfalls. *Ther. Drug Monit.* **2009**, *31*, 327–336. [CrossRef]
18. International Council for Harmonisation (ICH) of Technical Requirements for Pharmaceuticals for Human Use, Guideline M10 on Bioanalytical Method Validation. 2019. Available online: https://www.ema.europa.eu/en/documents/scientific-guideline/draft-ich-guideline-m10-bioanalytical-method-validation-step-2b_en.pdf (accessed on 18 August 2020).
19. Royo, J.L.; Hidalgo, M.; Ruiz, A. Pyrosequencing protocol using a universal biotinylated primer for mutation detection and SNP genotyping. *Nat. Protoc.* **2007**, *2*, 1734–1739. [CrossRef]
20. Moore, C.M.; Jacobson, S.A.; Fingerlin, T.E. Power and Sample size calculation for genetic association studies in the presence of genetic model misspecification. *Hum. Hered.* **2020**, *28*, 1–16. [CrossRef]
21. Mikkelsen, T.S.; Thorn, C.F.; Yang, J.J.; Ulrich, C.M.; French, D.; Zaza, G.; Dunnenberger, H.M.; Marsh, S.; McLeod, H.L.; Giacomini, K.; et al. PharmGKB summary: Methotrexate pathway. *Pharmacogenet. Genom.* **2011**, *21*, 679–686. [CrossRef]
22. Thorn, C.F.; Klein, T.E.; Altman, R.B. PharmGKB: The Pharmacogenomics Knowledge Base. *Methods Mol. Biol.* **2014**, *1015*, 311–320.
23. Eektimmerman, F.; Allaart, C.F.; Hazes, J.M.; Broeder, A.D.; Fransen, J.; Swen, J.J.; Guchelaar, H.-J. Validation of a clinical pharmacogenetic model to predict methotrexate nonresponse in rheumatoid arthritis patients. *Pharmacogenomics* **2019**, *20*, 85–93. [CrossRef]

24. Salazar, J.; Moya, P.; Altes, A.; Díaz-Torné, C.; Casademont, J.; Cerdà-Gabaroi, D.; Corominas, H.; Baiget, M. Polymorphisms in genes involved in the mechanism of action of methotrexate: Are they associated with outcome in rheumatoid arthritis patients? *Pharmacogenomics* **2014**, *15*, 1079–1090. [CrossRef] [PubMed]
25. Kumar, P.; Banik, S. Pharmacotherapy options in rheumatoid arthritis. *Clin. Med. Insights Arthritis Musculoskelet. Disord.* **2013**, *6*, 35–43. [CrossRef]
26. Conway, R.; Carey, J.J. Risk of liver disease in methotrexate treated patients. *World J. Hepatol.* **2017**, *9*, 1092–1100. [CrossRef] [PubMed]
27. Kremer, J.M. Toward a better understanding of methotrexate. *Arthritis Rheum.* **2004**, *50*, 1370–1382. [CrossRef] [PubMed]
28. Jolivet, J.; Schilsky, R.L.; Bailey, B.D.; Drake, J.C.; Chabner, B.A. Synthesis, retention, and biological activity of methotrexate polyglutamates in cultured human breast cancer cells. *J. Clin. Investig.* **1982**, *70*, 351–360. [CrossRef]
29. Kurkó, J.; Besenyei, T.; Laki, J.; Glant, T.T.; Mikecz, K.; Szekanecz, Z. Genetics of rheumatoid arthritis—A comprehensive review. *Clin. Rev. Allergy Immunol.* **2013**, *45*, 170–179. [CrossRef]
30. Dervieux, T.; Fürst, D.; Lein, D.O.; Capps, R.; Smith, K.; Walsh, M.; Kremer, J.M. Polyglutamation of methotrexate with common polymorphisms in reduced folate carrier, aminoimidazole carboxamide ribonucleotide transformylase, and thymidylate synthase are associated with methotrexate effects in rheumatoid arthritis. *Arthritis Rheum.* **2004**, *50*, 2766–2774. [CrossRef]
31. Tian, H.; Cronstein, B.N. Understanding the mechanisms of action of methotrexate: Implications for the treatment of rheumatoid arthritis. *Bull. NYU Hosp. Jt. Dis.* **2007**, *65*, 168–173.
32. Chan, E.S.L.; Cronstein, B.N. Molecular action of methotrexate in inflammatory diseases. *Arthritis Res.* **2002**, *4*, 266–273. [CrossRef]
33. Winograd, B.; Lippens, R.J.; Oosterbaan, M.J.; Dirks, M.J.; Vree, T.B.; van der Kleijn, E. Renal excretion and pharmacokinetics of methotrexate and 7-hydroxy-methotrexate following a 24-h high dose infusion of methotrexate in children. *Eur. J. Clin. Pharmacol.* **1986**, *30*, 231–238. [CrossRef]
34. Baggott, J.E.; Morgan, S.L. Methotrexate catabolism to 7-hydroxymethotrexate in rheumatoid arthritis alters drug efficacy and retention and is reduced by folic acid supplementation. *Arthritis Rheum.* **2009**, *60*, 2257–2261. [CrossRef] [PubMed]
35. Bird, H.A. Sulphasalazine, sulphapyridine or 5-aminosalicylic acid—Which is the active moiety in rheumatoid arthritis? *Br. J. Rheumatol.* **1995**, *34* (Suppl. 2), 16–19. [CrossRef] [PubMed]
36. Box, S.A.; Pullar, T. Sulphasalazine in the treatment of rheumatoid arthritis. *Br. J. Rheumatol.* **1997**, *36*, 382–386. [CrossRef] [PubMed]
37. Kumar, P.J.; Clark, M.L. *Kumar & Clark Clinical Medicine*, 6th ed.; Saunders: Edinburgh, UK, 2002.

Journal of *Personalized Medicine*

MDPI

Review

Modelling Neuromuscular Diseases in the Age of Precision Medicine

Alfina A. Speciale, Ruth Ellerington, Thomas Goedert and Carlo Rinaldi *

Department of Paediatrics, University of Oxford, Oxford OX1 3QX, UK; ambra.speciale@paediatrics.ox.ac.uk (A.A.S.); ruth.ellerington@paediatrics.ox.ac.uk (R.E.); tg5g18@soton.ac.uk (T.G.)

* Correspondence: carlo.rinaldi@paediatrics.ox.ac.uk; Tel.: +44-(0)1865-272148; Fax: +44-(0)1865-282840

Received: 23 September 2020; Accepted: 15 October 2020; Published: 17 October 2020

Abstract: Advances in knowledge resulting from the sequencing of the human genome, coupled with technological developments and a deeper understanding of disease mechanisms of pathogenesis are paving the way for a growing role of precision medicine in the treatment of a number of human conditions. The goal of precision medicine is to identify and deliver effective therapeutic approaches based on patients' genetic, environmental, and lifestyle factors. With the exception of cancer, neurological diseases provide the most promising opportunity to achieve treatment personalisation, mainly because of accelerated progress in gene discovery, deep clinical phenotyping, and biomarker availability. Developing reproducible, predictable and reliable disease models will be key to the rapid delivery of the anticipated benefits of precision medicine. Here we summarize the current state of the art of preclinical models for neuromuscular diseases, with particular focus on their use and limitations to predict safety and efficacy treatment outcomes in clinical trials.

Keywords: neuromuscular diseases; translational research; disease models; precision medicine

1. Introduction

Neuromuscular diseases are a broad and heterogeneous group of conditions characterized by an impairment in one or more components of the motor unit, defined as the motor neuron and the muscle fibres it innervates. Whilst most are individually rare, collectively neuromuscular diseases are significantly prevalent, with a cumulative prevalence of approximately 100–200 cases per 100,000 individuals worldwide [1], accounting for a substantial proportion of population-wide health care costs [2]. Very few treatments currently exist to treat these diseases. Nevertheless, as research progressively disentangles their pathogenic mechanisms, many opportunities are finally starting to land in the clinic.

Precision medicine refers to a treatment approach wherein the most appropriate treatment for an individual is chosen based on their specific disease manifestation, alongside their genetic/epigenetic information and other features such as their microbiome, age, nutrition, and lifestyle. The clinical and genetic heterogeneity of neuromuscular diseases make them ideal candidates for personalized therapeutic approaches, with many individuals suffering from rare or ultrarare diseases that cannot be treated by conventional blanket approach treatment. One example is Duchenne muscular dystrophy (DMD), the most prevalent childhood-onset muscular dystrophy, where progressive muscle degeneration and weakness is caused by mutations in the *DMD* gene, leading to loss of dystrophin protein production [3]. The vast majority of DMD patients carry an exon deletion (~65%) or a duplication (~10%) of one or multiple exons and these mutations tend to manifest in regions of vulnerability between exons 2 and 20 and exons 45 and 55 [4–6]. In addition, small mutations (insertions, deletions, nonsense mutations and splice site mutations) account for the remaining ~25% mutations and occur throughout the length of the gene [4]. Excision of specific exons, or exon skipping, by use of antisense oligonucleotides (AON) to allow restoration of the disrupted reading frame and therefore production of a shortened but functional dystrophin protein, has surfaced as a promising therapy for DMD [7]. Therefore, diagnosis by genetic sequencing has become a crucial tool in determining eligibility for these treatments, as multiple AON products need to address the large series of mutations carried by DMD subjects.

While presenting new challenges for researchers, precision medicine is rapidly taking the lead in the pursuit of radically transforming health care. Choosing the appropriate disease model that recapitulates the complexity and heterogeneity of patients is therefore paramount to understand disease mechanisms and increase the chances of success of translating a treatment opportunity into a safe and effective marketed drug.

In this review, we aim to discuss the currently available tools used to model neuromuscular diseases and to evaluate their utility and applicability to personalized medical research and therapeutic development (Table 1).

2. Cellular Models

2.1. Myoblasts

Primary myoblasts (activated satellite cells) obtained from human subjects or animal models typically go through multiple rounds of cell division until reaching confluence in growth media, followed by iterations of cellular fusions to form multinuclear myotubes and eventually terminal differentiation [8]. Due to several inherent traits of human-derived muscle cells, including the slower growth rate as well as the flattened morphology, primary human myotubes typically exhibit poorer contractile activity than their mouse counterparts in response to electric stimulation [9]. Obtaining a substantial number of satellite cells from skeletal muscle biopsies of patients is markedly limited by the restricted proliferative capability of activated satellite cells in culture. In order to overcome this limitation, myogenic conversion of non-muscle primary cells, such as primary human and murine fibroblasts from skin, has been widely employed, mainly using transduction of *MyoD* gene (myogenic differentiation), a master regulator of skeletal muscle differentiation [10]. In order to increase proliferative capacity, transduction with both telomerase-expressing and cyclin-dependent kinase 4-expressing vectors has been used to produce immortalized human muscle stem-cell lines from patients with different muscle diseases such as DMD, limb-girdle muscular dystrophy type 2B, facioscapulohumeral muscular dystrophy, oculopharyngeal muscular dystrophy and congenital muscular dystrophy [11]. These immortalized cultures have been extensively used both to study disease mechanism and to test treatment strategies.

2.2. Induced Pluripotent Stem Cells (iPSCs)

The development of induced pluripotent stem cell (iPSC) technology has brought a great paradigm shift in the field of precision medicine [12] and now they have a prominent role as a tool for disease modelling and drug screening. Moreover, they are highly expandable, are free from the ethical issues linked to the use of embryonic stem cells (ESCs), and their source of cells easily accessible.

Two major strategies have been recently developed to differentiate PSCs into satellite-like cells. The first involved overexpressing PAX7, the master transcription factor for satellite cells, in an inducible fashion [13]. After being generated from human embryonic stem cells and iPSCs, these cells showed capability for in vitro expansion and differentiation, as well as engraftment and myofibre formation in immunodeficient mice [13,14]. The second strategy involved the use of a small molecule, and consists of glycogen synthase kinase 3 beta (GSK3beta) inhibition, in order to activate the Wnt pathway, as well as treatment with fibroblast growth factor 2 (FGF2) in a minimal medium [15–20]. Alternative protocols have used bone morphogenic protein 4 (BMP4) inhibition to promote differentiation into the myogenic lineage [21–23], or Notch signalling inhibitor DAPT [24]. Purified by fluorescence-activated cell sorting (FACS) [15,19,24], partially purified, or unpurified [16,17,20,21,23], cell mixtures are then plated.

Table 1. Key features of the various models used for neuromuscular diseases.

Models.	Myoblasts	Stem Cells Derived Cultures	Organoids	Muscle-on-a-Chip	Mouse Models	Other Animal Models	Computational Models
Parameter		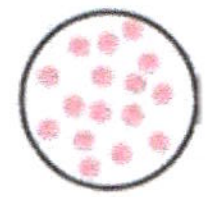	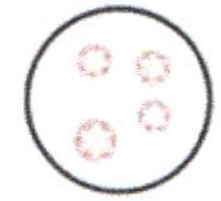	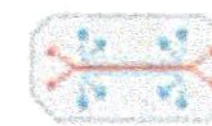			
Production complexity	Low	Medium, depends on protocol	Medium/High	High, requires engineered chambers	High	High/Medium	Medium/High
On-platform assay	Easy access to readouts, individual cell analyses possible	Medium difficulty, individual cell analyses possible	Medium/High difficulty, analyses possible at the tissue level	Medium/High, organ function analyses possible	Low difficulty, analyses possible both at cellular and tissue levels	Low difficulty, analyses possible both at cellular and tissue levels	Can account for contributions only of known variables
Duration of experiments	Minutes to days	Days to weeks	Days to weeks	Days to weeks, depends on platform design	Weeks to months	Days to weeks	Days
Variability and clinical relevance	Low variability and relevance	High variability and relevance	High variability and relevance	Low variability, high relevance	Low variability, high relevance	Low variability, moderate relevance	Low variability but requires in vivo confirmation
Level of control over variables	High	Medium/Low	Low	High	High/Medium	High/Medium	High
Biodistribution and toxicology studies	Useful for initial toxicology	Useful for initial toxicology	Useful for initial toxicology and limited biodistribution studies	Useful for initial toxicology and limited biodistribution studies	Useful for toxicology, biodistribution and life span assays	Useful for toxicology and life span assays	Useful to predict outcomes
High throughput feasibility	High	Medium/High	Medium/High, depends on tissues	Medium, depends on platform design	Low	High	High
Precision medicine potential	Low, easy to manipulate but homogenous cultures	High, takes into account individual variability	High, depends on tissues and is subject to model validation	High, allows high level of control and personalization	High, depends on model availability	Medium/High, easy to manipulate but low translational relevance	High when used in combination with in vitro/vivo methods

By generating an in vitro DMD model from patient-derived iPS cells, Shoji et al. noted excess Ca^{2+} influx in DMD myocytes when compared to control myocytes in response to stimulation via electricity. This was alleviated by restoring dystrophin expression via exon skipping, therefore establishing a model that recapitulates early DMD pathogenesis and is appropriate for assessing the efficacy of exon-skipping drugs by phenotypic assay [25]. IPSC models of several other neuromuscular diseases are currently available, including Miyoshi myopathy, a muscle disease caused by the mutation in dysferlin [26], Pompe disease, a paediatric disease caused by lysosomal glycogen accumulation in skeletal muscle that leads to muscle weakness [27], and myotonic dystrophy type 1, a multisystem disorder that affects skeletal and smooth muscle caused by a CTG trinucleotide repeat expansion in the non-coding region of the *DMPK* gene [28]. Overall, the introduction of iPSC technology has allowed scientists to model diseases directly from patients' cells, this being a cornerstone for personalized medicine. However, if they are planned to be used for personalized cell therapy, several issues remain to be addressed, including alterations in the differentiation efficiency, line-to-line variability, and risk of tumorigenicity.

2.3. Urine-Derived Stem Cells

In addition to representing an ideal source of cells for generating iPSCs, with a reprogramming efficiency approximately 100-fold higher than that of fibroblasts [29], urine stem cells (USCs) can also be induced into myogenic lineage by direct MyoD1 reprogramming [30]. Muscle differentiation can be further enhanced by adding 3-deazaneplanocin A hydrochloride [31]. These cells carry pluripotency markers such as CD29, CD105, CD166, CD90, and CD13 [32], and are able to self-renew and differentiate into the mesodermal, endodermal and ectodermal lineage [33]. Direct reprogramming of these cells, which can be easily isolated by centrifugation method and standard cell culture, has been recently shown to efficiently and reproducibly establish human myogenic cells from patients with DMD and limb-girdle muscular dystrophy (LGMD) type 2 [30]. Upon further molecular characterisation, this cost-effective and efficient in vitro model system shows great potential for more efficient drug development and targeted therapies development for neuromuscular diseases.

2.4. Skeletal Muscle Organoids

As the use of human iPSCs for tissue engineering and disease modelling expands, iPSC-derived organoids are rapidly becoming a powerful tool for modelling human organogenesis, homeostasis, injury repair and disease aetiology [34]. These miniature 3D tissues are generated using a combination of signposted differentiation, morphogenetic processes, and the embryonic organogenesis mimicking intrinsically driven self-assembly of cells, resulting in architecture and function remarkably similar to their in vivo counterparts. By using natural or synthetic scaffolds to create the artificial tissue [35], these models account for the cell–cell and cell–extracellular matrix interactions as well as the mechanical and/or chemical cues [36,37]. The development of physiologically relevant 3D in vitro models holds great promise to provide more economic, scalable and reproducible means of testing drugs and therapies for successful clinical translation. Few studies have reported methods to engineer human skeletal muscle tissue [38–43]. Induced myogenic progenitor cells derived from multiple human iPSC lines have been shown to form functional skeletal muscle tissues and are able to survive, progressively vascularize, and maintain functionality when implanted into the hindlimb muscle or dorsal window chamber in immunocompromised mice [44]. Isogenic human iPSC-derived 3D artificial muscles from patients affected by DMD, limb-girdle type 2D, and lamin A/C (LMNA)-related muscular dystrophies have been recently generated, recapitulating several pathogenic hallmarks in these diseases and also showing potential for muscle engraftment [45]. These studies have indicated that generation of fully functional artificial muscles require the contribution from other cellular lineages, for example vascular cells and motor neurons [45–49]. The major challenges the field is currently facing are mainly related to improving organoids' scalability as well as their complexity and maturity. Recent success in growing brain organoids using multiwell spinning bioreactors represents a significant step towards high-throughput drug screening via large-scale organoid generation [50]. These models resemble more closely foetal than adult tissue, therefore optimisation of protocols is essential before being able to advance these tissues into replacement therapy. Bearing in mind the speed at which the field has advanced over the past few years, the range of possible future applications of this platform in the study of human diseases and in regenerative medicine is expected to rapidly expand.

2.5. Muscle on Chip

Advancement in culturing models with mixed culture capabilities, together with the latest developments in 3D printing, microfluidics and microfabrication engineering, has led to the rapid expansion of organ-on-chip technologies. These platforms have recently attracted substantial interest due to their potential to be informative at multiple stages of the drug discovery process, while offering new ways to model disease states and perform mechanistic investigations in vitro. The critical and defining features of these platforms are the 3D structure, the possibility of integration of multiple cell types to reflect tissue physiology, and the presence of relevant biomechanical forces [51]. Organ on chips have been adapted for the human gut [52], heart [53], blood–brain barrier [54], and kidney [55]. Human primary myogenic cells have been engineered to form 3D myobundles, which respond to electrical stimuli and undergo dose-dependent hypertrophy or myopathy in response to pharmacological stimulation [40]. The decreased muscle regeneration capacity and weakness observed in DMD patients have been recapitulated in a human dystrophic skeletal muscle on a chip [56]. Using a 3D photo-patterning approach, other researchers have developed a skeletal muscle platform by confining a cell-laden gelatin network around two hydrogel pillars, which serve as anchoring sites for the cells, as the muscle tissues form and mature [57]. In other instances, neurons and rhabdomyocytes, both originating from mouse embryonic cells, have been differentiated in a 3D hydrogel culture, to effectively constitute a neuromuscular unit on a chip [58].

Tissue engineering requires a deep understanding of the functional interplay of cell types and the effect of the scaffold on cellular architecture, as well as careful characterising and validation of the model for the purpose of study. Additionally, due to safety concerns around the potential for unexpected toxic side effects, the biocompatibility of the materials to be used must be well profiled [51].

As iPSCs or adult stem cells taken from mass production of tissue organoids are increasingly employed as a source of cells for these platforms, organ on a chip represents an ideal tool for precision medicine.

2.6. Other

Sources in addition to the muscle-derived cells or reprogrammed cells can be employed to model muscle diseases. For example, melanocytes from DMD patients show the same morphological alterations as DMD muscle-derived cells [59]. Cultured melanocytes from skin biopsies have been shown to be a useful alternative to muscle biopsies for the mRNA-based molecular diagnosis of DMD [60]. Additionally, in the case of Ullrich congenital muscular dystrophy (UCMD) and Bethlem myopathy (BM), diseases caused by mutations in collagen VI genes [61], patients' derived melanocytes recapitulated the mitochondrial dysfunction and ultrastructural alterations that are found in patient myoblasts [62].

3. Animal Models

3.1. Mouse Models

A large fraction of currently available therapies have been developed with the help of animal models, especially mice, mainly due to the high similarity in sequence homology and organ physiology to humans, as well as cost-effective husbandry. Additionally, the external environment in mice studies can be well controlled and monitored and studies using inbred mice allow resampling isogenic individuals, therefore minimising variability.

Nevertheless, many differences remain: mice are smaller in size, have a markedly reduced lifespan and an increased heart rate, just to name a few. Approximately 1% of human genes are not present in the mouse genome [63], while the differences in the promoter regions, non-coding sequences, and RNA splicing might be even more marked, accounting for species-specific disparities in gene expression that in some cases can affect disease phenotype [64,65]. Overall these considerations, together with the realisation that treatments in mice have frequently resulted in disappointing outcomes in clinical trials, have recently called into question the translational potential of findings in mouse models [66].

One way of making mouse models for studying human diseases more suitable is to follow approaches pioneered over 30 years ago, which comprise incorporating human DNA into the mouse genome (genetic humanisation) and/or engrafting human cells and tissue into mouse tissues (cellular humanisation) [67–70]. Genetic humanisation can be achieved through a variety of methods, most commonly by injection of plasmids or artificial chromosome vectors into the mouse zygotes. Transgenic models have substantially contributed to advancing the understanding of human disease and have helped develop treatment strategies. One notorious major breakthrough in biomedical research using transgenic mice carrying the human *SMN2* gene led to the recent clinical approval of an AON, able to block an intronic splicing silencer in human *SMN2* [71], increasing full-length *SMN2* isoform expression, which compensates for the loss of *SMN1* that causes spinal muscular atrophy [72–75].

However, some key features must be considered: the cDNA or genomic DNA used to generate the transgenic mice tend to integrate randomly in multiple copies and thus overexpress the protein of interest. Overexpression of wild-type proteins may give a dose-dependent phenotype not related to the disease mutation, like in the case of the androgen receptor [76], and RNA binding proteins, such as TAR DNA-binding protein 43 [77]. The rise of genome engineering technology has revolutionized the field of molecular biology by allowing the generation of physiological, humanized knock-in mice models by precise editing [78,79]. Most DMD preclinical studies have been carried out in the mdx mouse that carries a nonsense point mutation in DMD exon 23 [80], which is only one out of the thousands of possible variations in this gene present in DMD patients. Despite a lack of dystrophin expression, these mice do not exhibit dilated cardiomyopathy or a shortened lifespan. To improve upon this model, a number of double knock-out mouse models have been created, such as mice deficient in both dystrophin and its homolog utrophin, which show decreased cardiac function and survival [81]. In recent years by using clustered regularly interspaced short palindromic repeat (CRISPR)-based editing, many new DMD mouse models carrying deletions, frameshifting mutations, a point mutation, and a mutant version of the human *DMD* gene have been generated [82–88], making testing of exon skipping strategies targeting different parts of the DMD transcript possible. It is worth considering that recent studies to assess the effects of disease-causing mutations or environmental stimuli in different mouse strains found a strong influence of the genetic background on phenotypic responses [89], highlighting the importance of genetic diversity of animal models in biomedical research.

It is becoming more and more evident that choosing the right model is critical. Depending on the specific research question, often combining different strains is the most appropriate way to minimize the risks of a lack of reproducibility of translational research. Despite the obvious differences between mice and humans, genetic mouse models have allowed us to look at the effects of a mutation at a system level. Combining genetic engineering, which has made genetic modifications of endogenous targets possible, with the use of genetic with cellular humanisation, we now have powerful tools to study human pathophysiology in vivo, in cell-autonomous and non-cell-autonomous contexts [90], as well as excellent preclinical models to identify and test the pharmacodynamic and pharmacokinetic properties of a treatment strategy, from gene therapy to small-molecule and cell replacement [91]. Overall, these considerations further support the use of 'mouse precision medicine' as a better prototype for future mouse studies.

3.2. Drosophila Melanogaster

Drosophila melanogaster can serve as a useful model of human neuromuscular disease, since flies have a neural circuitry, albeit much simpler than in humans, as well as multinucleated muscle cells and neuromuscular junctions (NMJ). The mechanisms of synaptic transmission seen at the NMJ in humans are conserved in Drosophila, with a key difference being that Drosophila uses glutamate, not acetylcholine, as the neurotransmitter. The ability to genetically manipulate Drosophila is useful when trying to better understand how certain myopathies occur. Moreover, their short life span and large progeny make flies a good system for carrying out large-scale genetic screens. Drosophila has helped us understand more about the NMJ, and in particular, the role that the dystrophin–glycoprotein complex plays (DGC). Like in mammals, the Drosophila gene of dystrophin also encodes multiple isoforms, which contain highly conserved domains and are mainly expressed in the muscle and the nervous system [92–94]. Studies into DGC function at the NMJ of Drosophila have shown that it plays an important role in the retrograde control of neurotransmitter release, neuronal migration and muscle stability and thus may help explain how neuromuscular pathology can occur. Removal of a dystrophin isoform (DLP2) in Drosophila, which is normally located at the post-synapse, has been shown to lead to an increase in presynaptic neurotransmitter release, causing increased muscle depolarisation, thus indicating a role of dystrophin in regulating presynaptic neurotransmitter release [95]. Previous work has shown that by studying sensory neurons (photoreceptor cells) in Drosophila [96], a lot can be learnt about axon guidance and target recognition. Perturbation of dystrophin and dystroglycan in photoreceptor cells led to disrupted axon guidance, similar to neuronal defects seen in human muscular dystrophy patients. Drosophila not only aids us in understanding the role that certain proteins play at the synapse of the NMJ, but also serves as a good model for studying age-dependent progression of muscular dystrophy. The reduction in levels of expression of dystrophin isoforms in Drosophila using RNAi led to muscle degeneration in larval and adult flies [95], thus potentially providing a useful model to help us understand Duchenne muscular dystrophy pathogenesis in humans.

3.3. Zebrafish

The zebrafish (Danio rerio) has become a useful organism for studying neuromuscular genetic disorders [97]. Comparison to the human reference genome has shown that approximately 70% of human genes have at least one zebrafish orthologue [98], and dozens of mutant zebrafish lines have already been generated to model the most common human myopathies [99–101]. As vertebrates, they possess desirable attributes, including small size, rapid development, and genetic tractability [97]. Zebrafish embryos are transparent, develop externally and can be easily genetically manipulated [102], making this model ideal for phenotypic high-throughput screening platform to investigate drug efficacy in a whole-organism context. The most commonly adopted screening criteria for assessing neuromuscular phenotype are spontaneous coiling, ability to hatch on time, swimming behaviour, and birefringence assay [103]. Compared to target-based drug discovery, a phenotype-driven approach offers several key advantages [104], such as rapid identification of compounds that have poor bioavailability, exhibit toxicity or off-target effects. By screening small-molecule libraries in the dystrophin-null zebrafish (sapje model), aminophylline, a non-selective phosphodiesterase inhibitor, was found to improve survival rate in animals, restore normal muscle structure and up-regulate the cAMP-dependent PKA pathway without affecting dystrophin expression [105]. In the sapje model, the mitochondrial defects present in DMD patients were recapitulated, making it an optimal model for the disease, and it was used to assess the effect of the cyclophilin inhibitor alisporivir treatment in vivo, resulting in an improvement in the morphology of mitochondria and myofibrils, and in mitochondrial respiration [106]. A zebrafish model showing severe myopathy has also been generated for UCMD via a deletion in the *col6a1* gene through the injection of an antisense morpholino [107]. Here, defects in the mitochondria permeability transition pore (mPTP) were corrected with the cyclophilin inhibitor NIM811 treatment [108]. In another study, the zebrafish model was used to test mitochondrial respiratory capacity after treatment with stable analogues of mPTP inhibitors [109]. Additionally, the zebrafish model has also provided insight into functional aspects of disease pathogenesis for several muscle conditions: for example, studies in zebrafish relatively relaxed (ryr) mutant, a model of RYR1-related myopathies [110], have contributed to identifying oxidative stress as an important disease mechanism in RYR1-related myopathies [111].

3.4. *Caenorhabditis Elegans*

With 40% of human disease genes having a nematode ortholog [112], and a fully sequenced genome [113], C. elegans is a valuable model to investigate several human physiological and pathological mechanisms. Studies of sarcomere maintenance and function in striated muscle led to the first identification of many conserved proteins, including twitchin, unc-89 (obscurin), unc-112 (kindlin), unc-45 (myosin chaperone) and unc-78 (AIP1) [114]. Using a large-scale screens in a C. elegans model of muscular dystrophy, carrying mutations in the dys-1 and the hlh-1 genes, which are respectively the homolog for the mammalian dystrophin and *MyoD* gene [115], compounds such as prednisone and serotonin have been shown to be effective in reducing muscle degeneration [116,117]. The obvious advantages of using this scalable and high-throughput model are counterbalanced by the limited phenotypic analyses, such as counting the number of times a worm bends in a C-shaped fashion in liquid in one minute, although new automated methods of quantifying muscle contraction and relaxation kinetics are emerging [118].

4. Computational Models

In silico models are becoming an increasingly useful tool for investigating muscle function and in helping us to understand which key players cause muscle pathology. These models integrate published experimental data, thus allowing us to encompass the many variables linked to pathology in a single model, enabling the study of multifaceted diseases. In doing these studies, one may understand better the underlying interactions between different disease mechanisms that lead to pathology, which may prove harder to do in live experiments. Over the last twenty years, big steps have been made in the computational modelling of muscle. A recent development has been the creation of agent-based models (ABMs), which allow us to assess what roles different biological agents play in muscle pathology, both at cellular and systems levels. For example, the use of ABMs for DMD has indicated a link between low satellite stem cell counts and impaired muscle regeneration symptom [119]. ABMs can also be used to predict the outcomes of given scenarios based on the rules derived from the literature, as well as having certain parameters that cannot be measured experimentally. This system can even add software agents that mimic certain biological cells into the simulation, with the aim of helping us to better understand their cellular interactions. This has been carried out in studies showing that fibroblasts can affect a muscle's susceptibility to disuse-induced atrophy [120].

However, these models do have their limitations: the simulated model is not a full replicate of the muscle cell and its microenvironment, as it only accounts for the contribution of known variables, which renders this model system not fully translatable to the in vivo situation.

5. Conclusions

The increasing availability of genetic and phenotypic information on patients with neuromuscular diseases, coupled with the unprecedented opportunity to manipulate eukaryotic genomes to generate disease models to study these diseases, has the potential to accelerate the translation of new therapeutic opportunities from preclinical settings into medical practice. Among the models available to researchers, 3D cultures and muscle on chips are best suited for precision medicine applications, due to their structural complexity and opportunity for genetic and environmental manipulation. However, as it becomes increasingly evident that we need to abandon the concept of 'one drug fits all', modelling every disease-associated variant for preclinical applications is likely to be unattainable and in many cases unnecessary. Achieving model precision is critical in translational research as long as it provides predictive validity, which is the ultimate goal of preclinical work, and may further be enhanced by using multiple models to capture the spectrum of mechanisms and testing therapies in diverse genetic backgrounds that more closely reflect the human population as a whole. This may be particularly true in complex diseases, where multiple risk loci concur to the development of a specific condition or to the treatment response.

Author Contributions: Literature review, writing and original draft preparation, A.A.S.; writing, review and editing, A.A.S., R.E., T.G., and C.R. All authors have read and agreed to the published version of the manuscript.

Funding: This research received no external funding.

Conflicts of Interest: The authors declare no conflict of interest.

References

1. Deenen, J.C.; Horlings, C.G.; Verschuuren, J.J.; Verbeek, A.L.; van Engelen, B.G. The Epidemiology of Neuromuscular Disorders: A Comprehensive Overview of the Literature. *J. Neuromuscul. Dis.* **2015**, *2*, 73–85. [CrossRef] [PubMed]
2. Olesen, J.; Gustavsson, A.; Svensson, M.; Wittchen, H.U.; Jonsson, B.; CDBE2010 Study Group; European Brain Council. The economic cost of brain disorders in Europe. *Eur. J. Neurol.* **2012**, *19*, 155–162. [CrossRef]
3. Blake, D.J.; Weir, A.; Newey, S.E.; Davies, K.E. Function and genetics of dystrophin and dystrophin-related proteins in muscle. *Physiol. Rev.* **2002**, *82*, 291–329. [CrossRef] [PubMed]
4. Aartsma-Rus, A.; Van Deutekom, J.C.; Fokkema, I.F.; Van Ommen, G.J.; Den Dunnen, J.T. Entries in the Leiden Duchenne muscular dystrophy mutation database: An overview of mutation types and paradoxical cases that confirm the reading-frame rule. *Muscle Nerve* **2006**, *34*, 135–144. [CrossRef] [PubMed]
5. Beggs, A.H.; Koenig, M.; Boyce, F.M.; Kunkel, L.M. Detection of 98% of DMD/BMD gene deletions by polymerase chain reaction. *Hum. Genet.* **1990**, *86*, 45–48. [CrossRef] [PubMed]
6. Liechti-Gallati, S.; Koenig, M.; Kunkel, L.M.; Frey, D.; Boltshauser, E.; Schneider, V.; Braga, S.; Moser, H. Molecular deletion patterns in Duchenne and Becker type muscular dystrophy. *Hum. Genet.* **1989**, *81*, 343–348. [CrossRef]
7. Dzierlega, K.; Yokota, T. Optimization of antisense-mediated exon skipping for Duchenne muscular dystrophy. *Gene Ther.* **2020**, *27*, 407–416. [CrossRef]
8. Koide, M.; Hagiwara, Y.; Tsuchiya, M.; Kanzaki, M.; Hatakeyama, H.; Tanaka, Y.; Minowa, T.; Takemura, T.; Ando, A.; Sekiguchi, T.; et al. Retained Myogenic Potency of Human Satellite Cells from Torn Rotator Cuff Muscles Despite Fatty Infiltration. *Tohoku J. Exp. Med.* **2018**, *244*, 15–24. [CrossRef]
9. Chen, W.; Nyasha, M.R.; Koide, M.; Tsuchiya, M.; Suzuki, N.; Hagiwara, Y.; Aoki, M.; Kanzaki, M. In vitro exercise model using contractile human and mouse hybrid myotubes. *Sci. Rep.* **2019**, *9*, 11914. [CrossRef]
10. Lattanzi, L.; Salvatori, G.; Coletta, M.; Sonnino, C.; Cusella De Angelis, M.G.; Gioglio, L.; Murry, C.E.; Kelly, R.; Ferrari, G.; Molinaro, M.; et al. High efficiency myogenic conversion of human fibroblasts by adenoviral vector-mediated MyoD gene transfer. An alternative strategy for ex vivo gene therapy of primary myopathies. *J. Clin. Investig.* **1998**, *101*, 2119–2128. [CrossRef]
11. Mamchaoui, K.; Trollet, C.; Bigot, A.; Negroni, E.; Chaouch, S.; Wolff, A.; Kandalla, P.K.; Marie, S.; Di Santo, J.; St Guily, J.L.; et al. Immortalized pathological human myoblasts: Towards a universal tool for the study of neuromuscular disorders. *Skelet Muscle* **2011**, *1*, 34. [CrossRef]
12. Takahashi, K.; Yamanaka, S. Induction of pluripotent stem cells from mouse embryonic and adult fibroblast cultures by defined factors. *Cell* **2006**, *126*, 663–676. [CrossRef] [PubMed]
13. Darabi, R.; Arpke, R.W.; Irion, S.; Dimos, J.T.; Grskovic, M.; Kyba, M.; Perlingeiro, R.C. Human ES- and iPS-derived myogenic progenitors restore DYSTROPHIN and improve contractility upon transplantation in dystrophic mice. *Cell Stem Cell* **2012**, *10*, 610–619. [CrossRef] [PubMed]
14. Magli, A.; Incitti, T.; Kiley, J.; Swanson, S.A.; Darabi, R.; Rinaldi, F.; Selvaraj, S.; Yamamoto, A.; Tolar, J.; Yuan, C.; et al. PAX7 Targets, CD54, Integrin alpha9beta1, and SDC2, Allow Isolation of Human ESC/iPSC-Derived Myogenic Progenitors. *Cell Rep.* **2017**, *19*, 2867–2877. [CrossRef]
15. Borchin, B.; Chen, J.; Barberi, T. Derivation and FACS-mediated purification of PAX3+/PAX7+ skeletal muscle precursors from human pluripotent stem cells. *Stem Cell Rep.* **2013**, *1*, 620–631. [CrossRef] [PubMed]

16. Caron, L.; Kher, D.; Lee, K.L.; McKernan, R.; Dumevska, B.; Hidalgo, A.; Li, J.; Yang, H.; Main, H.; Ferri, G.; et al. A Human Pluripotent Stem Cell Model of Facioscapulohumeral Muscular Dystrophy-Affected Skeletal Muscles. *Stem Cells Transl. Med.* **2016**, *5*, 1145–1161. [CrossRef] [PubMed]
17. Shelton, M.; Metz, J.; Liu, J.; Carpenedo, R.L.; Demers, S.P.; Stanford, W.L.; Skerjanc, I.S. Derivation and expansion of PAX7-positive muscle progenitors from human and mouse embryonic stem cells. *Stem Cell Rep.* **2014**, *3*, 516–529. [CrossRef]
18. Shelton, M.; Kocharyan, A.; Liu, J.; Skerjanc, I.S.; Stanford, W.L. Robust generation and expansion of skeletal muscle progenitors and myocytes from human pluripotent stem cells. *Methods* **2016**, *101*, 73–84. [CrossRef]
19. van der Wal, E.; Bergsma, A.J.; van Gestel, T.J.M.; In't Groen, S.L.M.; Zaehres, H.; Arauzo-Bravo, M.J.; Scholer, H.R.; van der Ploeg, A.T.; Pijnappel, W. GAA Deficiency in Pompe Disease Is Alleviated by Exon Inclusion in iPSC-Derived Skeletal Muscle Cells. *Mol. Ther. Nucleic Acids* **2017**, *7*, 101–115. [CrossRef]
20. Xu, C.; Tabebordbar, M.; Iovino, S.; Ciarlo, C.; Liu, J.; Castiglioni, A.; Price, E.; Liu, M.; Barton, E.R.; Kahn, C.R.; et al. A zebrafish embryo culture system defines factors that promote vertebrate myogenesis across species. *Cell* **2013**, *155*, 909–921. [CrossRef]
21. Chal, J.; Oginuma, M.; Al Tanoury, Z.; Gobert, B.; Sumara, O.; Hick, A.; Bousson, F.; Zidouni, Y.; Mursch, C.; Moncuquet, P.; et al. Differentiation of pluripotent stem cells to muscle fiber to model Duchenne muscular dystrophy. *Nat. Biotechnol.* **2015**, *33*, 962–969. [CrossRef]
22. Chal, J.; Al Tanoury, Z.; Hestin, M.; Gobert, B.; Aivio, S.; Hick, A.; Cherrier, T.; Nesmith, A.P.; Parker, K.K.; Pourquie, O. Generation of human muscle fibers and satellite-like cells from human pluripotent stem cells in vitro. *Nat. Protoc.* **2016**, *11*, 1833–1850. [CrossRef] [PubMed]
23. Swartz, E.W.; Baek, J.; Pribadi, M.; Wojta, K.J.; Almeida, S.; Karydas, A.; Gao, F.B.; Miller, B.L.; Coppola, G. A Novel Protocol for Directed Differentiation of C9orf72-Associated Human Induced Pluripotent Stem Cells Into Contractile Skeletal Myotubes. *Stem Cells Transl. Med.* **2016**, *5*, 1461–1472. [CrossRef]
24. Choi, I.Y.; Lim, H.; Estrellas, K.; Mula, J.; Cohen, T.V.; Zhang, Y.; Donnelly, C.J.; Richard, J.P.; Kim, Y.J.; Kim, H.; et al. Concordant but Varied Phenotypes among Duchenne Muscular Dystrophy Patient-Specific Myoblasts Derived using a Human iPSC-Based Model. *Cell Rep.* **2016**, *15*, 2301–2312. [CrossRef]
25. Shoji, E.; Sakurai, H.; Nishino, T.; Nakahata, T.; Heike, T.; Awaya, T.; Fujii, N.; Manabe, Y.; Matsuo, M.; Sehara-Fujisawa, A. Early pathogenesis of Duchenne muscular dystrophy modelled in patient-derived human induced pluripotent stem cells. *Sci. Rep.* **2015**, *5*, 12831. [CrossRef] [PubMed]
26. Liu, J.; Aoki, M.; Illa, I.; Wu, C.; Fardeau, M.; Angelini, C.; Serrano, C.; Urtizberea, J.A.; Hentati, F.; Hamida, M.B.; et al. Dysferlin, a novel skeletal muscle gene, is mutated in Miyoshi myopathy and limb girdle muscular dystrophy. *Nat. Genet.* **1998**, *20*, 31–36. [CrossRef] [PubMed]
27. Yoshida, T.; Awaya, T.; Jonouchi, T.; Kimura, R.; Kimura, S.; Era, T.; Heike, T.; Sakurai, H. A Skeletal Muscle Model of Infantile-onset Pompe Disease with Patient-specific iPS Cells. *Sci. Rep.* **2017**, *7*, 13473. [CrossRef]
28. Ueki, J.; Nakamori, M.; Nakamura, M.; Nishikawa, M.; Yoshida, Y.; Tanaka, A.; Morizane, A.; Kamon, M.; Araki, T.; Takahashi, M.P.; et al. Myotonic dystrophy type 1 patient-derived iPSCs for the investigation of CTG repeat instability. *Sci. Rep.* **2017**, *7*, 42522. [CrossRef]
29. Ousterout, D.G.; Kabadi, A.M.; Thakore, P.I.; Majoros, W.H.; Reddy, T.E.; Gersbach, C.A. Multiplex CRISPR/Cas9-based genome editing for correction of dystrophin mutations that cause Duchenne muscular dystrophy. *Nat. Commun.* **2015**, *6*, 6244. [CrossRef]
30. Kim, E.Y.; Page, P.; Dellefave-Castillo, L.M.; McNally, E.M.; Wyatt, E.J. Direct reprogramming of urine-derived cells with inducible MyoD for modeling human muscle disease. *Skelet Muscle* **2016**, *6*, 32. [CrossRef]
31. Takizawa, H.; Hara, Y.; Mizobe, Y.; Ohno, T.; Suzuki, S.; Inoue, K.; Takeshita, E.; Shimizu-Motohashi, Y.; Ishiyama, A.; Hoshino, M.; et al. Modelling Duchenne muscular dystrophy in MYOD1-converted urine-derived cells treated with 3-deazaneplanocin A hydrochloride. *Sci. Rep.* **2019**, *9*, 3807. [CrossRef]
32. He, W.; Zhu, W.; Cao, Q.; Shen, Y.; Zhou, Q.; Yu, P.; Liu, X.; Ma, J.; Li, Y.; Hong, K. Generation of Mesenchymal-Like Stem Cells From Urine in Pediatric Patients. *Transplant. Proc.* **2016**, *48*, 2181–2185. [CrossRef]
33. Zhang, Y.; McNeill, E.; Tian, H.; Soker, S.; Andersson, K.E.; Yoo, J.J.; Atala, A. Urine derived cells are a potential source for urological tissue reconstruction. *J. Urol.* **2008**, *180*, 2226–2233. [CrossRef] [PubMed]
34. Passier, R.; Orlova, V.; Mummery, C. Complex Tissue and Disease Modeling using hiPSCs. *Cell Stem Cell* **2016**, *18*, 309–321. [CrossRef]
35. Langer, R.; Vacanti, J.P. Tissue engineering. *Science* **1993**, *260*, 920–926. [CrossRef]

36. Discher, D.E.; Janmey, P.; Wang, Y.L. Tissue cells feel and respond to the stiffness of their substrate. *Science* **2005**, *310*, 1139–1143. [CrossRef]
37. Schmeichel, K.L.; Bissell, M.J. Modeling tissue-specific signaling and organ function in three dimensions. *J. Cell Sci.* **2003**, *116*, 2377–2388. [CrossRef] [PubMed]
38. Chiron, S.; Tomczak, C.; Duperray, A.; Laine, J.; Bonne, G.; Eder, A.; Hansen, A.; Eschenhagen, T.; Verdier, C.; Coirault, C. Complex interactions between human myoblasts and the surrounding 3D fibrin-based matrix. *PLoS ONE* **2012**, *7*, e36173. [CrossRef]
39. Fuoco, C.; Rizzi, R.; Biondo, A.; Longa, E.; Mascaro, A.; Shapira-Schweitzer, K.; Kossovar, O.; Benedetti, S.; Salvatori, M.L.; Santoleri, S.; et al. In vivo generation of a mature and functional artificial skeletal muscle. *EMBO Mol. Med.* **2015**, *7*, 411–422. [CrossRef] [PubMed]
40. Madden, L.; Juhas, M.; Kraus, W.E.; Truskey, G.A.; Bursac, N. Bioengineered human myobundles mimic clinical responses of skeletal muscle to drugs. *eLife* **2015**, *4*, e04885. [CrossRef]
41. Powell, C.; Shansky, J.; Del Tatto, M.; Forman, D.E.; Hennessey, J.; Sullivan, K.; Zielinski, B.A.; Vandenburgh, H.H. Tissue-engineered human bioartificial muscles expressing a foreign recombinant protein for gene therapy. *Hum. Gene Ther.* **1999**, *10*, 565–577. [CrossRef] [PubMed]
42. Quarta, M.; Cromie, M.; Chacon, R.; Blonigan, J.; Garcia, V.; Akimenko, I.; Hamer, M.; Paine, P.; Stok, M.; Shrager, J.B.; et al. Bioengineered constructs combined with exercise enhance stem cell-mediated treatment of volumetric muscle loss. *Nat. Commun.* **2017**, *8*, 15613. [CrossRef]
43. Tchao, J.; Kim, J.J.; Lin, B.; Salama, G.; Lo, C.W.; Yang, L.; Tobita, K. Engineered Human Muscle Tissue from Skeletal Muscle Derived Stem Cells and Induced Pluripotent Stem Cell Derived Cardiac Cells. *Int. J. Tissue Eng.* **2013**, *2013*, 198762. [CrossRef] [PubMed]
44. Rao, L.; Qian, Y.; Khodabukus, A.; Ribar, T.; Bursac, N. Engineering human pluripotent stem cells into a functional skeletal muscle tissue. *Nat. Commun.* **2018**, *9*, 126. [CrossRef] [PubMed]
45. Maffioletti, S.M.; Sarcar, S.; Henderson, A.B.H.; Mannhardt, I.; Pinton, L.; Moyle, L.A.; Steele-Stallard, H.; Cappellari, O.; Wells, K.E.; Ferrari, G.; et al. Three-Dimensional Human iPSC-Derived Artificial Skeletal Muscles Model Muscular Dystrophies and Enable Multilineage Tissue Engineering. *Cell Rep.* **2018**, *23*, 899–908. [CrossRef] [PubMed]
46. Christov, C.; Chretien, F.; Abou-Khalil, R.; Bassez, G.; Vallet, G.; Authier, F.J.; Bassaglia, Y.; Shinin, V.; Tajbakhsh, S.; Chazaud, B.; et al. Muscle satellite cells and endothelial cells: Close neighbors and privileged partners. *Mol. Biol. Cell* **2007**, *18*, 1397–1409. [CrossRef] [PubMed]
47. Ecob-Prince, M.S.; Jenkison, M.; Butler-Browne, G.S.; Whalen, R.G. Neonatal and adult myosin heavy chain isoforms in a nerve-muscle culture system. *J. Cell Biol.* **1986**, *103*, 995–1005. [CrossRef]
48. Kostallari, E.; Baba-Amer, Y.; Alonso-Martin, S.; Ngoh, P.; Relaix, F.; Lafuste, P.; Gherardi, R.K. Pericytes in the myovascular niche promote post-natal myofiber growth and satellite cell quiescence. *Development* **2015**, *142*, 1242–1253. [CrossRef]
49. Perry, L.; Flugelman, M.Y.; Levenberg, S. Elderly Patient-Derived Endothelial Cells for Vascularization of Engineered Muscle. *Mol. Ther.* **2017**, *25*, 935–948. [CrossRef]
50. Qian, X.; Nguyen, H.N.; Song, M.M.; Hadiono, C.; Ogden, S.C.; Hammack, C.; Yao, B.; Hamersky, G.R.; Jacob, F.; Zhong, C.; et al. Brain-Region-Specific Organoids Using Mini-bioreactors for Modeling ZIKV Exposure. *Cell* **2016**, *165*, 1238–1254. [CrossRef]
51. Low, L.A.; Mummery, C.; Berridge, B.R.; Austin, C.P.; Tagle, D.A. Organs-on-chips: Into the next decade. *Nat. Rev. Drug Discov.* **2020**. [CrossRef] [PubMed]
52. Kim, H.J.; Huh, D.; Hamilton, G.; Ingber, D.E. Human gut-on-a-chip inhabited by microbial flora that experiences intestinal peristalsis-like motions and flow. *Lab Chip* **2012**, *12*, 2165–2174. [CrossRef] [PubMed]
53. Maoz, B.M.; Herland, A.; Henry, O.Y.F.; Leineweber, W.D.; Yadid, M.; Doyle, J.; Mannix, R.; Kujala, V.J.; FitzGerald, E.A.; Parker, K.K.; et al. Organs-on-Chips with combined multi-electrode array and transepithelial electrical resistance measurement capabilities. *Lab Chip* **2017**, *17*, 2294–2302. [CrossRef]
54. Herland, A.; van der Meer, A.D.; FitzGerald, E.A.; Park, T.E.; Sleeboom, J.J.; Ingber, D.E. Distinct Contributions of Astrocytes and Pericytes to Neuroinflammation Identified in a 3D Human Blood-Brain Barrier on a Chip. *PLoS ONE* **2016**, *11*, e0150360. [CrossRef]

55. Musah, S.; Mammoto, A.; Ferrante, T.C.; Jeanty, S.S.F.; Hirano-Kobayashi, M.; Mammoto, T.; Roberts, K.; Chung, S.; Novak, R.; Ingram, M.; et al. Mature induced-pluripotent-stem-cell-derived human podocytes reconstitute kidney glomerular-capillary-wall function on a chip. *Nat. Biomed. Eng.* **2017**, *1*. [CrossRef] [PubMed]
56. Nesmith, A.P.; Wagner, M.A.; Pasqualini, F.S.; O'Connor, B.B.; Pincus, M.J.; August, P.R.; Parker, K.K. A human in vitro model of Duchenne muscular dystrophy muscle formation and contractility. *J. Cell Biol.* **2016**, *215*, 47–56. [CrossRef] [PubMed]
57. Agrawal, G.; Aung, A.; Varghese, S. Skeletal muscle-on-a-chip: An in vitro model to evaluate tissue formation and injury. *Lab Chip* **2017**, *17*, 3447–3461. [CrossRef]
58. Uzel, S.G.; Platt, R.J.; Subramanian, V.; Pearl, T.M.; Rowlands, C.J.; Chan, V.; Boyer, L.A.; So, P.T.; Kamm, R.D. Microfluidic device for the formation of optically excitable, three-dimensional, compartmentalized motor units. *Sci. Adv.* **2016**, *2*, e1501429. [CrossRef]
59. Pellegrini, C.; Zulian, A.; Gualandi, F.; Manzati, E.; Merlini, L.; Michelini, M.E.; Benassi, L.; Marmiroli, S.; Ferlini, A.; Sabatelli, P.; et al. Melanocytes–A novel tool to study mitochondrial dysfunction in Duchenne muscular dystrophy. *J. Cell Physiol.* **2013**, *228*, 1323–1331. [CrossRef]
60. Tyers, L.; Davids, L.M.; Wilmshurst, J.M.; Esterhuizen, A.I. Skin cells for use in an alternate diagnostic method for Duchenne muscular dystrophy. *Neuromuscul. Disord.* **2018**, *28*, 553–563. [CrossRef]
61. Lampe, A.K.; Bushby, K.M. Collagen VI related muscle disorders. *J. Med. Genet.* **2005**, *42*, 673–685. [CrossRef] [PubMed]
62. Zulian, A.; Tagliavini, F.; Rizzo, E.; Pellegrini, C.; Sardone, F.; Zini, N.; Maraldi, N.M.; Santi, S.; Faldini, C.; Merlini, L.; et al. Melanocytes from Patients Affected by Ullrich Congenital Muscular Dystrophy and Bethlem Myopathy have Dysfunctional Mitochondria That Can be Rescued with Cyclophilin Inhibitors. *Front. Aging Neurosci.* **2014**, *6*, 324. [CrossRef]
63. Waterston, R.H.; Lindblad-Toh, K.; Birney, E.; Rogers, J.; Abril, J.F.; Agarwal, P.; Agarwala, R.; Ainscough, R.; Alexandersson, M.; An, P.; et al. Initial sequencing and comparative analysis of the mouse genome. *Nature* **2002**, *420*, 520–562. [CrossRef] [PubMed]
64. Lee, Y.; Rio, D.C. Mechanisms and Regulation of Alternative Pre-mRNA Splicing. *Annu. Rev. Biochem.* **2015**, *84*, 291–323. [CrossRef]
65. Deveson, I.W.; Brunck, M.E.; Blackburn, J.; Tseng, E.; Hon, T.; Clark, T.A.; Clark, M.B.; Crawford, J.; Dinger, M.E.; Nielsen, L.K.; et al. Universal Alternative Splicing of Noncoding Exons. *Cell Syst.* **2018**, *6*, 245–255.e5. [CrossRef]
66. Seok, J.; Warren, H.S.; Cuenca, A.G.; Mindrinos, M.N.; Baker, H.V.; Xu, W.; Richards, D.R.; McDonald-Smith, G.P.; Gao, H.; Hennessy, L.; et al. Genomic responses in mouse models poorly mimic human inflammatory diseases. *Proc. Natl. Acad. Sci. USA* **2013**, *110*, 3507–3512. [CrossRef]
67. Brundin, P.; Nilsson, O.G.; Strecker, R.E.; Lindvall, O.; Astedt, B.; Bjorklund, A. Behavioural effects of human fetal dopamine neurons grafted in a rat model of Parkinson's disease. *Exp. Brain Res.* **1986**, *65*, 235–240. [CrossRef]
68. Gordon, J.W.; Ruddle, F.H. Integration and stable germ line transmission of genes injected into mouse pronuclei. *Science* **1981**, *214*, 1244–1246. [CrossRef]
69. Gumpel, M.; Lachapelle, F.; Gansmuller, A.; Baulac, M.; Baron van Evercooren, A.; Baumann, N. Transplantation of human embryonic oligodendrocytes into shiverer brain. *Ann. N. Y. Acad. Sci.* **1987**, *495*, 71–85. [CrossRef] [PubMed]
70. Stromberg, I.; Bygdeman, M.; Goldstein, M.; Seiger, A.; Olson, L. Human fetal substantia nigra grafted to the dopamine-denervated striatum of immunosuppressed rats: Evidence for functional reinnervation. *Neurosci. Lett.* **1986**, *71*, 271–276. [CrossRef]
71. Schoch, K.M.; Miller, T.M. Antisense Oligonucleotides: Translation from Mouse Models to Human Neurodegenerative Diseases. *Neuron* **2017**, *94*, 1056–1070. [CrossRef] [PubMed]
72. Hua, Y.; Sahashi, K.; Rigo, F.; Hung, G.; Horev, G.; Bennett, C.F.; Krainer, A.R. Peripheral SMN restoration is essential for long-term rescue of a severe spinal muscular atrophy mouse model. *Nature* **2011**, *478*, 123–126. [CrossRef]
73. Passini, M.A.; Bu, J.; Richards, A.M.; Kinnecom, C.; Sardi, S.P.; Stanek, L.M.; Hua, Y.; Rigo, F.; Matson, J.; Hung, G.; et al. Antisense oligonucleotides delivered to the mouse CNS ameliorate symptoms of severe spinal muscular atrophy. *Sci. Transl. Med.* **2011**, *3*, 72ra18. [CrossRef] [PubMed]

74. Porensky, P.N.; Mitrpant, C.; McGovern, V.L.; Bevan, A.K.; Foust, K.D.; Kaspar, B.K.; Wilton, S.D.; Burghes, A.H. A single administration of morpholino antisense oligomer rescues spinal muscular atrophy in mouse. *Hum. Mol. Genet.* **2012**, *21*, 1625–1638. [CrossRef] [PubMed]
75. Williams, J.H.; Schray, R.C.; Patterson, C.A.; Ayitey, S.O.; Tallent, M.K.; Lutz, G.J. Oligonucleotide-mediated survival of motor neuron protein expression in CNS improves phenotype in a mouse model of spinal muscular atrophy. *J. Neurosci.* **2009**, *29*, 7633–7638. [CrossRef] [PubMed]
76. Coome, L.A.; Swift-Gallant, A.; Ramzan, F.; Melhuish Beaupre, L.; Brkic, T.; Monks, D.A. Neural androgen receptor overexpression affects cell number in the spinal nucleus of the bulbocavernosus. *J. Neuroendocrinol.* **2017**, *29*, 12515. [CrossRef] [PubMed]
77. De Giorgio, F.; Maduro, C.; Fisher, E.M.C.; Acevedo-Arozena, A. Transgenic and physiological mouse models give insights into different aspects of amyotrophic lateral sclerosis. *Dis. Model. Mech.* **2019**, *12*. [CrossRef]
78. Cong, L.; Ran, F.A.; Cox, D.; Lin, S.; Barretto, R.; Habib, N.; Hsu, P.D.; Wu, X.; Jiang, W.; Marraffini, L.A.; et al. Multiplex genome engineering using CRISPR/Cas systems. *Science* **2013**, *339*, 819–823. [CrossRef]
79. Mali, P.; Esvelt, K.M.; Church, G.M. Cas9 as a versatile tool for engineering biology. *Nat. Methods* **2013**, *10*, 957–963. [CrossRef]
80. Sicinski, P.; Geng, Y.; Ryder-Cook, A.S.; Barnard, E.A.; Darlison, M.G.; Barnard, P.J. The molecular basis of muscular dystrophy in the mdx mouse: A point mutation. *Science* **1989**, *244*, 1578–1580. [CrossRef]
81. Deconinck, A.E.; Rafael, J.A.; Skinner, J.A.; Brown, S.C.; Potter, A.C.; Metzinger, L.; Watt, D.J.; Dickson, J.G.; Tinsley, J.M.; Davies, K.E. Utrophin-dystrophin-deficient mice as a model for Duchenne muscular dystrophy. *Cell* **1997**, *90*, 717–727. [CrossRef]
82. Kim, K.; Ryu, S.M.; Kim, S.T.; Baek, G.; Kim, D.; Lim, K.; Chung, E.; Kim, S.; Kim, J.S. Highly efficient RNA-guided base editing in mouse embryos. *Nat. Biotechnol.* **2017**, *35*, 435–437. [CrossRef] [PubMed]
83. Amoasii, L.; Long, C.; Li, H.; Mireault, A.A.; Shelton, J.M.; Sanchez-Ortiz, E.; McAnally, J.R.; Bhattacharyya, S.; Schmidt, F.; Grimm, D.; et al. Single-cut genome editing restores dystrophin expression in a new mouse model of muscular dystrophy. *Sci. Transl. Med.* **2017**, *9*, 418. [CrossRef] [PubMed]
84. Young, C.S.; Mokhonova, E.; Quinonez, M.; Pyle, A.D.; Spencer, M.J. Creation of a Novel Humanized Dystrophic Mouse Model of Duchenne Muscular Dystrophy and Application of a CRISPR/Cas9 Gene Editing Therapy. *J. Neuromuscul. Dis.* **2017**, *4*, 139–145. [CrossRef] [PubMed]
85. Koo, T.; Lu-Nguyen, N.B.; Malerba, A.; Kim, E.; Kim, D.; Cappellari, O.; Cho, H.Y.; Dickson, G.; Popplewell, L.; Kim, J.S. Functional Rescue of Dystrophin Deficiency in Mice Caused by Frameshift Mutations Using Campylobacter jejuni Cas9. *Mol. Ther.* **2018**, *26*, 1529–1538. [CrossRef]
86. Min, Y.L.; Li, H.; Rodriguez-Caycedo, C.; Mireault, A.A.; Huang, J.; Shelton, J.M.; McAnally, J.R.; Amoasii, L.; Mammen, P.P.A.; Bassel-Duby, R.; et al. CRISPR-Cas9 corrects Duchenne muscular dystrophy exon 44 deletion mutations in mice and human cells. *Sci. Adv.* **2019**, *5*, eaav4324. [CrossRef] [PubMed]
87. Egorova, T.V.; Zotova, E.D.; Reshetov, D.A.; Polikarpova, A.V.; Vassilieva, S.G.; Vlodavets, D.V.; Gavrilov, A.A.; Ulianov, S.V.; Buchman, V.L.; Deykin, A.V. CRISPR/Cas9-generated mouse model of Duchenne muscular dystrophy recapitulating a newly identified large 430 kb deletion in the human DMD gene. *Dis. Model. Mech.* **2019**, *12*. [CrossRef]
88. Amoasii, L.; Li, H.; Zhang, Y.; Min, Y.L.; Sanchez-Ortiz, E.; Shelton, J.M.; Long, C.; Mireault, A.A.; Bhattacharyya, S.; McAnally, J.R.; et al. In vivo non-invasive monitoring of dystrophin correction in a new Duchenne muscular dystrophy reporter mouse. *Nat. Commun.* **2019**, *10*, 4537. [CrossRef]
89. Sittig, L.J.; Carbonetto, P.; Engel, K.A.; Krauss, K.S.; Barrios-Camacho, C.M.; Palmer, A.A. Genetic Background Limits Generalizability of Genotype-Phenotype Relationships. *Neuron* **2016**, *91*, 1253–1259. [CrossRef] [PubMed]
90. Espuny-Camacho, I.; Arranz, A.M.; Fiers, M.; Snellinx, A.; Ando, K.; Munck, S.; Bonnefont, J.; Lambot, L.; Corthout, N.; Omodho, L.; et al. Hallmarks of Alzheimer's Disease in Stem-Cell-Derived Human Neurons Transplanted into Mouse Brain. *Neuron* **2017**, *93*, 1066–1081. [CrossRef] [PubMed]
91. Xu, D.; Peltz, G. Can Humanized Mice Predict Drug "Behavior" in Humans? *Annu. Rev. Pharmacol. Toxicol.* **2016**, *56*, 323–338. [CrossRef]
92. Greener, M.J.; Roberts, R.G. Conservation of components of the dystrophin complex in Drosophila. *FEBS Lett.* **2000**, *482*, 13–18. [CrossRef]
93. Neuman, S.; Kaban, A.; Volk, T.; Yaffe, D.; Nudel, U. The dystrophin / utrophin homologues in Drosophila and in sea urchin. *Gene* **2001**, *263*, 17–29. [CrossRef]

94. Dekkers, L.C.; van der Plas, M.C.; van Loenen, P.B.; den Dunnen, J.T.; van Ommen, G.J.; Fradkin, L.G.; Noordermeer, J.N. Embryonic expression patterns of the Drosophila dystrophin-associated glycoprotein complex orthologs. *Gene Exp. Patterns* **2004**, *4*, 153–159. [CrossRef] [PubMed]
95. van der Plas, M.C.; Pilgram, G.S.; Plomp, J.J.; de Jong, A.; Fradkin, L.G.; Noordermeer, J.N. Dystrophin is required for appropriate retrograde control of neurotransmitter release at the Drosophila neuromuscular junction. *J. Neurosci.* **2006**, *26*, 333–344. [CrossRef]
96. Shcherbata, H.R.; Yatsenko, A.S.; Patterson, L.; Sood, V.D.; Nudel, U.; Yaffe, D.; Baker, D.; Ruohola-Baker, H. Dissecting muscle and neuronal disorders in a Drosophila model of muscular dystrophy. *EMBO J.* **2007**, *26*, 481–493. [CrossRef]
97. Lin, Y.Y. Muscle diseases in the zebrafish. *Neuromuscul. Disord.* **2012**, *22*, 673–684. [CrossRef] [PubMed]
98. Howe, K.; Clark, M.D.; Torroja, C.F.; Torrance, J.; Berthelot, C.; Muffato, M.; Collins, J.E.; Humphray, S.; McLaren, K.; Matthews, L.; et al. The zebrafish reference genome sequence and its relationship to the human genome. *Nature* **2013**, *496*, 498–503. [CrossRef]
99. Li, M.; Hromowyk, K.J.; Amacher, S.L.; Currie, P.D. Muscular dystrophy modeling in zebrafish. *Methods Cell Biol.* **2017**, *138*, 347–380. [CrossRef]
100. Gibbs, E.M.; Horstick, E.J.; Dowling, J.J. Swimming into prominence: The zebrafish as a valuable tool for studying human myopathies and muscular dystrophies. *FEBS J.* **2013**, *280*, 4187–4197. [CrossRef]
101. Goody, M.F.; Carter, E.V.; Kilroy, E.A.; Maves, L.; Henry, C.A. "Muscling" Throughout Life: Integrating Studies of Muscle Development, Homeostasis, and Disease in Zebrafish. *Curr. Top. Dev. Biol.* **2017**, *124*, 197–234. [CrossRef] [PubMed]
102. Steffen, L.S.; Guyon, J.R.; Vogel, E.D.; Beltre, R.; Pusack, T.J.; Zhou, Y.; Zon, L.I.; Kunkel, L.M. Zebrafish orthologs of human muscular dystrophy genes. *BMC Genom.* **2007**, *8*, 79. [CrossRef] [PubMed]
103. Widrick, J.J.; Kawahara, G.; Alexander, M.S.; Beggs, A.H.; Kunkel, L.M. Discovery of Novel Therapeutics for Muscular Dystrophies using Zebrafish Phenotypic Screens. *J. Neuromuscul. Dis.* **2019**, *6*, 271–287. [CrossRef] [PubMed]
104. Kell, D.B. Finding novel pharmaceuticals in the systems biology era using multiple effective drug targets, phenotypic screening and knowledge of transporters: Where drug discovery went wrong and how to fix it. *FEBS J.* **2013**, *280*, 5957–5980. [CrossRef]
105. Kawahara, G.; Karpf, J.A.; Myers, J.A.; Alexander, M.S.; Guyon, J.R.; Kunkel, L.M. Drug screening in a zebrafish model of Duchenne muscular dystrophy. *Proc. Natl. Acad. Sci. USA* **2011**, *108*, 5331–5336. [CrossRef]
106. Schiavone, M.; Zulian, A.; Menazza, S.; Petronilli, V.; Argenton, F.; Merlini, L.; Sabatelli, P.; Bernardi, P. Alisporivir rescues defective mitochondrial respiration in Duchenne muscular dystrophy. *Pharmacol. Res.* **2017**, *125*, 122–131. [CrossRef]
107. Telfer, W.R.; Busta, A.S.; Bonnemann, C.G.; Feldman, E.L.; Dowling, J.J. Zebrafish models of collagen VI-related myopathies. *Hum. Mol. Genet.* **2010**, *19*, 2433–2444. [CrossRef]
108. Zulian, A.; Rizzo, E.; Schiavone, M.; Palma, E.; Tagliavini, F.; Blaauw, B.; Merlini, L.; Maraldi, N.M.; Sabatelli, P.; Braghetta, P.; et al. NIM811, a cyclophilin inhibitor without immunosuppressive activity, is beneficial in collagen VI congenital muscular dystrophy models. *Hum. Mol. Genet.* **2014**, *23*, 5353–5363. [CrossRef]
109. Sileikyte, J.; Devereaux, J.; de Jong, J.; Schiavone, M.; Jones, K.; Nilsen, A.; Bernardi, P.; Forte, M.; Cohen, M.S. Second-Generation Inhibitors of the Mitochondrial Permeability Transition Pore with Improved Plasma Stability. *ChemMedChem* **2019**, *14*, 1771–1782. [CrossRef]
110. Hirata, H.; Watanabe, T.; Hatakeyama, J.; Sprague, S.M.; Saint-Amant, L.; Nagashima, A.; Cui, W.W.; Zhou, W.; Kuwada, J.Y. Zebrafish relatively relaxed mutants have a ryanodine receptor defect, show slow swimming and provide a model of multi-minicore disease. *Development* **2007**, *134*, 2771–2781. [CrossRef]
111. Dowling, J.J.; Arbogast, S.; Hur, J.; Nelson, D.D.; McEvoy, A.; Waugh, T.; Marty, I.; Lunardi, J.; Brooks, S.V.; Kuwada, J.Y.; et al. Oxidative stress and successful antioxidant treatment in models of RYR1-related myopathy. *Brain* **2012**, *135*, 1115–1127. [CrossRef] [PubMed]
112. Culetto, E.; Sattelle, D.B. A role for Caenorhabditis elegans in understanding the function and interactions of human disease genes. *Hum. Mol. Genet.* **2000**, *9*, 869–877. [CrossRef]
113. Consortium, C.E.S. Genome sequence of the nematode C. elegans: A platform for investigating biology. *Science* **1998**, *282*, 2012–2018. [CrossRef]

114. Ono, S. The Caenorhabditis elegans unc-78 gene encodes a homologue of actin-interacting protein 1 required for organized assembly of muscle actin filaments. *J. Cell Biol.* **2001**, *152*, 1313–1319. [CrossRef] [PubMed]
115. Gieseler, K.; Grisoni, K.; Segalat, L. Genetic suppression of phenotypes arising from mutations in dystrophin-related genes in Caenorhabditis elegans. *Curr. Biol.* **2000**, *10*, 1092–1097. [CrossRef]
116. Gaud, A.; Simon, J.M.; Witzel, T.; Carre-Pierrat, M.; Wermuth, C.G.; Segalat, L. Prednisone reduces muscle degeneration in dystrophin-deficient Caenorhabditis elegans. *Neuromuscul. Disord.* **2004**, *14*, 365–370. [CrossRef]
117. Carre-Pierrat, M.; Mariol, M.C.; Chambonnier, L.; Laugraud, A.; Heskia, F.; Giacomotto, J.; Segalat, L. Blocking of striated muscle degeneration by serotonin in C. elegans. *J. Muscle Res. Cell Motil.* **2006**, *27*, 253–258. [CrossRef]
118. Hwang, H.; Barnes, D.E.; Matsunaga, Y.; Benian, G.M.; Ono, S.; Lu, H. Muscle contraction phenotypic analysis enabled by optogenetics reveals functional relationships of sarcomere components in Caenorhabditis elegans. *Sci. Rep.* **2016**, *6*, 19900. [CrossRef]
119. Virgilio, K.M.; Martin, K.S.; Peirce, S.M.; Blemker, S.S. Agent-based model illustrates the role of the microenvironment in regeneration in healthy and mdx skeletal muscle. *J. Appl. Physiol.* **2018**, *125*, 1424–1439. [CrossRef]
120. Martin, K.S.; Blemker, S.S.; Peirce, S.M. Agent-based computational model investigates muscle-specific responses to disuse-induced atrophy. *J. Appl. Physiol.* **2015**, *118*, 1299–1309. [CrossRef]

Journal of
Personalized Medicine

MDPI

Review

An Omics View of Emery–Dreifuss Muscular Dystrophy

Nicolas Vignier and Antoine Muchir *

INSERM, Center of Research in Myology, Institute of Myology, Sorbonne University, 75013 Paris, France; n.vignier@institut-myologie.org

* Correspondence: a.muchir@institut-myologie.org; Tel.: +33-1-4216-5705

Received: 13 May 2020; Accepted: 12 June 2020; Published: 15 June 2020

Abstract: Recent progress in Omics technologies has started to empower personalized healthcare development at a thorough biomolecular level. Omics have subsidized medical breakthroughs that have started to enter clinical proceedings. The use of this scientific know-how has surfaced as a way to provide a more far-reaching view of the biological mechanisms behind diseases. This review will focus on the discoveries made using Omics and the utility of these approaches for Emery–Dreifuss muscular dystrophy.

Keywords: *LMNA*; Emery–Dreifuss muscular dystrophy; Omics

---

1. Introduction

To understand the complexity of systems biology, Omics' technologies adopt a holistic view. In this, in opposition to hypothesis-generating experiments, no rationale is known, but instead, biological inputs are acquired and analyzed to delineate a hypothesis that can be then tested. Omics technology can be used not only to decipher physiological conditions but also in disease states, where they have a key role in diagnosis, as well as promoting our knowledge of the development of diseases [1]. Omics approaches to conditions, such as muscular dystrophies, are being used for small molecule therapy discovery by isolating innovative targets for drug development [2]. The scope of this review is to provide an overview of the Omics approaches and their application in Emery–Dreifuss muscular dystrophy research.

2. Emery–Dreifuss Muscular Dystrophy

Muscular dystrophies are characterized by the progressive weakness and degeneration of the skeletal muscle system, which may or may not be associated with cardiac impairment, leading to loss of mobility, and swallowing and respiratory difficulties. Death originates from respiratory defects or heart failure. Muscular dystrophies are a heterogeneous group of inherited disorders, and they differ in the distribution of affected muscles, the rate of muscle weakness progression and the age of onset [3]. The development of molecular genetic mapping techniques has shown that these disorders are genetically heterogeneous, and more than 50 genes have been identified as causing muscular dystrophies [4].

In the 1960s, an X-linked muscular dystrophy associated with contractures, which was first diagnosed as a benign variant of Duchenne muscular dystrophy, was reported [5,6]. In the 1980s, Alan Emery re-investigated the original family and reported that cardiomyopathy was a significant feature of the disease, which was thereafter called Emery–Dreifuss muscular dystrophy (EDMD) [7]. In EDMD, the onset of symptoms occurs within the first decade of life [7]. Contractures of the elbows, neck extensor muscles and Achilles' tendons appear to be the first symptoms of the disease, and occur before muscle weakness and wasting. The progressive muscle degeneration begins during the end of the second decade of life, in a humeroperoneal distribution. Cardiac alteration begins during the teenage years, with no link to the severity of the muscular dystrophy [7–10]. Over time, dilated cardiomyopathy develops, and is associated with severe ventricular tachydysrrhythmias. Sudden cardiac death is frequent, and an implantable defibrillator can be lifesaving [10,11].

3. Application of Omics Approaches for Emery-Dreifuss Muscular Dystrophy

3.1. Omics and Diagnosis

In the 1990s, a positional cloning study—a technique for the positioning of a trait-associated gene within the genome—showed that mutations in *EMD*, and *LMNA* cause the X-linked [12] and autosomal dominant [13] forms of EDMD, respectively [14]. EMD encodes emerin, which is a transmembrane protein of the nuclear envelope. *LMNA* encodes nuclear lamins A and C, which are intermediate filament proteins associated with the nuclear envelope. Until recently, genetic screening for EDMD was performed with Sanger sequencing of the exons and intron-exon regions of both the *EMD* and *LMNA* genes. Since then, diagnosis of EDMD has been made available if the clinical signs are suggestive or if a family member is known to have EDMD. However, classical DNA sequencing methods have many shortcomings (time and cost) hampering their use for diagnosis. By contrast, as introduced in 2005, next-generation sequencing is a potent novel technology and a cost-effective method that has completely revolutionized the field [15,16]. Full exome sequencing is routinely used in clinical diagnostic laboratories to identify pathogenic variants in a given patient at a reasonable cost [17–20]. A new next-generation sequencing approach to identify potential new candidate EDMD genes has also recently been tested [21].

Subsequently, *LMNA* mutations have been shown to cause other striated muscle diseases, i.e., dilated cardiomyopathy [22], limb-girdle muscular dystrophy type 1B [23] and congenital muscular dystrophy [24]. Limb-girdle muscular dystrophy type 1B and dilated cardiomyopathy can occur in the same families as subjects with EDMD, and can be therefore be considered variants of the same clinical entity. This supports the concept that modifier genes arouse the severity and the peculiar symptoms. To map the modifier locus, microsatellite markers were genotyped in a large French family, where patients carrying the same *LMNA* mutation exhibited phenotypic variability [25]. The linked DNA region harbors two candidate modifier genes, *DES* and *MYL1*, encoding desmin and light chain of myosin, respectively, thus providing insights for the natural history and the physiopathology of EDMD.

3.2. Omics and Abnormal Cellular Signaling

The mechanisms by which mutations in *EMD* and *LMNA* cause muscular dystrophies are poorly understood. A few models have been proposed to explain the physiopathology of EDMD [26]. The model called the 'mechanical stress hypothesis' relies on the premise that striated muscle is steadily subjected to mechanical strains. Abnormalities in nuclear envelope composition may imply a weakening of the nucleus, which could represent an initial step in the chain of events leading to EDMD.

The ambition of transcriptomics studies is to pinpoint genome-wide changes and to expose coordinately organized gene networks [27,28]. Gene expression profiles associated with EDMD have been studied in several experimental models using various technologies. Mouse models have been helpful in deciphering mechanisms involved in the pathogenesis of EDMD, as well as for opening pharmacological therapies perspectives. The development of *Lmna*$^{-/-}$ mice by Sullivan and colleagues was the first animal model of the disease [29]. The mice expressing a truncated peptide, lamin A delta8-11 [30], develop cardiomyopathy and skeletal muscle wasting reminiscent of human EDMD. Then, other mouse models of EDMD were generated. These are knock-in mice that express A-type lamins with the p.H222P [31], p.deltaK32 [32] and p.N195K [33] residue substitutions. Arimura et al. developed *Lmna* knock-in mice carrying the p.H222P mutation that was identified in the human *LMNA* gene in an EDMD family [31]. This mutation was also chosen because it putatively dramatically altered the coil-coiled organization of A-type lamins, based on in silico analysis. This was the first *Lmna* mouse model mimicking human EDMD from the gene mutation to the clinical symptoms. Two separate groups have generated *Emd* null mice [34,35]. These mice have subtle motor coordination abnormalities, with a prolongation of atrioventricular conduction time in *Emd* null mice [35]. Notwithstanding, these animal models may not reflect the "natural" human condition in terms of physiological mechanism and genetic outlook. Induced pluripotent stem cell (iPSC) technology represents a means to surmount these shortcomings, allowing the generation of any cell type through peculiar differentiation protocols [36]. Tissue-specific in vitro models of EDMD have been created from iPSCs that recapitulate traits of the disease [37–42].

To explore the pathogenesis of cardiomyopathy associated with EDMD, we carried out a genome-wide RNA expression analysis of hearts from *Lmna*$^{p.H222P/H222P}$ mice and *Emd* knockout mice, two mouse models of EDMD. This analysis revealed changes in the expression of genes encoding proteins in mitogen-activated protein (MAP) kinases, Wnt/β-catenin, AKT/mTOR and transforming growth factor (Tgf)-β signaling pathways in the mutated models [43–48]. Using RNA-sequencing technology on *Lmna*$^{-/-}$ mice, Auguste and colleagues showed that the FOXO signaling pathway impacted different signaling pathways, i.e., NFκB, TNFα, P53 and OxPHOS signaling pathways and biological processes, i.e., apoptosis, sustaining the cardiac phenotype associated with EDMD [49]. Furthermore, whole genome expression analysis of the primary cells of EDMD patients showed aberrant activity of unfolded protein response signaling [50]. In a cardiac specific expression of *Lmna* p.D300N, the Marian group showed using bulk RNA sequencing strategy that an increase of DNA damage response/TP53 pathway was contributing to the pathogenesis of cardiomyopathy associated with EDMD [51]. All these datasets led to the hypothesis of a model of how abnormalities of A-type lamins and emerin may lead to EDMD [52]. Using a transcriptomic approach from regenerating skeletal muscle from emerin-deficient mice, Melcon et al. [34] have shown delayed myogenic differentiation, which is regulated by *Rb* and *MyoD* genes. Since then, small molecules have been used by others to rescue the impaired myogenic differentiation in emerin deficiency, which could represent a potential strategy for improving the muscle wasting phenotype seen in EDMD [53]. Hence, abnormalities in satellite cell behavior may be responsible in part for the skeletal muscle disease in EDMD.

The mechanisms that bridge the *LMNA* genetic defects to malignant arrhythmias [54] are unknown. To better understand this phenotype in EDMD, Dr. Wu's group modeled the disease in vitro using patient-specific iPSC-derived cardiomyocytes (iPSC-CMs) carrying *LMNA* frameshift p.K117fs mutation [42]. They showed an abnormal activation of the platelet-derived growth factor (PDGF) signaling in *LMNA* p.K117fs iPSC-CMs. The inhibition of the PDGF signaling improves the arrhythmic phenotype of mutated iPSC-CMs, opening novel therapeutic perspectives for the treatment of EDMD.

All these studies contributed to a better knowledge of the functional and molecular mechanisms of the disease. We can expect that new findings will design applications of iPSC-models to pharmacological testing in striated muscle-specific contexts [55], making the technology available to patients.

3.3. Omics and Chromatin Regulation

Among the models proposed to explain how EDMD phenotypes arise, the 'gene-expression model', posits an effect of mutated lamin A/C on the transcription activity of genes and/or pathways that could impact striated muscle-homeostasis process. According to this model, it has been described that A-type lamins interact with heterochromatin regions called Lamin-Associated Domains (LADs). These LADs play a role for chromatin organization and gene expression regulation [56–59]. It has been shown that the LADs are re-organized in EDMD steering modifications of the epigenetic program, ultimately driving the loss of myogenic differentiation [60]. This has been recently shown in an elegant manner by Bianchi and colleagues on a murine model of EDMD [61]. The authors described an abnormal positioning of polycomb proteins, which are epigenetic repressors involved in cell identity. This causes impairment in self-renewal, deficiency of cell identity and the early exhaustion of the quiescent satellite cell pool, demonstrating that muscular dystrophy in EDMD can be partially caused by epigenetic dysfunctions of muscle stem cells [61]. Mewborn and colleagues showed that the *LMNA* p.E161K mutation perturbed the positioning and compaction of chromosomal domains in primary fibroblasts, resulting in an altered gene expression profile [62]. Another study focused on the organization of LADs in EDMD using a multi-Omics approach (Chip-Seq/RNA-sequencing) in explanted hearts from five patients carrying *LMNA* mutations [62]. LADs were redistributed, ensuing in a functional chromatin state in mutated hearts, suggesting the loss of specific functional chromatin binding. This aberrant distribution impacted both the gene expression profile and CpG methylation. An integrated analysis showed the combined role of LADs and CpG methylation in the regulation of gene expression, and identified numerous transcription factors involved in biological processes such as cell death/survival, cell cycle and metabolism [63]. Bertero and colleagues showed at the same time, using genome-wide chromosome conformation capture (HiC) analyses on iPSC-CMs carrying *LMNA* p.R225X mutation [64], a slight chromatin compartment dysregulation (around 1% of the genome). RNA-sequencing of these altered chromatin domains revealed the abnormal up-regulation of only a handful of genes, among which was *CACNA1A*. This latter encodes a subunit of a calcium channel, whose abnormal expression might partially explain the electrophysiological and contractile aberrations observed in the mutant hiPSC-CMs. The authors showed that pharmacological treatments to prevent both the electrophysiological and contractile alterations helped improve the abnormal phenotypes described in the mutated iPSC-CMs. This suggests that *CACNA1A* may be a good therapeutic target to reverse the cardiac abnormalities in EDMD patients. However, the work by Bertero et al. showed only minor alterations in chromatin compartmentalization in mutated hiPSC-CMs, a finding that challenges the aberrant gene-expression model [64].

3.4. Omics and Metabolism

Contracting incessantly, the heart demands a lot of energy to ensure optimal contractile function. Research has demonstrated that the high requirements of the heart are satisfied by a preference for the oxidation of fatty acids. Studies have demonstrated that the failing heart deviates from its inherent profile and relies heavily on glucose metabolism, primarily achieved by acceleration in glycolysis. To gain further insight into the molecular mechanism ruling this disease, scientists have studied the cardiac metabolic rates in *Lmna*$^{-/-}$ mice. West and colleagues described a targeted metabolomics assay that quantifies metabolites relevant to cardiac metabolism [65]. The assay demonstrates that the *Lmna*$^{-/-}$ mouse heart has decreased metabolites associated with the citric acid cycle and fatty acid oxidation [65]. This corroborated another study, which showed that activated AKT/mTOR signaling reduces tolerance to energy deficits in hearts from *Lmna*$^{p.H222P/H222P}$ mice [45]. A rapamycin analog that blocks AKT/mTOR activity has been used to prevent the progression of cardiomyopathy in *Lmna*$^{p.H222P/H222P}$ mice [45]. These works highlighted that the heart is unable to compensate for increased or fluctuating energy demand and, over time, develops dilated cardiomyopathy in EDMD. This metabolic remodeling probably represents an adaptive cardio-protective mechanism that can help improve contractile function, thus slowing the progression of EDMD and improving prognosis. As such, metabolic modulators, which have the potential to shift myocardial substrate utilization from fatty acids toward glucose metabolism, may have a place in the management of patients. Some of these modulators have already been investigated as treatment for cardiomyopathy, with some beneficial effects [66,67]. It would be relevant to test their efficacy in EDMD.

Moreover, West and colleagues showed increased responses to oxidative stress and reactive oxygen species (ROS) exposure in *Lmna*$^{-/-}$ mouse hearts [65]. ROS are small, short-lived signaling molecules that mediate various cellular responses. Based on this, we recently showed that N-acetyl cysteine treatment reduces cardiac oxidative stress injury and ameliorates contractile dysfunction in *Lmna*$^{p.H222P/H222P}$ mice [68].

3.5. Omics and Biomarkers

Omics are powerful tools to identify diseases' molecular biomarkers. Molecular biomarkers are molecules with particular biophysical properties, the quantities of which are measured in biological samples, and are of critical importance in the support of the clinical diagnosis of pathology, the monitoring of its time course and the evaluation of the impact of therapeutic approaches. Molecular biomarkers have to be quantified from patients in the least invasive manner possible. Circulating molecular biomarkers in body fluids, i.e., blood, plasma, serum or urine, are thus the main interest. Proteomic, transcriptomic and metabolomic analysis have been driven in many diseases to identify such biomarkers [69,70].

To identify circulating microRNA as molecular biomarkers for EDMD, the microRNA transcriptome (miRnome) from plasma of *Lmna*$^{p.H222P/H222P}$ mice was screened [71]. A specific and distinctive microRNA expression profile was identified, and three of the dystromirs (mir-133b, mir-133a and mir-1) were downregulated in these mice. Furthermore, two microRNAs were upregulated (mir-146b and mir-200a) and six microRNAs were downregulated (mir-130a, mir-133a, mir-133b, mir-1, mir-151-3p, mir-339-3p) in *Lmna*$^{p.H222P/H222P}$ mice compared with wild type animals [71].

4. Conclusions

Omics technologies have hastened the identification of genetic mutations associated with EDMD, and have unveiled the existence of rare variants and modifier genes that might establish the phenotypical heterogeneity of the disease. Transcriptomic studies have uncovered alterations in signaling mechanisms causing some of the symptoms in EDMD, but are restricted to experimental models (in vitro and in vivo) with technical and theoretical shortcomings. Furthermore, although the number of parameters being measured has increased with Omics technologies, the number of biological and methodological replicates has not. In addition, because of the large number of measurements and the limited number of subjects, unique problems arise in Omics studies involving statistics and bias. A single Omics technique will only capture changes in a subset of biological cascades; it cannot provide a systemic understanding of the complexity of systems biology. The integration of multiple Omics data sets promises a substantial improvement, through an increase of information and, especially, systemic understanding. Therefore, much work is needed before using this research in the clinic. All of this will depend on integrative collaborations among physicians and scientists that will be essential for major breakthroughs for both the diagnosis and treatment of EDMD.

Author Contributions: Writing—Original Version, A.M.; Writing—Review & Editing, A.M. and N.V.; Funding Acquisition, A.M. All authors have read and agreed to the published version of the manuscript.

Funding: This work was supported by the Association Française contre les Myopathies, the Institut National de la Santé et de la Recherche Médicale and Sorbonne Université.

Conflicts of Interest: The authors declare no conflict of interest.

References

1. Karczewski, K.J.; Snyder, M.P. Integrative omics for health and disease. *Nat. Rev. Genet.* **2018**, *19*, 299–310. [CrossRef]
2. Matthews, H.; Hanison, J.; Nirmalan, N. "Omics"-Informed Drug and Biomarker Discovery: Opportunities, Challenges and Future Perspectives. *Proteomes* **2016**, *4*, 28. [CrossRef]
3. Mercuri, E.; Bönnemann, C.G.; Muntoni, F. Muscular dystrophies. *Lancet* **2019**, *394*, 2025–2038. [CrossRef]
4. Benarroch, L.; Bonne, G.; Rivier, F.; Hamroun, D. The 2020 version of the gene table of neuromuscular disorders (nuclear genome). *Neuromuscul. Disord.* **2019**, *29*, 980–1018. [CrossRef] [PubMed]
5. Dreifuss, F.E.; Hogan, G.R. Survival in x-chromosomal muscular dystrophy. *Neurology* **1961**, *11*, 734–737. [CrossRef]
6. Emery, A.E.; Dreifuss, F.E. Unusual type of benign x-linked muscular dystrophy. *J. Neurol. Neurosurg. Psychiatry* **1966**, *29*, 338–342. [CrossRef] [PubMed]
7. Emery, A.E. X-linked muscular dystrophy with early contractures and cardiomyopathy (Emery-Dreifuss type). *Clin. Genet.* **1987**, *32*, 360–367. [CrossRef] [PubMed]
8. Waters, D.D.; Nutter, D.O.; Hopkins, L.C.; Dorney, E.R. Cardiac features of an unusual X-linked humeroperoneal neuromuscular disease. *N. Engl. J. Med.* **1975**, *293*, 1017–1022. [CrossRef]

9. Bialer, M.G.; Mcdaniel, N.L.; Kelly, T.E. Progression of cardiac disease in emery-dreifuss muscular dystrophy. *Clin. Cardiol. Int. J. Cardiovasc. Dis.* **1991**, *14*, 411. [CrossRef] [PubMed]
10. Bécane, H.M.; Bonne, G.; Varnous, S.; Muchir, A.; Ortega, V.; Hammouda, E.H.; Urtizberea, J.A.; Lavergne, T.; Fardeau, M.; Eymard, B.; et al. High incidence of sudden death with conduction system and myocardial disease due to lamins A and C gene mutation. *Pacing Clin. Electrophysiol.* **2000**, *23*, 1661–1666. [CrossRef]
11. van Berlo, J.H.; de Voogt, W.G.; van der Kooi, A.J.; van Tintelen, J.P.; Bonne, G.; Yaou, R.B.; Duboc, D.; Rossenbacker, T.; Heidbüchel, H.; de Visser, M.; et al. Meta-analysis of clinical characteristics of 299 carriers of LMNA gene mutations: Do lamin A/C mutations portend a high risk of sudden death? *J. Mol. Med.* **2005**, *83*, 79–83. [CrossRef] [PubMed]
12. Bione, S.; Maestrini, E.; Rivella, S.; Mancini, M.; Regis, S.; Romeo, G.; Toniolo, D. Identification of a novel X-linked gene responsible for Emery-Dreifuss muscular dystrophy. *Nat. Genet.* **1994**, *8*, 323–327. [CrossRef] [PubMed]
13. Bonne, G.; Barletta, M.R.D.; Varnous, S.; Bécane, H.-M.; Hammouda, E.-H.; Merlini, L.; Muntoni, F.; Greenberg, C.R.; Gary, F.; Urtizberea, J.-A.; et al. Mutations in the gene encoding lamin A/C cause autosomal dominant Emery-Dreifuss muscular dystrophy. *Nat. Genet.* **1999**, *21*, 285–288. [CrossRef] [PubMed]
14. Bonne, G.; Mercuri, E.; Muchir, A.; Urtizberea, A.; Bécane, H.M.; Recan, D.; Merlini, L.; Wehnert, M.; Boor, R.; Reuner, U.; et al. Clinical and molecular genetic spectrum of autosomal dominant Emery-Dreifuss muscular dystrophy due to mutations of the lamin A/C gene. *Ann. Neurol.* **2000**, *48*, 170–180. [CrossRef]
15. Ropers, H.-H. New perspectives for the elucidation of genetic disorders. *Am. J. Hum. Genet.* **2007**, *81*, 199–207. [CrossRef]
16. Voelkerding, K.V.; Dames, S.A.; Durtschi, J.D. Next-generation sequencing: From basic research to diagnostics. *Clin. Chem.* **2009**, *55*, 641–658. [CrossRef]
17. Roncarati, R.; Viviani Anselmi, C.; Krawitz, P.; Lattanzi, G.; von Kodolitsch, Y.; Perrot, A.; di Pasquale, E.; Papa, L.; Portararo, P.; Columbaro, M.; et al. Doubly heterozygous LMNA and TTN mutations revealed by exome sequencing in a severe form of dilated cardiomyopathy. *Eur. J. Hum. Genet.* **2013**, *21*, 1105–1111. [CrossRef]
18. Park, H.-Y. Hereditary Dilated Cardiomyopathy: Recent Advances in Genetic Diagnostics. *Korean Circ. J.* **2017**, *47*, 291–298. [CrossRef]
19. Fu, Y.; Eisen, H.J. Genetics of Dilated Cardiomyopathy. *Curr. Cardiol. Rep.* **2018**, *20*, 121. [CrossRef]
20. Park, J.; Levin, M.G.; Haggerty, C.M.; Hartzel, D.N.; Judy, R.; Kember, R.L.; Reza, N.; Regeneron Genetics Center; Ritchie, M.D.; Owens, A.T.; et al. A genome-first approach to aggregating rare genetic variants in LMNA for association with electronic health record phenotypes. *Genet. Med.* **2020**, *22*, 102–111. [CrossRef]
21. Meinke, P.; Kerr, A.R.W.; Czapiewski, R.; Heras, J.I.D.L.; Dixon, C.R.; Harris, E.; Kölbel, H.; Muntoni, F.; Schara, U.; Straub, V.; et al. A multistage sequencing strategy pinpoints novel candidate alleles for Emery-Dreifuss muscular dystrophy and supports gene misregulation as its pathomechanism. *EBioMedicine* **2020**, *51*. [CrossRef] [PubMed]
22. Fatkin, D.; MacRae, C.; Sasaki, T.; Wolff, M.R.; Porcu, M.; Frenneaux, M.; Atherton, J.; Vidaillet, H.J., Jr.; Spudich, S.; De Girolami, U.; et al. Missense Mutations in the Rod Domain of the Lamin A/C Gene as Causes of Dilated Cardiomyopathy and Conduction-System Disease. *NEJM* **1999**, *341*, 1715–1724. [CrossRef] [PubMed]
23. Muchir, A.; Bonne, G.; van der Kooi, A.J.; van Meegen, M.; Baas, F.; Bolhuis, P.A.; de Visser, M.; Schwartz, K. Identification of mutations in the gene encoding lamins A/C in autosomal dominant limb girdle muscular dystrophy with atrioventricular conduction disturbances (LGMD1B). *Hum. Mol. Genet.* **2000**, *9*, 1453–1459. [CrossRef] [PubMed]
24. Quijano-Roy, S.; Mbieleu, B.; Bönnemann, C.G.; Jeannet, P.-Y.; Colomer, J.; Clarke, N.F.; Cuisset, J.-M.; Roper, H.; De Meirleir, L.; D'Amico, A.; et al. De novo LMNA mutations cause a new form of congenital muscular dystrophy. *Ann. Neurol.* **2008**, *64*, 177–186. [CrossRef]
25. Granger, B.; Gueneau, L.; Drouin-Garraud, V.; Pedergnana, V.; Gagnon, F.; Ben Yaou, R.; Du Montcel, S.T.; Bonne, G. Modifier locus of the skeletal muscle involvement in Emery-Dreifuss muscular dystrophy. *Hum. Genet.* **2011**, *129*, 149–159. [CrossRef]
26. Worman, H.J.; Bonne, G. "Laminopathies": A wide spectrum of human diseases. *Exp. Cell Res.* **2007**, *313*, 2121–2133. [CrossRef]

27. Raghavachari, N. Microarray technology: Basic methodology and application in clinical research for biomarker discovery in vascular diseases. *Methods Mol. Biol.* **2013**, *1027*, 47–84. [CrossRef]
28. Matkovich, S.J.; Zhang, Y.; Van Booven, D.J.; Dorn, G.W. Deep mRNA sequencing for in vivo functional analysis of cardiac transcriptional regulators: Application to Galphaq. *Circ. Res.* **2010**, *106*, 1459–1467. [CrossRef]
29. Sullivan, T.; Escalante-Alcalde, D.; Bhatt, H.; Anver, M.; Bhat, N.; Nagashima, K.; Stewart, C.L.; Burke, B. Loss of A-type lamin expression compromises nuclear envelope integrity leading to muscular dystrophy. *J. Cell Biol.* **1999**, *147*, 913–920. [CrossRef]
30. Jahn, D.; Schramm, S.; Schnölzer, M.; Heilmann, C.J.; de Koster, C.G.; Schütz, W.; Benavente, R.; Alsheimer, M. A truncated lamin A in the Lmna−/−mouse line. *Nucleus* **2012**, *3*, 463–474. [CrossRef]
31. Arimura, T.; Helbling-Leclerc, A.; Massart, C.; Varnous, S.; Niel, F.; Lacène, E.; Fromes, Y.; Toussaint, M.; Mura, A.-M.; Keller, D.I.; et al. Mouse model carrying H222P-Lmna mutation develops muscular dystrophy and dilated cardiomyopathy similar to human striated muscle laminopathies. *Hum. Mol. Genet.* **2005**, *14*, 155–169. [CrossRef]
32. Bertrand, A.T.; Renou, L.; Papadopoulos, A.; Beuvin, M.; Lacène, E.; Massart, C.; Ottolenghi, C.; Decostre, V.; Maron, S.; Schlossarek, S.; et al. DelK32-lamin A/C has abnormal location and induces incomplete tissue maturation and severe metabolic defects leading to premature death. *Hum. Mol. Genet.* **2012**, *21*, 1037–1048. [CrossRef] [PubMed]
33. Mounkes, L.C.; Kozlov, S.V.; Rottman, J.N.; Stewart, C.L. Expression of an LMNA-N195K variant of A-type lamins results in cardiac conduction defects and death in mice. *Hum. Mol. Genet.* **2005**, *14*, 2167–2180. [CrossRef]
34. Melcon, G.; Kozlov, S.; Cutler, D.A.; Sullivan, T.; Hernandez, L.; Zhao, P.; Mitchell, S.; Nader, G.; Bakay, M.; Rottman, J.N.; et al. Loss of emerin at the nuclear envelope disrupts the Rb1/E2F and MyoD pathways during muscle regeneration. *Hum. Mol. Genet.* **2006**, *15*, 637–651. [CrossRef] [PubMed]
35. Ozawa, R.; Hayashi, Y.K.; Ogawa, M.; Kurokawa, R.; Matsumoto, H.; Noguchi, S.; Nonaka, I.; Nishino, I. Emerin-lacking mice show minimal motor and cardiac dysfunctions with nuclear-associated vacuoles. *Am. J. Pathol.* **2006**, *168*, 907–917. [CrossRef]
36. Takahashi, K.; Yamanaka, S. A decade of transcription factor-mediated reprogramming to pluripotency. *Nat. Rev. Mol. Cell Biol.* **2016**, *17*, 183–193. [CrossRef] [PubMed]
37. Liu, G.-H.; Suzuki, K.; Qu, J.; Sancho-Martinez, I.; Yi, F.; Li, M.; Kumar, S.; Nivet, E.; Kim, J.; Soligalla, R.D.; et al. Targeted gene correction of laminopathy-associated LMNA mutations in patient-specific iPSCs. *Cell Stem Cell* **2011**, *8*, 688–694. [CrossRef] [PubMed]
38. Lee, Y.-K.; Lau, Y.-M.; Cai, Z.-J.; Lai, W.-H.; Wong, L.-Y.; Tse, H.-F.; Ng, K.-M.; Siu, C.-W. Modeling Treatment Response for Lamin A/C Related Dilated Cardiomyopathy in Human Induced Pluripotent Stem Cells. *J. Am. Heart Assoc.* **2017**, *6*. [CrossRef] [PubMed]
39. Siu, C.-W.; Lee, Y.-K.; Ho, J.C.-Y.; Lai, W.-H.; Chan, Y.-C.; Ng, K.-M.; Wong, L.-Y.; Au, K.-W.; Lau, Y.-M.; Zhang, J.; et al. Modeling of lamin A/C mutation premature cardiac aging using patient-specific induced pluripotent stem cells. *Aging (Albany NY)* **2012**, *4*, 803–822. [CrossRef] [PubMed]
40. Steele-Stallard, H.B.; Pinton, L.; Sarcar, S.; Ozdemir, T.; Maffioletti, S.M.; Zammit, P.S.; Tedesco, F.S. Modeling Skeletal Muscle Laminopathies Using Human Induced Pluripotent Stem Cells Carrying Pathogenic LMNA Mutations. *Front. Physiol.* **2018**, *9*, 1332. [CrossRef]
41. Salvarani, N.; Crasto, S.; Miragoli, M.; Bertero, A.; Paulis, M.; Kunderfranco, P.; Serio, S.; Forni, A.; Lucarelli, C.; Dal Ferro, M.; et al. The K219T-Lamin mutation induces conduction defects through epigenetic inhibition of SCN5A in human cardiac laminopathy. *Nat. Commun.* **2019**, *10*, 2267. [CrossRef] [PubMed]
42. Lee, J.; Termglinchan, V.; Diecke, S.; Itzhaki, I.; Lam, C.K.; Garg, P.; Lau, E.; Greenhaw, M.; Seeger, T.; Wu, H.; et al. Activation of PDGF pathway links LMNA mutation to dilated cardiomyopathy. *Nature* **2019**, *572*, 335–340. [CrossRef] [PubMed]
43. Muchir, A.; Pavlidis, P.; Bonne, G.; Hayashi, Y.K.; Worman, H.J. Activation of MAPK in hearts of EMD null mice: Similarities between mouse models of X-linked and autosomal dominant Emery Dreifuss muscular dystrophy. *Hum. Mol. Genet.* **2007**, *16*, 1884–1895. [CrossRef] [PubMed]
44. Muchir, A.; Pavlidis, P.; Decostre, V.; Herron, A.J.; Arimura, T.; Bonne, G.; Worman, H.J. Activation of MAPK pathways links LMNA mutations to cardiomyopathy in Emery-Dreifuss muscular dystrophy. *J. Clin. Investig.* **2007**, *117*, 1282–1293. [CrossRef]

45. Choi, J.C.; Muchir, A.; Wu, W.; Iwata, S.; Homma, S.; Morrow, J.P.; Worman, H.J. Temsirolimus activates autophagy and ameliorates cardiomyopathy caused by lamin A/C gene mutation. *Sci. Transl. Med.* **2012**, *4*, 144ra102. [CrossRef]
46. Muchir, A.; Wu, W.; Choi, J.C.; Iwata, S.; Morrow, J.; Homma, S.; Worman, H.J. Abnormal p38α mitogen-activated protein kinase signaling in dilated cardiomyopathy caused by lamin A/C gene mutation. *Hum. Mol. Genet.* **2012**, *21*, 4325–4333. [CrossRef]
47. Chatzifrangkeskou, M.; Le Dour, C.; Wu, W.; Morrow, J.P.; Joseph, L.C.; Beuvin, M.; Sera, F.; Homma, S.; Vignier, N.; Mougenot, N.; et al. ERK1/2 directly acts on CTGF/CCN2 expression to mediate myocardial fibrosis in cardiomyopathy caused by mutations in the lamin A/C gene. *Hum. Mol. Genet.* **2016**, *25*, 2220–2233. [CrossRef]
48. Le Dour, C.; Macquart, C.; Sera, F.; Homma, S.; Bonne, G.; Morrow, J.P.; Worman, H.J.; Muchir, A. Decreased WNT/β-catenin signalling contributes to the pathogenesis of dilated cardiomyopathy caused by mutations in the lamin a/C gene. *Hum. Mol. Genet.* **2017**, *26*, 333–343. [CrossRef]
49. Auguste, G.; Gurha, P.; Lombardi, R.; Coarfa, C.; Willerson, J.T.; Marian, A.J. Suppression of Activated FOXO Transcription Factors in the Heart Prolongs Survival in a Mouse Model of Laminopathies. *Circ. Res.* **2018**, *122*, 678–692. [CrossRef]
50. West, G.; Gullmets, J.; Virtanen, L.; Li, S.-P.; Keinänen, A.; Shimi, T.; Mauermann, M.; Heliö, T.; Kaartinen, M.; Ollila, L.; et al. Deleterious assembly of the lamin A/C mutant p.S143P causes ER stress in familial dilated cardiomyopathy. *J. Cell. Sci.* **2016**, *129*, 2732–2743. [CrossRef]
51. Chen, S.N.; Lombardi, R.; Karmouch, J.; Tsai, J.-Y.; Czernuszewicz, G.; Taylor, M.R.G.; Mestroni, L.; Coarfa, C.; Gurha, P.; Marian, A.J. DNA Damage Response/TP53 Pathway Is Activated and Contributes to the Pathogenesis of Dilated Cardiomyopathy Associated With LMNA (Lamin A/C) Mutations. *Circ. Res.* **2019**, *124*, 856–873. [CrossRef] [PubMed]
52. Worman, H.J.; Fong, L.G.; Muchir, A.; Young, S.G. Laminopathies and the long strange trip from basic cell biology to therapy. *J. Clin. Investig.* **2009**, *119*, 1825–1836. [CrossRef] [PubMed]
53. Bossone, K.A.; Ellis, J.A.; Holaska, J.M. Histone acetyltransferase inhibition rescues differentiation of emerin-deficient myogenic progenitors. *Muscle Nerve* 2020. [CrossRef]
54. Kumar, S.; Baldinger, S.H.; Gandjbakhch, E.; Maury, P.; Sellal, J.-M.; Androulakis, A.F.A.; Waintraub, X.; Charron, P.; Rollin, A.; Richard, P.; et al. Long-Term Arrhythmic and Nonarrhythmic Outcomes of Lamin A/C Mutation Carriers. *J. Am. Coll. Cardiol.* **2016**, *68*, 2299–2307. [CrossRef] [PubMed]
55. Blondel, S.; Jaskowiak, A.-L.; Egesipe, A.-L.; Le Corf, A.; Navarro, C.; Cordette, V.; Martinat, C.; Laabi, Y.; Djabali, K.; de Sandre-Giovannoli, A.; et al. Induced pluripotent stem cells reveal functional differences between drugs currently investigated in patients with hutchinson-gilford progeria syndrome. *Stem Cells Transl. Med.* **2014**, *3*, 510–519. [CrossRef] [PubMed]
56. Shimi, T.; Pfleghaar, K.; Kojima, S.; Pack, C.-G.; Solovei, I.; Goldman, A.E.; Adam, S.A.; Shumaker, D.K.; Kinjo, M.; Cremer, T.; et al. The A- and B-type nuclear lamin networks: Microdomains involved in chromatin organization and transcription. *Genes Dev.* **2008**, *22*, 3409–3421. [CrossRef]
57. Guelen, L.; Pagie, L.; Brasset, E.; Meuleman, W.; Faza, M.B.; Talhout, W.; Eussen, B.H.; de Klein, A.; Wessels, L.; de Laat, W.; et al. Domain organization of human chromosomes revealed by mapping of nuclear lamina interactions. *Nature* **2008**, *453*, 948–951. [CrossRef]
58. Zullo, J.M.; Demarco, I.A.; Piqué-Regi, R.; Gaffney, D.J.; Epstein, C.B.; Spooner, C.J.; Luperchio, T.R.; Bernstein, B.E.; Pritchard, J.K.; Reddy, K.L.; et al. DNA sequence-dependent compartmentalization and silencing of chromatin at the nuclear lamina. *Cell* **2012**, *149*, 1474–1487. [CrossRef]
59. Solovei, I.; Wang, A.S.; Thanisch, K.; Schmidt, C.S.; Krebs, S.; Zwerger, M.; Cohen, T.V.; Devys, D.; Foisner, R.; Peichl, L.; et al. LBR and lamin A/C sequentially tether peripheral heterochromatin and inversely regulate differentiation. *Cell* **2013**, *152*, 584–598. [CrossRef]
60. Perovanovic, J.; Dell'Orso, S.; Gnochi, V.F.; Jaiswal, J.K.; Sartorelli, V.; Vigouroux, C.; Mamchaoui, K.; Mouly, V.; Bonne, G.; Hoffman, E.P. Laminopathies disrupt epigenomic developmental programs and cell fate. *Sci. Transl. Med.* **2016**, *8*, 335ra58. [CrossRef]
61. Bianchi, A.; Mozzetta, C.; Pegoli, G.; Lucini, F.; Valsoni, S.; Rosti, V.; Petrini, C.; Cortesi, A.; Gregoretti, F.; Antonelli, L.; et al. Dysfunctional polycomb transcriptional repression contributes to lamin A/C-dependent muscular dystrophy. *J. Clin. Investig.* **2020**, *130*, 2408–2421. [CrossRef] [PubMed]

62. Mewborn, S.K.; Puckelwartz, M.J.; Abuisneineh, F.; Fahrenbach, J.P.; Zhang, Y.; MacLeod, H.; Dellefave, L.; Pytel, P.; Selig, S.; Labno, C.M.; et al. Altered Chromosomal Positioning, Compaction, and Gene Expression with a Lamin A/C Gene Mutation. *PLoS ONE* **2010**, *5*, e14342. [CrossRef] [PubMed]
63. Cheedipudi, S.M.; Matkovich, S.J.; Coarfa, C.; Hu, X.; Robertson, M.J.; Sweet, M.; Taylor, M.; Mestroni, L.; Cleveland, J.; Willerson, J.T.; et al. Genomic Reorganization of Lamin-Associated Domains in Cardiac Myocytes Is Associated With Differential Gene Expression and DNA Methylation in Human Dilated Cardiomyopathy. *Circ. Res.* **2019**, *124*, 1198–1213. [CrossRef]
64. Bertero, A.; Fields, P.A.; Smith, A.S.T.; Leonard, A.; Beussman, K.; Sniadecki, N.J.; Kim, D.-H.; Tse, H.-F.; Pabon, L.; Shendure, J.; et al. Chromatin compartment dynamics in a haploinsufficient model of cardiac laminopathy. *J. Cell Biol.* **2019**, *218*, 2919–2944. [CrossRef] [PubMed]
65. West, J.A.; Beqqali, A.; Ament, Z.; Elliott, P.; Pinto, Y.M.; Arbustini, E.; Griffin, J.L. A targeted metabolomics assay for cardiac metabolism and demonstration using a mouse model of dilated cardiomyopathy. *Metabolomics* **2016**, *12*, 59. [CrossRef] [PubMed]
66. Beadle, R.M.; Williams, L.K.; Kuelh, M.; Bowater, S.; Abozguia, K.; Leyva, F.; Yousef, Z.; Wagenmakers, A.J.M.; Thies, F.; Horowitz, J.; et al. Improvement in cardiac energetics by perhexiline in heart failure due to dilated cardiomyopathy. *J. Am. Coll. Cardiol. Heart Fail.* **2015**, *3*, 202–211. [CrossRef] [PubMed]
67. Tuunanen, H.; Engblom, E.; Naum, A.; Någren, K.; Scheinin, M.; Hesse, B.; Airaksinen, J.; Nuutila, P.; Iozzo, P.; Ukkonen, H.; et al. Trimetazidine, a metabolic modulator, has cardiac and extracardiac benefits in idiopathic dilated cardiomyopathy. *Circulation* **2008**, *118*, 1250–1258. [CrossRef] [PubMed]
68. Morales Rodriguez, B.; Khouzami, L.; Decostre, V.; Varnous, S.; Pekovic-Vaughan, V.; Hutchison, C.J.; Pecker, F.; Bonne, G.; Muchir, A. N-acetyl cysteine alleviates oxidative stress and protects mice from dilated cardiomyopathy caused by mutations in nucelar A-type lamins gene. *Hum. Mol. Genet.* **2018**, *27*, 3353–3360. [CrossRef]
69. Olivier, M.; Asmis, R.; Hawkins, G.A.; Howard, T.D.; Cox, L.A. The Need for Multi-Omics Biomarker Signatures in Precision Medicine. *Int. J. Mol. Sci.* **2019**, *20*, 4781. [CrossRef] [PubMed]
70. Murphy, S.; Zweyer, M.; Mundegar, R.R.; Swandulla, D.; Ohlendieck, K. Proteomic serum biomarkers for neuromuscular diseases. *Expert Rev. Proteom.* **2018**, *15*, 277–291. [CrossRef] [PubMed]
71. Vignier, N.; Amor, F.; Fogel, P.; Duvallet, A.; Poupiot, J.; Charrier, S.; Arock, M.; Montus, M.; Nelson, I.; Richard, I.; et al. Distinctive Serum miRNA Profile in Mouse Models of Striated Muscular Pathologies. *PLoS ONE* **2013**, *8*, e55281. [CrossRef] [PubMed]

Journal of
*Personalized
Medicine*

Review

What Can Machine Learning Approaches in Genomics Tell Us about the Molecular Basis of Amyotrophic Lateral Sclerosis?

Christina Vasilopoulou [1], Andrew P. Morris [2], George Giannakopoulos [3,4], Stephanie Duguez [1] and William Duddy [1,*]

1 Northern Ireland Centre for Stratified Medicine, Altnagelvin Hospital Campus, Ulster University, Londonderry BT47 6SB, UK; Vasilopoulou-C@ulster.ac.uk (C.V.); s.duguez@ulster.ac.uk (S.D.)
2 Centre for Genetics and Genomics Versus Arthritis, Centre for Musculoskeletal Research, Manchester Academic Health Science Centre, University of Manchester, Manchester M13 9PT, UK; andrew.morris-5@manchester.ac.uk
3 Institute of Informatics and Telecommunications, NCSR Demokritos, 153 10 Aghia Paraskevi, Greece; ggianna@iit.demokritos.gr
4 Science For You (SciFY) PNPC, TEPA Lefkippos-NCSR Demokritos, 27, Neapoleos, 153 41 Ag. Paraskevi, Greece
* Correspondence: w.duddy@ulster.ac.uk; Tel.: +44-28-71-675-686

Received: 30 October 2020; Accepted: 23 November 2020; Published: 26 November 2020

Abstract: Amyotrophic Lateral Sclerosis (ALS) is the most common late-onset motor neuron disorder, but our current knowledge of the molecular mechanisms and pathways underlying this disease remain elusive. This review (1) systematically identifies machine learning studies aimed at the understanding of the genetic architecture of ALS, (2) outlines the main challenges faced and compares the different approaches that have been used to confront them, and (3) compares the experimental designs and results produced by those approaches and describes their reproducibility in terms of biological results and the performances of the machine learning models. The majority of the collected studies incorporated prior knowledge of ALS into their feature selection approaches, and trained their machine learning models using genomic data combined with other types of mined knowledge including functional associations, protein-protein interactions, disease/tissue-specific information, epigenetic data, and known ALS phenotype-genotype associations. The importance of incorporating gene-gene interactions and cis-regulatory elements into the experimental design of future ALS machine learning studies is highlighted. Lastly, it is suggested that future advances in the genomic and machine learning fields will bring about a better understanding of ALS genetic architecture, and enable improved personalized approaches to this and other devastating and complex diseases.

Keywords: Amyotrophic Lateral Sclerosis; machine learning; genome-wide association studies; GWAS; genomics; ALS pathology; gene prioritization

1. Introduction

Amyotrophic Lateral Sclerosis (ALS) is a progressively fatal, late-onset motor neuron disorder that is predominately characterised by the loss of upper and lower motor neurons. Progressive muscle atrophy in ALS patients leads to swallowing difficulties, paralysis and ultimately to death from neuromuscular respiratory failure [1–3]. ALS is the most common type of motor neuron disorder, and has peak onset at 54–67 years old, although it can affect individuals of any age [2–5]. Patients typically survive 2–5 years after the first symptoms occur, with 5–10% surviving more than 10 years [1,2,6]. A population-based study of estimated ALS incidence in 10 countries found that prevalence could increase more than 31% from 2015 to 2040 [4]. Thus, there is an increasing need to understand ALS pathology and the molecular pathways involved, towards prevention or successful therapeutic intervention.

There are two major classifications among ALS patients, based on family history: 5–10% of cases are genetically linked, and are classified as *familial*, having one or more relatives that suffer from ALS, while 90% are classified as *sporadic*, in which a familial history is not established, and where a genetic cause is usually not identified [7]. However, the distinction between the two categories is not always simple, with familial ALS-associated mutations also being present among sporadic ALS cases [3]. The extent and form of genetic contribution to sporadic ALS remains unclear, but genetic factors are considered to play an important role in the disease pathology [3,8]. Further investigation of the genetic architecture of both familial and sporadic cases is necessary.

In recent years, advances in high-throughput technologies have enabled the discovery of multiple Single Nucleotide Polymorphisms (SNPs) that are associated with ALS, mainly by the application of the Genome-Wide Association Study (GWAS) approach. GWAS aims to identify SNPs and other types of genetic variation (such as structural variants, copy number variations and multiple nucleotide polymorphisms) that are more frequent in patients than in people without the disease [9]. Statistical tests are carried out for disease association across genetic markers numbering from hundreds of thousands up to millions, depending on the genomic analytical platform. The most popular genotype-phenotype association studies use statistical models such as logistic or linear regression, depending on whether the trait is binary (i.e., case-control studies, such as ALS versus healthy controls) or quantitative (e.g., different scales of height). GWAS has been successful in discovering tens of thousands of significant genotype-phenotype associations in a large spectrum of diseases and traits, such as schizophrenia, anorexia nervosa, body-mass index (BMI), type 2 diabetes, and ALS [10–13]. Over the past decade, the discovery of significant genotype-phenotype associations has provided new insights into disease susceptibility, pathology, prevention, drug design and personalized medical approaches [11,14,15].

Rapid recent technological advances and great efforts in the field have led to the genomic profiling of large ALS cohorts, providing new insights into the pathology of ALS [12]. Initiatives such as Project MinE and dbGaP have contributed to the systematic release of ALS GWAS data [16,17]. The ALSoD publicly available database for genes that are implicated in ALS records 126 genes, with a subset having been reproduced in multiple studies [18]. As of July 2020, the GWAS catalogue has published 317 variants and risk allele associations with ALS [10].

The scope of this review covers genome-wide association studies that employ machine learning approaches with the aim to understand ALS pathology through gene prioritization. A search of PubMed and Google Scholar for the terms "amyotrophic lateral sclerosis", "GWAS" and "machine learning" yielded 420 results, of which 7 research papers were identified as falling into this scope. Machine learning studies that refer to the estimation of ALS heritability, to drug repurposing/prediction for ALS, or survival analyses, are not considered.

The review is structured as follows: first, knowledge from relevant literature about ALS pathology and genetic architecture is summarized, then the central challenges and limitations of traditional GWAS studies are introduced in the context of ALS, while the third section provides a brief overview of key machine learning concepts along with a description and comparison of published feature selection and machine learning approaches using ALS GWAS datasets. The main contribution of the review is to outline the challenges of ALS genomic studies, summarizing and comparing how these have been addressed by the collected research papers. Finally, the further use of machine learning as a method to understand ALS pathology is advocated.

1.1. Current Knowledge of Molecular Pathways Implicated by the Functions of Known ALS-Linked Genes

Our current knowledge of the aetiology and the genetic architecture of ALS is still elusive. Genetic mutations, environmental contributions, epigenetic changes and DNA damage are hypothesized as potential causal factors that ultimately lead to motor neuron death [19,20]. Variants in more than 30 genes are recognized as monogenic causes of ALS [12,19,21–23]. The most frequent monogenic cause in European populations is the intronic hexanucleotide GGGGCC (G4C2) repeat expansion (HRE) in the *C9orf72* gene [24,25]. Other genes linked to ALS with high reproducibility include Cu/Zn superoxide dismutase 1 *SOD1*, fused in sarcoma *FUS*, and transactive response DNA-binding protein of 43 kD *TARDBP/TDP-43* [19]. The discovery of risk gene mutations has helped to unravel the molecular mechanisms of ALS, and may lead ultimately to targeted therapy and stratified drug discovery [19,26–28].

Numerous studies have been published aimed at explaining motor neuron death, investigating the functional effects of specific mutations of known risk-associated genes such as *C9orf72, FUS, SOD1* and *TDP-43* [23,27]. Recent systematic reviews from our group have summarised the molecular pathways and biomarkers for which there are strong supporting evidence in ALS [19,26,27]. The molecular pathways affected in ALS can be grouped as follows (see [27] for detailed review):

- Mitochondrial dysfunction as a direct or indirect consequence of ALS-associated gene mutations *CHCHD10, FUS, SOD1, C9orf72 and TDP-43* can lead to an increase in oxidative stress, an increase in cytosolic calcium, ATP deficiency and/or stimulation of pro-apoptotic pathways [20,22,27,29,30].
- Oxidative stress can also be derived from a stimulation of NADPH oxidase, as observed with *ATXN2* mutations [31], or from deficiency in the elimination of Reactive Oxygen Species (ROS) as observed with some *SOD1* mutations in familial cases [27,32,33]. It then may contribute to DNA damage. Interestingly, other mutations on the ALS-associated genes *NEK1* [34], *SETX* [35] and *C21orf2* [36] are suspected to alter the DNA repair machinery, leading to an accumulation of oxidative damage over time. Consequently, these events could ultimately lead to motor neuron death [34,37].
- Disrupted axonal transport has been directly linked to a mutation in the C-terminal of the ALS-associated gene, *KIF5A* [12,38], and to mutations in genes encoding for neurofilaments (*NEFH*), microtubules and motor proteins (*PFN1,TUBA4A, DCTN1*) [27,39]. Consequently, organelle transport, protein degradation, and RNA transport are affected, disrupting cellular homeostasis. Similarly, axonal transport disruptions have been observed in fALS patients harboring mutations in non-cytoskeletal-related genes such as *SOD1* [38].
- Protein degradation is suspected to be a key pathway that is defective in ALS. This can be a direct consequence of mutations in ALS-associated genes involved in proteasome activity and the autophagy pathway, such as *UBQLN2, VCP, SQSTM1/P62, OPTN, FIG4, Spg11,* or *TBK1* [40], and may lead to an accumulation of misfolded and non-functional proteins [27,41]. It can also be an indirect consequence of other mutations leading to the formation of protein aggregates such as SOD1, FUS, TDP43, C9orf72-derived DPR - aggregates that in turn impair the proteasome and autophagic degradation pathways [42], thus exacerbating the accumulation of misfolded proteins. Consequently, the blockade of autophagy pathways may affect vesicle secretion [43,44]. Interestingly, some ALS-associated genes are known to be directly or indirectly involved in exosome biogenesis such as *CHMP2B* [45] or *C9orf72* [46], respectively.
- Glutamate-mediated excitotoxicity has been suggested to cause motor neuron deterioration, and could be an indirect consequence of ALS-associated gene mutations such as in *SOD1* or *C9orf72*, resulting in an elevated level of glutamate in the cerebrospinal fluid of patients [47–49].
- RNA processing and metabolism is another key pathway affected in ALS. For example, mutations to RNA-binding proteins encoded by *FUS, TDP-43, hnRNPA1, hnRNPA2B1,* and *MATR3,* result in altered mRNA splicing, RNA nucleocytoplasmic transport and translation [27,50–55], as well as in the generation and accumulation of toxic stress granules [56]. Similarly, accumulation of toxic RNA foci can be observed in motor neurons in the context of *C9orf72* mutations, and may lead to the sequestration of splicing proteins, thus affecting RNA maturation and translation [57]. The biogenesis of microRNA is also directly affected by mutated *FUS,*

TDP-43, or *C9orf72*-mediated DPRs, thus having an impact on the expression of genes involved in motor neuron survival [27,58].

Understanding the functional processes that drive ALS pathology has proven to be a difficult and complex task, compounded by the heterogeneity that characterises the disease. The gene products of the 30 or more known ALS-associated genes interact with each other, are implicated in multiple molecular pathways, and result in multiple disease phenotypes, making functional curation and interpretation complex [19,27]. In addition, these monogenic causes in ALS occur only in ~15% of sporadic ALS and ~66% of familial ALS patients, so that more than 80% of the ALS population do not currently have any known ALS-associated mutations [19,21]. Nonetheless, acquiring an in-depth understanding of the molecular mechanisms and the genetic architecture of ALS could potentially lead to the identification of multiple patient strata and therefore targeted therapies to be applied to different subgroups of ALS patients.

1.2. The Genetic Architecture of ALS

The genetic contribution to familial and sporadic ALS has not been fully explained by genotype-phenotype discoveries [8,25], and the known Mendelian causes of ALS represent only a small proportion of the ALS population [19,21]. Nonetheless, estimates of heritability are high in sporadic ALS patients - for example, 61% in a twin meta-analysis study - suggesting that genetic factors are strongly represented in sporadic ALS and that further investigation may yet identify novel causal variants and/or multilocus interactions that could account for this high estimated heritability [59].

So far, evidence supports a model implicating rare variants (minor allele frequency <1%) along with non-genetic causes, such as environmental factors [3,36,60,61]. Large GWAS efforts suggest a genetic architecture for ALS that falls somewhere in the middle of the spectrum of genetic pathology in terms of effect size and prevalence of risk variants-i.e., an *intermediate genetic architecture*, lying between conditions such as schizophrenia which have many common variants each imparting a small increase to disease risk, and conditions such as Huntington's disease which are caused by rare large-effect variants located in a single gene [3,20,62,63].

Many ALS-associated variants, particularly for *C9orf72*, also contribute to other conditions such as frontotemporal dementia (FTD) and cerebellar disease, suggesting that ALS is a multi-system syndrome [3,60,61]. ALS has an established overlap with other neurodegenerative and neuropsychiatric disorders, investigation of which could lead to insights into the understanding of pathology [3,5,25,60,64,65]. An example of this is the degree of overlap between familial ALS (~40%) and familial FTD (~25%) patients that carry the G6C4 expansion of *C9orf72* [65,66]. *C9orf72* hexanucleotide expansion has been associated to multiple traits including Alzheimer's and Parkinson's diseases, ataxia, chorea and schizophrenia [21,67–69]. A population-based GWAS study reported a higher prevalence of psychosis, suicidal behaviour, and schizophrenia, in Irish ALS kindreds, which was associated with the C9orf72 repeat expansion, based on an aggregation analysis [64]. Further evidence for a shared susceptibility to ALS was provided by the greater occurrence of dementia among first-degree relatives of ALS patients [69]. Several studies have suggested that the genetic overlap between ALS and other neurodegenerative and neuropsychiatric disorders could also be explained by the presence of ALS-associated pleiotropic variants that influence multiple, and in some cases quite distinct, phenotypic traits [70–72]. One study that supports this hypothesis is that of O'Brien et al., which shows that first-degree and second-degree relatives of Irish ALS patients have a significantly higher prevalence of schizophrenia and neuropsychiatric diseases than healthy controls, including obsessive-compulsive disorder, psychotic illness, and autism-the authors performed k-means clustering and calculated the relative risk to estimate aggregation [71,73–75].

Further investigation is needed to achieve a deep understanding of ALS heritability and genetic architecture, incorporating pleiotropic gene effects into experimental design.

2. ALS-Specific GWAS Challenges and Limitations

So far, numerous ALS GWAS studies have been published, aiming to identify novel ALS-associated variants through standard genotype-phenotype analyses. The first was published in 2007, providing genomic data for 276 cases and 271 controls [76]. Technological advances have provided the opportunity for studies with a higher number of genotyped ALS cohorts. The largest release of ALS genomic data was published in 2018 by Nicolas et al., and identified *KIF5A* as a novel ALS-associated gene; the study included a publicly-available large meta-analysis dataset of 10,031,630 imputed SNPs of 20,806 ALS and 59,804 controls as well as providing controlled access to "raw" genomic data including SNP-arrays of 12,188 cases and 3,292 controls [12,17]. Despite that hundreds of ALS-associated variants have been recorded in public databases such as the GWAS Catalog [10], these associations show very little reproducibility across different studies and have not been able to explain a large percentage of ALS heritability [3,36]; a phenomenon which is generally known as the "missing heritability" paradox [77]. It has been proposed that SNPs contribute ~8.5% of the overall heritability of ALS, although it should be noted that such estimates consider only linear single-marker effects of SNPs [36,77]. Here we outline some general GWAS limitations in the context of ALS, as well as potential reasons why standard GWAS phenotype-genotype analysis is unlikely to fully explain the genetic architecture of ALS.

A first general challenge in large scale genomic analyses is to ensure a high quality of the genotype data, so that the downstream results of the experimental design reflect true biology and not artifacts. Therefore, the collected genomic data first need to pass a comprehensive Quality Control (QC) pipeline including multiple sample and variant QC steps [78–81]. One challenge is that each dataset has its own specific features, thus there are not fixed thresholds for each quality-control step. For this reason, each study needs to follow a data-driven approach, taking into consideration the distribution of each data metric. However, there are some good practices in QC that may be generally applicable to most studies [78,81]. For example, it is typical to follow a procedure first filtering out low quality samples then removing poor quality markers, the order of this ensuring that as many genetic markers as possible are kept in the final dataset. However, overly strict thresholds can lead to the loss of a substantial proportion of samples, reducing study power. Another challenge is to ensure homogeneity of the collected samples in terms of ancestry. This QC step is carried out by analysing the population structure to remove ethnic outliers, and by accounting for confounding factors in later stages of the analysis, such as a potential inner population sub-structure, usually using the first few Principal Components, after performing a Principal Component Analysis on the homogeneous sample cohort. Also, it is very important to check for duplicated samples and, in non-family GWAS analyses, ensure that all samples are unrelated so that specific genotypes are not over-represented (and thereby contributing a bias to the subsequent analysis). Identity-by-descent (IBD) is a metric that corrects for such bias and takes into account the number of variants that a pair of individuals share.

GWAS is a single marker analysis treating each variant association as an independent event that contributes to the phenotype. Due to this, it is a standard practice for results to be corrected under the strict multiple testing threshold ($p < 5 \times 10^{-8}$) of the Bonferroni correction in order to control for false positive discoveries (Family-wise type I errors). This threshold derives from the hypothesis of 1,000,000 independent markers being tested under a significance level of 5%. Particularly in low sample size studies this correction can result in a loss of power of the analysis, which may then fail to capture a portion of potential risk variants that do not pass the significance threshold (Family-wise type II errors) [15,82].

Univariate analyses such as GWAS that test trait association for one locus at a time are not able to capture multilocus interactions-a phenomenon called epistasis-and the interaction of the environment with the genome; events that could potentially account for the missing heritability of ALS and explain the disease pathology [83,84]. The term *epistasis* was introduced in genetics over a century ago by Bateson et al. [85], and genetic and evolutionary biology studies have highlighted the importance of gene-gene interactions not only in the genetic architecture of an organism but also in evolution [77,86,87]. Epistasis represents non-additive events in the genome including interactions among two or more loci that have an effect on the phenotype [88]. Several studies have highlighted the role of epistasis in pathology, showing that SNP interactions provide a stronger association to the disease than the participating SNPs do individually [77,84,89,90]. To understand pathology in a complex disease such as ALS, it may be necessary to identify complex genetic interactions, including epistatic interations [77,87]. Nevertheless, the study of multilocus interactions poses a number of challenges, in particular the need for a high computational power as the number of tested interactions is extremely high even in pairwise combinations. As such, multivariate computational approaches and appropriate machine learning methods may be able to capture the potentially complex relationships among risk variants in ALS [77,90,91].

GWAS is more successfully employed under a "common disease-common variant" hypothesis, being of particular use in common diseases such as schizophrenia which are driven by many risk alleles each with high frequency [92]. In contrast, ALS is a heterogeneous disease likely comprised of multiple strata each resulting from combinations of different rare mutations and other factors. As a result, stratum-specific mutations may each have very small effects that are diluted and thus not captured by GWAS [3,36]. The majority of GWAS analyses have used SNP-arrays as they have until recently had a lower experimental cost in comparison to sequencing of the exome or the whole genome. SNP-array analyses can typically capture the effect of only common variants to the phenotype whereas sequencing analyses identify both common and rare variants. In most SNP-array GWAS studies, variants with Minor Allele Frequency (MAF) of <1–5% are removed from subsequent analysis as they are generally more difficult to genotype and therefore are considered potential false positives [15,78]. Nevertheless, whole genome sequencing, custom designed exome sequencing arrays, rare variant burden analyses and imputation approaches using large reference panels (such as the Haplotype Reference Consortium, which contains 64,976 haplotypes), face this challenge by recovering both rare (up to 0.1% MAF) and common variants that SNP-array platforms do not usually contain [15,93–95]. However, there is still a proportion of low frequency minor allele effects on the phenotype that cannot yet be detected by GWAS approaches and that could also potentially explain some of the missing heritability in ALS [3,36].

Lastly, another common GWAS challenge in complex diseases is the difficulty to distinguish causal variants from other non-disease-associated variants that are in high linkage disequilibrium [15]. Linkage disequilibrium describes the phenomenon where an allele of a variant is inherited together with the alleles of other variants [9]. These alleles of other variants are highly correlated and will have very similar GWAS signals with the truly causal SNP. The majority of disease-related variants are located in cis-regulatory regions of the genome [96], and given our limited knowledge of non-coding genomic loci, it is even more challenging for those to discern causal SNPs from the noise. Our difficulty to identify the causal variants in complex diseases among a pool of statistically significant associated variants adds to the challenge of identifying molecular processes that could have a significant impact on the disease.

Advanced machine learning prediction models trained in ALS genomic data could overcome the aforementioned challenges, moving towards better insights into disease causality and ultimately to a personalized understanding of ALS [15,97]. In Figure 1, we describe the basic steps of an ALS machine learning experimental design in order to discover ALS-associated novel loci or combinations of loci, as well as the main challenges of each step. Each of the main challenges is addressed in successive chapters of the review, as we describe and compare the experimental design of the collected gene prioritization studies. Some of the challenges in Figure 1 have already been mentioned, such as the need for a *large sample size* that could increase the power of the study, a *comprehensive quality control pipeline* to assure high quality genomic data, as well as the *curse of dimensionality* which is a very common problem in genomic studies that include an extremely high number of features and especially in studies that focus on multilocus interactions.

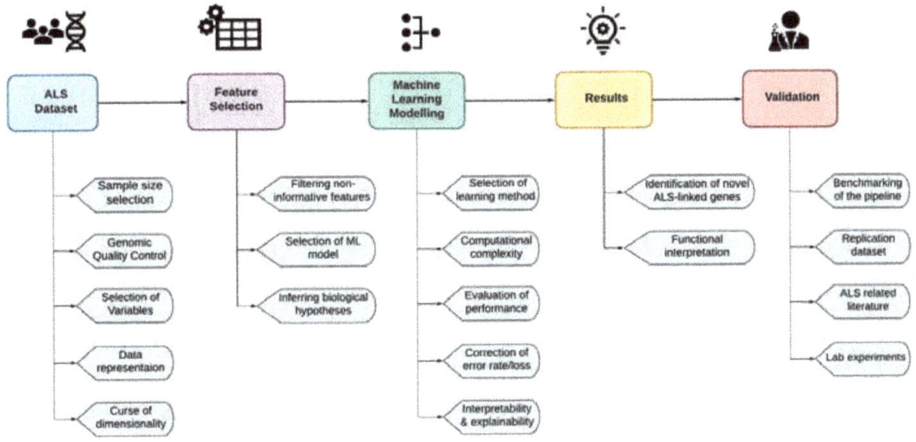

Figure 1. The main challenges of an ALS machine learning experimental design.

3. Facing the Challenges

In this chapter, we outline gene prioritization approaches by which published research studies have employed machine learning methods in order to identify and rank novel ALS-linked genes, SNPs and multilocus interactions. First, a short introduction is made to some basic machine learning concepts that will be useful in later discussion of the machine learning approaches that have been used. Then, details are provided about the data representation, feature curation, and selection methods that each study chose for their experimental design and, finally, the machine learning methods and the overall experimental designs of each study are compared, as well as considering their results.

3.1. A Brief Overview of Machine Learning Concepts

Based on the task and the type of learning there are three main machine-learning categories: supervised, unsupervised and semi-supervised algorithms. Supervised learning methods aim to make predictions on unknown instances (e.g., a sample or a gene) based on known labels (e.g., ALS/non ALS) [98]. For instance, classification is a supervised machine learning approach, which trains a classifier using labeled data e.g., samples/genes including a case/control label (training set) and predicts the class of an unknown sample or gene (testing set) based on specific rules and patterns that the classifier learned during training and testing.

In classification, it has always been a challenge to identify the input features that are most informative and maximally affect the prediction. This domain of research is described as *feature selection* research [99] and has been traditionally connected to statistical methods. More recently, with the advent of deep neural networks, *explainability* and *interpretability* of machine learning suggestions, predictions and decisions has become an even more acute problem. This is due to the complex nature of the network itself, which does not clearly illustrate the connection between input (features) and output (prediction) in a humanly understandable manner. Thus, a number of recent studies aim for *explainability* in deep models [100], including visualized explanations [101].

On the other hand, unsupervised learning is performed on unlabelled data, with an ultimate purpose to identify interesting patterns or novel sub-groupings of the data. Clustering research offers a well-established set of unsupervised algorithms which identify patterns in a group of instances (e.g., Genes, SNPs, ALS patients), most commonly based on notions of distance (or similarity) between instances. Distance in such cases can be measured with metrics such as Euclidean distance or the Pearson Correlation Coefficient [74].

Lastly, in semi-supervised learning the prediction is carried out in positive and unlabelled data. Semi-supervised learning is applied when information is available only on instances of a single class (usually called "positive" instances, e.g., already known-ALS genes) but there is not sufficient information to label the rest of the instances as "negative". Semi-supervised learning methods can be quite challenging when the aim is to predict novel disease-associated instances since the classifier is trained treating potentially novel instances as "negative". Semi-supervised learning methods are particularly common in gene prioritization studies.

An *instance* in a machine learning task is an entity that the classifier is trained to predict, for example a gene in a gene prioritization algorithm or an ALS sample in an ALS/non-ALS patient classification task. An instance is described by a number of *features* (e.g., gene functional annotations, SNP genotypes etc.), typically represented as a vector, termed *feature vector*. All X instances (e.g., samples or genes/SNPs) need to have the same number of Z features (e.g., genetic mutations or functional annotations), leading to a two-dimensional matrix K that has a size of K = X * Z. Each instance has a specific location on a Z-dimensional space (where Z is the number of features) and each feature has, accordingly, a specific location in X-dimensional space (where X is the number of instances).

There are a wide variety of metrics that can be used to evaluate the performance of a machine learning model, the choice of which depends on the nature of the task. The most popular metrics include Accuracy, Precision, Recall, F1-score, Specificity and ROC (Receiver Operating Characteristic) curves. Accuracy represents the fraction of true positive and true negative predictions out of the total number of the model predictions. Precision expresses how many positive predictions of the model were truly positive. Recall (True Positive Rate) explains the percentage of the missed true positive predictions. Recall and Precision are both very important metrics that need to be taken into account for the evaluation of a model's performance. F1-score calculates the harmonic mean of those two metrics, hence the higher the F1-score is, the better the model performed. Specificity expresses the fraction of the negative predicted instances that were actually negative. The ROC curve is a plot of True Positive Rate against False Positive Rate expressed as 1-Specificity. The Area Under the Curve (AUC) of ROC is employed to calculate how well the model performed - the closer the AUC is to 1, the better the model performed.

3.2. Recent Feature Preparation and Selection Approaches in ALS Genomic Studies

Modern analytical genomic platforms and imputation methods have led to the profiling of samples containing up to tens of millions of genetic markers. This vast amount of genomic information makes machine learning modelling more complicated and demanding, as it is likely that only a small minority, if any, of markers are truly causal and associated to the disease. Thus, a first challenge in the genomic machine learning experimental design is to deal with the *curse of dimensionality* using appropriate feature selection methods [102]. Below we describe and group the feature selection and dataset curation approaches that have been used by each study, also summarized in Table 1.

Table 1. Details of the feature selection approaches that have been followed by each study prior to the machine learning experiments. The "Data/Instances" column summarises the primary number of instances before further filtering, and details about the datasets that have been used by each study. The "Features" column refers to the initial number of features in each dataset before feature selection and after quality control (the latter being applicable only in studies using genotype data). The "Epistasis" column indicates whether the respective study has incorporated epistatic events (i.e., multi-locus interactions) as a feature selection method. The "Regulatory Elements" column describes studies that have filtered their initial dataset by including only non-coding regulatory regions. Studies that have used prior ALS-related information (e.g., already known ALS-associated SNPs/genes, known functional information, filtering based on an ALS versus Control genotype-phenotype association analysis *p*-value etc.) in order to select and reduce their initial instance and feature space are indicated under "ALS-linked knowledge". Lastly, we indicate machine learning methods that were used to select only highly informative features based on specific criteria. ML: Machine Learning, SNP: Single Nucleotide Polymorphism, MDR: Multifactor Dimensionality Reduction, CNN: Convolutional Neural Network, PPIs: Protein- Protein Interactions, DHS: DNase I hypersensitive sites, TFBS: Transcription Factor Binding Sites, PCA: Principal Component Analysis, t-SNE: t-distributed Stochastic Neighboring Embedding, UMAP: Uniform Manifold Approximation and Projection.

Study	Data/Instances	Features	Genomic Structure	Epistasis	Cis-Regulatory Elements	ALS-Linked Knowledge	ML Methods
Vitsios et al. [103]	18,626 coding genes (label:positive/unlabelled)	1,249 gene-annotations: generic, disease- and tissue -specific features	No	No	No	Yes	PCA, t-SNE, UMAP
Yousefian et al. [104]	8,697,640 SNP *p*-values of 14,791 ALS cases and 26,898 controls [36]	2,252 functional features: DHS mapping data, histone modifications, target gene functions, and TFBS	Yes	No	Yes	Yes	None
Bean et al. [105]	ALS-linked gene lists: DisGeNet: 101 genes ALSoD: 126 genes, ClinVar: 44 genes, Manual list: 40 genes Union: 199 genes	PPIs, disease-gene associations and functional annotations	No	No	No	Yes	None
Yin et al. [90]	4511 cases and 7397 controls [16]	823,504 SNPs from 7,9, 17 and 22 chromosomes	Yes	No	Yes	Yes	CNN
Kim et al. [91]	SNP pairwise interactions	550,000 SNPs of 276 cases/271 controls and 211 cases/211 controls [76,106]	No	Yes	No	Yes	MDR
Greene et al.[107]	SNP pairwise interactions	210,382 SNPs of 276 cases/271 controls and 211 cases/211 controls [76,106]	No	Yes	No	No	MDR
Sha et al. [108]	SNP pairwise interactions	555,352 SNPs of 276 cases/271 controls [76]	No	No	No	Yes	None

The use of feature selection methods are a common strategy to reduce the high number of initial features for machine learning models [102]. As described in Table 2, four of the seven studies used a machine learning model as a feature selection method, in some cases using specific hypotheses or along with a combination of other strategies that will be described later in this sub-chapter. Mantis-ml is one example of a multi-step gene prioritisation framework which extracts a heterogeneous set of 1249 gene-annotation features mined from a large collection of databases in order to discover and rank novel disease-related genes [103]. Each instance in this model consists of 18,626 coding genes which are labelled as positive (seeds) or unlabelled, depending on known association to the disease, based on information retrieved from the Human Phenotype Ontology (HPO) [109]. A number of pre-processing steps were applied that included filtering of highly correlated pairs of features, removing features with missing data, and imputing certain features with a low missing rate. Some exploratory analyses are then performed (i.e., heat maps, variable distributions, etc.) in the original feature space and then three dimensionality reduction methods are automatically applied: principal component analysis (PCA), t-distributed stochastic neighboring embedding (t-SNE) [110], and uniform manifold approximation and projection (UMAP) [111], in order to identify any interesting pattern(s) and linear/non-linear relationships among the features. The gene pool is randomly split into K balanced datasets containing both positive and unlabelled genes. Mantis-ml can use the Boruta feature selection algorithm which labels the features (given a decision threshold) as "confirmed"/"tentative"/"rejected" by assessing the contribution of each feature to the prediction [103], although the entire feature space is used by default.

Table 2. The biological hypotheses incorporated to the experimental design and main focuses of the collected studies.

	Genomic Structure	Epistasis	Cis-Regulatory Elements	ALS-Linked Knowledge	Functional Annotations
Vitsios et al. [103]	No	No	No	Yes	Yes
Yousefian et al. [104]	Yes	No	Yes	Yes	Yes
Bean et al. [105]	No	No	No	Yes	Yes
Yin et al. [90]	Yes	Yes	Yes	No	No
Kim et al. [91]	No	Yes	Yes	Yes	Yes
Greene et al. [107]	No	Yes	No	No	No
Sha et al. [108]	No	Yes	No	Yes	No

Several recent studies have focused on uncovering gene-gene interactions (epistasis) that could potentially explain some part of the missing heritability of complex genetic traits such as ALS. However, modern genomic studies that aim to model epistatic events face considerable challenges, demanding large computational and statistical power for the analysis of a large number of combinations among millions of genotyped loci (even when considering only pairwise interactions). Two of the studies -Greene et al. and Kim et al.- included epistatic events in their feature selection approach [91,107]. Both used the wrapper method Multifactor Dimensionality Deduction (MDR)-a non-parametric, model-free approach which reduces the feature space of multilocus combinations by creating new single variables pooled from multiple SNP genotypes [87,112]- and then estimates statistically significant ALS-risk pairwise interactions of SNPs. Both studies used the same datasets and pre-processing steps to identify pairwise SNP interactions that are significantly associated with ALS. The datasets included two sporadic ALS cohorts along with healthy controls containing 276 cases versus 271 controls in the detection dataset and 211 cases versus 211 controls in the replication dataset [76,106]. SNPs were filtered using a 0.2 Minor Allele Frequency cut-off and a less than 90% call rate. Lastly, Greene et al. considered only independent SNPs in further analysis, leading to a dataset of 210,382 SNPs.

Another strategy to reduce feature space is to include only regulatory elements as features for the subsequent machine learning experiments. Two of the studies focus only on the effect of noncoding regulatory elements in ALS pathology. It has been shown that disease-related variants are mostly located in cis-regulatory elements of the genome marked by DNase I hypersensitive sites (DHSs)-zones of the genome that have been associated with elevated levels of transcriptional activity [96]. In 2020, Yousefian et al. investigated the effect of noncoding variants in ALS [104]. Firstly, they applied a p-value threshold ($p < 5 \times 10^{-4}$) on SNPs from a previously published large ALS meta-analysis GWAS dataset. The authors constructed association blocks by identifying lead SNPs having strong ALS GWAS p-value associations and being at least 1 Mb apart from each other, then selecting also the top 30 ALS-associated SNPs located upstream and downstream of each lead SNP; leading ultimately to 274 association blocks [104]. They enriched their selected association blocks with functional and epigenetic information including DHS profile data, histone modifications, functional gene-sets from the KEGG database and TF binding sites collected from TRANSFAC and JASPAR databases [104,113–115]. After the functional enrichment of the features, they constructed a binary feature matrix representing whether a SNP within an association block was associated or not with a particular functional feature [104]. The second ALS study that included only noncoding regulatory elements was published in 2019 by Yin et al. who proposed Promoter-CNN as a feature selection method; a convolutional neural network model (comprised of 2 convolutional layers and two deep layers) that reduces the initial feature space by selecting only the top 8 highest performing promoter regions among variants located on chromosomes 7, 9, 17 and 22 [90]. They assessed the performance of Promoter-CNN using a 9-fold cross validation. Each promoter region of each individual was represented by a window of 64 genomic features having a value of 0,1,2 to represent the genotype at each of these 64 loci, utilizing the genomic structure of the regions [90]. The aforementioned studies employed the genomic structure in order to build the association blocks and the promoter regions (see Table 1).

A popular strategy that has been used to select the initial data and then further reduce the feature and instance space is ALS-associated knowledge (see Table 1). The most relevant study that falls into this category is Bean et al. which modelled only previously known ALS-linked gene lists, mining information from the literature and from disease databases, as well as including a manually curated set of ALS-associated genes [105]. In order to reduce the feature space they performed an enrichment test on all features and, for the predictive model, kept only those features that are significantly enriched in the mechanism(s) of the disease [116]. Another example is Yin et al. which before applying their Promoter-CNN model as a feature selection method, they first limited their feature space by studying only non-additive phenomena of multiple promoters located on four specific chromosomes (7, 9, 17 and 22), those chromosomes have being selected based on the amounts of missing heritability that have been previously identified in ALS [36,90]. Bean et al., Kim et al., Yousefian et al. and Sha et al., implement multi- step algorithms in which one of the initial steps reduces the feature space by keeping only the highest performing genes/SNPs, by assessing the enrichment of genes in ALS [105] or by applying a specific threshold of single-marker association analysis to ALS [91,104,108].

3.3. Experimental Design and Results of the ALS Gene Prioritization Approaches

In this section we will focus on the approach and rationale of the collected studies that aim to understand the pathology of ALS using machine learning and probabilistic models on genomic data. We also briefly consider the main findings of each study. Although the majority of the collected studies aim to answer more than one research question, in this section we group studies by methodology, based on common features of their experimental design that we considered to be the main focus in each study. In order to avoid confusion, in Table 2 we describe the main biological focus of each study. We investigate the reproducibility of the results of each analysis, comparing it with related literature as well as mentioning putative novel ALS discoveries in the Discussion section.

The identified studies (see Table 1) all fall under the gene prioritization umbrella category. In Table 3, we provide brief information about the machine learning models that have been used in these studies (after feature selection and filtering approaches have been applied, as discussed in the previous section, see Table 1), as well as details about the assessment and the performance of the models.

Four of the studies included epistatic events in their experimental design, testing the hypothesis that multilocus interactions have an effect on ALS susceptibility (see Table 2). Three studies that fall into this category, all use the same ALS detection dataset in order to discover pairwise SNP interactions [91,107,108]. Greene et al. and Kim et al. use multifactor dimensionality reduction (MDR); a wrapper method which performs both a feature selection and classification to predict pairwise combinations of SNPs as high-risk and low-risk for ALS [91,107]. As proposed by [112], MDR reduces multilocus dimensions into a one-dimensional multilocus variable, the prediction performance of which is evaluated in classification tasks (ALS versus healthy controls) by cross validation and permutation tests. Out of pairwise combinations among 210,382 SNPs, Greene et al. reported the pair of SNPs rs4363506 and rs6014848 to have the highest accuracy (Acc: 0.6551) and a p-value of 0.048 after permutation testing. Their replication dataset showed a lower accuracy (Acc: 0.5821) but with a higher statistical significance (p-value < 0.021) [107]. Kim et al. chose the best performing MDR model for each SNP, then they mapped SNPs to genes (including neighboring regulatory elements) and investigated their enrichment using Gene Ontology functional terms [91]. Unfortunately, they did not report any specific high performing pairs of SNPs in terms of MDR accuracy, nor specific genes. The statistical significance of each MDR model was estimated using permutation, with the highest enriched Gene Ontology gene-sets being "Regulation of Cellular Component Organization and Biogenesis" (*p*-values: 0.010 and 0.014), and "Actin Cytoskeleton" (*p*-values: 0.040 and 0.046). The *p*-values for each gene-set refer to the detection and replication dataset respectively, after multiple testing. The third study is proposed by Sha et al. implementing a two-stage probabilistic algorithm that attempts to predict two-locus combinations that are associated to ALS using eight epistatic and nine multiplicative two-locus predicting models. The first step was a single-marker association test using the x^2 test statistic, keeping only the 1000 SNPs having the strongest *p*-values. The associations of all the two-locus combinations of those 1000 variants were then tested using the above-mentioned seventeen models. Three SNPs were discovered participating to two two-locus combinations: rs4363506 with rs3733242 (*p*-value = 0.032) and rs4363506 with rs16984239 (*p*-value = 0.042). Reported *p*-values were adjusted for multiple testing correction using permutation. They also performed Multifactor dimensionality reduction and Combinatorial Searching Method to identify high performing two-locus interactions, but none of the results reached the significance threshold (see Table 3). Single locus analysis was not able to capture the three SNPs that participate to the bi-locus interaction.

Table 3. The machine learning approaches followed by each study. Acc: Accuracy, SVM: Support Vector Machines, LR: Logistic Regression, RF: Random Forest, DNN: Deep Neural Network, SVC: Support Vector Classifier (SVC), CNN: Convolutional Neural Network, AUC: Area under the receiver operator characteristic curve, CSM: Combinatorial Searching Method, Chr: Chromosome.

Study	Machine Learning Models	Model Assessment	Performance
Vitsios et al. [103]	Stochastic Semi-supervised Learning: Stacking, DNN, Gradient Boosting, RF, SVC, XGBoost, ExtraTrees Classifiers	10-fold cross validation	Stacking: Avg AUC: 0.767, DNN: Avg AUC: 0.774, Gradient Boosting: Avg AUC: 0.79, RF: Avg AUC: 0.798, SVC: Avg AUC: 0.801, XGBoost: Avg AUC: 0.805, ExtraTrees: Avg AUC: 0.814
Yousefian et al. [104]	Convolutional Neural Network	Autoencoder pre-training, Chr 1-10: training set, Chr 11–14: testing set and Chr 15–22: validation set	CNN: AUC: 0.96 F1-score: 0.83
Bean et al. [105]	Knowledge graph edge prediction model [116]	5-fold cross validation	Fold-Change Enrichment and random guess baseline: ALSoD: 23.33 (23.53), ClinVar: 30.05 (15.64), DisGeNet: 55.90 (81.66), Manual: 84.54 (13.27), Union: 8.92 (4.28)
Yin et al. [90]	Deep Neural Network (ALS-Net), Logistic Regression, SVM, Random Forest and Adaboost	9-fold cross validation	ALS-Net: Acc: 0.769 F1-score: 0.797 LR: Acc: 0.739 F1-score:0.728 SVM: Acc: 0.725 F1-score:0.694 RM: Acc: 0.596 F1-score:0.381 Adaboost: Acc: 0.661 F1-score:0.625 (+PromoterCNN and all 4 chromosomes combined)
Kim et al. [91]	Multifactor dimensionality reduction; using a naïve Bayes classifier	1000 permutation tests	Critical Acc: 0.629 and 0.640 (replication dataset)
Greene et al. [107]	Multifactor dimensionality reduction	1000 permutation tests	Best SNP pairwise model: Acc: 0.6551 and 0.5821 (replication dataset); with $p < 0.048$ and $p < 0.021$ (replication dataset)
Sha et al. [108]	Two-locus probabilistic models, Multifactor dimensionality reduction, Combinatorial Searching Method	1000 permutation tests	Two-locus models: rs4363506-rs3733242: $p = 0.032$ rs4363506-rs16984239: $p = 0.042$ MDR model: rs4363506-rs12680546: $p = 0.156$ CSM model: rs4363506-rs12680546: $p = 0.2$

A far more complex machine learning model for ALS patient classification, studying non-additive interactions of multilocus promoters among four chromosomes, was published in 2019 by Yin et al. [90]. Building upon previous knowledge that the majority of the disease-associated variants in GWAS are cis-regulatory elements, this study showed that using only the highest performing promoter regions of 4 chromosomes as features provides enough information for successful classification of the ALS genomic profile versus healthy controls [90]. Specifically, they applied a prediction model for putative ALS-related genotypes located in promoter regions, using a case-control genomic dataset containing 4,511 cases and 7,397 controls from the Dutch cohort of ProjectMinE. A two-level pipeline was constructed in which, as a first step, deep neural networks select the eight highest performing promoter regions of chromosomes 7, 9, 17 and 22 having the highest accuracy in ALS status prediction (Promoter-CNN model) and then the selected promoter regions of each individual are combined for a final classification task (ALS-Net model). A 9-fold cross validation was used to train both models. The authors compared their ALS-Net deep learning model to other classification models using both the pre-selected promoter regions combined from all four chromosomes by Promoter-CNN models, and markers from individual chromosomes. The compared classification models included a logistic regression Polygenic Risk Score (PRS) approach [117], Support Vector Machines (SVM) [118], Random Forest [119] and AdaBoost [120,121]. The two-tier deep learning model showed very promising results identifying both already reported associated ALS genes and new putative ALS markers. The results showed that ALS-Net combined with Promoter-CNN pre-selected promoter regions produced better performance (Acc: 0.769, F1-score: 0.797) than the logistic regression Polygenic Risk Score (PRS) approach (Acc: 0.739, F1-score: 0.728) [117], Support Vector Machines (SVM) (Acc: 0.725, F1-score: 0.694) [118], Random Forest (Acc: 0.596, F1-score: 0.381) [119], or AdaBoost (Acc: 0.661, F1-score: 0.625) [120,121]. They highlight that their Promoter-CNN model is a successful feature selection method keeping only the highest performing promoters from each chromosome individually, advancing the performance of the subsequently tested classifiers [90]. The findings indicate that combining genomic information from all four chromosomes improves the performance for the majority of the models, supporting further the hypothesis that non-additive events take place in ALS pathology.

Another recent study that builds upon the hypothesis that cis-regulatory noncoding variants can have an important effect on ALS pathology, applies a functional SNP prioritization framework using convolutional neural networks (CNN) to make ALS rare noncoding risk-variant predictions [104]. The authors build upon a previously published deep learning CNN-based model that used functional features in order to predict causal regulatory elements in complex diseases [122]. They tested their proposed method on a large GWAS meta-analysis cohort including 8,697,640 SNP p-values for 14,791 ALS patients and 26,898 healthy controls [12]. The functional SNP prioritization framework followed a multi-step procedure starting from (a) collecting the upstream and downstream flanking regions of the 274 highest GWAS ALS-associated variants, then (b) functionally annotating the variants (including DNase I hypersensitive sites (DHSs), histone modifications, target gene functions, and transcription factor binding sites (TFBS)) and finally (c) training the CNN model on these association blocks with uncertain class labels using chromosomes 1–10 as a training set, chromosomes 11–14 as a testing test and finally 15–22 as a validation set [104]. The CNN model used two convolutional layers; the first layer measured how well each individual SNP matched the pattern of 50 functional features using a rectified linear unit (ReLU), and the second level output prediction scores for each SNP with a 0–1 range, with values close to one indicating that there are common regulatory patterns embedded for a particular SNP. The CNN model shows a high predictive performance (AUC = 0.96 and F1 = 0.83). A random forest classification for the ALS cell-type specificity showed that a high portion of their ALS selected features have neuronal cell-type specificity within Trancriptional Factor binding sites. The proposed framework highlights two potentially functional ALS-risk variants rs2370964 (chromosome 3, located in enhancer site of *CX3CR1*) and rs3093720 (chromosome 17, intron variant in *TNFAIP1*). An eQTL analysis was performed to investigate the effect of these two variants on the expression of other genes. The analysis showed that the two noncoding variants may impact ALS risk by affecting the expression levels of *CX3CR1* and *TNFAIP1*. The *CX3CR1* gene deletion has been associated with microglia neurotoxicity and neuron loss in transgenic ALS mice [123,124]. TNFAIP1 is an apoptotic protein which has been associated to neurotoxicity [125]. The rs2370964 variant also affects the *CTCF* and *NFAT* binding sites [104]. Mutations in *CTCF* gene has also been related to microglial dysfunction, among other effects [126]. NFAT is a transcription factor that is involved with the regulation of pro-inflammatory responses in cultured murine microglia [127]. The rs3093720 SNP affects the *NR3C1* binding site, a gene which has been associated with neurodegeneration and multiple sclerosis [128].

The accumulation of large amounts of ALS multi-omic data, functional annotations, and tissue-specific information, has provided a great opportunity and challenge for researchers to combine these in ways that could potentially lead to stronger machine learning models. Two collected studies combine known ALS genes and ALS-related information mined from a variety of databases to predict novel disease-specific genes and rank them. The first one is, Mantis-ml, a recently published multi-step disease agnostic gene prioritisation pipeline which employs known disease-associated genes to predict scores for putative novel genes based on feature pattern similarity [103]. As mentioned in Section 3.2, the disease-related gene information is extracted from multiple databases and resources, including tissue- and disease- specific data [103]. Depending on the users' disease-related queries, the pipeline follows automatic feature selection and pre-processing as well as an exploratory data analysis on disease-related features. A repeated stochastic semi-supervised model is used to iteratively predict disease-related probabilities for each gene, and then rank each gene based on the mean prediction probability of all iterations. The starting point for the modelling is the labelling of an entire coding gene pool (18,626 genes) based on disease relevance retrieved from the Human Phenotype Ontology (HPO), with positive and unlabelled genes then being split into random balanced datasets [109]. They evaluate the performance of their classifier using a stratified 10-fold random split in every balanced dataset, which is followed by testing using the out-of-bag k-fold method. The model generates gene prediction probabilities belonging to the respective testing set of each k-prediction cycle. Lastly, the model calculates an aggregated prediction probability combined from all iteration cycles. The classification performance of Mantis-ml was assessed using seven different supervised models (gradient boosting, random forest, extra trees, extreme gradient boosting, support vector classifier, deep neural networks, and a stacking classifier) with 10 stochastic iterations and 10-fold cross validation in three diseases. All models showed similar performance (AUC: 0.83–0.85), with Extra Trees having the highest mean Area Under Curve for ALS (AUC = 0.814) (see Table 3). For ALS, 77 positively labelled genes were selected, having an average AUC of 0.814 (combined from 7 classifiers). Among the top 50 genes there were two already known ALS associated genes, *FUS* and *L1CAM*. Specifically, "MGI mouse knockout feature" was ranked as the top feature for ALS, including human orthologue mouse genes that have been associated with survival and developmental pathways. Unlabelled genes (i.e., genes not annotated to ALS in the HPO) were also identified as being predictive of ALS. Some of the top novel predicted genes among 5 out of 6 classifiers are *SYNE1, ALDH5A1, ABCA1, DNMT3A, NF2, SZT2, ACADVL, MED12, TSC2, EP400, RYR2, VCL* and *BBS2*. *SYNE1* causes recessive ataxia and has been associated with motor neuron degeneration and ALS [129]; ALDH5A1 has been identified as significantly down-regulated protein in ALS murine models [130]; ABCA1 has been linked with damage of neuromuscular junctions and identified in significant clusters of altered frontal cortex genes in ALS samples [131,132]; a DNMT3A isoform has been identified in synapses and in mitochondria and has been associated with degeneration in motor neurons in ALS patients and abnormal expression levels in skeletal muscle and spinal cord of presymptomatic ALS mice [133,134].

Another study that combined genomic data with other types of data sources to increase the power of the machine learning gene prioritization method was that of Bean et al., integrating functional annotations, known ALS-gene associations, and protein-protein interactions [105]. Protein-protein interaction networks have been useful to decipher new disease mechanisms, as proteins that are encoded by disease-related genes are likely to interact with proteins that are implicated in similar pathologies [135]. These authors used a previously published knowledge graph-based completion model [116], which is trained combining protein-protein interactions data mined from Intact [136,137], known disease-gene lists from DisGeNet [138,139] and functional gene-sets from the Gene Ontology [140] in order to make predictions of novel ALS-linked genes [105] (see Table 1). The algorithm starts by building a knowledge graph containing known ALS data represented as nodes and their interrelationships as edges. The aim of the model is to predict the missing edges of the graph that could represent novel ALS genes, providing a predictive score to each. This is a similarity score deriving from the profile which is built by the trained knowledge graph model comparing the ALS-known genes to the rest of the genes. They train 5 models for 5 input sets of ALS-linked genes mined from the ALS Online Database (ALSoD), which is intended to host all known ALS- associated genetic variants [18], from the ClinVar database of curated clinical variants [141], from DisGeNet [138,139], and a manual curated ALS gene list generated by the authors [105] (see Table 1). For the training and testing of the model, 5-fold cross validation was used to estimate how well the model would have predicted known ALS genes. As seen in Table 3, all models-except the DisGeNet list- performed very well in each fold above the random baseline, with the manual list being on top. In total, the 5 models predicted 45, 176, 192, 327 and 575 novel ALS for the Manual list, DisGeNet, ClinVar, ALSoD and union ALS-known lists, respectively. All predicted novel genes of the manual list were also present in the other 4 lists. The authors also tested the functional enrichment of the predicted genes following an overrepresentation analysis using Gene Ontology terms, with all gene-sets having statistically significant enrichment to ALS-specific biological processes, like mitochondrial activity, endosome transport and vesicular trafficking, lipid metabolism and others. To validate the relevance of the predicted ALS genes, a gene-set and gene-level analysis was performed using MAGMA [142] on a large ALS meta-analysis GWAS dataset (European cohort, including 20,806 cases and 59,804 controls) [12], keeping only the variants that mapped to the putative ALS-genes. Only ClinVar predicted genes had statistically significant results (p-value = 0.038), followed by the Manual model which did not pass the Bonferonni correction but it was close with a p-value of 0.060.

4. Discussion

Here we identified gene prioritization machine learning studies aimed at the understanding of genomic data in ALS, outlining the main challenges faced by such studies. We compared these studies in terms of their feature selection methods, experimental design, machine learning performance and their biological results. In Figure 2, we summarize some of the key decision making steps taken by machine learning approaches in ALS genomics, and examples of some possible choices at each step.

In gene prioritization studies, the initial number of potentially "novel" SNPs/genes or, more problematically, the potential novel multilocus combinations, can be so high as to make computational analysis unfeasible (see Figure 1). In this context, it was intriguing to group and compare the chosen dimensionality reduction approaches in each study. As seen in Table 1, most studies handled this problem quite differently. More than half used a machine learning method for feature selection, along with one or more biological hypotheses for additional filtering of the input variables. The most common approach of biological hypothesis filtering was to infer ALS-specific knowledge early in the experimental design. This is both an advantage and a potential disadvantage as it makes the results more likely to have biological relevance but at the same time risks introducing bias to later stages of the machine learning approach. An example of a feature selection method that fell into this category is an early SNP-filtering approach based on a specific threshold of ALS GWAS p-values. This approach succeeds in a straightforward way to reduce the feature space, but comes at the cost that potential GWAS false positives could be inferred and/or that true positives (not captured by GWAS) might be removed from further analysis. This may be especially problematic in the capturing of epistatic events, as traditional GWAS analysis is a linear single-marker analysis, so filtering based on single SNP-disease association p-values could risk losing putative significant multi-locus interactions in later stages of the analysis [143,144]. However, a large cohort size could increase the power of a standard GWAS analysis utilized as a feature selection method in a machine learning study.

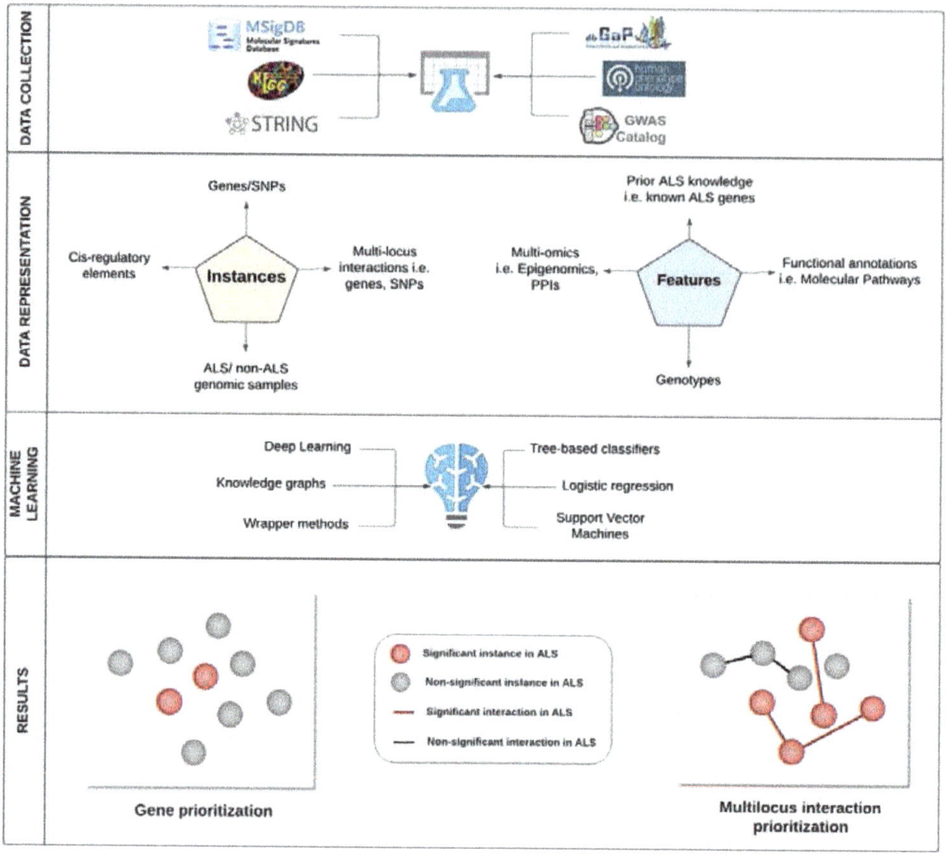

Figure 2. Some of the key decision making steps taken by machine learning approaches in ALS genomics, and examples of some possible choices at each step. These included data collection, data representation, selection of the Machine Learning algorithm for the classification task, and the types of result obtained by the model. The studies collected information from a variety of databases in order to mine, among other things, data on genotypes (e.g., dbGaP, GWAS Catalog, gnomAD), functional annotations (e.g., KEGG), and Protein-Protein Interactions (e.g., STRING). Depending on the purpose of the experimental design, the collected studies modelled genes, SNPs, cis-regulatory regions, multilocus interactions, and/or ALS/non-ALS patients. Each instance was described using features such as Genomic, Epigenomic, and Proteomic data, functional annotations, and/or prior ALS-related knowledge (e.g., ALS gene-sets). Various machine learning algorithms were selected for the classification tasks. Lastly, we visualize the distinction between the modelling results of gene prioritization and multilocus interaction prioritization studies. Gene prioritization studies aim to identify significant ALS associated instances (e.g., SNPs, genes and cis-regulatory regions), whereas multilocus interaction prioritization studies aim to discover significant interactions among multiple loci. Lastly, we note that an ALS versus non-ALS sample classification experiment can be used to prioritize genes if the interpretability of the chosen model permits the identification of informative genomic features.

Deep learning seems a promising machine learning method for ALS gene prioritization studies, as well as in ALS patient classification. Deep learning methods are known to perform well in regulatory genomic classification tasks, being able to incorporate information from the structure of the genomic data and capturing non-linear relationships and patterns of multilocus interactions [145,146]. This statement is further validated by comparing the performance of three collected ALS gene prioritization studies. More specifically, three recent studies that employ deep learning as at least one of their classification methods had very good predictive performance [90,103,104], as summarized in Table 2. Yin et al. showed that deep neural networks perform better than other methods in classifying ALS versus healthy controls, not only in terms of accuracy (0.769) and F1-score (0.797), but also with an excellent recall of 0.908, using a 9-fold cross-validation [90]. Moreover, all of the benchmarked classification methods showed improved performance when promoter selection Convolutional Neural Network models (Promoter-CNN) where incorporated as an extra feature selection stage with the classification models. Yousefian et al. constructed a semi-supervised Convolutional Neural Network model in order to predict ALS-associated non-coding variants using epigenetic features. The model showed an excellent performance achieving an AUC of 0.96 and F1-score of 0.83 [104]. This study did not include benchmarking against other machine learning models, and interestingly the association blocks of non-coding variants that were constructed for training, testing, and validation sets of the model were separated into chromosome numbers 1–10, 11–14 and 15–22, respectively, rather than the cross validation methods typically employed by other studies. Lastly, Vitsios et al. benchmarked their proposed multi-step non-disease specific gene prioritization pipeline assessing the performance of seven classifiers using 10-fold cross validation in three diseases. In ALS, all classifiers achieved very similar performance with an average AUC ranging from 0.767 to 0.814, with Deep Neural Networks achieving an average AUC of 0.774 and Extra Trees classifier being at the top [103]. Even though, the majority of the ALS collected studies show that deep learning yields very good classification results, one very popular challenge in such learning algorithms is the lack of explainability in the model's results, identifying which features are the most informative to the classification task.

In terms of reproducibility, comparing the highest ranked genes and SNPs, as well as the reported statistically significant functional pathways, it is noteworthy that the sALS-associated variant rs4363506 (initially identified by Schymick et al. with an empirical p-value = 10^{-6} [76]) was found by both Greene et al. and Sha et al. to have statistically significant pairwise interactions with other variants. Specifically, Greene et al. reported rs4363506 to interact with rs6014848 (Acc of 0.6551 and a p-value < 0.048 in the detection dataset, and Acc of 0.5821 and p-value < 0.021 in the replication dataset) and Sha et al. found that rs4363506 participated to two separate two-locus interactions: one with rs3733242 (p-value = 0.032) and another with rs16984239 (p-value = 0.042) [107]. These were the only significant pairwise combinations identified by either study. The replicated SNP rs4363506 is an intergenic variant (chr10:127476239, GRCh38.p12) located between *DOCK1* dedicator of cytokinesis 1 and *NPS* neuropeptide S [147]. *DOCK1* is implicated in neural growth and is a member of the KEGG pathway term "Regulation of actin cytoskeleton" and *NPS* in "positive regulation of synaptic transmission, glutamatergic" (GO:0051966) and "regulation of synaptic transmission, GABAergic" (GO:0032228), among others, all processes that are linked to ALS (see Section 1.1) [113,148]. This is consistent with other work implicating the actin cytoskeleton in ALS, including the finding of the third study, Kim et al.-investigating the functional enrichment of pairwise interactions in sporadic ALS-of statistical significance of the Gene Ontology term "Actin Cytoskeleton" (p-value = 0.040). However, it should be kept in mind that all 3 of these studies are analyses of the same primary dataset.

There could be several reasons for the disparity in the results of Greene and Sha et al. (i.e., the identification of different interaction partner SNPs to rs4363506). One reason could be that different quality control methods and thresholds are applied to the genomic datasets. In addition, Greene et al. investigated the pairwise combinations of 210,382 uncorrelated SNPs (keeping only independent SNPs in terms of linkage disequilibrium) using their proposed MDRGPU model, whereas Sha et al. tested the pairwise combinations of the top 1000 SNPs with a significant

ALS-association *p*-value, before testing for significant interactions using a series of different two-locus probabilistic models. The significant results of these studies also differed in the design of the predictive model used, with Greene et al. using Multifactor Dimensionality Reduction (MDR)-a non-parametric method combining feature engineering and then classification which does not make any assumption about the underlying genetic mechanisms-and Sha et al. using two-locus probabilistic parametric models which tested specific hypotheses about the type of the genotype interaction, taking into account the order of the genotypes in terms of penetrance of high-risk variants. Asides from the two-locus probabilistic models, Sha et al. did in fact also test (separately) an MDR and a Combinatorial Searching Method-unlike Greene et al., these analyses did not identify any significant pairwise combinations but, interestingly, SNP rs4363506 was present in a pairwise combination in both models that almost passed the threshold of statistical significance. Lastly, Kim et al. followed a multi-level approach starting from SNP pairwise interaction feature selection, moving to testing genes associations and gene-set functional associations to ALS. Unfortunately, they did not report any results about the MDR predicted pairwise interactions, hence we cannot make a direct comparison with the other two studies.

From the comparison of the most statistically significant genes and SNPs predicted from the optimally performing machine learning models of Mantis-ml [103], the knowledge graph completion model [105], the Promoter-CNN model [90], and the non-coding variant CNN model [104], we note several points. First, we compared the top 50 ALS genes predicted from the Mantis-ml model using Extra Trees-the highest performing classifier of the ALS data [103]-with the top 45 performing genes predicted by the knowledge graph-based machine learning approach using the highest performing model trained on the manually curated ALS-linked list [105]. *SLC1A2* (solute carrier family 1 member 2) was the only gene that was predicted by both approaches. SLC1A2 protein is the dominant transporter that clears the extracellular neurotransmitter glutamate in the synapses, expressed by astrocytes [149]. The down-regulation of *SLC1A2* has previously been associated with excitotoxicity leading to motor neuron degeneration and therefore contributing to ALS pathology, as described in the introduction (see Section 1.1) [47,48,150]. The Mantis-ml ALS model top predicted genes also had an overlap with the genes associated with the top performing promoters predicted by the Promoter-CNN model [90], but only in terms of shared protein families: within the top 8 promoter regions that Promoter-CNN selected for chromosomes 7, 9, 17, and 22, there were two promoter regions that were associated with genes *LAMB4* (laminin subunit beta 4) and *TRIM16* (tripartite motif containing 16); while among Mantis-ml highest predicted ALS genes were *LAMB3* (laminin subunit beta 3) and *TRIM28* (tripartite motif containing 28). The Laminin family contains heterotrimeric glycoproteins of the extracellular matrix that are associated with processes such as adhesion, survival, neuronal development and proliferation [151]. *LAMB4* is implicated in tissue development and cell migration, and has been associated with different types of cancer [140,152]. Interestingly, a laminin-4 isoform is expressed in neuromuscular junctions and has been associated with muscular dystrophy [151]. A recent study of *LAMB3* upregulation implicates this gene in cell apoptotic, proliferating and metastatic events in patients that suffer from pancreatic cancer [153]. *TRIM16* has been associated with autophagy, degradation of protein aggregates and ubiquitination of misfolded proteins; pathways that have been previously associated with ALS (see Section 1.1) [154]. Finally, *TRIM28* encodes a co-repressor protein which is expressed in the human brain and is a major regulator of transposable elements [155]. Elevated transcription of transposable elements has been linked with neurological disorders, including ALS, as well as binding to the ALS-associated RNA-processing protein TDP-43 [156].

It is noteworthy that the vast majority of the overlapping genes that we identified among all the collected studies are implicated in previously known ALS-associated functional pathways (as we outlined in Section 1.1) as well as the majority of the highest predicted novel genes (as described in Section 3.3). Nevertheless, we did not observe any further overlap among the other studies in terms of ALS-predicted genes. Limited reproducibility among the four studies could be due to multiple factors that derive from a number of differences in their experimental design and the focus of each study (see Tables 1–3). As summarised in Table 1, these four studies all use different instances, features and feature selection methods. As far as the machine learning models are concerned, the best performing machine learning models which were utilized for comparison were different. However, the low reproducibility could also derive from a more general challenge of gene prioritization studies, which concerns the difficulty of identifying the truly causal genes out of a usually large pool of novel predicted genes that pass a chosen significance threshold [157]. The difficulty of reproducibility is also emphasized in Bean et al. where 5 known ALS-linked gene lists mined from different databases, and one which was manually curated, were used, with each yielding very different results even with the same model, highlighting that the methodology and results of each study should be compared with caution [105]. Lastly, we need to mention that due to the high number of potential novel genes among all the benchmarked models of each study, we only compared the top genes/SNPs predicted by the highest performing model in each case. Hence, due to this limitation, we acknowledge the possibility of a larger existing overlap of the top genes predicted by the rest of the well performing models that were benchmarked among the studies. These challenges, makes the identification of the ALS implicated functional pathways even harder as well as the task of investigating reproducibility in different studies (see Figure 1) [157,158].

We also note that several studies incorporated prior ALS biological knowledge and functional annotations into the general experimental design (see Table 2), and the trained models showed a very good predictive performance (see Table 3). Specifically, as seen in Table 1 all of the considered studies except Greene et al. used an ALS-linked knowledge feature selection method to reduce their feature space as well as the number of instances (i.e., genes and SNPs). Moreover, Bean et al., Vitsios et al. and Yousefian et al. include ALS- specific and generic biological knowledge into their initial feature space, such as tissue/disease-specific features and known ALS disease-gene associations, as well as gene functional annotations, epigenetic features and protein-protein interactions (see Table 1). The most characteristic study that falls into this category is the one of Bean et al., where the instances were only ALS-linked lists from various resources, and the model was an ALS knowledge-based graph which uses the neighborhood of genes and enrichment tests to define significance of association, trained using PPIs, ALS-gene associations and functional annotations, showing the highest performance using a manually curated ALS-linked gene list.

Multilocus interactions may have a significant role in ALS and should be considered in future ALS genomic studies [90,91,107]. Each of the studies that investigated multilocus interactions has highlighted that the statistically significant variants could not be replicated in single-locus analyses [90,91,107,108]. Yin et al. carried out the most complex study in this category, using a large ALS cohort (see Table 1) followed by a thorough quality control -this was the first ALS multilocus study to investigate complex non-additive events in such a large scale of input variables. The results provided further support for the involvement of non-additive genetic interactions in ALS, showing that combining the genomic structure from multiple cis-regulatory elements (in this case promoters located in different chromosomes) yields very promising results in ALS patient classification [36,90]. Also, related literature supports that the "missing heritability" in genetic traits could be uncovered and explained to a significant degree by a network of gene-gene interactions which is not taken into account in the methods typically used to estimate the proportion of heritability that is missing [77,144,159]. Even though, the prediction of novel disease-specific gene-gene interactions poses, as described in previous chapters, a greater number of challenges than single-marker GWAS, it offers the potential to understand the heritability and genetic architecture of complex traits like ALS disease in greater depth. These challenges could be faced using machine learning approaches.

Machine Learning is a rapidly evolving field that has great potential in helping us to understand the complexity of ALS genomics, and how this relates to molecular pathways. However, further advances are needed in GWAS machine learning approaches in order to fully uncover the underlying mechanisms of this deadly disease which ultimately may lead us to successful personalized disease and drug-targeting prediction approaches.

Author Contributions: Conceptualization, C.V., S.D. and W.D.; data curation, C.V.; writing—original draft preparation, C.V.; writing—review and editing, C.V., A.P.M., G.G., S.D. and W.D.; supervision, W.D.; project administration, S.D.; funding acquisition, W.D. and S.D. All authors have read and agreed to the published version of the manuscript.

Funding: This work was financed by the European Union Regional Development Fund (ERDF) EU Sustainable Competitiveness Programme for N. Ireland, Northern Ireland Public Health Agency (HSC R&D) & Ulster University. C.V. was the recipient of a DfE international scholarship from Ulster University.

Acknowledgments: We thank George Paliouras for kind and helpful advice.

Conflicts of Interest: The authors declare no conflict of interest.

References

1. Niedermeyer, S.; Murn, M.; Choi, P.J. Respiratory Failure in Amyotrophic Lateral Sclerosis. *Chest* **2019**, *155*, 401–408. [CrossRef] [PubMed]
2. Chiò, A.; Logroscino, G.; Traynor, B.; Collins, J.; Simeone, J.; Goldstein, L.; White, L. Global Epidemiology of Amyotrophic Lateral Sclerosis: A Systematic Review of the Published Literature. *Neuroepidemiology* **2013**, *41*, 118–130. [CrossRef] [PubMed]
3. Al-Chalabi, A.; Van Den Berg, L.H.; Veldink, J. Gene discovery in amyotrophic lateral sclerosis: Implications for clinical management. *Nat. Rev. Neurol.* **2017**, *13*, 96. [CrossRef] [PubMed]
4. Arthur, K.C.; Calvo, A.; Price, T.R.; Geiger, J.T.; Chiò, A.; Traynor, B.J. Projected increase in amyotrophic lateral sclerosis from 2015 to 2040. *Nat. Commun.* **2016**, *7*, 1–6. [CrossRef] [PubMed]
5. Rowland, L.P.; Shneider, N.A. Amyotrophic Lateral Sclerosis. *N. Engl. J. Med.* **2001**, *344*, 1688–1700. [CrossRef] [PubMed]
6. Chiò, A.; Logroscino, G.; Hardiman, O.; Swingler, R.; Mitchell, D.; Beghi, E.; Traynor, B.G. Prognostic factors in ALS: A critical review. *Amyotroph. Lateral Scler.* **2009**, *10*, 310–323. [CrossRef]
7. Nicaise, C.; Mitrecic, D.; Pochet, R. Brain and spinal cord affected by amyotrophic lateral sclerosis induce differential growth factors expression in rat mesenchymal and neural stem cells. *Neuropathol. Appl. Neurobiol.* **2011**, *37*, 179–188. [CrossRef]
8. McLaughlin, L.R.; Vajda, A.; Hardiman, O. Heritability of amyotrophic lateral sclerosis insights from disparate numbers. *Jama Neurol.* **2015**, *72*, 857–858. [CrossRef]
9. Bush, W.S.; Moore, J.H. Chapter 11: Genome-Wide Association Studies. *PLoS Comput. Biol.* **2012**, *8*, e1002822. [CrossRef]
10. MacArthur, J.; Bowler, E.; Cerezo, M.; Gil, L.; Hall, P.; Hastings, E.; Junkins, H.; McMahon, A.; Milano, A.; Morales, J.; et al. The new NHGRI-EBI Catalog of published genome-wide association studies (GWAS Catalog). *Nucleic Acids Res.* **2017**, *45*, D896–D901. [CrossRef]
11. Klein, R.J.; Xu, X.; Mukherjee, S.; Willis, J.; Hayes, J. Successes of Genome-wide association studies. *Cell* **2010**, *142*, 350–351. [CrossRef] [PubMed]
12. Nicolas, A.; Kenna, K.; Renton, A.E.; Ticozzi, N.; Faghri, F.; Chia, R.; Dominov, J.A.; Kenna, B.J.; Nalls, M.A.; Keagle, P.; et al. Genome-wide Analyses Identify KIF5A as a Novel ALS Gene. *Neuron* **2018**, *97*, 1268–1283. [CrossRef] [PubMed]
13. Zhao, W.; Rasheed, A.; Tikkanen, E.; Lee, J.J.; Butterworth, A.S.; Howson, J.M.; Assimes, T.L.; Chowdhury, R.; Orho-Melander, M.; Damrauer, S.; et al. Identification of new susceptibility loci for type 2 diabetes and shared etiological pathways with coronary heart disease. *Nat. Genet.* **2017**, *49*, 1450–1457. [CrossRef] [PubMed]
14. Duncan, L.; Yilmaz, Z.; Gaspar, H.; Walters, R.; Goldstein, J.; Anttila, V.; Bulik-Sullivan, B.; Ripke, S.; Thornton, L.; Hinney, A.; et al. Significant locus and metabolic genetic correlations revealed in genome-wide association study of anorexia nervosa. *Am. J. Psychiatry* **2017**, *174*, 850–858. [CrossRef] [PubMed]
15. Tam, V.; Patel, N.; Turcotte, M.; Bossé, Y.; Paré, G.; Meyre, D. Benefits and limitations of genome-wide association studies. *Nat. Rev. Genet.* **2019**, *20*, 467–484. [CrossRef]
16. Van Rheenen, W.; Pulit, S.L.; Dekker, A.M.; Al Khleifat, A.; Brands, W.J.; Iacoangeli, A.; Kenna, K.P.; Kavak, E.; Kooyman, M.; McLaughlin, R.L.; et al. Project MinE: Study design and pilot analyses of a large-scale whole-genome sequencing study in amyotrophic lateral sclerosis. *Eur. J. Hum. Genet.* **2018**, *26*, 1537–1546. [CrossRef]

17. Mailman, M.D.; Feolo, M.; Jin, Y.; Kimura, M.; Tryka, K.; Bagoutdinov, R.; Hao, L.; Kiang, A.; Paschall, J.; Phan, L.; et al. The NCBI dbGaP database of genotypes and phenotypes. *Nat. Genet.* **2007**, *39*, 1181–1186. [CrossRef]
18. Abel, O.; Powell, J.F.; Andersen, P.M.; Al-Chalabi, A. ALSoD: A user-friendly online bioinformatics tool for amyotrophic lateral sclerosis genetics. *Hum. Mutat.* **2012**, *33*, 1345–1351. [CrossRef]
19. Vijayakumar, U.G.; Milla, V.; Stafford, M.Y.C.; Bjourson, A.J.; Duddy, W.; Duguez, S.M.R. A systematic review of suggested molecular strata, biomarkers and their tissue sources in ALS. *Front. Neurol.* **2019**, *10*, 400. [CrossRef]
20. Hardiman, O.; Al-Chalabi, A.; Chio, A.; Corr, E.M.; Logroscino, G.; Robberecht, W.; Shaw, P.J.; Simmons, Z.; Van Den Berg, L.H. Amyotrophic lateral sclerosis. *Nat. Rev. Dis. Primers* **2017**, *3*. [CrossRef]
21. Turner, M.R.; Al-Chalabi, A.; Chio, A.; Hardiman, O.; Kiernan, M.C.; Rohrer, J.D.; Rowe, J.; Seeley, W.; Talbot, K. Genetic screening in sporadic ALS and FTD. *J. Neurol. Neurosurg. Psychiatry* **2017**, *88*. [CrossRef] [PubMed]
22. Chia, R.; Chiò, A.; Traynor, B.J. Novel genes associated with amyotrophic lateral sclerosis: Diagnostic and clinical implications. *Lancet Neurol.* **2018**, *17*, 94–102. [CrossRef]
23. Volk, A.E.; Weishaupt, J.H.; Andersen, P.M.; Ludolph, A.C.; Kubisch, C. Current knowledge and recent insights into the genetic basis of amyotrophic lateral sclerosis. *Med. Genet.* **2018**, *30*, 252–258. [CrossRef] [PubMed]
24. Zou, Z.Y.; Zhou, Z.R.; Che, C.H.; Liu, C.Y.; He, R.L.; Huang, H.P. Genetic epidemiology of amyotrophic lateral sclerosis: A systematic review and meta-analysis. *J. Neurol. Neurosurg. Psychiatry* **2017**, *88*, 540–549. [CrossRef] [PubMed]
25. Connolly, O.; Le Gall, L.; McCluskey, G.; Donaghy, C.G.; Duddy, W.J.; Duguez, S. A Systematic Review of Genotype–Phenotype Correlation across Cohorts Having Causal Mutations of Different Genes in ALS. *J. Pers. Med.* **2020**, *10*, 58. [CrossRef] [PubMed]
26. Morgan, S.; Duguez, S.; Duddy, W. Personalized Medicine and Molecular Interaction Networks in Amyotrophic Lateral Sclerosis (ALS): Current Knowledge. *J. Pers. Med.* **2018**, *8*, 44. [CrossRef]
27. Gall, L.L.; Anakor, E.; Connolly, O.; Vijayakumar, U.G.; Duguez, S. Molecular and cellular mechanisms affected in ALS. *J. Pers. Med* **2020**, *10*, 101. [CrossRef]
28. Volonté, C.; Morello, G.; Spampinato, A.G.; Amadio, S.; Apolloni, S.; D'Agata, V.; Cavallaro, S. Omics-based exploration and functional validation of neurotrophic factors and histamine as therapeutic targets in ALS. *Ageing Res. Rev.* **2020**, *62*, 101121. [CrossRef]
29. Deng, J.; Yang, M.; Chen, Y.; Chen, X.; Liu, J.; Sun, S.; Cheng, H.; Li, Y.; Bigio, E.H.; Mesulam, M.; et al. FUS Interacts with HSP60 to Promote Mitochondrial Damage. *PLoS Genet.* **2015**, *11*, 1005357. [CrossRef]
30. Gupta, R.; Lan, M.; Mojsilovic-Petrovic, J.; Choi, W.H.; Safren, N.; Barmada, S.; Lee, M.J.; Kalb, R. The proline/arginine dipeptide from hexanucleotide repeat expanded C9ORF72 inhibits the proteasome. *eNeuro* **2017**, *4*, 249–265. [CrossRef]
31. Elden, A.C.; Kim, H.J.; Hart, M.P.; Chen-Plotkin, A.S.; Johnson, B.S.; Fang, X.; Armakola, M.; Geser, F.; Greene, R.; Lu, M.M.; et al. Ataxin-2 intermediate-length polyglutamine expansions are associated with increased risk for ALS. *Nature* **2010**, *466*, 1069–1075. [CrossRef]
32. Chang, Y.; Kong, Q.; Shan, X.; Tian, G.; Ilieva, H.; Cleveland, D.W.; Rothstein, J.D.; Borchelt, D.R.; Wong, P.C.; Lin, C.L.G. Messenger RNA oxidation occurs early in disease pathogenesis and promotes motor neuron degeneration in ALS. *PLoS ONE* **2008**, *3*, e2849. [CrossRef]
33. Rosen, D.R.; Siddique, T.; Patterson, D.; Figlewicz, D.A.; Sapp, P.; Hentati, A.; Donaldson, D.; Goto, J.; O'Regan, J.P.; Deng, H.X.; et al. Mutations in Cu/Zn superoxide dismutase gene are associated with familial amyotrophic lateral sclerosis. *Nature* **1993**, *362*, 59–62. [CrossRef]
34. Fang, X.; Lin, H.; Wang, X.; Zuo, Q.; Qin, J.; Zhang, P. The NEK1 interactor, C21ORF2, is required for efficient DNA damage repair. *Acta Biochim. Biophys. Sin.* **2015**, *47*, 834–841. [CrossRef] [PubMed]
35. Chen, Y.Z.; Bennett, C.L.; Huynh, H.M.; Blair, I.P.; Puls, I.; Irobi, J.; Dierick, I.; Abel, A.; Kennerson, M.L.; Rabin, B.A.; et al. DNA/RNA helicase gene mutations in a form of juvenile amyotrophic lateral sclerosis (ALS4). *Am. J. Hum. Genet.* **2004**, *74*, 1128–1135. [CrossRef] [PubMed]

36. Van Rheenen, W.; Shatunov, A.; Dekker, A.M.; McLaughlin, R.L.; Diekstra, F.P.; Pulit, S.L.; Van Der Spek, R.A.; Võsa, U.; De Jong, S.; Robinson, M.R.; et al. Genome-wide association analyses identify new risk variants and the genetic architecture of amyotrophic lateral sclerosis. *Nat. Genet.* **2016**, *48*, 1043–1048. [CrossRef] [PubMed]
37. Higelin, J.; Catanese, A.; Semelink-Sedlacek, L.L.; Oeztuerk, S.; Lutz, A.K.; Bausinger, J.; Barbi, G.; Speit, G.; Andersen, P.M.; Ludolph, A.C.; et al. NEK1 loss-of-function mutation induces DNA damage accumulation in ALS patient-derived motoneurons. *Stem Cell Res.* **2018**, *30*, 150–162. [CrossRef] [PubMed]
38. De vos, K.J.; Chapman, A.L.; Tennant, M.E.; Manser, C.; Tudor, E.L.; Lau, K.F.; Brownlees, J.; Ackerley, S.; Shaw, P.J.; Mcloughlin, D.M.; et al. Familial amyotrophic lateral sclerosis-linked SOD1 mutants perturb fast axonal transport to reduce axonal mitochondria content. *Hum. Mol. Genet.* **2007**, *16*, 2720–2728. [CrossRef]
39. Puls, I.; Jonnakuty, C.; LaMonte, B.H.; Holzbaur, E.L.; Tokito, M.; Mann, E.; Floeter, M.K.; Bidus, K.; Drayna, D.; Oh, S.J.; et al. Mutant dynactin in motor neuron disease. *Nat. Genet.* **2003**, *33*, 455–456. [CrossRef]
40. Oakes, J.A.; Davies, M.C.; Collins, M.O. TBK1: A new player in ALS linking autophagy and neuroinflammation. *Mol. Brain* **2017**, *10*, 1–10. [CrossRef]
41. Taylor, J.P.; Brown, R.H.; Cleveland, D.W. Decoding ALS: From genes to mechanism. *Nature* **2016**, *539*, 197–206. [CrossRef] [PubMed]
42. Wen, X.; Tan, W.; Westergard, T.; Krishnamurthy, K.; Markandaiah, S.S.; Shi, Y.; Lin, S.; Shneider, N.A.; Monaghan, J.; Pandey, U.B.; et al. Antisense proline-arginine RAN dipeptides linked to C9ORF72-ALS/FTD form toxic nuclear aggregates that initiate invitro and invivo neuronal death. *Neuron* **2014**, *84*, 1213–1225. [CrossRef]
43. Silverman, J.M.; Christy, D.; Shyu, C.C.; Moon, K.M.; Fernando, S.; Gidden, Z.; Cowan, C.M.; Ban, Y.; Greg Stacey, R.; Grad, L.I.; et al. CNS-derived extracellular vesicles from superoxide dismutase 1 (SOD1)G93A ALS mice originate from astrocytes and neurons and carry misfolded SOD1. *J. Biol. Chem.* **2019**, *294*, 3744–3759. [CrossRef] [PubMed]
44. Buratta, S.; Tancini, B.; Sagini, K.; Delo, F.; Chiaradia, E.; Urbanelli, L.; Emiliani, C. Lysosomal exocytosis, exosome release and secretory autophagy: The autophagic- and endo-lysosomal systems go extracellular. *Int. J. Mol. Sci.* **2020**, *21*, 2576. [CrossRef] [PubMed]
45. Parkinson, N.; Ince, P.G.; Smith, M.O.; Highley, R.; Skibinski, G.; Andersen, P.M.; Morrison, K.E.; Pall, H.S.; Hardiman, O.; Collinge, J.; et al. ALS phenotypes with mutations in CHMP2B (charged multivesicular body protein 2B). *Neurology* **2006**, *67*, 1074–1077. [CrossRef] [PubMed]
46. Blanc, L.; Vidal, M. New insights into the function of Rab GTPases in the context of exosomal secretion. *Small GTPases* **2018**, *9*, 95–106. [CrossRef] [PubMed]
47. Laslo, P.; Lipski, J.; Nicholson, L.F.; Miles, G.B.; Funk, G.D. GluR2 AMPA Receptor Subunit Expression in Motoneurons at Low and High Risk for Degeneration in Amyotrophic Lateral Sclerosis. *Exp. Neurol.* **2001**, *169*, 461–471. [CrossRef]
48. Spreux-Varoquaux, O.; Bensimon, G.; Lacomblez, L.; Salachas, F.; Pradat, P.F.; Le Forestier, N.; Marouan, A.; Dib, M.; Meininger, V. Glutamate levels in cerebrospinal fluid in amyotrophic lateral sclerosis: A reappraisal using a new HPLC method with coulometric detection in a large cohort of patients. *J. Neurol. Sci.* **2002**, *193*, 73–78. [CrossRef]
49. Milanese, M.; Zappettini, S.; Onofri, F.; Musazzi, L.; Tardito, D.; Bonifacino, T.; Messa, M.; Racagni, G.; Usai, C.; Benfenati, F.; et al. Abnormal exocytotic release of glutamate in a mouse model of amyotrophic lateral sclerosis. *J. Neurochem.* **2011**, *116*, 1028–1042. [CrossRef]
50. Schwartz, J.C.; Ebmeier, C.C.; Podell, E.R.; Heimiller, J.; Taatjes, D.J.; Cech, T.R. FUS binds the CTD of RNA polymerase II and regulates its phosphorylation at Ser2. *Genes Dev.* **2012**, *26*, 2690–2695. [CrossRef]
51. Buratti, E.; Baralle, F.E. Characterization and Functional Implications of the RNA Binding Properties of Nuclear Factor TDP-43, a Novel Splicing Regulator of CFTR Exon 9. *J. Biol. Chem.* **2001**, *276*, 36337–36343. [CrossRef]
52. Leblond, C.S.; Gan-Or, Z.; Spiegelman, D.; Laurent, S.B.; Szuto, A.; Hodgkinson, A.; Dionne-Laporte, A.; Provencher, P.; de Carvalho, M.; Orrù, S.; et al. Replication study of MATR3 in familial and sporadic amyotrophic lateral sclerosis. *Neurobiol. Aging* **2016**, *37*, 17–209. [CrossRef] [PubMed]
53. Jutzi, D.; Akinyi, M.V.; Mechtersheimer, J.; Frilander, M.J.; Ruepp, M.D. The emerging role of minor intron splicing in neurological disorders. *Cell Stress* **2018**, *2*, 40–54. [CrossRef]

54. Scotti, M.M.; Swanson, M.S. RNA mis-splicing in disease. *Nat. Rev. Genet.* **2016**, *17*, 19. [CrossRef] [PubMed]
55. Johnson, J.O.; Pioro, E.P.; Boehringer, A.; Chia, R.; Feit, H.; Renton, A.E.; Pliner, H.A.; Abramzon, Y.; Marangi, G.; Winborn, B.J.; et al. Mutations in the Matrin 3 gene cause familial amyotrophic lateral sclerosis. *Nat. Neurosci.* **2014**, *17*, 664–666. [CrossRef] [PubMed]
56. Vance, C.; Scotter, E.L.; Nishimura, A.L.; Troakes, C.; Mitchell, J.C.; Kathe, C.; Urwin, H.; Manser, C.; Miller, C.C.; Hortobágyi, T.; et al. ALS mutant FUS disrupts nuclear localization and sequesters wild-type FUS within cytoplasmic stress granules. *Hum. Mol. Genet.* **2013**, *22*, 2676–2688. [CrossRef] [PubMed]
57. Kumar, V.; Hasan, G.M.; Hassan, M.I. Unraveling the role of RNA mediated toxicity of C9orf72 repeats in C9-FTD/ALS. *Front. Neurosci.* **2017**, *11*, 711. [CrossRef] [PubMed]
58. Kawahara, Y.; Mieda-Sato, A. TDP-43 promotes microRNA biogenesis as a component of the Drosha and Dicer complexes. *Proc. Natl. Acad. Sci. USA* **2012**, *109*, 3347–3352. [CrossRef]
59. Al-Chalabi, A.; Fang, F.; Hanby, M.F.; Leigh, P.N.; Shaw, C.E.; Ye, W.; Rijsdijk, F. An estimate of amyotrophic lateral sclerosis heritability using twin data. *J. Neurol. Neurosurg. Psychiatry* **2010**, *81*, 1324–1326. [CrossRef]
60. Dion, P.A.; Daoud, H.; Rouleau, G.A. Genetics of motor neuron disorders: New insights into pathogenic mechanisms. *Nat. Rev. Genet.* **2009**, *10*, 769–782. [CrossRef]
61. Andersen, P.M.; Al-Chalabi, A. Clinical genetics of amyotrophic lateral sclerosis: What do we really know? *Nat. Rev. Neurol.* **2011**, *7*, 603–615. [CrossRef] [PubMed]
62. Myers, R.H. Huntington's Disease Genetics. *NeuroRx* **2004**, *1*, 255–262. [CrossRef] [PubMed]
63. Loh, P.R.; Bhatia, G.; Gusev, A.; Finucane, H.K.; Bulik-Sullivan, B.K.; Pollack, S.J.; Lee, H.; Wray, N.R.; Kendler, K.S.; O'donovan, M.C.; et al. Contrasting genetic architectures of schizophrenia and other complex diseases using fast variance-components analysis. *Nat. Genet.* **2015**, *47*, 1385. [CrossRef] [PubMed]
64. Byrne, S.; Heverin, M.; Elamin, M.; Bede, P.; Lynch, C.; Kenna, K.; MacLaughlin, R.; Walsh, C.; Al Chalabi, A.; Hardiman, O. Aggregation of neurologic and neuropsychiatric disease in amyotrophic lateral sclerosis kindreds: A population-based case-control cohort study of familial and sporadic amyotrophic lateral sclerosis. *Ann. Neurol.* **2013**, *74*, 699–708. [CrossRef]
65. Renton, A.E.; Chiò, A.; Traynor, B.J. State of play in amyotrophic lateral sclerosis genetics. *Nat. Neurosci.* **2014**, *17*, 17–23. [CrossRef]
66. Majounie, E.; Renton, A.E.; Mok, K.; Dopper, E.G.; Waite, A.; Rollinson, S.; Chiò, A.; Restagno, G.; Nicolaou, N.; Simon-Sanchez, J.; et al. Frequency of the C9orf72 hexanucleotide repeat expansion in patients with amyotrophic lateral sclerosis and frontotemporal dementia: A cross-sectional study. *Lancet Neurol.* **2012**, *11*, 323–330. [CrossRef]
67. Majounie, E.; Abramzon, Y.; Renton, A.E.; Perry, R.; Bassett, S.S.; Pletnikova, O.; Troncoso, J.C.; Hardy, J.; Singleton, A.B.; Traynor, B.J. Repeat expansion in C9ORF72 in Alzheimer's disease. *N. Engl. J. Med.* **2012**, *366*, 283. [CrossRef]
68. Lesage, S.; Le Ber, I.; Condroyer, C.; Broussolle, E.; Gabelle, A.; Thobois, S.; Pasquier, F.; Mondon, K.; Dion, P.A.; Rochefort, D.; et al. C9orf72 repeat expansions are a rare genetic cause of parkinsonism. *Brain* **2013**, *136*, 385–391. [CrossRef]
69. Majoor-Krakauer, D.; Ottman, R.; Johnson, W.G.; Rowland, L.P. Familial aggregation of amyotrophic lateral sclerosis, dementia, and Parkinson's disease: Evidence of shared genetic susceptibility. *Neurology* **1994**, *44*, 1872–1877. [CrossRef]
70. Karch, C.M.; Wen, N.; Fan, C.C.; Yokoyama, J.S.; Kouri, N.; Ross, O.A.; Höglinger, G.; Müller, U.; Ferrari, R.; Hardy, J.; et al. Selective genetic overlap between amyotrophic lateral sclerosis and diseases of the frontotemporal dementia spectrum. *JAMA Neurol.* **2018**, *75*, 860–875. [CrossRef]
71. O'Brien, M.; Burke, T.; Heverin, M.; Vajda, A.; McLaughlin, R.; Gibbons, J.; Byrne, S.; Pinto-Grau, M.; Elamin, M.; Pender, N.; et al. Clustering of neuropsychiatric disease in first-degree and second-degree relatives of patients with amyotrophic lateral sclerosis. *JAMA Neurol.* **2017**, *74*, 1425–1430. [CrossRef] [PubMed]
72. Stearns, F.W. One hundred years of pleiotropy: A retrospective. *Genetics* **2010**, *186*, 767–773. [CrossRef] [PubMed]
73. Wu, J.; Liu, H.; Xiong, H.; Cao, J.; Chen, J. K-means-based consensus clustering: A unified view. *IEEE Trans. Knowl. Data Eng.* **2015**, *27*, 155–169. [CrossRef]
74. Wagstaff, K.; Cardie, C.; Rogers, S.; Schroedl, S. Constrained K-means Clustering with Background Knowledge. *Proc. Eighteenth Int. Conf. Mach. Learn.* **2001**, *1*, 577–584.

75. MacQueen, J.B. Some methods for classification and analysis of multivariate observations. In *Proceedings of the Fifth Berkeley Symposium on Mathematical Statistics and Probability;* University of California Press: Berkeley, CA, USA, 1967; pp. 281–297.
76. Schymick, J.C.; Scholz, S.W.; Fung, H.C.; Britton, A.; Arepalli, S.; Gibbs, J.R.; Lombardo, F.; Matarin, M.; Kasperaviciute, D.; Hernandez, D.G.; et al. Genome-wide genotyping in amyotrophic lateral sclerosis and neurologically normal controls: First stage analysis and public release of data. *Lancet Neurol.* **2007**, *6*, 322–328. [CrossRef]
77. Zuk, O.; Hechter, E.; Sunyaev, S.R.; Lander, E.S. The mystery of missing heritability: Genetic interactions create phantom heritability. *Proc. Natl. Acad. Sci. USA* **2012**, *109*, 1193–1198. [CrossRef]
78. Anderson, C.A.; Pettersson, F.H.; Clarke, G.M.; Cardon, L.R.; Morris, P.; Zondervan, K.T. Data quality control in genetic case-control association studies. *Nat. Protoc.* **2011**, *5*, 1564–1573. [CrossRef]
79. Marees, A.T.; de Kluiver, H.; Stringer, S.; Vorspan, F.; Curis, E.; Marie-Claire, C.; Derks, E.M. A tutorial on conducting genome-wide association studies: Quality control and statistical analysis. *Int. J. Methods Psychiatr. Res.* **2018**, *27*, 1–10. [CrossRef]
80. Verma, S.S.; de Andrade, M.; Tromp, G.; Kuivaniemi, H.; Pugh, E.; Namjou-Khales, B.; Mukherjee, S.; Jarvik, G.P.; Kottyan, L.C.; Burt, A.; et al. Imputation and quality control steps for combining multiple genome-wide datasets. *Front. Genet.* **2014**, *5*, 1–15. [CrossRef]
81. Laurie, C.C.; Doheny, K.F.; Mirel, D.B.; Pugh, E.W.; Bierut, L.J.; Bhangale, T.; Boehm, F.; Caporaso, N.E.; Cornelis, M.C.; Edenberg, H.J.; et al. Quality control and quality assurance in genotypic data for genome-wide association studies. *Genet. Epidemiol.* **2011**, *34*, 591–602. [CrossRef]
82. Dudbridge, F.; Gusnanto, A. Estimation of significance thresholds for genomewide association scans. *Genet. Epidemiol.* **2008**, *32*, 227–234. [CrossRef]
83. Culverhouse, R.; Suarez, B.K.; Lin, J.; Reich, T. A perspective on epistasis: Limits of models displaying no main effect. *Am. J. Hum. Genet.* **2002**, *70*, 461–471. [CrossRef]
84. Hemani, G.; Knott, S.; Haley, C. An Evolutionary Perspective on Epistasis and the Missing Heritability. *PLoS Genet.* **2013**, *9*, 1003295. [CrossRef]
85. Bateson, W.; Saunders, E.; Punnett, R.; Sons, C.H.U.H. Reports to the Evolution Committee of the Royal Society, Report II. London. *R. Soc.* **1905**, *2*, 5–131.
86. de Visser, J.A.G.M.; Cooper, T.F.; Elena, S.F. The causes of epistasis. *Proc. R. Soc. B Biol. Sci.* **2011**, *278*, 3617–3624. [CrossRef]
87. Pan, Q.; Hu, T.; Moore, J.H. *Epistasis, Complexity, and Multifactor Dimensionality Reduction*; Humana Press: Totowa, NJ, USA, 2013; pp. 465–477. [CrossRef]
88. Churchill, G.A. Epistasis. In *Brenner's Encyclopedia of Genetics*, 2nd ed.; Elsevier Inc.: Amsterdam, The Netherlands, 2013; pp. 505–507. [CrossRef]
89. Goudey, B.; Rawlinson, D.; Wang, Q.; Shi, F.; Ferra, H.; Campbell, R.M.; Stern, L.; Inouye, M.T.; Ong, C.S.; Kowalczyk, A. GWIS–model-free, fast and exhaustive search for epistatic interactions in case-control GWAS. *BMC Genom.* **2013**, *14* (Suppl. S3), 1–18. [CrossRef]
90. Yin, B.; Balvert, M.; Van Der Spek, R.A.; Dutilh, B.E.; Bohté, S.; Veldink, J.; Schönhuth, A. Using the structure of genome data in the design of deep neural networks for predicting amyotrophic lateral sclerosis from genotype. *Bioinformatics* **2019**, *35*, i538–i547. [CrossRef]
91. Kim, N.C.; Andrews, P.C.; Asselbergs, F.W.; Frost, H.R.; Williams, S.M.; Harris, B.T.; Read, C.; Askland, K.D.; Moore, J.H. Gene ontology analysis of pairwise genetic associations in two genome-wide studies of sporadic ALS. *BioData Min.* **2012**, *5*, 9. [CrossRef]
92. Reich, D.E.; Lander, E.S. On the allelic spectrum of human disease. *Trends Genet.* **2001**, *17*, 502–510. [CrossRef]
93. McCarthy, S.; Das, S.; Kretzschmar, W.; Delaneau, O.; Wood, A.R.; Teumer, A.; Kang, H.M.; Fuchsberger, C.; Danecek, P.; Sharp, K.; et al. A reference panel of 64,976 haplotypes for genotype imputation. *Nat. Genet.* **2016**, *48*, 1279–1283. [CrossRef]
94. Naj, A.C. Genotype Imputation in Genome-Wide Association Studies. *Curr. Protoc. Hum. Genet.* **2019**, *102*, e84. [CrossRef]
95. Pistis, G.; Porcu, E.; Vrieze, S.I.; Sidore, C.; Steri, M.; Danjou, F.; Busonero, F.; Mulas, A.; Zoledziewska, M.; Maschio, A.; et al. Rare variant genotype imputation with thousands of study-specific whole-genome sequences: Implications for cost-effective study designs. *Eur. J. Hum. Genet.* **2015**, *23*, 975–983. [CrossRef]

96. Maurano, M.T.; Humbert, R.; Rynes, E.; Thurman, R.E.; Haugen, E.; Wang, H.; Reynolds, A.P.; Sandstrom, R.; Qu, H.; Brody, J.; et al. Systematic Localization of Common Disease-Associated Variation in Regulatory DNA. *Science* **2012**, *337*, 1190–1195. [CrossRef]
97. Myszczynska, M.A.; Ojamies, P.N.; Lacoste, A.M.; Neil, D.; Saffari, A.; Mead, R.; Hautbergue, G.M.; Holbrook, J.D.; Ferraiuolo, L. Applications of machine learning to diagnosis and treatment of neurodegenerative diseases. *Nat. Rev. Neurol.* **2020**, *16*, 440–456. [CrossRef]
98. Deo, R.C. Machine learning in medicine. *Circulation* **2015**, *132*, 1920–1930. [CrossRef]
99. Chandrashekar, G.; Sahin, F. A survey on feature selection methods. *Comput. Electr. Eng.* **2014**, *40*, 16–28. [CrossRef]
100. Adadi, A.; Berrada, M. Peeking Inside the Black-Box: A Survey on Explainable Artificial Intelligence (XAI). *IEEE Access* **2018**, *6*, 52138–52160. [CrossRef]
101. Zhang, Q.S.; Zhu, S.C. Visual interpretability for deep learning: A survey. *Front. Inf. Technol. Electron. Eng.* **2018**, *19*, 27–39. [CrossRef]
102. Mirza, B.; Wang, W.; Wang, J.; Choi, H.; Chung, N.C.; Ping, P. Machine learning and integrative analysis of biomedical big data. *Genes* **2019**, *10*, 87. [CrossRef]
103. Vitsios, D.; Petrovski, S. Mantis-ml: Disease-Agnostic Gene Prioritization from High-Throughput Genomic Screens by Stochastic Semi-supervised Learning. *Am. J. Hum. Genet.* **2020**, *106*, 659–678. [CrossRef]
104. Yousefian-Jazi, A.; Sung, M.K.; Lee, T.; Hong, Y.H.; Choi, J.K.; Choi, J. Functional fine-mapping of noncoding risk variants in amyotrophic lateral sclerosis utilizing convolutional neural network. *Sci. Rep.* **2020**, *10*, 12872. [CrossRef] [PubMed]
105. Bean, D.M.; Al-Chalabi, A.; Dobson, R.J.B.; Iacoangeli, A. A Knowledge-Based Machine Learning Approach to Gene Prioritisation in Amyotrophic Lateral Sclerosis. *Genes* **2020**, *11*, 668. [CrossRef] [PubMed]
106. Cronin, S.; Berger, S.; Ding, J.; Schymick, J.C.; Washecka, N.; Hernandez, D.G.; Greenway, M.J.; Bradley, D.G.; Traynor, B.J.; Hardiman, O. A genome-wide association study of sporadic ALS in a homogenous Irish population. *Hum. Mol. Genet.* **2008**, *17*, 768–774. [CrossRef] [PubMed]
107. Greene, C.S.; Sinnott-Armstrong, N.A.; Himmelstein, D.S.; Park, P.J.; Moore, J.H.; Harris, B.T. Multifactor dimensionality reduction for graphics processing units enables genome-wide testing of epistasis in sporadic ALS. *Bioinformatics* **2010**, *26*, 694–695. [CrossRef]
108. Sha, Q.; Zhang, Z.; Schymick, J.C.; Traynor, B.J.; Zhang, S. Genome-wide association reveals three SNPs associated with sporadic amyotrophic lateral sclerosis through a two-locus analysis. *BMC Med Genet.* **2009**, *10*, 86. [CrossRef]
109. Köhler, S.; Vasilevsky, N.A.; Engelstad, M.; Foster, E.; McMurry, J.; Aymé, S.; Baynam, G.; Bello, S.M.; Boerkoel, C.F.; Boycott, K.M.; et al. The human phenotype ontology in 2017. *Nucleic Acids Res.* **2017**, *45*, D865–D876. [CrossRef]
110. Van Der Maaten, L.; Hinton, G. Visualizing Data using t-SNE. *J. Mach. Learn. Res.* **2008**, *9*, 2579–2605.
111. McInnes, L.; Healy, J.; Saul, N.; Großberger, L. UMAP: Uniform Manifold Approximation and Projection. *J. Open Source Softw.* **2018**, *3*, 861. [CrossRef]
112. Ritchie, M.D.; Hahn, L.W.; Roodi, N.; Bailey, L.R.; Dupont, W.D.; Parl, F.F.; Moore, J.H. Multifactor-dimensionality reduction reveals high-order interactions among estrogen-metabolism genes in sporadic breast cancer. *Am. J. Hum. Genet.* **2001**, *69*, 138–147. [CrossRef]
113. Kanehisa, M.; Furumichi, M.; Tanabe, M.; Sato, Y.; Morishima, K. KEGG: New perspectives on genomes, pathways, diseases and drugs. *Nucleic Acids Res.* **2017**, *45*, D353–D361. [CrossRef]
114. Matys, V.; Fricke, E.; Geffers, R.; Gößling, E.; Haubrock, M.; Hehl, R.; Hornischer, K.; Karas, D.; Kel, A.E.; Kel-Margoulis, O.V.; et al. TRANSFAC®: Transcriptional regulation, from patterns to profiles. *Nucleic Acids Res.* **2003**, *31*, 374–378. [CrossRef]
115. Bryne, J.C.; Valen, E.; Tang, M.H.E.; Marstrand, T.; Winther, O.; Da piedade, I.; Krogh, A.; Lenhard, B.; Sandelin, A. JASPAR, the open access database of transcription factor-binding profiles: New content and tools in the 2008 update. *Nucleic Acids Res.* **2008**, *36*, D102–D106. [CrossRef]
116. Bean, D.M.; Wu, H.; Dzahini, O.; Broadbent, M.; Stewart, R.; Dobson, R.J. Knowledge graph prediction of unknown adverse drug reactions and validation in electronic health records. *Sci. Rep.* **2017**, *7*. [CrossRef]

117. Dudbridge, F. Power and Predictive Accuracy of Polygenic Risk Scores. *PLoS Genet.* **2013**, *9*, 1003348. [CrossRef]
118. Boser, B.E.; Guyon, I.M.; Vapnik, V.N. Training algorithm for optimal margin classifiers. In *Proceedings of the Fifth Annual ACM Workshop on Computational Learning Theory*; ACM: New York, NY, USA, 1992; pp. 144–152. [CrossRef]
119. Breiman, L. Random forests. *Mach. Learn.* **2001**, *45*, 5–32. [CrossRef]
120. Freund, Y.; Schapire, R.E. A Short Introduction to Boosting. *J. Jpn. Soc. Artif. Intell.* **1999**, *14*, 771–780.
121. Friedman, J.; Hastie, T.; Tibshirani, R. Additive logistic regression: A statistical view of boosting (With discussion and a rejoinder by the authors). *Ann. Stat.* **2000**, *28*, 337–407. [CrossRef]
122. Lee, T.; Sung, M.K.; Lee, S.; Yang, W.; Oh, J.; Kim, J.Y.; Hwang, S.; Ban, H.J.; Choi, J.K. Convolutional neural network model to predict causal risk factors that share complex regulatory features. *Nucleic Acids Res.* **2019**, *47*, 146. [CrossRef]
123. Cardona, A.E.; Pioro, E.P.; Sasse, M.E.; Kostenko, V.; Cardona, S.M.; Dijkstra, I.M.; Huang, D.R.; Kidd, G.; Dombrowski, S.; Dutta, R.; et al. Control of microglial neurotoxicity by the fractalkine receptor. *Nat. Neurosci.* **2006**, *9*, 917–924. [CrossRef]
124. Ransohoff, R.M.; Cardona, A.E. The myeloid cells of the central nervous system parenchyma. *Nature* **2010**, *468*, 253–262. [CrossRef]
125. Liu, N.; Yu, Z.; Xun, Y.; Li, M.; Peng, X.; Xiao, Y.; Hu, X.; Sun, Y.; Yang, M.; Gan, S.; et al. TNFAIP1 contributes to the neurotoxicity induced by Aβ25-35 in Neuro2a cells. *BMC Neurosci.* **2016**, *17*, 51. [CrossRef] [PubMed]
126. McGill, B.E.; Barve, R.A.; Maloney, S.E.; Strickland, A.; Rensing, N.; Wang, P.L.; Wong, M.; Head, R.; Wozniak, D.F.; Milbrandt, J. Abnormal microglia and enhanced inflammation-related gene transcription in mice with conditional deletion of Ctcf in Camk2a-Cre-expressing neurons. *J. Neurosci.* **2018**, *38*, 200–219. [CrossRef] [PubMed]
127. Nagamoto-Combs, K.; Combs, C.K. Microglial phenotype is regulated by activity of the transcription factor, NFAT (nuclear factor of activated T cells). *J. Neurosci.* **2010**, *30*, 9641–9646. [CrossRef] [PubMed]
128. Limviphuvadh, V.; Tanaka, S.; Goto, S.; Ueda, K.; Kanehisa, M. The commonality of protein interaction networks determined in neurodegenerative disorders (NDDs). *Bioinformatics* **2007**, *23*, 2129–2138. [CrossRef] [PubMed]
129. Mademan, I.; Harmuth, F.; Giordano, I.; Timmann, D.; Magri, S.; Deconinck, T.; Claaßen, J.; Jokisch, D.; Genc, G.; Di Bella, D.; et al. Multisystemic SYNE1 ataxia: Confirming the high frequency and extending the mutational and phenotypic spectrum. *Brain* **2016**, *139*, e46. [CrossRef]
130. Bergemalm, D.; Forsberg, K.; Jonsson, P.A.; Graffmo, K.S.; Brä Nnströ M §, T.; Andersen, P.M.; Antti, H.; Marklund, S.L. Changes in the Spinal Cord Proteome of an Amyotrophic Lateral Sclerosis Murine Model Determined by Differential In-gel Electrophoresis. *Mol. Cell. Proteom.* **2009**, *8*, 1306–1317. [CrossRef] [PubMed]
131. Saris, C.G.; Horvath, S.; van Vught, P.W.; van Es, M.A.; Blauw, H.M.; Fuller, T.F.; Langfelder, P.; DeYoung, J.; Wokke, J.H.; Veldink, J.H.; et al. Weighted gene co-expression network analysis of the peripheral blood from Amyotrophic Lateral Sclerosis patients. *BMC Genom.* **2009**, *10*, 405. [CrossRef] [PubMed]
132. Andrés-Benito, P.; Moreno, J.; Aso, E.; Povedano, M.; Ferrer, I. Amyotrophic lateral sclerosis, gene deregulation in the anterior horn of the spinal cord and frontal cortex area 8: Implications in frontotemporal lobar degeneration. *Aging* **2017**, *9*, 823–851. [CrossRef]
133. Martin, L.J.; Wong, M. Aberrant Regulation of DNA Methylation in Amyotrophic Lateral Sclerosis: A New Target of Disease Mechanisms. *Neurotherapeutics* **2013**, *10*, 722–733. [CrossRef]
134. Wong, M.; Gertz, B.; Chestnut, B.A.; Martin, L.J. Mitochondrial DNMT3A and DNA methylation in skeletal muscle and CNS of transgenic mouse models of ALS. *Front. Cell. Neurosci.* **2013**, *7*, 279. [CrossRef]
135. Gandhi, T.K.; Zhong, J.; Mathivanan, S.; Karthick, L.; Chandrika, K.N.; Mohan, S.S.; Sharma, S.; Pinkert, S.; Nagaraju, S.; Periaswamy, B.; et al. Analysis of the human protein interactome and comparison with yeast, worm and fly interaction datasets. *Nat. Genet.* **2006**, *38*, 285–293. [CrossRef]
136. Licata, L.; Orchard, S. The MIntAct Project and Molecular Interaction Databases. *Methods Mol. Biol.* **2016**, *1415*, 55–69. [CrossRef]

137. Orchard, S.; Ammari, M.; Aranda, B.; Breuza, L.; Briganti, L.; Broackes-Carter, F.; Campbell, N.H.; Chavali, G.; Chen, C.; Del-Toro, N.; et al. The MIntAct project - IntAct as a common curation platform for 11 molecular interaction databases. *Nucleic Acids Res.* **2014**, *42*, D358–D363. [CrossRef]
138. Piñero, J.; Ramírez-Anguita, J.M.; Saüch-Pitarch, J.; Ronzano, F.; Centeno, E.; Sanz, F.; Furlong, L.I. The DisGeNET knowledge platform for disease genomics: 2019 update. *Nucleic Acids Res.* **2020**, *48*, D845–D855. [CrossRef]
139. Piñero, J.; Bravo, À.; Queralt-Rosinach, N.; Gutiérrez-Sacristán, A.; Deu-Pons, J.; Centeno, E.; García-García, J.; Sanz, F.; Furlong, L.I. DisGeNET: A comprehensive platform integrating information on human disease-associated genes and variants. *Nucleic Acids Res.* **2016**, *45*, D833–D839. [CrossRef]
140. Ashburner, M.; Ball, C.A.; Blake, J.A.; Botstein, D.; Butler, H.; Cherry, J.M.; Davis, A.P.; Dolinski, K.; Dwight, S.S.; Eppig, J.T.; et al. Gene ontology: Tool for the unification of biology. *Nat. Genet.* **2000**, *25*, 25–29. [CrossRef]
141. Landrum, M.J.; Lee, J.M.; Riley, G.R.; Jang, W.; Rubinstein, W.S.; Church, D.M.; Maglott, D.R. ClinVar: Public archive of relationships among sequence variation and human phenotype. *Nucleic Acids Res.* **2014**, *42*, D980–D985. [CrossRef]
142. de Leeuw, C.A.; Mooij, J.M.; Heskes, T.; Posthuma, D. MAGMA: Generalized Gene-Set Analysis of GWAS Data. *PLoS Comput. Biol.* **2015**, *11*, 1–19. [CrossRef]
143. Moore, J.H.; Williams, S.M. Epistasis: Methods and Protocols. *Epistasis Methods Protoc.* **2014**, *1253*, 1–346. [CrossRef]
144. Chattopadhyay, A.; Lu, T.P. Gene-gene interaction: The curse of dimensionality. *Ann. Transl. Med.* **2019**, *7*, 813. [CrossRef]
145. Angermueller, C.; Pärnamaa, T.; Parts, L.; Stegle, O. Deep learning for computational biology. *Mol. Syst. Biol.* **2016**, *12*, 878. [CrossRef] [PubMed]
146. Bellot, P.; de los Campos, G.; Pérez-Enciso, M. Can deep learning improve genomic prediction of complex human traits? *Genetics* **2018**, *210*, 809–819. [CrossRef] [PubMed]
147. Sherry, S.T.; Ward, M.H.; Kholodov, M.; Baker, J.; Phan, L.; Smigielski, E.M.; Sirotkin, K. DbSNP: The NCBI database of genetic variation. *Nucleic Acids Res.* **2001**, *29*, 308–311. [CrossRef] [PubMed]
148. Gaudet, P.; Livstone, M.S.; Lewis, S.E.; Thomas, P.D. Phylogenetic-based propagation of functional annotations within the Gene Ontology consortium. *Brief. Bioinform.* **2011**, *12*, 449–462. [CrossRef] [PubMed]
149. De Felice, B.; Guida, M.; Guida, M.; Coppola, C.; De Mieri, G.; Cotrufo, R. A miRNA signature in leukocytes from sporadic amyotrophic lateral sclerosis. *Gene* **2012**, *508*, 35–40. [CrossRef]
150. Maragakis, N.J.; Dykes-Hoberg, M.; Rothstein, J.D. Altered Expression of the Glutamate Transporter EAAT2b in Neurological Disease. *Ann. Neurol.* **2004**, *55*, 469–477. [CrossRef]
151. Tzu, J.; Marinkovich, M.P. Bridging structure with function: Structural, regulatory, and developmental role of laminins. *Int. J. Biochem. Cell Biol.* **2008**, *40*, 199–214. [CrossRef]
152. Choi, M.R.; An, C.H.; Yoo, N.J.; Lee, S.H. Laminin gene LAMB4 is somatically mutated and expressionally altered in gastric and colorectal cancers. *APMIS* **2015**, *123*, 65–71. [CrossRef]
153. Zhang, H.; Pan, Y.Z.; Cheung, M.; Cao, M.; Yu, C.; Chen, L.; Zhan, L.; He, Z.W.; Sun, C.Y. LAMB3 mediates apoptotic, proliferative, invasive, and metastatic behaviors in pancreatic cancer by regulating the PI3K/Akt signaling pathway. *Cell Death Dis.* **2019**, *10*, 230. [CrossRef]
154. Jena, K.K.; Kolapalli, S.P.; Mehto, S.; Nath, P.; Das, B.; Sahoo, P.K.; Ahad, A.; Syed, G.H.; Raghav, S.K.; Senapati, S.; et al. TRIM16 controls assembly and degradation of protein aggregates by modulating the p62-NRF2 axis and autophagy. *EMBO J.* **2018**, *37*, e98358. [CrossRef]
155. Grassi, D.A.; Jönsson, M.E.; Brattås, P.L.; Jakobsson, J. TRIM28 and the control of transposable elements in the brain. *Brain Res.* **2019**, *1705*, 43–47. [CrossRef] [PubMed]
156. Li, W.; Jin, Y.; Prazak, L.; Hammell, M.; Dubnau, J. Transposable Elements in TDP-43-Mediated Neurodegenerative Disorders. *PLoS ONE* **2012**, *7*, e44099. [CrossRef]
157. Nicholls, H.L.; John, C.R.; Watson, D.S.; Munroe, P.B.; Barnes, M.R.; Cabrera, C.P. Reaching the End-Game for GWAS: Machine Learning Approaches for the Prioritization of Complex Disease Loci. *Front. Genet.* **2020**, *11*, 350. [CrossRef]

158. Ritchie, M.D.; Van Steen, K. The search for gene-gene interactions in genome-wide association studies: Challenges in abundance of methods, practical considerations, and biological interpretation. *Ann. Transl. Med.* **2018**, *6*, 157. [CrossRef] [PubMed]
159. Ritchie, M.D. Using Biological Knowledge to Uncover the Mystery in the Search for Epistasis in Genome-Wide Association Studies. *Ann. Hum. Genet.* **2011**, *75*, 172–182. [CrossRef]

Journal of Personalized Medicine

Editorial

Understanding Neuromuscular Health and Disease: Advances in Genetics, Omics, and Molecular Function

William J. Duddy * and Stephanie Duguez *

Northern Ireland Center for Stratified/Personalized Medicine, Ulster University, Glenshane Road, Derry-Londonderry BT47 6SB, UK
* Correspondence: w.duddy@ulster.ac.uk (W.J.D.); s.duguez@ulster.ac.uk (S.D.)

The field of neuromuscular research has seen considerable recent advances in the molecular and cellular understanding of muscle biology, and the treatment of neuromuscular disease. These advances are at the forefront of modern molecular methodologies, often integrating across wet-lab cell and tissue models, dry-lab computational approaches, and clinical studies. The continuing development and application of multiomics methods offer particular challenges and opportunities, not least in the potential for personalized medicine. More than 500 different genes are known to be associated with neuromuscular disorders [1], and the identification of causative mutations has allowed the development of personalized therapies [1,2]. However, even if great progress has been made during the last two decades in different subgroups of neuromuscular disorders, there are still numerous challenges to resolve, such as the optimization of therapeutic knock-down strategies [3], targeting specific muscles and/or tissues of the nervous system [1], identifying genetic modifiers that can impair a therapeutic strategy [2], targeting common pathways being affected in different patient subgroups for a given disease [4,5], or understanding the impact of neuromuscular disorders on other tissues that could be affected but may be understudied.

This Special Issue, entitled "Understanding Neuromuscular Health and Disease: Advances in Genetics, Omics, and Molecular Function", encompasses some 15 publications from colleagues working on a diverse range of neuromuscular diseases, including Duchenne muscular dystrophy [6–9], facioscapulohumeral dystrophy [3,10,11], amyotrophic lateral sclerosis [4,5,12], spinal muscular atrophy [2], Emery–Dreifuss muscular dystrophy [13], and rheumatoid arthritis [14]. Looking across diseases, several themes are recurrent, such as the efforts to identify genotype–phenotype correlations in DMD [6,7,9] and ALS [4,5], the quest for effective biomarkers in many neuromuscular conditions [2,8,10,14], and the use of genomic and multi-omic approaches towards better ways to identify biomarkers and to understand disease [10,12,13].

The search for genotype–phenotype correlations can be aimed at the improved understanding of disease [4,5,7,9], but may also be relevant to potential therapeutic outcomes [6]. Of relevance to this Special Issue are genotype–phenotype correlations in DMD [6,7,9] and in ALS [4,5]. It is interesting to contrast the state of these investigations in these two conditions, differences which are related to the underlying genetics: DMD being due to mutations at a single gene, and therefore correlations being sought between clinical outcomes and specific mutation patterns within that gene [6,7,9]; ALS being a disease of unclear aetiology for the majority of patients and the focus of these genotype–phenotype investigations thus being on the relationship of different genes and their functional roles to the implicated mechanisms [4] or clinical outcomes of the condition [5].

The use of genomics and multi-omics approaches is a theme which itself cuts across the aims of current research, from the overlay of multiple omics data to achieve a global perspective and new understanding of Emery–Dreifuss muscular dystrophy [13], to the identification of novel circulating miRNA and protein biomarkers for FSHD using multi-omics [10], through to the application of machine learning to the genomics of ALS, which

Citation: Duddy, W.J.; Duguez, S. Understanding Neuromuscular Health and Disease: Advances in Genetics, Omics, and Molecular Function. *J. Pers. Med.* **2021**, *11*, 438. https://doi.org/10.3390/jpm11050438

Received: 30 April 2021
Accepted: 18 May 2021
Published: 20 May 2021

is aimed at understanding the molecular basis of this disease (and is also relevant to genotype–phenotype correlations) [12]. Deciphering the pathways and gene mutations involved in neuromuscular diseases may allow for the development of computational models helping our understanding of muscle pathologies, which could enable preclinical studies of neuromuscular diseases in the context of personalized medicine [15].

Aside from therapeutic strategy development, the use of biomarkers may be critical as disease trackers for the development of effective therapeutics (for example, in FSHD [10]), but also to the personalized tailoring of existing treatments (in DMD [8], and in rheumatoid arthritis [14]), and may prove useful in a broad sense for improved stratification, diagnosis, and/or treatment (e.g., in adult SMA [2]).

We hope that studies such as these, that integrate modern molecular methodologies across cell and tissue models, computational approaches, and clinical studies, will continue to drive progress towards improved neuromuscular health and treatments for these often severe diseases.

Funding: Not applicable.

Institutional Review Board Statement: Not applicable.

Informed Consent Statement: Not applicable.

Data Availability Statement: Not applicable.

Acknowledgments: We wish to thank all the authors who contributed their work, and in so doing made the Special Issue a success. We also wish to thank the staff of *JPM* for their excellent support throughout the editorial process.

Conflicts of Interest: The authors declare no conflict of interest.

References

1. Buscara, L.; Gross, D.-A.; Daniele, N. Of rAAV and Men: From Genetic Neuromuscular Disorder Efficacy and Toxicity Preclinical Studies to Clinical Trials and Back. *J. Pers. Med.* **2020**, *10*, 258. [CrossRef] [PubMed]
2. Smeriglio, P.; Langard, P.; Querin, G.; Biferi, M.G. The Identification of Novel Biomarkers Is Required to Improve Adult SMA Patient Stratification, Diagnosis and Treatment. *J. Pers. Med.* **2020**, *10*, 75. [CrossRef] [PubMed]
3. Joubert, R.; Mariot, V.; Charpentier, M.; Concordet, J.P.; Dumonceaux, J. Gene Editing Targeting the DUX4 Polyadenylation Signal: A Therapy for FSHD? *J. Pers. Med.* **2020**, *11*, 7. [CrossRef] [PubMed]
4. Le Gall, L.; Anakor, E.; Connolly, O.; Vijayakumar, U.; Duddy, W.; Duguez, S. Molecular and Cellular Mechanisms Affected in ALS. *J. Pers. Med.* **2020**, *10*, 101. [CrossRef] [PubMed]
5. Connolly, O.; Le Gall, L.; McCluskey, G.; Donaghy, C.G.; Duddy, W.J.; Duguez, S. A Systematic Review of Genotype–Phenotype Correlation across Cohorts Having Causal Mutations of Different Genes in ALS. *J. Pers. Med.* **2020**, *10*, 58. [CrossRef] [PubMed]
6. Anwar, S.; He, M.; Lim, K.R.Q.; Maruyama, R.; Yokota, T. A Genotype-Phenotype Correlation Study of Exon Skip-Equivalent In-Frame Deletions and Exon Skip-Amenable Out-of-Frame Deletions across the DMD Gene to Simulate the Effects of Exon-Skipping Therapies: A Meta-Analysis. *J. Pers. Med.* **2021**, *11*, 46. [CrossRef] [PubMed]
7. Lim, K.R.Q.; Nguyen, Q.; Yokota, T. Genotype–Phenotype Correlations in Duchenne and Becker Muscular Dystrophy Patients from the Canadian Neuromuscular Disease Registry. *J. Pers. Med.* **2020**, *10*, 241. [CrossRef] [PubMed]
8. Tawalbeh, S.; Samsel, A.; Gordish-Dressman, H.; Hathout, Y.; Investigators, C.-D.; Dang, U.J. Comparison of Serum Pharmacodynamic Biomarkers in Prednisone-Versus Deflazacort-Treated Duchenne Muscular Dystrophy Boys. *J. Pers. Med.* **2020**, *10*, 164. [CrossRef] [PubMed]
9. Sheikh, O.; Yokota, T. Advances in Genetic Characterization and Genotype–Phenotype Correlation of Duchenne and Becker Muscular Dystrophy in the Personalized Medicine Era. *J. Pers. Med.* **2020**, *10*, 111. [CrossRef] [PubMed]
10. Heier, C.; Zhang, A.; Nguyen, N.; Tully, C.; Panigrahi, A.; Gordish-Dressman, H.; Pandey, S.; Guglieri, M.; Ryan, M.; Clemens, P.; et al. Multi-Omics Identifies Circulating miRNA and Protein Biomarkers for Facioscapulohumeral Dystrophy. *J. Pers. Med.* **2020**, *10*, 236. [CrossRef] [PubMed]
11. Sidlauskaite, E.; Le Gall, L.; Mariot, V.; Dumonceaux, J. DUX4 Expression in FSHD Muscles: Focus on Its mRNA Regulation. *J. Pers. Med.* **2020**, *10*, 73. [CrossRef] [PubMed]
12. Vasilopoulou, C.; Morris, A.P.; Giannakopoulos, G.; Duguez, S.; Duddy, W. What Can Machine Learning Approaches in Genomics Tell Us about the Molecular Basis of Amyotrophic Lateral Sclerosis? *J. Pers. Med.* **2020**, *10*, 247. [CrossRef] [PubMed]
13. Vignier, N.; Muchir, A. An Omics View of Emery–Dreifuss Muscular Dystrophy. *J. Pers. Med.* **2020**, *10*, 50. [CrossRef] [PubMed]

14. D'Cruz, L.G.; McEleney, K.G.; Tan, K.B.C.; Shukla, P.; Gardiner, P.V.; Connolly, P.; Conway, C.; Cobice, D.; Gibson, D.S. Clinical and Laboratory Associations with Methotrexate Metabolism Gene Polymorphisms in Rheumatoid Arthritis. *J. Pers. Med.* **2020**, *10*, 149. [CrossRef] [PubMed]
15. Speciale, A.A.; Ellerington, R.; Goedert, T.; Rinaldi, C. Modelling Neuromuscular Diseases in the Age of Precision Medicine. *J. Pers. Med.* **2020**, *10*, 178. [CrossRef] [PubMed]

Journal of *Personalized Medicine*

MDPI

Article

A Genotype-Phenotype Correlation Study of Exon Skip-Equivalent In-Frame Deletions and Exon Skip-Amenable Out-of-Frame Deletions across the *DMD* Gene to Simulate the Effects of Exon-Skipping Therapies: A Meta-Analysis

Saeed Anwar [1], Merry He [1], Kenji Rowel Q. Lim [1], Rika Maruyama [1] and Toshifumi Yokota [1,2,*]

[1] Department of Medical Genetics, Faculty of Medicine and Dentistry, University of Alberta, Edmonton, AB T6G 2H7, Canada; sanwar@ualberta.ca (S.A.); merry1@ualberta.ca (M.H.); kenjirow@ualberta.ca (K.R.Q.L.); yokotama@ualberta.ca (R.M.)

[2] The Friends of Garrett Cumming Research and Muscular Dystrophy Canada, HM Toupin Neurological Science Research Chair, Edmonton, AB T6G 2H7, Canada

* Correspondence: toshifum@ualberta.ca; Tel.: +1-(780)-492-1102

Abstract: Dystrophinopathies are caused by mutations in the *DMD* gene. Out-of-frame deletions represent most mutational events in severe Duchenne muscular dystrophy (DMD), while in-frame deletions typically lead to milder Becker muscular dystrophy (BMD). Antisense oligonucleotide-mediated exon skipping converts an out-of-frame transcript to an in-frame one, inducing a truncated but partially functional dystrophin protein. The reading frame rule, however, has many exceptions. We thus sought to simulate clinical outcomes of exon-skipping therapies for *DMD* exons from clinical data of exon skip-equivalent in-frame deletions, in which the expressed quasi-dystrophins are comparable to those resulting from exon-skipping therapies. We identified a total of 1298 unique patients with exon skip-equivalent mutations in patient registries and the existing literature. We classified them into skip-equivalent deletions of each exon and statistically compared the ratio of DMD/BMD and asymptomatic individuals across the *DMD* gene. Our analysis identified that five exons are associated with significantly milder phenotypes than all other exons when corresponding exon skip-equivalent in-frame deletion mutations occur. Most exon skip-equivalent in-frame deletions were associated with a significantly milder phenotype compared to corresponding exon skip-amenable out-of-frame mutations. This study indicates the importance of genotype-phenotype correlation studies in the rational design of exon-skipping therapies.

Keywords: dystrophinopathy; duchenne muscular dystrophy (DMD); becker muscular dystrophy (BMD); dystrophin; reading frame rule; exon skipping; skip-equivalent deletions

check for updates

Citation: Anwar, S.; He, M.; Lim, K.R.Q.; Maruyama, R.; Yokota, T. A Genotype-Phenotype Correlation Study of Exon Skip-Equivalent In-Frame Deletions and Exon Skip-Amenable Out-of-Frame Deletions across the *DMD* Gene to Simulate the Effects of Exon-Skipping Therapies: A Meta-Analysis. *J. Pers. Med.* **2021**, *11*, 46. https://doi.org/10.3390/jpm11010046

Received: 1 November 2020
Accepted: 22 December 2020
Published: 14 January 2021

1. Introduction

Dystrophinopathies are a spectrum of X-linked muscular dystrophies caused by mutations in the *DMD* gene encoding the dystrophin protein, which helps maintain the integrity of muscle membranes [1]. The most lethal end of this spectrum, Duchenne muscular dystrophy (DMD), generally arises from mutations that disrupt the translational reading frame and result in an absence of dystrophin [2]. Three other major conditions that belong to this spectrum are Becker muscular dystrophy (BMD, a mild form of DMD), intermediate muscular dystrophy (IMD, an intermediate form between DMD and BMD), and DMD-associated dilated cardiomyopathy (DCM, a type of nonischemic heart-disease) [2].

With 79 constitutive exons and (at least) seven alternatively-used exons, the *DMD* gene is one of the largest known genes in the human genome [2,3]. Because of the enormous length of the gene, it is highly vulnerable to mutations and, one out of three mutations are de novo in nature. Besides, the presence of two mutational hotspots, encompassing exons 3–22 and exons 45–55, within the coding sequence of the gene makes it more vulnerable to mutations [4–6]. To date, over five thousand *DMD* mutations have been reported. Large deletions involving deletion of one or more exons account for most cases (~68%), while duplications, other small mutations, and rarely deep intronic copy number variations cause the rest of the cases [7–9]. The major determinant of phenotypes is the reading frame rule, i.e., the mutations giving rise to an mRNA with disrupted reading frame (out-of-frame) result in a severe pathology, i.e., DMD. On the other hand, in-frame deletions result in the production of truncated yet (partially) functional dystrophins, causing a mild clinical phenotype called BMD [10]. The reading frame rule is accurately predictive in ~90% of DMD cases; however, not as consistent for BMD cases, with prediction rates ranging from 56–91% in different cohorts [8,11].

This reading frame rule provides the rationale for a therapeutic strategy, called therapeutic exon skipping using antisense oligonucleotides to transform DMD-related out-of-frame mRNAs into in-frame ones to produce truncated dystrophin, against severe and lethal cases of DMD. Recently, three exon-skipping oligonucleotide drugs, namely eteplirsen, golodirsen, and viltolarsen, received accelerated approval by the U.S. Food and Drug Administration (FDA), and a few other oligonucleotide-based drugs are in later stages of clinical trials [12–14]. As such, antisense oligonucleotide-mediated single exon skipping has great promise for treating DMD effectively. However, because of these drugs' highly specific nature, they apply to only a small portion of the DMD population, e.g., eteplirsen, golodirsen, and viltolarsen together are applicable for a total of around 20% of the entire DMD population in the U.S. [12,13]. Recently, skipping of multiple exons, e.g., exons 45–55, using mutation-tailored cocktails of antisense oligonucleotides have shown potential for clinical application, and the applicability of such cocktail drug was shown to reach over 65% of DMD patients with large deletions [6,15].

The reading frame rule, however, is not always an accurate predictor of clinical phenotypes [16]. As multiple exonic deletions are amenable to skipping of each exon, the quasi-dystrophins resulting from the various exon skipped transcripts may vary in stability, function, and phenotype. The DMD/BMD phenotypic ratio observed in individuals with confirmed in-frame deletions starting and/or ending at frameshifting exons have been examined to estimate the therapeutic outcomes from skipping several exons [17–19]. In this study, we employed meta-analysis on the published literature and databases containing data on the genotype-phenotype association. It provides an overview of clinical presentations in patients with each exon "skip-equivalent" deletion and determines the best estimate of each exon skip-amenable mutation's clinical outcome of exon-skipping therapies.

2. Materials and Methods

2.1. Clinical Database Search and Inclusion/Exclusion Criteria

The French UMD-DMD-France database and the eDystrophin are two online registries for individuals and families affected by dystrophinopathies [20,21]. From 30 March 2020 to 30 July 2020, we queried these two databases for relevant cases. We excluded patients with pending clinical phenotypes and female carriers from the analysis. If a record were present in both databases, the data were entered only once into the data tabulation sheet.

2.2. Literature Search and Inclusion/Exclusion Criteria

To collect more data on the clinical features of genetically confirmed DMD patients, a literature search of PubMed was conducted using the following query terms: Dystrophin*[Title/Abstract]) OR (Duchenne[Title/Abstract])) OR (Becker*[Title/Abstract])) OR (muscular dystrophy[Title/Abstract]) OR (mutation[Title/Abstract])) OR (large deletion[Title/Abstract])) OR (mutation spectrum[Title/Abstract])) OR (MLPA[Title/Abstract])) OR (ligation[Title/Abstract])) OR (CGH[Title/Abstract])) OR (comparative genomic hybridization[Title/Abstract])). We then also searched google scholar using an equivalent query.

Two researchers (S.A. and M.H.) independently reviewed titles and abstracts of the identified articles to determine that only relevant publications were included. Articles were excluded if, after review, it was evident that the article discussed a non-human disease model, diseases other than dystrophinopathies, a non-dystrophin study, or presents a meta-analysis. We excluded the manuscripts presenting inadequate genotype-phenotype information or discussing only female carriers. Additionally, we excluded the studies that were included in any of the two databases mentioned above. In addition, only the manuscripts reporting cases of dystrophinopathies detected by multiplex ligation-dependent probe amplification (MLPA), comparative genomic hybridization (CGH), or equivalent diagnostic procedures were included. No restrictions were made based on language, publication year, publication status, and the latest search date.

We identified a total of 612 unique articles by searching PubMed and Google Scholar, out of which 12 articles were finally selected based on our exclusion criteria after the title and abstract review. We then conducted a comprehensive review of these 12 articles to identify individuals with genetically confirmed large deletions in *dystrophin* (Figure 1, Figure S1, Table S1).

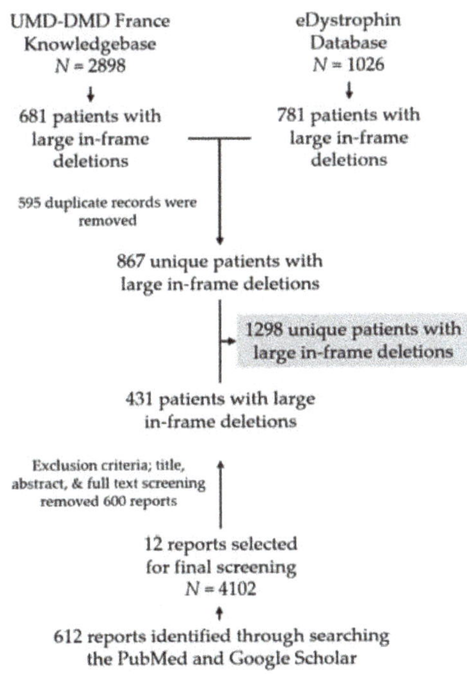

Figure 1. Flow chart showing database screening and literature search procedure to collect patients' clinical information with confirmed DMD large deletions. *N* indicates the number of individuals present in each data source.

2.3. Data Extraction and Analysis of the Data

Two authors (S.A. and M.H.) independently reviewed the full-text versions and, if available, the Supplementary Materials to extract clinical data of dystrophinopathy patients with in-frame deletions from the literature. Data of patients with deletions involving exons 1 and 79 were not included as deletions of these two exons would not produce truncated dystrophins. To assess duplicate records in multiple databases and literature, we utilized the following strategy. One of the investigators (S.A.) compared the partially de-identified information (identifiers, country origin, age at diagnosis, date of birth, and available mutation information, e.g., mutation start point and end point, and type of mutation), assuming any records with these same items represented the same patient [22,23]. Another author (K.R.Q.L.) scrutinized the inclusion of mutation information and the removal of duplicate records. We categorized the clinically confirmed patients into two groups: (i) DMD, and (ii) BMD and asymptomatic. We categorized BMD and asymptomatic cases together because exon-skipping therapies aim to convert lethal DMD cases into milder forms, whether BMD or asymptomatic, with in-frame deletions. In addition, those asymptomatic cases at the time of examination could become BMD at a later age. The asymptomatic cases, therefore, are combined with BMD in a single group.

2.4. Exon Skip-Amenable and Exon Skip-Equivalent Mutations

We defined exon skip-amenable mutation as follows; (1) for frame-shifting exons (i.e., exons 2, 6, 7, 8, 11, 12, 17, 18, 19, 20, 21, 22, 43, 44, 45, 46, 50, 51, 52, 53, 54, 55, 56, 57, 58, 59, 61, 62, 63, 65, 66, 67, 68, 69, 70, 75, 76, and 78), we included large out-of-frame deletion mutations that can be made in-frame by skipping of exon X as "exon X skip-amenable mutations". For example, exon 52 of DMD is a frameshifting exon, and skipping of an adjacent exon, exon 51 or exon 53, theoretically restores the reading frame (exon 51 skip-amenable and exon 53 skip-amenable), (2) for non-frame-shifting exons (all other exons between exons 2–78), we included nonsense mutations and small out-of-frame indels in exon Y as "exon Y skip-amenable mutations". For example, therapeutic skipping of exon 23 for a nonsense mutation in exon 23 would restore the reading frame, leading to truncated dystrophin protein expression (exon 23 skip-amenable). We included patients who naturally harbor large in-frame deletion mutations starting and/or ending at exon Z as "exon Z skip-equivalent mutation". Deletions starting at exon 1 and ending at exon 79 are not included.

The same overall definitions of exon skip-amenable and exon skip-equivalent mutations can be applied for multi-exon regions, e.g., exons 3–9 or 45–55. Mutations (i.e., large deletions, large duplications, small indels, and nonsense) whose disrupted reading frame can be restored by multi-exon skipping are considered multi-exon skip-amenable. Naturally, mutations involving both frame-shifting or non-frame-shifting exons are included in our counts. Mutations partly outside of exons 3–9 and 45–55 are not included in exons 3–9 and 45–55 skip-amenable mutations, respectively (e.g., del. ex 43–45, del. ex 51–57). On the other hand, naturally occurring mutations (e.g., in-frame large deletions of exons 3 to 9 and exons 45 to 55) that model multi-exon-skipped transcripts are considered multi-exon skip-equivalent.

2.5. Statistical Analysis

The odds of DMD for a given exon were compared to BMD and asymptomatic phenotype's odds using Fisher's exact test. The ratios of DMD to BMD/asymptomatic phenotypes associated with a specific mutation and all other mutations were compared using Fisher's exact test. The test-statistic values of multiple test comparisons were adjusted using the Benjamini–Hochberg method to control false discovery rates [24]. To identify if there were hotspot regions in the gene, we compared the frequency of in-frame deletions starting and/or ending at a given exonic region and all other exons using f-statistic. Additionally, we compared the phenotypic outcomes associated with out-of-frame mutations amenable to exon skipping (exon skip-amenable) and exon skip-equivalent in-frame deletions of each exon using Fisher's exact test. Since the databases included and the manuscripts reviewed were, in general, descriptive, case-specific or focus group studies rather than randomized ones, we could not do a formal risk of bias assessment on the data. All data were analyzed in Microsoft Excel (Office 365, 2019).

3. Results

3.1. Patient Pool and Deletional Patterns

The UMD-DMD France and eDystrophin databases contain clinical data of 681 and 781 individuals, respectively, with confirmed clinical outcomes with in-frame exon deletions within *dystrophin*. Data of 595 individuals (29 DMD, 556 BMD, and 10 asymptomatic individuals) was included in both databases. In addition, our literature search identified clinical data of 431 unique patients with in-frame deletions. The final patient pool we analyzed for clinical information contained information of 1298 individuals with confirmed clinical outcomes resulting from genetically diagnosed dystrophinopathy, including 277 (21.34%) individuals with DMD and 1021 (78.66%) with milder BMD (n = 976) or asymptomatic phenotypes (n = 45) (Figure S2A). There was a significant difference in the ratio of DMD to BMD and asymptomatic phenotypes between the records obtained from the two databases and literature searching (eDystrophin vs. literature search: p-value < 0.0001, 95% CI = 27.60–37.79; UMD-DMD France database vs. literature search: p-value < 0.001, 95% CI = 30.49–40.55; UMD-DMD France database vs. eDystrophin: p-value = 0.0519, 95% CI = −0.034–5.65) (Figure S2A).

Theoretically, there are 1408 potential large in-frame deletions possible across the *DMD* gene (Table S2). Our patient pool represented 180 (12.78%) of these potential in-frame deletions (Figure S2B). Of these theoretically possible large in-frame deletions, Δ 45–47 (n = 317, 23 DMD) and Δ 45–48 (n = 205, 19 DMD) were the most common deletional patterns reported among the patients. Two deletional hotspots were identified between exons 3 and 13, and 44 and 55, as 15.58% and 74.46% deletions started and/or ended within these two regions (p-value < 0.0001, for both regions) (Figures S2C and S3).

3.2. More BMD and Asymptomatic Phenotypes Were Associated with In-Frame Deletions

Large in-frame deletions were associated with 9.73%, 6.90%, and 42.46% cases with a severe DMD phenotype in the eDystrophin, UMD-DMD France databases, and the literature (Figures S4–S6). Overall, the reading frame rule was predictive for nearly 78.54% of the cases with large in-frame deletions (Figure 2A). However, exclusive of the literature's records, the prediction rate was 88.95% (Figures S4–S6). We identified 19 exons in-frame deletions, resulting in a significantly severer phenotype than average (Figure 2A,B). Additionally, in-frame deletions starting and/or ending at five exons, including exon 4, 45, 47, 48, and 55, are deemed to result in milder phenotypes (Figure 2A). In addition, we observed a lower incidence of DMD phenotype associated with in-frame deletions starting and/or ending at exons encoding the dystrophin protein's central rod domain (Figure S7). On the other hand, in-frame deletions starting and/or ending at the extreme ends of the protein was associated with more DMD phenotype (Figure S7).

3.3. Distribution of Phenotypes for Exon Skip-Amenable Mutations to Each Exon

We then intended to look at the clinical phenotypes associated with out-of-frame mutations that can be converted to in-frame by therapeutic skipping of one (or more) exons, i.e., skip-amenable mutations, using the patient data collected from the UMD-DMD France Knowledgebase. We determined the ratio of individuals with severe and milder phenotypes associated with exon skip-amenable mutations. Collectively, our analysis identified a total of 1149 individuals with single exon skip-amenable mutations, including 1120 (97.476%) patients with a severe phenotype, i.e., DMD (Figure 3).

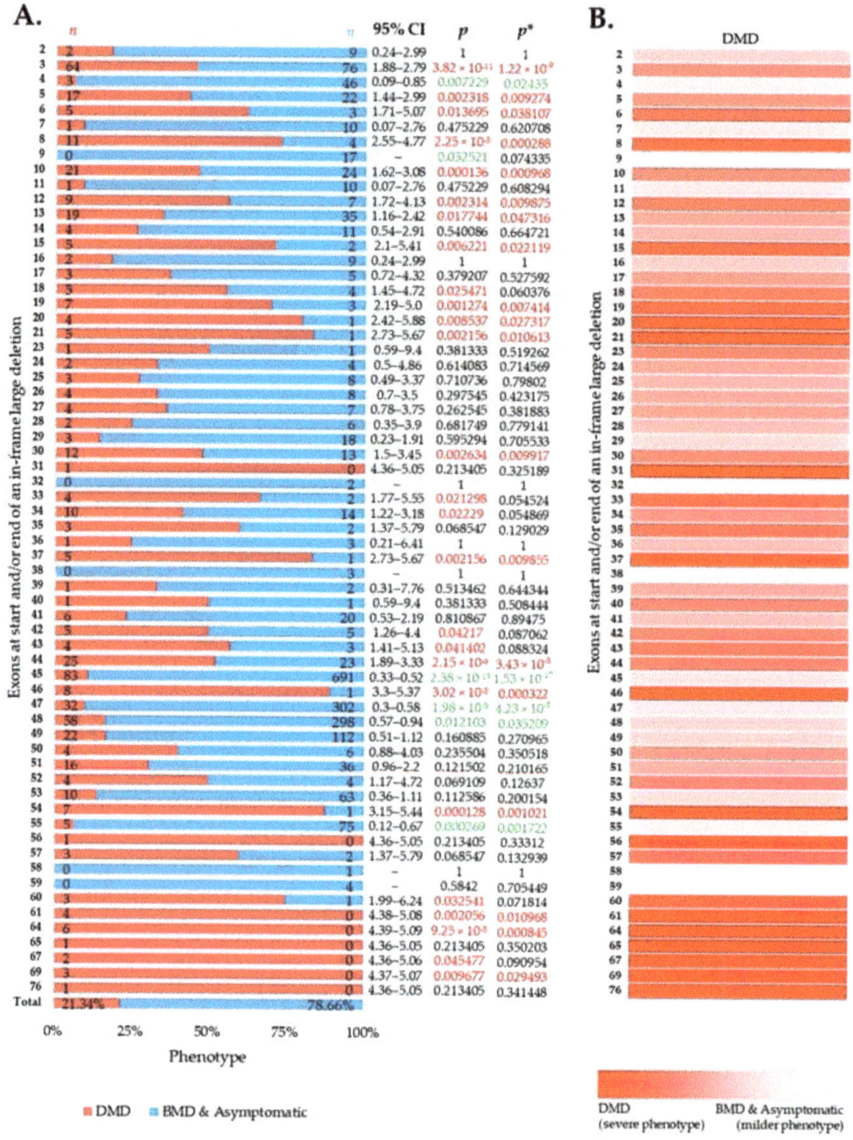

Figure 2. Clinical phenotypes of exon skip-equivalent in-frame *DMD* exon deletions. (**A**) Association between in-frame deletions starting and/or ending at each exon (exon skip-equivalent) and consequent phenotypes. Phenotypic ratios associated with in-frame deletions starting and/or ending at a given exon and all other exons were compared using Fisher's exact test. n indicates the number of individuals with DMD (red) and milder (blue; BMD and asymptomatic) phenotypes. Green and red color indicate a significantly lower and higher incidence of DMD phenotype for a given exon, respectively, as compared to the overall incidence rate. p = p-value, as calculated by Fisher's exact test; p^* = Benjamini–Hochberg adjusted p-value. (**B**) Heatmap showing the relative severity of the consequence of in-frame deletions starting and/or ending at specific exons.

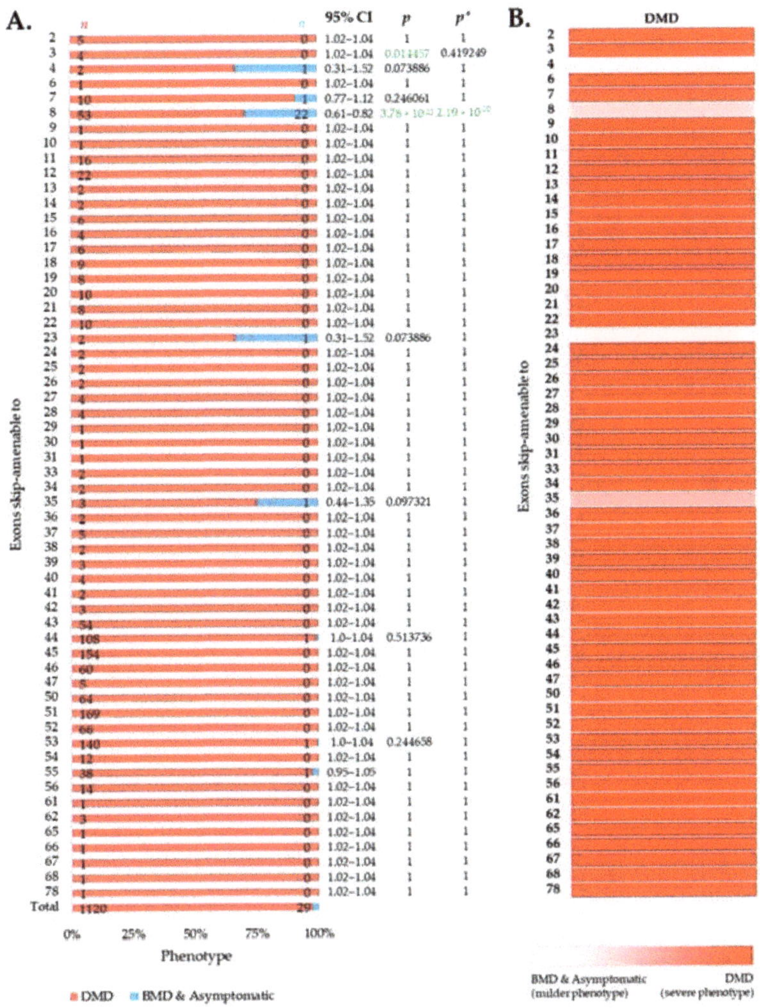

Figure 3. Phenotypic outcomes of exon skip-amenable mutations present in the UMD-DMD France database. (**A**) Distribution of phenotypes for each exon skip-amenable mutation. Phenotypic ratios associated with exon skip-amenable mutations for each exon vs. all other exons are compared using Fisher's exact test. *n* indicates the number of individuals with DMD (red) and milder (blue; BMD and asymptomatic) phenotypes. Green color indicates a significantly lower incidence of DMD phenotype. *p* = *p*-value, as calculated by Fisher's exact test; *p** = Benjamini–Hochberg adjusted *p*-value. (**B**) Heatmap showing the relative severity of the consequence of out-of-frame mutations amenable to skipping of each exon.

3.4. Comparison of Phenotypes of In-Frame Exon Skip-Equivalent and Out-of-Frame Exon Skip-Amenable Mutations

To simulate the clinical phenotype before and after exon-skipping therapy, we looked at the differences in phenotypic outcomes of in-frame exon skip-equivalent and out-of-frame exon skip-amenable mutations for each exon using the data present in the UMD-DMD France Knowledgebase. Our analysis identified 21 exons at which in-frame deletions start and/or end (exon skip-equivalent mutations) were associated with a significantly lower DMD incidence compared to corresponding exon skip-amenable mutations (Figure 4A,B). Additionally, exons 3–9 and exons 45–55 skip equivalent deletions were associated with a significantly lower incidence of a severe phenotype compared to exons 3–9 and exons 45–55 skip-amenable mutations (p-value < 0.0001).

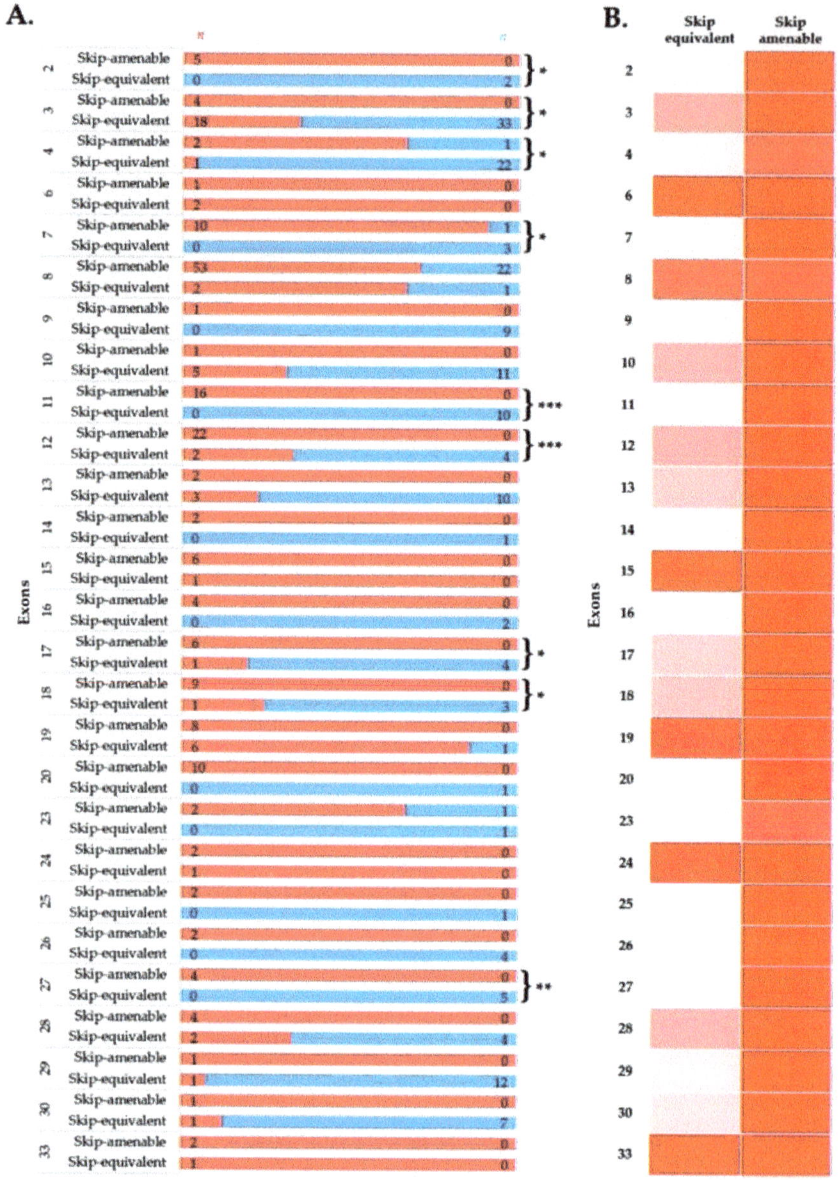

Figure 4. *Cont.*

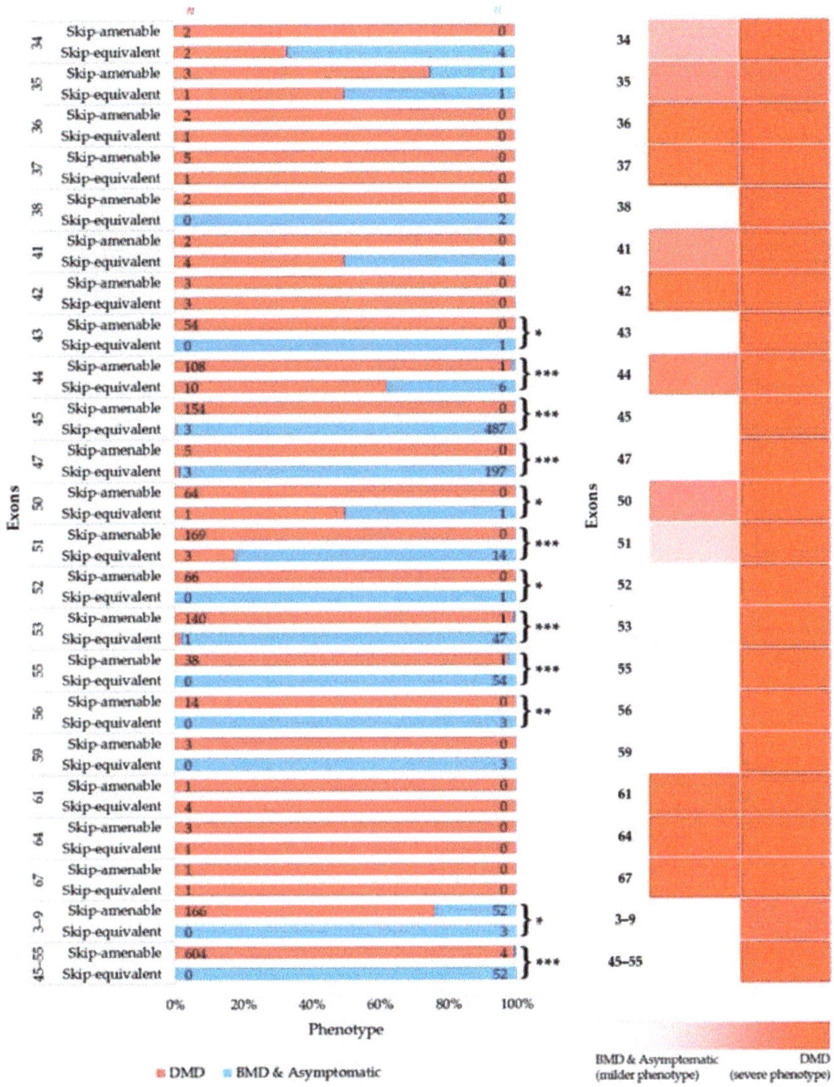

Figure 4. Comparison of clinical phenotypes associated with out-of-frame mutations amenable to exon skipping in UMD-DMD database and in-frame exon skip-equivalent deletions of each exon to simulate the effects of exon-skipping therapies. (**A**) Phenotypic outcomes associated with mutations amenable to exon skipping and exon skip-equivalent in-frame deletions of each exon and their consequent phenotypes. *n* indicates the number of individuals with DMD (red) and milder (blue; BMD and asymptomatic) phenotypes. Asterisks indicate that exon skip-equivalent in-frame deletions are associated with a significantly milder phenotype compared to corresponding exon skip-amenable out-of-frame mutations. We compared the incidence of DMD associated with exon skip-equivalent (or group of exons, e.g., exons 3–9, and exons 45–55) and mutations amenable to skipping each exon (or group of exons, e.g., exons 3–9, and exons 45–55). The statistical significance was calculated using Fisher's exact test. (**B**) Heatmap showing the relative severity of the consequence of exon skip-equivalent in-frame deletions and exon skip-amenable mutations. * $p < 0.05$, ** $p < 0.01$, *** $p < 0.001$

4. Discussion

The FDA approvals of eteplirsen, viltolarsen, and golodirsen are an inspiring development in treating DMD, although these drugs cumulatively can only be administered to approximately 20% of the total DMD population who are specifically amenable to exon 51 and exon 53 skipping [12,25,26]. Besides, the development of antisense therapies to skip additional exons could theoretically help to treat up to ~80% of individuals living with DMD [16,27]. Currently, clinical trials to evaluate exon 44, 45, 51, and 53 skipping antisense drugs are underway (clinicaltrials.gov: NCT02530905, NCT04179409, NCT02667483, NCT02500381, NCT04433234, NCT04129294). Besides, the development of mutation-tailored cocktail antisense drugs applicable for treating a large portion of the DMD population carrying out-of-frame deletions also showed potential [15,28].

This study focused on the phenotypes associated with large deletions involving one or more exons to result in in-frame transcripts. While 1408 large in-frame deletion patterns are theoretically possible within the *DMD* gene, this study identified only 180 (12.78%) of them in individuals in the searched sources of clinical data (Figure S2B and Table S1). The absence of clinical data on individuals with the remaining deletions, perhaps due to the deletions' extreme rarity or that the associated phenotypes are asymptomatic or very mild, that the mutations remain undetected and unreported.

The clinical data on individuals with large in-frame deletions in *DMD* present that nearly 80% (78.54%) have milder BMD or asymptomatic phenotypes (Figure S2A and Figure 2A). Nearly 1 out of 5 individuals (21.46%) developed severe phenotype, i.e., DMD, despite their predicted in-frame deletions. Interestingly, the occurrence of severe phenotype was over four-times higher among the individuals constituted from the literature (Figure S2A). This indicates the potential presence of reporting bias in the clinical reports [29]. To note, in the aggregate of the data collected from the two databases but without those obtained from the literature, 11.05% of individuals with in-frame large deletions developed severe phenotype, which is comparable with the numbers indicated by previous reports (Figures S5 and S6) [7,8,10,20,30].

A previous report studying the clinical phenotypes of *DMD* exon 51 skip-equivalent deletions identified 12% of patients had severe phenotypes despite their predicted in-frame deletions [31]. The individuals with DMD phenotype with in-frame deletions indicate the need for subtle knowledge of the reading-frame rule for dystrophinopathies. It also reflects the complex biology of dystrophinopathies and an array of different factors may influence the ultimate clinical phenotype. These factors may include but are not limited to the inherent leakiness of splicing seen in some exons, the effect of predicted frame-altering mutations on splicing signals, and the stability and tertiary structure of the resultant protein [16,32–34].

Irrespective of the data source, deletions starting and/or ending at some exons were associated with a more severe phenotype (Figure 2 and Figures S2–S6). Additionally, in-frame deletions starting and/or ending within the exons 3–13 hotspot region was associated with an elevated frequency of a severe phenotype, while those starting and/or ending within exons 44–55 region resulted in a significantly lower incidence rate for a DMD phenotype (Figure S2C and Figure 2). An explanation for why the occurrence of a severe phenotype is associated with these deletions is currently unclear. One possible reasoning could be that these deletions, although been predicted as in-frame, result in a less stable or non-functional protein, as a part of N-terminal hotspot encodes for the actin-binding domain. However, the deletion of certain exons involved with coding for portions of the central rod domain may have a less significant impact on skeletal muscle pathology [35], which is what is reflected in this study. Given the retrospective and meta-analytic nature of this study, we did not have access to do any muscle tissue biopsy to investigate the actual cause beneath the elevated frequency of severe phenotype associated with these deletions, and it would also be beyond the scope of this study. Further assessment in clinical status and muscle tissue analysis may help conclude the factors behind the findings of a severe phenotype associated with these exons, which could be interesting and may provide essential insights for therapeutic development.

Importantly, this study identified five exons, including exons 4, 45, 47, 48, and 55, at the start and/or end of in-frame deletions associated with a significantly lower frequency of a DMD phenotype than average (Figure 2A). A previous study of 4894 patients with in-frame and out-of-frame large deletions from the TREAT-NMD DMD Global database listed the top ten skippable *DMD* exons, including exons 8, 43, 44, 45, 50, 51, 52, 53, and 55, that would apply to the largest group of patients with DMD [7]. Our study identified that individuals with large in-frame deletions starting and/or ending at many of these exons had a significantly lower frequency of developing a severe phenotype than those having a mutation amenable to skipping those exons (Figures 2 and 4). It also suggests that the skipping of exons 3 to 9 and 45 to 55 could be promising targets for treating DMD (p-value < 0.0001).

Among the exons we identified as associated with a significantly lower incidence of DMD, exon 45 is already being thoroughly studied as a target for the development of exon-skipping therapies [16,36]. Casimersen, an antisense drug of phosphorodiamidate morpholino chemistry to treat DMD amenable to exon 45 skipping, has recently been accepted and placed under priority review by the FDA [37,38]. The other five exons are also deemed promising therapeutic targets of exon skipping, although some of them are not applicable to many patients.

While the present study included a reasonably comprehensive dataset, it has several limitations. First, a significant limitation is that the specific deletional patterns between DMD patients eligible for exon skipping with out-of-frame deletions and in-frame deletions of exon skip-equivalent deletion mutations are not always comparable. This is particularly the case for exon 44 skipping. Over two-thirds of the (67.89%) out-of-frame deletions amenable to exon 44 skipping start at exon 45, whereas ~98% of the in-frame deletions (exon 44 skip-equivalent) end at exon 44 (Figure S8). Second, the effects of in-frame deletions at the DNA level are not necessarily equivalent to the effects at the RNA level in antisense-treated cells. For example, in a DMD dog model, skipping of exon 8 using an antisense morpholino led to spontaneous skipping of exon 9 in addition to exon 8 [28]. As such, the effects of exon 8 skipping might be more relevant to exon 9 skip-equivalent in-frame deletion rather than exon 8 skip-equivalent in-frame deletion. Third, among the literature included, some studies only enrolled patients diagnosed with either DMD or BMD, potentially influencing the results. Fourth, some studies did not interrogate all *DMD* exons, leaving the possibility that, although unlikely, the patients may have harbored a second mutation. Fifth, the exceedingly high proportion of the DMD phenotype in the literature also raises the concern that there could be some reporting biases in play. Lastly, other factors, including the variability of antisense oligonucleotide-mediated exon skipping efficacy for different exons, also need to be considered to design exon-skipping therapy. Hence, these potential limitations should be considered with care when interpreting the data presented in this study.

5. Conclusions

This study highlights the genotype-phenotype association among individuals with large in-frame deletions starting and/or ending at different exons in *DMD*. While most exon skip-equivalent in-frame deletions are associated with a significantly milder phenotype compared to corresponding exon skip-amenable out-of-frame mutations, we identified several exon skip-equivalent in-frame deletions that are the most promising therapeutic targets. However, the phenotypic variability in individuals with specific in-frame exon deletions found in this study is suggestive of the issue that the response to exon-skipping therapy could be variable and may be impacted by multiple factors. This study largely indicates that genotype-phenotype correlation analysis can significantly contribute to the rational design of exon-skipping therapies; however, hints at the necessity of continued evaluation of genetic and other modifiers.

Supplementary Materials: The following are available online at https://www.mdpi.com/2075-4426/11/1/46/s1, Figure S1: Deletional pattern in the DMD gene. From left to right on the bottom row represents the number of exons (2–78) in the DMD gene. The black shaded area shows the range of each deletion in the *DMD* gene. Figure S2: Patient pool and dystrophin mutational patterns among the individuals with large in-frame deletions. Figure S3: Frequency of deletions starting and/or ending at a given exon was significantly high in a proximal (exons 3 to 13, p-value = 0.000015, f = 20.64776) and a distal (exons 44 to 55, p-value = 0.000069, f = 17.2277) region. Figure S4: Association between in-frame deletions (A) starting, (B) ending, and (C) starting and/or ending at each exon and consequent phenotypes in the eDystrophin database. Figure S5: Association between in-frame deletions (A) starting, (B) ending, and (C) starting and/or ending at each exon and consequent phenotypes in the UMD-DMD France database. Figure S6: Association between in-frame deletions (A) starting, (B) ending, and (C) starting and/or ending at each exon and consequent phenotypes in the records identified from existing literature; Figure S7: Overview of clinical phenotypes associated with in-frame deletions starting and/or ending at different exons and their corresponding dystrophin domains. Figure S8: Phenotypes associated with different exon 44 skip equivalent and amenable to exon 44 skipping mutations. Table S1: Clinical data obtained from literature searching. Table S2: List of all theoretically possible in-frame deletion within *dystrophin*.

Author Contributions: Conceptualization, T.Y.; methodology, S.A., M.H., K.R.Q.L., and R.M.; software, S.A., M.H., and R.M.; validation and scrutinization, S.A., M.H., R.M., and K.R.Q.L.; investigation, S.A. and M.H.; data curation, S.A., and M.H.; formal analysis, S.A., and M.H.; writing—original draft preparation, S.A.; writing—review and editing, S.A. and T.Y.; visualization, S.A.; supervision, T.Y.; project administration, T.Y.; funding acquisition, T.Y. All authors have read and agreed to the published version of the manuscript.

Funding: T.Y. is supported by the Muscular Dystrophy Canada, the Friends of Garrett Cumming Research Fund, the HM Toupin Neurological Science Research Fund, Canadian Institutes of Health Research (CIHR), Alberta Innovates: Health Solutions (AIHS), Jesse's Journey, and the Women and Children's Health Research Institute (WCHRI). S.A. is supported by scholarships from the Maternal and Child Health (MatCH) Program, and the Alberta Innovates Graduate Student Scholarship (AIGSS).

Institutional Review Board Statement: For each database and the literature included, each study reported that the investigators have obtained approval from the relevant Institutional Review Board. No data sharing is applicable to this study since no new data were created or analyzed in this study.

Informed Consent Statement: Not applicable.

Data Availability Statement: The data prepared for this study are available in the article or Supplementary Materials.

Acknowledgments: We thank Aurélie Nicolas and Sanjana Fatema Chowdhury for their help in data coding and identifying the redundant entries in the patient registries. We especially thank Nur-E Jannat Hoque (Department of Statistics, Biostatistics, and Informatics, University of Dhaka), Maryna Yaskina (Biostatistician, University of Alberta Faculty of Medicine & Dentistry—Women and Children's Health Research Institute), and Matthew Pietrosanu (Department of Mathematics & Statistical Sciences/Training and Consulting Centre, University of Alberta) for kindly reviewing the statistical analyses.

Conflicts of Interest: The authors declare no conflict of interest. The funders had no role in the design of the study; in the collection, analyses, or interpretation of data; in the writing of the manuscript, or in the decision to publish the results.

References

1. Koenig, M.; Beggs, A.H.; Moyer, M.; Scherpf, S.; Heindrich, K.; Bettecken, T.; Meng, G.; Müller, C.R.; Lindlöf, M.; Kaariainen, H.; et al. The molecular basis for duchenne versus becker muscular dystrophy: Correlation of severity with type of deletion. *Am. J. Hum. Genet.* **1989**, *45*, 498–506. [CrossRef] [PubMed]
2. Muntoni, F.; Torelli, S.; Ferlini, A. Dystrophin and mutations: One gene, several proteins, multiple phenotypes. *Lancet Neurol.* **2003**, *2*, 731–740. [CrossRef]
3. Roberts, R.G.; Coffey, A.J.; Bobrow, M.; Bentley, D.R. Exon Structure of the Human Dystrophin Gene. *Genomics* **1993**, *16*, 536–538. [CrossRef] [PubMed]

4. Buzin, C.H.; Feng, J.; Yan, J.; Scaringe, W.; Liu, Q.; Den Dunnen, J.; Mendell, J.R.; Sommer, S.S. Mutation rates in the dystrophin gene: A hotspot of mutation at a CpG dinucleotide. *Hum. Mutat.* **2005**, *25*, 177–188. [CrossRef] [PubMed]
5. Takeshima, Y.; Yagi, M.; Okizuka, Y.; Awano, H.; Zhang, Z.; Yamauchi, Y.; Nishio, H.; Matsuo, M. Mutation spectrum of the dystrophin gene in 442 Duchenne/Becker muscular dystrophy cases from one Japanese referral center. *J. Hum. Genet.* **2010**, *55*, 379–388. [CrossRef] [PubMed]
6. Echigoya, Y.; Lim, K.R.Q.; Nakamura, A.; Yokota, T. Multiple Exon Skipping in the Duchenne Muscular Dystrophy Hot Spots: Prospects and Challenges. *J. Pers. Med.* **2018**, *8*, 41. [CrossRef] [PubMed]
7. Bladen, C.L.; Salgado, D.; Monges, S.; Foncuberta, M.E.; Kekou, K.; Kosma, K.; Dawkins, H.; Lamont, L.; Roy, A.J.; Chamova, T.; et al. The TREAT-NMD DMD global database: Analysis of more than 7000 duchenne muscular dystrophy mutations. *Hum. Mutat.* **2015**, *36*, 395–402. [CrossRef]
8. Aartsma-Rus, A.; Van Deutekom, J.C.T.; Fokkema, I.F.; Van Ommen, G.-J.B.; Den Dunnen, J.T. Entries in the Leiden Duchenne muscular dystrophy mutation database: An overview of mutation types and paradoxical cases that confirm the reading-frame rule. *Muscle Nerve* **2006**, *34*, 135–144. [CrossRef]
9. White, S.J.; den Dunnen, J.T. Copy number variation in the genome; the human DMD gene as an example. *Cytogenet. Genome Res.* **2006**, *115*, 240–246. [CrossRef] [PubMed]
10. Monaco, A.P.; Bertelson, C.J.; Liechti-Gallati, S.; Moser, H.; Kunkel, L.M. An explanation for the phenotypic differences between patients bearing partial deletions of the DMD locus. *Genomics* **1988**, *2*, 90–95. [CrossRef]
11. Flanigan, K.M.; Dunn, D.M.; Von Niederhausern, A.; Soltanzadeh, P.; Gappmaier, E.; Howard, M.T.; Sampson, J.B.; Mendell, J.R.; Wall, C.; King, W.M.; et al. Mutational spectrum of DMD mutations in dystrophinopathy patients: Application of modern diagnostic techniques to a large cohort. *Hum. Mutat.* **2009**, *30*, 1657–1666. [CrossRef] [PubMed]
12. Lim, K.R.; Maruyama, R.; Yokota, T. Eteplirsen in the treatment of Duchenne muscular dystrophy. *Drug Des. Devel. Ther.* **2017**, *11*, 533–545. [CrossRef]
13. Heo, Y.A. Golodirsen: First Approval. *Drugs* **2020**, *80*, 329–333. [CrossRef]
14. Dhillon, S. Viltolarsen: First Approval. *Drugs* **2020**, *80*, 1027–1031. [CrossRef]
15. Echigoya, Y.; Lim, K.R.Q.; Melo, D.; Bao, B.; Trieu, N.; Mizobe, Y.; Maruyama, R.; Mamchaoui, K.; Tanihata, J.; Aoki, Y.; et al. Exons 45–55 Skipping Using Mutation-Tailored Cocktails of Antisense Morpholinos in the DMD Gene. *Mol. Ther.* **2019**, *27*, 2005–2017. [CrossRef] [PubMed]
16. Findlay, A.R.; Wein, N.; Kaminoh, Y.; Taylor, L.E.; Dunn, D.M.; Mendell, J.R.; King, W.M.; Pestronk, A.; Florence, J.M.; Mathews, K.D.; et al. Clinical phenotypes as predictors of the outcome of skipping around DMD exon 45. *Ann. Neurol.* **2015**, *77*, 668–674. [CrossRef] [PubMed]
17. Anthony, K.; Arechavala-Gomeza, V.; Ricotti, V.; Torelli, S.; Feng, L.; Janghra, N.; Tasca, G.; Guglieri, M.; Barresi, R.; Armaroli, A.; et al. Biochemical characterization of patients with in-frame or out-of-frame DMD deletions pertinent to exon 44 or 45 skipping. *JAMA Neurol.* **2014**, *71*, 32–40. [CrossRef]
18. Van Den Bergen, J.C.; Van Westrum, S.M.S.; Dekker, L.; Van Der Kooi, A.J.; De Visser, M.; Wokke, B.H.A.; Straathof, C.S.; Hulsker, M.A.; Aartsma-Rus, A.; Verschuuren, J.J.; et al. Clinical characterisation of Becker muscular dystrophy patients predicts favourable outcome in exon-skipping therapy. *J. Neurol. Neurosurg. Psychiatry* **2014**, *85*, 92–98. [CrossRef] [PubMed]
19. Nakamura, A.; Shiba, N.; Miyazaki, D.; Nishizawa, H.; Inaba, Y.; Fueki, N.; Maruyama, R.; Echigoya, Y.; Yokota, T. Comparison of the phenotypes of patients harboring in-frame deletions starting at exon 45 in the Duchenne muscular dystrophy gene indicates potential for the development of exon skipping therapy. *J. Hum. Genet.* **2017**, *62*, 459–463. [CrossRef]
20. Tuffery-Giraud, S.; Béroud, C.; Leturcq, F.; Yaou, R.B.; Hamroun, D.; Michel-Calemard, L.; Moizard, M.P.; Bernard, R.; Cossée, M.; Boisseau, P.; et al. Genotype-phenotype analysis in 2405 patients with a dystrophinopathy using the UMD-DMD database: A model of nationwide knowledgebase. *Hum. Mutat.* **2009**, *30*, 934–945. [CrossRef]
21. Nicolas, A.; Lucchetti-Miganeh, C.; Yaou, R.B.; Kaplan, J.C.; Chelly, J.; Leturcq, F.; Barloy-Hubler, F.; Le Rumeur, E. Assessment of the structural and functional impact of in-frame mutations of the DMD gene, using the tools included in the eDystrophin online database. *Orphanet J. Rare Dis.* **2012**, *7*, 45. [CrossRef] [PubMed]
22. North American Association of Central Cancer Registries Protocol for Assessing Duplicate Cases for CiNA, Registry Certification, and Evaluation of 2017 Cases. Available online: https://www.naaccr.org/wp-content/uploads/2018/09/3_Duplicate_protocol2019.pdf (accessed on 20 November 2020).
23. Garfinkel, S.L. *De-Identification of Personal Information*; National Institute of Standards and Technology: Gaithersburg, MD, USA, 2015.
24. Benjamini, Y.; Hochberg, Y. Controlling the False Discovery Rate: A Practical and Powerful Approach to Multiple Testing. *J. R. Stat. Soc. Ser. B* **1995**, *57*, 289–300. [CrossRef]
25. Roshmi, R.R.; Yokota, T. Viltolarsen for the treatment of Duchenne muscular dystrophy. *Drugs Today* **2019**, *55*, 627–639. [CrossRef] [PubMed]
26. Anwar, S.; Yokota, T. Golodirsen for Duchenne muscular dystrophy. *Drugs Today (Barc)* **2020**, *56*, 491–504. [CrossRef]
27. Aartsma-Rus, A.; Fokkema, I.; Verschuuren, J.; Ginjaar, I.; van Deutekom, J.; van Ommen, G.-J.; den Dunnen, J.T. Theoretic applicability of antisense-mediated exon skipping for Duchenne muscular dystrophy mutations. *Hum. Mutat.* **2009**, *30*, 293–299. [CrossRef]

28. Lim, K.R.Q.; Echigoya, Y.; Nagata, T.; Kuraoka, M.; Kobayashi, M.; Aoki, Y.; Partridge, T.; Maruyama, R.; Takeda, S.; Yokota, T. Efficacy of Multi-exon Skipping Treatment in Duchenne Muscular Dystrophy Dog Model Neonates. *Mol. Ther.* **2019**, *27*, 76–86. [CrossRef]
29. McGauran, N.; Wieseler, B.; Kreis, J.; Schüler, Y.-B.; Kölsch, H.; Kaiser, T. Reporting bias in medical research—A narrative review. *Trials* **2010**, *11*, 37. [CrossRef]
30. Lim, K.R.Q.; Nguyen, Q.; Yokota, T. Genotype–Phenotype Correlations in Duchenne and Becker Muscular Dystrophy Patients from the Canadian Neuromuscular Disease Registry. *J. Pers. Med.* **2020**, *10*, 241. [CrossRef]
31. Waldrop, M.A.; Yaou, R.B.; Lucas, K.K.; Martin, A.S.; O'Rourke, E.; Ferlini, A.; Ferlini, A.; Muntoni, F.; Leturcq, F.; Tuffery-Giraud, S.; et al. Clinical Phenotypes of DMD Exon 51 Skip Equivalent Deletions: A Systematic Review. *J. Neuromuscul. Dis.* **2020**, *7*, 217–229. [CrossRef]
32. Wang, R.T.; Barthelemy, F.; Martin, A.S.; Douine, E.D.; Eskin, A.; Lucas, A.; Lavigne, J.; Peay, H.; Khanlou, N.; Sweeney, L.; et al. DMD genotype correlations from the Duchenne Registry: Endogenous exon skipping is a factor in prolonged ambulation for individuals with a defined mutation subtype. *Hum. Mutat.* **2018**, *39*, 1193–1202. [CrossRef]
33. Flanigan, K.M.; Dunn, D.M.; von Niederhausern, A.; Soltanzadeh, P.; Howard, M.T.; Sampson, J.B.; Swoboda, K.J.; Bromberg, M.B.; Mendell, J.R.; Taylor, L.E.; et al. Nonsense mutation-associated Becker muscular dystrophy: Interplay between exon definition and splicing regulatory elements within the DMD gene. *Hum. Mutat.* **2011**, *32*, 299–308. [CrossRef] [PubMed]
34. Nicolas, A.; Raguénès-Nicol, C.; Ben Yaou, R.; Ameziane-Le Hir, S.; Chéron, A.; Vié, V.; Claustres, M.; Leturcq, F.; Delalande, O.; Hubert, J.-F.; et al. Becker muscular dystrophy severity is linked to the structure of dystrophin. *Hum. Mol. Genet.* **2015**, *24*, 1267–1279. [CrossRef] [PubMed]
35. Harper, S.Q. Molecular dissection of dystrophin identifies the docking site for nNOS. *Proc. Natl. Acad. Sci. USA* **2013**, *110*, 387–388. [CrossRef] [PubMed]
36. Rodrigues, M.; Yokota, T. An Overview of Recent Advances and Clinical Applications of Exon Skipping and Splice Modulation for Muscular Dystrophy and Various Genetic Diseases. *Methods Mol. Biol.* **2018**, *1828*, 31–55. [CrossRef]
37. Kuntz, N.; Wagner, K.; East, L.; Upadhyay, S.; Han, B.; Koenig, E.; Steiner, D.; Shieh, P. DMD—THERAPY. *Neuromuscul. Disord.* **2020**, *30*, S130–S131. [CrossRef]
38. Wexler, M. Casimersen, Exon 45 Skipping Duchenne Therapy, under FDA Priority Review. *Muscular Dystrophy News*, 27 August 2020; 4.

MDPI
St. Alban-Anlage 66
4052 Basel
Switzerland
Tel. +41 61 683 77 34
Fax +41 61 302 89 18
www.mdpi.com

Journal of Personalized Medicine Editorial Office
E-mail: jpm@mdpi.com
www.mdpi.com/journal/jpm

www.ingramcontent.com/pod-product-compliance
Lightning Source LLC
Chambersburg PA
CBHW061757210926
51850CB00006B/582
9783036516226